Systems & Control: Foundations & Applications

Systems & Control: Foundations & Applications

Han-Fu Chen
and Lei Guo

Identification and Stochastic Adaptive Control

1991

Birkhäuser
Boston · Basel · Berlin

Chen, Han-Fu
Institute of Systems Science
Academia Sinica
Peking University
Beijing 100080, China

Lei Guo
Institute of Systems Science
Academia Sinica
Peking University
Beijing 100080, China

Library of Congress Cataloging-in-Publication Data

Chen, Han-Fu
Identification and stochastic control / Han-Fu Chen and Lei Guo.
p. cm.
ISBN 0-8176-3597-1 (hard : alk. paper) : -- ISBN 3-7643-3597-1 (hard : alk. paper)
1. Adaptive control systems. 2. System identification. I. Guo, Lei, 1961-
TJ217.C475 1991 91-25633
629.8'36--dc20 CIP

Printed on acid-free paper.

ISBN 0-8176-3597-1
ISBN 3-7643-3597-1

Camera-ready copy prepared by the authors using Latex.
Printed and bound by Edwards Brothers, Ann Arbor, Michigan.
Printed in the USA.

9 8 7 6 5 4 3 2 1

Preface

Identifying the input-output relationship of a system or discovering the evolutionary law of a signal on the basis of observation data, and applying the constructed mathematical model to predicting, controlling or extracting other useful information constitute a problem that has been drawing a lot of attention from engineering and gaining more and more importance in econometrics, biology, environmental science and other related areas. Over the last 30-odd years, research on this problem has rapidly developed in various areas under different terms, such as time series analysis, signal processing and system identification. Since the randomness almost always exists in real systems and in observation data, and since the random process is sometimes used to model the uncertainty in systems, it is reasonable to consider the object as a stochastic system.

In some applications identification can be carried out off line, but in other cases this is impossible, for example, when the structure or the parameter of the system depends on the sample, or when the system is time-varying. In these cases we have to identify the system on line and to adjust the control in accordance with the model which is supposed to be approaching the true system during the process of identification. This is why there has been an increasing interest in identification and adaptive control for stochastic systems from both theorists and practitioners.

The basic model that is used to describe systems in this book is the so-called ARMAX model, the acronym standing for "autoregressive-moving average with exogenous input". If the system does not involve the control, then the discrete-time model is called the ARMA process, which is the research topic of a vast amount of publications, including the well-known monographs [BJ], [An2], [HDe] and so on. When the control term is present in the system although some results on identification and adaptive control for the stochastic system can be found in [GS], [Che2], [Ca], [KV], it is difficult to find a book that completely focuses on this subject that demonstrates both theoretical methods and analytical techniques applied in the field and shows not only the classical results but also the leading edge of research. This book is an attempt to supply this lack and to stimulate the research on and interest in an area full of opportunity.

This book is written for graduate students, teachers and research workers in the areas of systems and control, statistics, signal processing, econometrics and other related areas. For a systematic and self-contained description of the subject, we include some basic knowledge from related fields that is necessary for understanding some parts of the book, but we do not intend to provide a complete and comprehensive presentation of stochastic systems. Further, it is almost certain that some significant result may be missed in this book. First, this is because of inadequate knowledge or possible oversight, and secondly because of the authors' personal interest which is reflected to a certain extent in the selection of material here.

Chapter 1 is mainly aimed at those who are not familiar with probability theory. Not all statements are given with proof; the readers looking for details are referred to well-known references, e.g. [Do], [Lo], [CT], [Sh], [Chu2], among others.

The material in Chapter 2 is selected not for its importance in martingale theory, for which there exist a series of references, for example, [Do], [N], [St], [HH], [LSh], but for its basic role in estimating the weighted sums of martingale difference sequences. The coefficient and order estimates for linear stochastic systems are essentially based on estimates for sums of martingale difference sequences with single- and double-indexed weights respectively. Some of these developments on martingales can be found only in recent periodicals.

Chapter 3 gives some basic results of linear stochastic systems, but we restrict the presentation to the framework within which we can deal with adaptive control problems without any further referencing. The topic of this chapter, generally speaking, is known by researchers in the systems and control area, who, however, may find some generalization and specialization of well-known results that are not available in standard references.

Consistency of coefficient estimates for ARMAX models is relatively easy to prove under the persistent excitation (PE) condition on regressors and the strictly positive realness (SPR) condition on the noise model. Unfortunately, the PE condition in general is not satisfied by an adaptive control system, even if its control performance is optimized. In Chapter 4 the consistency of coefficient estimates is established under a weakened excitation condition. Consistency results without using the PE condition are important not only for themselves, but also in that they make it possible to simultaneously achieve both optimality of control performance and consistency of coefficient estimates. The moving average process with coefficients satisfying the SPR condition is "close" to the uncorrelated noise. Chapter 4 also provides consistency theorems for systems with noise model not satisfying the SPR condition.

The adaptive tracker is an important class of adaptive control systems. Its control purpose consists in forcing the system output to follow a given reference signal. In particular, if the reference signal is a constant, then the adaptive tracker is called the self-tuning regulator, which has obtained great success in applications. Chapter 5 gives an analysis for the convergence and optimality of the adaptive trackers with parameter estimates provided by SG or ELS algorithms. In practice adaptive trackers are applied mostly on the basis of the ELS rather than SG algorithm, but the analysis of ELS-based trackers is much more complicated than that of the SG-based ones. In this chapter a model reference adaptive control problem is also solved for the system with noise being a possibly unstable ARMA process.

Chapter 6 shows how the weakened excitation condition derived in Chapter 4 can be satisfied for the general feedback control system. It is demonstrated that a diminishing dither added to the control asymptotically has no effect on the control performance but gives sufficient excitation to make the estimate strongly consistent. This is a significant step towards simultaneous acquisition of consistency of estimates and optimality of performance for adaptive control systems, which is the topic of Chapter 8.

The order estimation problem for stationary ARMA processes is well studied in time series analysis. Chapter 7 discusses estimating the order and the time-delay for stochastic feedback control systems. Because of the control, the output of the system is no longer stationary; hence results in time series analysis cannot be directly applied to the order and time-delay estimation problem of an ARMAX model.

Using the method developed in Chapter 6 for systems with unknown parameters including order, time-delay and coefficients, Chapter 8 designs the optimal adaptive controls minimizing either the tracking error or the general quadratic loss function while keeping the consistency of parameter estimates. A quadratic cost turns out to be a tracking error if the weight for the control term in the loss function is set to equal zero. But for these two cases the optimal adaptive controls are structurally different from each other. The intrinsic connection between them is also demonstrated in this chapter.

Approximating a real system, which in general is of infinite order, by a system of finite order is of great importance in applications, but its solution is far from complete in theory. Restricted to the ARX(∞) system, Chapter 9 concerns only two aspects of this problem, namely, the transfer function approximation and the estimation of the noise process. The latter is crucial when an ARMAX model is directly fitted to the input-output data. The basic estimation method used is the increasing lag least squares algorithm.

As mentioned at the beginning, one of the motivations of adaptive control is to update the control in correspondence with the parameter drift during the system operation. Chapters 10 and 11 are concerned with the interesting but difficult topic: identification and adaptive control for stochastic systems with time-varying parameters. Chapter 10 analyzes two basic algorithms for tracking time-varying parameters and demonstrates that the averaged tracking error is dominated by the averaged parameter variation and the averaged system noise. This property is established under a reasonable "conditional richness" condition but under no other restriction on independence, stationarity or other statistical relationship. Chapter 11 solves the adaptive tracking problem for stochastic systems with a class of slowly time-varying parameters. The readers will find a series of problems that need further research in this important area.

Chapter 12 deals with continuous-time stochastic systems. Some results on estimation and adaptive control are presented. Although they are not as complete as those for discrete-time systems, we believe that they are useful for further study.

The authors are indebted to Dr. J.F. Zhang for his valuable discussion and kind help in preparing the book. The authors also gratefully acknowledge support from the National Natural Science Foundation of China.

Han-Fu Chen, Lei Guo

January, 1991
Beijing, China

Contents

CHAPTER 1

Probability Theory Preliminaries

In this chapter we introduce the main concepts and some basic results of probability theory for convenience of reference. They are presented mostly without proof, but are explained by examples for those who are not familiar with measure-theory-based probability theory. For detailed material we refer readers to standard texts, for example, [Do], [Chu2], [Lo], [CT] and [Sh].

1.1 Random Variables

We first define the probability space, where the random variables are defined.

Let Ω be the *basic space*, the point ω of which is called the elementary event or the sample.

Let $\mathcal{F}$ be a *σ-algebra* of sets in Ω, i.e. $\mathcal{F}$ is a class of sets in Ω satisfying the following conditions:

1. The whole space Ω is an element of $\mathcal{F}$, i.e. $\Omega \in \mathcal{F}$.

2. The complement A^c of any element (set) A in $\mathcal{F}$ belongs to $\mathcal{F}$, i.e. $A^c \in \mathcal{F}$ if $A \in \mathcal{F}$.

3. The countable union of elements in $\mathcal{F}$ belongs to $\mathcal{F}$, i.e.

$$\bigcup_{i=1}^{\infty} A_i \in \mathcal{F} \qquad if \quad A_i \in \mathcal{F}, \quad i = 1, 2, \ldots.$$

The element of $\mathcal{F}$ is called the measurable set or the random event, $(\Omega, \mathcal{F})$ is called the measurable space.

From properties 2 and 3 it is clear that $\bigcap_{i=1}^{\infty} A_i \in \mathcal{F}$ if $A_i \in \mathcal{F}$, $i = 1, 2, \ldots$, because $\bigcap_{i=1}^{\infty} A_i = \left(\bigcup_{i=1}^{\infty} A_i^c\right)^c$.

Let P be a set function defined on $\mathcal{F}$ with the following properties:

1. P is a non-negative function, i.e. $P(A) \geq 0, \forall A \in \mathcal{F}$.
2. The value of P at Ω is 1, i.e. $P(\Omega) = 1$.
3. P is countably additive, i.e.

$$P\left(\bigcup_{i=1}^{\infty} A_i\right) = \sum_{i=1}^{\infty} P(A_i)$$

if $A_i \in \mathcal{F}, \forall i = 1, 2, \ldots$ and the intersection of A_i with A_j is empty for each $i \neq j$, i.e. $A_i \cap A_j = \emptyset$ (empty set), $\forall i \neq j$.

A set function P satisfying the above-mentioned properties is called a *probability measure* on $\mathcal{F}$.

The triple $(\Omega, \mathcal{F}, P)$ consisting of a basic space, a σ-algebra $\mathcal{F}$ of sets in Ω and a probability measure P defined on $\mathcal{F}$ is called a *probability space*.

Example 1.1. Drawing cards from a pack with the same probability (1/52) for each outcome of a card can be modeled as follows: Ω is a finite set consisting of 52 points; each point is endowed with probability of 1/52 and $\mathcal{F}$ consists of all possible subpacks (including the whole pack) of cards.

Example 1.2. $\Omega = [0, 1]$, a real interval. $\mathcal{F}$ is composed of 4 elements:

$$\mathcal{F} = \left\{\emptyset, \Omega, [0, \frac{1}{3}), [\frac{1}{3}, 1]\right\}$$

and P is defined as

$$P\left([0, \frac{1}{3})\right) = p, \quad P\left([\frac{1}{3}, 1]\right) = q = 1 - p,$$

where $0 \leq p \leq 1$.

Example 1.3. $\Omega = (-\infty, \infty)$, $\mathcal{F} = \mathcal{B}$, the *Borel σ-algebra*, which is generated by open intervals in $(-\infty, \infty)$, i.e. $\mathcal{B}$ contains and only contains all open intervals in $(-\infty, \infty)$ and all sets obtained from the open intervals by countable operations of the union and intersection. The elements in $\mathcal{B}$ are called the Borel-measurable sets or Borel sets. By the way, it is worth noting that the Borel σ-algebra $\mathcal{B}^n$ in $\mathbb{R}^n$ can be defined in a similar way.

Let $F(\cdot)$ be a continuous function non-decreasing on (c, d) with $F(c) = 0$, $F(d) = 1$, where $-\infty \leq c$ and $d \leq \infty$. Then it can be shown that a probability measure $P(\cdot)$ on (a, b) can be defined by setting $P([a\,b]) = F(b) - F(a)$, $\forall\, a, b : d \geq b \geq a \geq c$. In particular, if $-\infty < c < d < \infty$ and $F(x) = \frac{x-c}{d-c}$, i.e. the measure is generated by the length (multiplied by the constant $\frac{1}{d-c}$) of intervals, then the measure is called the Borel measure.

For a probability space $(\Omega, \mathcal{F}, P)$ it is clear that the subsets of a measurable set A in $\mathcal{F}$ are not necessarily measurable; hence the probabilities for them may not be defined. It is natural to assume that all subsets of a measurable set of zero probability are measurable sets with probability 0. In other words, we add all subsets of the measurable sets with probability 0 to $\mathcal{F}$ and define their probability equal to 0. After such a treatment the probability space is called *completed*. The elements of the completed Borel σ-algebra are said to be Lebesgue measurable and the completed Borel measure the Lebesgue measure.

In the subsequence we always assume that the probability space $(\Omega, \mathcal{F}, P)$ is completed.

A real function $\xi = \xi(\omega)$ defined on $(\Omega, \mathcal{F})$ and finitely valued with the possible exception on a set of probability 0 is called *the random variable* if it is measurable, i.e. if for any Borel set $B \in \mathcal{B}$ the ω-set $\{\omega : \xi(\omega) \in B\}$ is $\mathcal{F}$ measurable:

$$\{\omega : \xi(\omega) \in B\} \in \mathcal{F}.$$

Example 1.4. The function

$$\xi(\omega) = \begin{cases} a, & \text{if } \omega \in [0, \frac{1}{3}), \\ b, & \text{if } \omega \in [\frac{1}{3}, 1] \end{cases}$$

with a and b being any constants is a random variable in the probability space given in Example 1.2.

Example 1.5. Any continuous function defined on $(-\infty, \infty)$ is a random variable in the probability space given in Example 1.3.

In the sequel, by a relationship between random quantities, we always mean that it holds for any ω with possible exception of a set with probability 0, but sometimes we omit writing "a.s." or "with probability one".

For any random variable ξ the *distribution function* $F_\xi(\cdot)$ is defined as

$$F_\xi(x) \triangleq P(\omega : \xi(\omega) < x), \qquad \forall\, x \in \mathbb{R}.$$

It is easy to see that $F_\xi(\cdot)$ is a non-decreasing and left-continuous function.

If $F_\xi(\cdot)$ is differentiable, then its derivative $f_\xi(x) = \dfrac{dF_\xi(x)}{dx}$ is called the density of ξ.

By an l-dimensional random vector $\xi = [\xi^1 \cdots \xi^l]^\tau$ we mean that each of its component ξ^i, $i = 1, \ldots l$ is a random variable.

The distribution function $F_\xi(x^1, \cdots, x^l)$ of ξ and the density $f_\xi(x^1, \cdots, x^l)$, if it exists, are similarly defined:

$$F_\xi(x^1, \cdots, x^l) = P(\omega : \ \xi^1 < x^1, \cdots, \xi^l < x^l);$$

$$F_\xi(x^1, \cdots, x^l) = \int_{-\infty}^{x^1} \cdots \int_{-\infty}^{x^l} f_\xi(\lambda^1, \cdots, \lambda^l) d\lambda^1 \cdots \lambda^l.$$

Example 1.6. The distribution function for ξ given in Example 1.4 is

$$F_\xi(x) = \begin{cases} 0, & x \le a, \\ p, & a < x \le b, \\ 1, & b < x. \end{cases}$$

In this Example $F_\xi(x)$ is stepwise and ξ has no density.

Example 1.7. The function

$$\frac{1}{\sqrt{2\pi}\sigma} exp\left\{ -\frac{(x-\mu)^2}{2\sigma^2} \right\}, \qquad \forall x \in I\!R^1, \quad \sigma > 0$$

or in the n-dimensional case

$$\frac{1}{(2\pi)^{\frac{n}{2}} (det\, R)^{(\frac{1}{2})}} exp\left\{ -\frac{1}{2}(x-\mu)^\tau R^{-1}(x-\mu) \right\}, \quad R > 0, \quad x, \mu \in I\!R^n$$

is called the Gaussian (or normal) density, and the corresponding random variable is said to have the Gaussian (normal) distribution.

This distribution will be generalized to the case $R \ge 0$ in Section 1.3.

1.2 Expectation

The concept of the integral for continuous functions can be extended to that for random variables. The integral of a random variable ξ over the whole space Ω if it exists is called the mathematical the expectation or simply expectation of ξ and is denoted by $E\xi$. We now give its precise definition.

We first define $E\xi$ for a non-negative ξ.

Let $\xi \ge 0$ a.s. and set

$$A_{ni} = \left\{ \omega : \ i2^{-n} < \xi \le (i+1)2^{-n} \right\}.$$

Define

$$E\xi \triangleq \int_\Omega \xi \, dP \triangleq \lim_{n\to\infty} \left[\sum_{i=0}^{n2^n-1} i2^{-n} P(A_{ni}) + nP(\xi > n) \right].$$

Clearly, the right-hand side always exists but may be infinite.

For a general random variable ξ we first split it into its positive

$$\xi^+ \triangleq max(\xi, 0)$$

and negative

$$\xi^- \triangleq max(-\xi, 0)$$

parts: $\xi = \xi^+ - \xi^-$.

Since both $E\xi^+$ and $E\xi^-$ are well-defined it is natural to define $E\xi$ as

$$E\xi = E\xi^+ - E\xi^-$$

if at least one of $E\xi^+$ and $E\xi^-$ is finite.

If $E|\xi| = E\xi^+ + E\xi^- < \infty$, then ξ is called integrable.

It can be shown that in the case where $E\xi$ exists $E\xi$ can be expressed as a Lebesgue-Stieltjes integral

$$E\xi \triangleq \int_\Omega \xi dP = \int_{-\infty}^{\infty} x dF_\xi(x)$$

where $F_\xi(x)$ denotes the distribution function of ξ.

In Example 1.4 we have

$$E\xi = ap + bq,$$

while for the n-dimensional Gaussian distribution

$$E\xi = \int_{-\infty}^{\infty} \cdots \int_{-\infty}^{\infty} x \frac{1}{(2\pi)^{\frac{n}{2}} (det\, R)^{(\frac{1}{2})}} exp\left\{ -\frac{1}{2}(x-\mu)^\tau R^{-1}(x-\mu) \right\} dx_1 \cdots dx_n = \mu,$$

where $x = [x_1 \ \cdots \ x_n]^\tau$.

We now give fundamental theorems on taking limits under the expectation sign. For this we first define several convergence types of random variables.

Let ξ and ξ_n $n = 1, 2, \cdots$ be random variables.

We say that 1) ξ_n converges to ξ with probability 1 or almost surely

$$\xi_n \xrightarrow[n\to\infty]{} \xi \quad a.s.$$

if $P(\xi_n \to \xi) = 1$; 2) ξ_n converges to ξ in probability $\xi_n \xrightarrow{P} \xi$ if for any $\varepsilon > 0$

$$P(|\xi_n - \xi| > \varepsilon) \xrightarrow[n \to \infty]{} 0;$$

3) ξ_n weakly converges to ξ, if the distribution function $F_{\xi_n}(x)$ of ξ_n converges to that of ξ

$$F_{\xi_n}(x) \xrightarrow[n \to \infty]{} F_\xi(x)$$

at any x where $F_\xi(x)$ is continuous and 4) ξ_n converges to ξ in the mean square sense if

$$E|\xi_n - \xi|^2 \xrightarrow[n \to \infty]{} 0.$$

For the cases of 2) and 3) we respectively denote ξ as $\xi = p - \lim \xi_n$ and $\xi = l \cdot i \cdot m \cdot \xi_n$.

The convergence a.s. implies the convergence in probability which in turn implies the weak convergence. Also, the convergence in the mean square sense implies the convergence in probability.

Theorem 1.1 *(On Monotone Convergence). If $\xi_n \uparrow \xi$, $\xi_n \geq \eta$ a.s. and $E\eta^- < \infty$ ($E\eta^+ < \infty$), then $E\xi_n \uparrow E\xi$ ($E\xi_n \downarrow E\xi$).*

Theorem 1.2 *(Fatou's Lemma). If $\xi_n \geq \eta$ ($\xi_n \leq \eta$), $n = 1, 2, \cdots$ for some random variable η with $E\eta^- < \infty$ ($E\eta^+ < \infty$), then*

$$E \liminf_{n\to\infty} \xi_n \leq \liminf_{n\to\infty} E\xi_n$$

$$(\limsup_{n\to\infty} E\xi_n \leq \limsup_{n\to\infty} \xi_n)$$

Theorem 1.3 *(On Dominated Convergence). If $\xi_n \xrightarrow[n \to \infty]{P} \xi$ and there exists an integrable random variable η such that $|\xi_n| \leq \eta$, then $E|\xi| < \infty$, $E\xi_n \xrightarrow[n \to \infty]{} E\xi$ and*

$$E|\xi_n - \xi| \xrightarrow[n \to \infty]{} 0.$$

A sequence $\{\xi_n\}$ of random variables is called uniformly integrable (u.i.) if

$$\sup_n E\left(|\xi_n| I_{[|\xi_n| > c]}\right) \xrightarrow[c \to \infty]{} 0.$$

A necessary and sufficient condition for $\{\xi_n\}$ to be u.i. is that there is a function $G(t)$, $t \geq 0$, which is positive, increasing and convex downward, such that

$$\lim_{t\to\infty} G(t)/t = \infty \quad \text{and} \quad \sup_n EG(|\xi_n|) < \infty.$$

Theorem 1.4. *Assume* $\xi_n \xrightarrow{P} \xi$, $\xi_n \geq 0$, $E\xi_n < \infty$. *Then the following statements are equivalent: 1)* $E\xi_n \to E\xi < \infty$; *2)* $E|\xi_n - \xi| \xrightarrow[n\to\infty]{} 0$; *3)* $\{\xi_n\}$ *is uniformly integrable.*

We now list inequalities which we shall refer to.

Chebyshev's Inequality. Let ξ be a random variable. Then for any $\varepsilon > 0$

$$P(|\xi| \geq \varepsilon) \leq \frac{E|\xi|}{\varepsilon}.$$

Jensen's Inequality. Let ξ be a random variable with $E|\xi| < \infty$. Then for any convex Borel measurable function $g(\cdot)$

$$g(E\xi) \leq Eg(\xi).$$

Lyapunov's Inequality. Let ξ be a random variable and let $0 < s < t$. Then

$$(E|\xi|^s)^{\frac{1}{s}} \leq (E|\xi|^t)^{\frac{1}{t}}.$$

Hölder's Inequality. Let $1 < p < \infty$, $1 < q < \infty$, $\frac{1}{p} + \frac{1}{q} = 1$. If for random variables ξ and η $E|\xi|^p < \infty$ and $E|\eta|^q < \infty$, then

$$E|\xi\eta| \leq (E|\xi|^p)^{\frac{1}{p}} (E|\eta|^q)^{\frac{1}{q}}.$$

For the special case $p = q = 2$ the Hölder's inequality is called Cauchy's inequality or Schwarz's inequality.

Minkowski's Inequality. If $E|\xi|^p < \infty$ and $E|\eta|^p < \infty$ for some $p \in [1, \infty)$, then

$$(E|\xi + \eta|^p)^{\frac{1}{p}} \leq (E|\xi|^p)^{\frac{1}{p}} + (E|\eta|^p)^{\frac{1}{p}}.$$

C_r-inequality.

$$\left(\sum_{i=1}^{n} |\xi_i|\right)^r \leq C_r \sum_{i=1}^{n} |\xi_i|^r,$$

where

$$C_r = \begin{cases} 1, & if\ r \leq 1, \\ n^{r-1}, & if\ r \geq 1. \end{cases}$$

1.3 Conditional Expectation

Let $(\Omega, \mathcal{F}, P)$ be a probability space, $\mathcal{F}_1$ be a sub-σ-algebra of $\mathcal{F}$: $\mathcal{F}_1 \subset \mathcal{F}$, (i.e. $\mathcal{F}_1$ is a σ-algebra and any measurable set A in $\mathcal{F}_1$ is measurable with respect to $\mathcal{F}$).

In the sequel we denote by I_A the indicator of a set A:

$$I_A = \begin{cases} 1, & \omega \in A, \\ 0, & \omega \notin A \end{cases}$$

and define

$$\int_A \xi dP = E\xi I_A$$

for a random variable ξ for which $E\xi I_A$ exists.

It can be shown (Radon-Nikodym Theorem) that for any random variable ξ with $E\xi$ defined (i.e. either $E\xi^+ < \infty$ or $E\xi^- < \infty$) there is a unique (up to sets of probability zero) $\mathcal{F}_1$-measurable random variable denoted by $E(\xi|\mathcal{F}_1)$ or $E^{\mathcal{F}_1}\xi$ such that

$$\int_A \xi dP = \int_A E(\xi|\mathcal{F}_1)dP \quad \text{a.s. for any } A \in \mathcal{F}_1. \tag{1.1}$$

Definition. The random variable satisfying (1.1) is called the conditional expectation of ξ given $\mathcal{F}_1$.

Definition. Let $\mathcal{F}^\eta$ be the σ-algebra generated by a random variable η (i.e. the smallest σ-algebra containing all sets of the form $\{\omega\colon \eta(\omega) \in B, B \in \mathcal{B}\}$) and let $E\xi$ exist. $E(\xi|\mathcal{F}^\eta)$ is called the conditional expectation of ξ given η and is denoted by $E(\xi|\eta)$ or $E^\eta\xi$.

Definition. If $A \in \mathcal{F}$, $\mathcal{F}_1 \subset \mathcal{F}$, then $E(I_A|\mathcal{F})$ $(E(I_A|\eta))$ is called the conditional probability of A given $\mathcal{F}_1$ (η) and denoted by $P(A|\mathcal{F}_1)$ or $P^{\mathcal{F}_1}A$ $(P(A|\eta)$ or $P^\eta A)$.

Example 1.8. Let $\Omega = [0,1]$, $\mathcal{F}$ be the Borel σ-algebra and P be the Lebesgue measure. If $\xi(\omega) = \omega$, $\mathcal{F}_1 = \{\Omega, \emptyset, [0, \frac{1}{3}), [\frac{1}{3}, 1]\}$, then the random variable

$$\eta(\omega) = \begin{cases} \frac{1}{6}, & 0 \le \omega < \frac{1}{3}, \\ \frac{2}{3}, & \frac{1}{3} \le \omega \le 1 \end{cases}$$

is obviously $\mathcal{F}_1$-measurable and satisfies (1.1) $\int_A \xi dP = \int_A \eta dP$, $\forall A \in \mathcal{F}_1$. By uniqueness we have

$$E(\xi|\mathcal{F}_1) = \eta.$$

Example 1.9. Let $\mathcal{F}_1 \subset \mathcal{F}$ and let $\mathcal{F}_1$ have a countable decomposition, i.e. let $\mathcal{F}_1$ be generated by $\{A_1, A_2, \cdots\}$ with $A_i \cap A_j = \emptyset$, $\forall i \neq j$ and $P(A_i) > 0$, $\forall i$. Then

$$E(\xi|\mathcal{F}_1) = \sum_{i=1}^{\infty} \frac{E(\xi I_{A_i})}{P(A_i)} I_{A_i}.$$

We now list the properties of the conditional expectation. Assume that the expectations concerned in the sequel are defined.

1. $E[(a\xi + b\eta)|\mathcal{F}_1] = aE(\xi|\mathcal{F}_1) + bE(\eta|\mathcal{F}_1)$, for constants a and b.
2. If $\xi \leq \eta$, then $E(\xi|\mathcal{F}_1) \leq E(\eta|\mathcal{F}_1)$.
3. Let ζ be a random variable. There exists a Borel measurable function $f(\cdot)$ such that $E(\xi|\zeta) = f(\zeta)$.
4. $E(E(\xi|\mathcal{F}_1)) = E\xi$.
5. $E(\eta^\tau \xi|\mathcal{F}_1) = \eta^\tau E(\xi|\mathcal{F}_1)$, if η is $\mathcal{F}_1$-measurable and $|\eta| < \infty$ *a.s.*, $E(|\xi||\mathcal{F}_1) < \infty$ *a.s.*.
6. $E^{\mathcal{F}_1}E^{\mathcal{F}_2}\xi = E^{\mathcal{F}_1}\xi$, if $\mathcal{F}_1 \subset \mathcal{F}_2 \subset \mathcal{F}$.
7. $E^{\mathcal{F}_1}\xi = E\xi$ if $\mathcal{F}_1 = (\Omega, \emptyset)$.
8. $P^{\mathcal{F}_1}(A) \geq 0$, $\quad P^{\mathcal{F}_1}(\Omega) = 1$, and

$$P^{\mathcal{F}_1}\left(\bigcup_{i=1}^{\infty} A_i\right) = \sum_{i=1}^{\infty} P^{\mathcal{F}_1}(A_i)$$

   if $A_i \cap A_j = \emptyset$, $\forall i \neq j$.

Theorems 1.1-1.3 can be extended from expectation to the conditional expectation.

Theorem 1.5. *If* $\xi_n \uparrow \xi$ $(\xi_n \downarrow \xi)$ *a.s. and* $E\xi_1^- < \infty$ $(E\xi_1^+ < \infty)$, *then*

$$E(\xi_n|\mathcal{F}_1) \uparrow E(\xi|\mathcal{F}_1) \quad (E(\xi_n|\mathcal{F}_1) \downarrow E(\xi|\mathcal{F}_1)) \quad a.s.$$

Theorem 1.6. *Chebyshev's inequality, Jensen's inequality, Lyapunov's inequality, Hölder's inequality, Minkowski's inequality and* C_r*-inequality remain valid if* E *is replaced by* $E^{\mathcal{F}_1}$.

Theorem 1.7.

1. If $\xi_n \le \eta$, $E\eta < \infty$, then

$$\limsup_{n\to\infty} E^{\mathcal{F}_1}\xi_n \le E^{\mathcal{F}_1} \limsup_{n\to\infty} \xi_n \quad a.s.$$

If, in addition, $\xi_n \downarrow \xi$ a.s., then

$$E^{\mathcal{F}_1}\xi_n \downarrow E^{\mathcal{F}_1}\xi \quad a.s.$$

2. If $\xi_n \ge \eta$, $E\eta > -\infty$, then

$$E^{\mathcal{F}_1}(\liminf_{n\to\infty} \xi_n) \le \liminf_{n\to\infty} E^{\mathcal{F}_1}\xi_n \quad a.s.$$

If, in addition, $\xi_n \uparrow \xi$ a.s., then

$$E^{\mathcal{F}_1}\xi_n \uparrow E^{\mathcal{F}_1}\xi \quad a.s.$$

3. If $|\xi_n| \le \eta$, $E\eta < \infty$, $\xi_n \to \xi$ a.s., then

$$E^{\mathcal{F}_1}\xi_n \to E^{\mathcal{F}_1}\xi \quad a.s.$$

1.4 Independence, Characteristic Functions

Let $A_i \in \mathcal{F}$, $i = 1, 2, \ldots$. If for any set of indices $\{i_1, \ldots, i_k\}$

$$P\left(\bigcap_{j=1}^{k} A_{i_j}\right) = \prod_{j=1}^{k} P(A_{i_j}),$$

then random events $\{A_i\}$ are called mutually independent. Sub-σ-algebra $\mathcal{F}_i$, $i = 1, 2, \ldots$ of $\mathcal{F}$ are called mutually independent if the random events $A_1, \ldots, A_k$ are mutually independent whenever $A_j \in \mathcal{F}_j$ for $j = 1, \ldots, k$, $\forall k \ge 1$.

Random vectors ξ_i, $i = 1, 2, \ldots$ are called mutually independent if σ-algebras $\mathcal{F}^{\xi_i}$ generated by ξ_i are mutually independent.

Theorem 1.8. *Let $f(\lambda, \mu)$ be a measurable function defined on the Borel space $(\mathbb{R}^l \times \mathbb{R}^m, \mathcal{B}^l \times \mathcal{B}^m)$. If the l-dimensional random vector ξ is independent of the m-dimensional random vector η, then*

$$E(f(\xi, \eta)|\xi) = g(\xi)$$

where

$$g(\lambda) = Ef(\lambda, \eta), \quad \text{if } Ef(\lambda, \eta) \text{ exists for all } \lambda \in \mathbb{R}^l \text{ in the range of } \xi.$$

Corollary 1.1. *If $f(\xi, \eta) = \eta$, then $g(\lambda) = E\eta$, i.e. $E(\eta|\xi) = E\eta$ when η is independent of ξ.*

Definition. Let $F(x)$ be the distribution function of an n-dimensional random vector ξ. The function

$$\varphi(\lambda) \triangleq Eexp\{i\lambda^\tau \xi\} = \int_{-\infty}^{\infty} \cdots \int_{-\infty}^{\infty} exp\{i\lambda^\tau x\} dF(x), \quad \lambda \in I\!R^n \tag{1.2}$$

is called the characteristic function of ξ.

The characteristic function $\varphi(\lambda)$ uniquely defines the distribution function $F(x)$. In fact, we have the inversion formula, which in the one-dimensional case ($n = 1$) has the form

$$F(b) - F(a) = \lim_{c \to \infty} \frac{1}{2\pi} \int_{-c}^{c} \frac{e^{-i\lambda a} - e^{-i\lambda b}}{i\lambda} \varphi(\lambda) d\lambda \tag{1.3}$$

for any a and b being the continuity points of $F(\cdot)$.

Theorem 1.9. *Let $f_\xi(\lambda)$, $f_\eta(\mu)$ and $f_\zeta(\gamma)$ be characteristic functions of random vectors ξ, η and $\zeta = [\xi^\tau \quad \eta^\tau]^\tau$ respectively, where $\gamma = [\lambda^\tau \quad \eta^\tau]^\tau$. ξ is independent of η if and only if $f_\zeta(\gamma) = f_\xi(\lambda) \cdot f_\eta(\mu)$.*

For the Gaussian distribution (see Example 1.7) the characteristic function is

$$e^{i\lambda^\tau \mu} exp\left\{-\frac{\lambda^\tau R \lambda}{2}\right\}. \tag{1.4}$$

Since the characteristic function and the distribution function of a random vector is in one-to-one correspondence, (1.4) may be taken as a definition of the Gaussian distribution. To be precise, a random vector ξ is called normally distributed if its characteristic function is expressed as (1.4). In this case we write $\xi \in N(\mu, R)$. It can be shown that $E\xi = \mu$, $E(\xi - \mu)(\xi - \mu)^\tau = R$. It is worth noting that the present definition for Gaussian distribution is more general than that given in Example 1.7 because here R is allowed to be degenerate.

Definition. Two random vectors ξ and η with $E\|\xi\|^2 < \infty$, $E\|\eta\|^2 < \infty$ are called uncorrelated if $E(\xi - E\xi)(\eta - E\eta)^\tau = 0$.

It is clear that if ξ is independent of η, then ξ and η are uncorrelated. The converse, in general, is not true. However, if $\zeta = [\xi^\tau \ \eta^\tau]^\tau$ is Gaussian and if ξ and η are uncorrelated, then they must be independent. This is because the covariance of ζ is $R_\zeta = \begin{bmatrix} R_\xi & 0 \\ 0 & R_\eta \end{bmatrix}$ by uncorrelatedness, and the characteristic function for ζ is

$$e^{i\gamma^T E\zeta} exp\left\{-\frac{1}{2}\gamma^T R_\zeta \gamma\right\}$$
$$= \left(e^{i\lambda^T E\xi} exp\left\{-\frac{1}{2}\lambda^T R_\xi \lambda\right\}\right)\left(e^{i\mu^T E\eta} exp\left\{-\frac{1}{2}\mu^T R_\eta \mu\right\}\right),$$

$\gamma = [\lambda^T \quad \mu^T]^T$. Then the desired independence follows from Theorem 1.9.

The method of characteristic functions is a powerful tool for analyzing the limiting behavior of the distribution for sums of mutually independent random variables, ζ_i, and yields a set of classical limit theorems. This is because, by independence, the characteristic function for $\sum_{i=1}^{n} \zeta_i$ is the product of characteristic functions of ζ_i, i.e.

$$Ee^{i\lambda^T \sum_{j=1}^{n} \zeta_j} = \prod_{j=1}^{n} Ee^{i\lambda^T \zeta_j},$$

and the limiting behavior of $\sum_{j=1}^{n} log\, Ee^{i\lambda^T \zeta_j}$ is relatively easy to analyze. For example, $\zeta_j \in N(0, R_j)$, then

$$Ee^{i\lambda^T \zeta_j} = exp\left\{-\frac{1}{2}\lambda^T R_j \lambda\right\}$$

$$\sum_{j=1}^{n} log\, Ee^{i\lambda^T \zeta_j} = -\frac{1}{2}\lambda^T \sum_{j=1}^{n} R_j \lambda.$$

This means that the characteristic function of the sum of independent Gaussian random variables $\sum_{j=1}^{n} \zeta_j$ is $exp\left\{-\frac{1}{2}\lambda^T R \lambda\right\}$ with $R = \sum_{j=1}^{n} R_j \geq 0$. Hence $\sum_{j=1}^{n} \zeta_j$ remains normally distributed.

Definition. Random vectors ξ and η are called conditionally independent given ζ if

$$P^\zeta(\xi < x, \eta < y) = P^\zeta(\xi < x) P^\zeta(\eta < y), \quad \forall x, y$$

where the inequality $\xi < x$ between vectors means inequalities between their components.

The function $f_\xi^\zeta \triangleq E(exp\{i\lambda^\tau \xi\}|\zeta)$ is called the conditional characteristic function of ξ given ζ and it uniquely defines the conditional distribution function $F_\xi^\zeta(x) \triangleq P^\zeta(\xi < x)$ of ξ given ζ.

Definition. Let $|\xi| < \infty$ a.s. ξ is called conditionally Gaussian given ζ if there exist an $\mathcal{F}^\zeta$-measurable random vector $\hat{\xi}$ and an $\mathcal{F}^\zeta$-measurable random matrix $P = P^\tau \geq 0$ a.s. such that

$$E(exp\{i\lambda^\tau \xi\}|\zeta) = e^{i\lambda^\tau \hat{\xi}} e^{-\lambda^\tau P \lambda}, \tag{1.5}$$

for any constant vector λ.

When (1.5) holds, then $\hat{\xi} = E(\xi|\zeta)$ and

$$P = E^\zeta[(\xi - E^\zeta \xi)(\xi - E^\zeta \xi)^\tau]$$

are respectively the conditional mean and the conditional covariance of ξ given ζ, and $P(\xi \in B|\zeta)$ is a Gaussian measure on the Borel σ-algebra $\mathcal{B}$, $B \in \mathcal{B}$. Calculating the conditional characteristic function by using the conditional distribution shows that $A(z)x + b(z)$ is conditionally Gaussian given z if x is conditionally Gaussian given z, and $A(\cdot)$ and $b(\cdot)$ are Borel measurable functions with $\|A(z)\| < \infty$ a.s. and $\|b(z)\| < \infty$ a.s.

Let $f_\xi^\zeta(\lambda)$, $f_\eta^\zeta(\mu)$ and $f_\pi^\zeta(\gamma)$ be the conditional characteristic functions of ξ, η and π, respectively, given ζ with

$$\pi = [\xi^\tau \quad \eta^\tau]^\tau, \qquad \gamma = [\lambda^\tau \quad \mu^\tau]^\tau, \quad \lambda \in \mathbb{R}^n, \quad \mu \in \mathbb{R}^n.$$

Theorem 1.10. *1) Given ζ, ξ and η are conditionally independent if and only if*

$$f_\pi^\zeta(\gamma) = f_\xi^\zeta(\lambda) f_\eta^\zeta(\mu).$$

2) If η and $[\xi^\tau \quad \zeta^\tau]^\tau$ are independent, then ξ and η are conditionally independent given ζ.

3) If ξ and η are conditionally independent given ζ, then for any Borel set B

$$P(\{\xi \in B\}|\zeta, \eta) = P(\{\xi \in B\}|\zeta)$$

and

$$E(\xi|\zeta, \eta) = E(\xi|\zeta).$$

Note that the last equality may not hold if only independency between ξ and η is assumed.

1.5 Random Processes

Let $(\Omega, \mathcal{F}, P)$ be a probability space, $\mathcal{B}(T)$ the σ-algebra of Borel sets on $T = [0, \infty)$.

Definition. An l-dimensional continuous-time stochastic process is a function $\xi_t(\omega)$ defined on $(\Omega \times T, \mathcal{F} \times \mathcal{B}(T))$ and taking values in $(\mathbb{R}^l, \mathcal{B}^l)$. By $\mathcal{F} \times \mathcal{B}(T)$ we mean that the minimal σ-algebra containing all sets of form $F \times B$, $\forall F \in \mathcal{F}$, $\forall B \in \mathcal{B}(T)$. If $\xi_t(\omega)$ is defined only at discrete time $t = 0$, 1, 2, ..., then it is called a discrete-time stochastic process or a random sequence.

For a fixed ω, $\xi_t(\omega)$ is a function of t and is called a trajectory of the stochastic process.

Definition. If for any Borel set $B \in \mathcal{B}^l$

$$\{(\omega, t) : \ \xi_t(\omega) \in B\} \in \mathcal{F} \times \mathcal{B}(T),$$

then $\xi_t(\omega)$ is called the measurable stochastic process.

If a.s. trajectories of ξ_t are left (or right) continuous, then the process is called the left- (right-) continuous process. In this case ξ_t is measurable.

Definition. Let $\{\mathcal{F}_t\}$ be a family of nondecreasing σ-algebras, i.e. $\mathcal{F}_s \subseteq \mathcal{F}_t$, $\forall s \leq t$. If ξ_t is $\mathcal{F}_t$-measurable $\forall t \in T$, then ξ_t is called $\mathcal{F}_t$-adapted.

When ξ_t is $\mathcal{F}_t$-adapted, we write it as $\{\xi_t, \mathcal{F}_t\}$.

If ξ_t is a measurable process, $E\|\xi_t\| < \infty$, $\forall t \in T$ and $\{\mathcal{F}_t\}$ is a family of nondecreasing σ-algebras, then $E(\xi_t|\mathcal{F}_t)$ can be chosen to be measurable. In the sequel we always assume that $E(\xi_t|\mathcal{F}_t)$ is so chosen.

Theorem 1.11 *(Fubini). If ξ_t is a measurable stochastic process, then almost all its trajectories are Borel measurable functions of t. In addition, if $E\xi_t$ exists $\forall t \in T$, then it is also a measurable function. Further, if*

$$\int_S E\|\xi_t\| dt < \infty,$$

then

$$\int_S \|\xi_t\| dt < \infty, \quad a.s.$$

and

$$E \int_S \xi_t dt = \int_S E\xi_t dt$$

where S is any measurable set in T.

Martingales

An adapted process $\{\xi_t, \mathcal{F}_t\}$ with $E|\xi_t| < \infty$, $\forall t \in T$ is called a

$$\begin{cases} \text{martingale,} & \text{if } E(\xi_t|\mathcal{F}_s) = \xi_s, \\ \text{supermartingale,} & \text{if } E(\xi_t|\mathcal{F}_s) \leq \xi_s, \\ \text{submartingale,} & \text{if } E(\xi_t|\mathcal{F}_s) \geq \xi_s \end{cases}$$

for any $s \le t$, $\forall s, t \in T$.

Example 1.10. Let $\{\zeta_i\}$ be a sequence of mutually independent random variables with $E\zeta_i = 0, \forall i \ge 1$. Then $\{\xi_n,\ \mathcal{F}_n^\zeta\}$ is a martingale, where $\xi_n = \sum_{i=1}^n \zeta_i$, $\mathcal{F}_n^\zeta$ is a σ-algebra generated by $(\zeta_i,\ i = 1, ..., n)$. In fact

$$
\begin{aligned}
E(\xi_n|\mathcal{F}_m^\zeta) = & \ E\left[\left(\xi_m + \sum_{m+1}^n \zeta_i\right)|\mathcal{F}_m^\zeta\right] = \xi_m + E\left(\sum_{m+1}^n \zeta_i|\mathcal{F}_m^\zeta\right) \\
= & \ \xi_m + E\sum_{m+1}^n \zeta_i = \xi_m
\end{aligned}
$$

where the last but one equality follows from the Corollary to Theorem 1.7.

Markov Process

Definition. Let $\mathcal{F}_s^\xi$ be the σ-algebra generated by $(\xi_\lambda,\ \lambda \le s)$, and let A belong to the σ-algebra generated by $(\xi_\lambda,\ \lambda \ge t)$. A stochastic process is called a Markov process if

$$
P(A|\mathcal{F}_s^\xi) = P(A|\xi_s), \quad \textit{for any } t \textit{ and } s: \ \forall t \ge s. \tag{1.6}
$$

This means that the process in the future is conditionally independent of the past given the present.

Let B be a Borel set. The Markov property (1.6) is equivalent to

$$
P(\xi_t \in B|\mathcal{F}_s^\xi) = P(\xi_t \in B|\xi_s), \quad \textit{for any } t \textit{ and } s: \ \forall t \ge s. \tag{1.7}
$$

Example 1.11. Let $\phi(t, x, y)$ be a measurable function with respect to t, $t \ge 0$, $x \in \mathbb{R}^n$ and $x \in \mathbb{R}^m$. Let ξ_0, w_1, w_2, ... , be mutually independent. Then

$$
\xi_{t+1} = \phi(t+1, \xi_t, w_{t+1}), \quad t \ge 0
$$

defines a Markov process, for which the Markov property follows from Theorem 1.8:

$$
\begin{aligned}
& P\left(\phi(t+1, \xi_t, w_{t+1}) \in B|\xi_0, \ ..., \ \xi_t\right) \\
= & \ P\left(\phi(t+1, x, w_{t+1}) \in B\right)|_{x=\xi_t} \\
= & \ P\left(\phi(t+1, \xi_t, w_{t+1}) \in B|\xi_t\right)
\end{aligned}
$$

where B is an arbitrary Borel set.

By Property 6 of the conditional expectation we have that for a Markov process $\xi_t,\ t \in T$

$$P(\xi_t \in B|\xi_r) = E\{P(\xi_t \in B|\xi_r,\ \xi_s)|\xi_r\} \tag{1.8}$$

for any $r < s < t$. Then by the Markov property from (1.8) we obtain

$$P(\xi_t \in B|\xi_r) = E\{P(\xi_t \in B|\xi_s)|\xi_r\}, \quad r < s < t. \tag{1.9}$$

Equation (1.9) is called the Kolmogorov equation and can be written via the transition probabilities

$$P_{rt}(x, B) = \int P_{rs}(x,\ dy)\ P_{st}(y\,B) \tag{1.10}$$

where

$$P_{rt}(x, B) = P(\xi_t \in B|\xi_r)|_{\xi_r = x}.$$

If $P_{r+u,\,t+u}(x, B) = P_{r,t}(x, B)$ for all r, t, u, x and B, then the Markov process is said to be stationary. In this case the transition probability $P_{r,t}(x, B)$ depends on the difference of t and r and can be written as $P_t(x, B)$.

For the stationary Markov process the Kolmogorov equation has the form

$$P_{s+t}(x, B) = \int P_s(x,\ dy) P_t(y,\ B). \tag{1.11}$$

Definition. A Markov process that moves through a countable set of states is called a Markov chain.

For a stationary Markov chain the transition probabilities

$$P(\xi_{m+1} = k|\xi_m = j)$$

do not depend on m and can be written as P_{ij}. In this case the Kolmogorov equation takes the form

$$P_{ij} = \sum_k P_{ik} P_{kj}.$$

For more on Markov processes the reader is referred to [Dy], [Chu1].

Stationary Process

Stationarity defined for Markov processes is characterized by the fact that the joint distribution for $[\xi_{t_1}\ \xi_{t_2}\ \dots\ \xi_{t_n}]^T$ is the same as that for $[\xi_{t_1+t}\ \xi_{t_2+t}\ \dots\ \xi_{t_n+t}]^T$. This property is called stationarity in the strict sense. There is also stationarity in the wide sense.

Definition. A process ξ_t is called a wide-sense stationary process if $E\xi_t$ is a constant vector, $E\|\xi_t\|^2 < \infty$ and

$$E(\xi_t - E\xi_t)(\xi_{t+s} - E\xi_{t+s})^* \triangleq R(s)$$

is independent of t, where "$*$" means taking the complex conjugate and transpose. R_s is called the correlation function of ξ_t. It is clear that

$$R(-s) = R(s)^*.$$

Example 1.12. Let $\{\zeta_i\}$ $(i = 0, \pm1, \pm2, ...)$ be a sequence of uncorrelated random vectors such that $E\zeta_i = 0$, $E|\zeta_i|^2 \geq 1$, $E\zeta_i\zeta_j^* = 0$, $i \neq j$. Then $\xi_t = \sum_{i=-\infty}^{\infty} c_{t-i}\zeta_i$ is a wide-sense stationary process for any sequence of complex numbers c_i such that $\sum_{i=-\infty}^{\infty} |c_i|^2 < \infty$. It is easy to see that

$$E\xi_t\xi_s^* = \sum_{i=-\infty}^{\infty} c_{t-s+i}c_i^*.$$

Using the Bochner theorem from the theory of Fourier transformation it is known that there is a nondecreasing function $F(x)$ with $F(\infty) - F(-\infty) = E|\xi_t|^2$ such that

$$R(s) = \int_{-\infty}^{\infty} e^{ixs}dF(x).$$

$F(x)$ is called the spectral function of ξ_t.

In the discrete-time case the correlation function can be presented as

$$R(s) = \int_{-\pi}^{\pi} e^{ixs}dF(x), \quad s = 0, \pm1, \pm2, ...$$

where $F(x)$ is a nondecreasing function defined on $[-\pi, \pi)$.

If $f(x) \triangleq \frac{dF(x)}{dx}$ exists for all x, then $f(x)$ is called the spectral density of $\{\xi_t\}$.

1.6 Stochastic Integral

A stochastic process η_t is said to have orthogonal increments if $E|\eta_t|^2 < \infty$, $-\infty < t < \infty$ and

$$E(\eta_{t_2} - \eta_{s_2})(\eta_{t_1} - \eta_{s_1})^* = 0$$

for any $s_1 < t_1 \leq s_2 < t_2$.

We now define the integral

$$\int_{-\infty}^{\infty} f(t) d\eta_t \tag{1.12}$$

for $f(t)$ being a deterministic function belonging to $L_2(dF)$, i.e.

$$\int_{-\infty}^{\infty} |f(t)|^2 dF(t) < \infty \tag{1.13}$$

where $F(t)$ is a nondecreasing function such that

$$E|\eta_t - \eta_s|^2 = F(t) - F(s). \tag{1.14}$$

If $f(t)$ is a step-wise function

$$f(t) = \sum_{i=1}^{n-1} c_i I_{(t_i, t_{i+1}]}(t), \quad t_1 < t_2 < \dots < t_n,$$

then (1.12) is defined as

$$\int_{-\infty}^{\infty} f(t) d\eta_t \triangleq \sum_{i=1}^{n-1} c_i \left(\eta_{t_{i+1}+0} - \eta_{t_i-0}\right) \tag{1.15}$$

where

$$I_{(t_i, t_{i+1}]}(t) = \begin{cases} 1, & if\ t \in (t_i, t_{i+1}], \\ 0, & otherwise \end{cases}$$

is called the indicator of $(t_i, t_{i+1}]$ and c_i may be complex.

For any $f(t)$ satisfying (1.13) there is a sequence of step-functions $\{f_n(t)\}$ so that

$$\int_{-\infty}^{\infty} |f(t) - f_n(t)|^2 dF(t) \xrightarrow[n \to \infty]{} 0.$$

It can be shown that there is a unique (a.s.) random variable x such that

$$E|x - \int_{-\infty}^{\infty} f_n(t) d\eta_t|^2 \xrightarrow[n \to \infty]{} 0.$$

We define

$$\int_{-\infty}^{\infty} f(t) d\eta_t \triangleq \underset{n \to \infty}{l \cdot i \cdot m \cdot} \int_{-\infty}^{\infty} f_n(t) d\eta_t (= x) \tag{1.16}$$

It is easy to establish that

$$E \int_{-\infty}^{\infty} f(t) d\eta_t \left(\int_{-\infty}^{\infty} g(t) d\eta_t\right)^* = \int_{-\infty}^{\infty} f(t) g^*(t) dF(t) \tag{1.17}$$

if $f(t), g(t) \in L_2(dF)$.

Theorem 1.12 *(The Spectral Representation). For any p-dimensional wide-sense stationary process ξ_t there exists a p-dimensional orthogonal increment process η_t such that*

$$\xi_t = \int e^{its} d\eta_s \tag{1.18}$$

where the integration is carried out over $(-\infty, \infty)$ for the continuous-time process ξ_t and over $[-\pi, \pi]$ for the discrete-time process.

Theorem 1.13 *(Ergodic Theorem). For the wide-sense stationary process ξ_t the time average converges to a random variable η_0 in the mean square sense, i.e.*

$$\begin{cases} \underset{T \to \infty}{l \cdot i \cdot m \cdot} \frac{1}{T} \sum_{i=1}^{T} \xi_i = \eta_0, & \textit{in the discrete} - \textit{time case}, \\ \underset{T \to \infty}{l \cdot i \cdot m \cdot} \frac{1}{T} \int_0^T \xi_t dt = \eta_0, & \textit{in the continuous} - \textit{time case}. \end{cases} \tag{1.19}$$

We say that ξ_t is ergodic if $\eta_0 = E\xi_t$.

For detailed material on stationary processes we refer readers to [Do], [Ro], [Ca].

The integral we just discussed is with respect to an orthogonal increment process with arbitrary distribution and its integrand is a deterministic function. We now define Ito's stochastic integral which is with respect to a normally distributed process called the Wiener process and its integrand may be a stochastic process.

Definition. An l-dimensional process w_t that is adapted to a nondecreasing family of σ-algebras $\mathcal{F}_t$, $t \geq 0$ is called a Wiener process if $w_0 = 0$, $Ew_t = 0$, $E\|w_t\|^2 < \infty$, $\forall t \geq 0$,

$$E\left[(w_t - w_s)(w_t - w_s)^\tau \,|\mathcal{F}_s\right] = (t-s)I, \quad t \geq s \tag{1.20}$$

and if $\{w_t, \mathcal{F}_t\}$ is a continuous martingale, where I denotes an identity matrix.

Theorem 1.14. *A Wiener process is an independent increment process, i.e. $w_{t_1} - w_{s_1}$ is independent of $w_{t_2} - w_{s_2}$ for $s_1 < t_1 \leq s_2 < t_2$. Further, any of its increment $w_t - w_s$ is Gaussian and the iterated logarithm law holds*

$$\limsup_{t \to \infty} \frac{\|w_t\|^2}{2lt \log t \log t} = 1 \quad a.s. \tag{1.21}$$

where l is the dimension of w_t.

The Wiener process is also called Brownian motion. Clearly, it is also an orthogonal increment process. We now define classes of integrands.

We say that $\{\xi_t, \mathcal{F}_t\} \in \mathcal{M}_T$, if

$$E\int_0^T \|\xi_t\|^2 dt < \infty$$

and $\{\xi_t, \mathcal{F}_t\} \in \mathcal{P}_T$, if

$$P\left(\int_0^T \|\xi_t\|^2 dt < \infty\right) = 1.$$

Let $\{t_i\}$ be a partition of $[0, T]$ with $0 = t_0 < t_1 < \ldots < t_n = T$ and let ξ_i be $\mathcal{F}_{t_i}$-measurable. We say that

$$\xi_t = \xi_0 I_{[0]} + \sum_{i=0}^{n-1} \xi_i I_{(t_i, t_{i+1}]}$$

is a simple process.

Like (1.15) the Ito stochastic integral is defined for simple processes as

$$\int_0^t \xi_s dw_s \triangleq \sum_{i=0}^{m} \xi_i (w_{t_{i+1}} - w_{t_i}) + \xi_{m+1}(w_t - w_{t_m+1}), \quad t_{m+1} < t \le t_{m+2},$$

where $\{w_t, \mathcal{F}_t\}$ is a Wiener process.

For $\xi_t \in \mathcal{M}_T$ there is a sequence of simple processes $\xi_t^n \in \mathcal{M}_T$ and a random variable η independent of the selection of ξ_t^n such that

$$E\int_0^T |\xi_t - \xi_t^n|^2 dt \xrightarrow[n \to \infty]{} 0$$

and

$$\lim_{n\to\infty} E\left(\eta - \int_0^T \xi_t^n dw_t\right)^2 = 0$$

The Ito stochastic integral for $\xi_t \in \mathcal{M}_T$ is defined by

$$\int_0^T \xi_t dw_t \triangleq \underset{n \to \infty}{l \cdot i \cdot m \cdot} \int_0^T \xi_t^n dw_t (= \eta)$$

and

$$\int_0^t \xi_s dw_s \triangleq \int_0^T I_{[0,t]} \xi_s dw_s.$$

Further, for any $\xi_t \in \mathcal{P}_T$ there exist $\xi_t^n \in \mathcal{M}_T$ and a random process ξ_t independent of the selection of $\{\xi_t^n\}$ such that

$$\int_0^T |\xi_t - \xi_t^n|^2 dt \xrightarrow[n \to \infty]{P} 0$$

and

$$\int_0^T \xi_t^n dw_t \xrightarrow[n \to \infty]{P} \zeta.$$

The Ito stochastic integral for $\xi_t \in \mathcal{P}_T$ is then defined by

$$\int_0^T \xi_t dw_t \triangleq p - \lim_{n\to\infty} \int_0^T \xi_t^n dw_t (= \zeta).$$

The Ito stochastic integral has the following properties.

1. $\int_0^t (a\xi_s^1 + b\xi_s^2) dw_s = a \int_0^t \xi_s^1 dw_s + b \int_0^t \xi_s^2 dw_s, \quad \forall t \in [0,T]$
for ξ_s^1, $\xi_s^2 \in \mathcal{P}_T$ and a, b being constants.

2. $\int_0^t \xi_s dw_s$ is continuous in $t \in [0,T]$ for $\xi_s \in \mathcal{P}_T$.

3. $\int_0^t \xi_s dw_s = \int_0^u \xi_s dw_s + \int_u^t \xi_s dw_s, \ 0 \le u \le t$
for $\xi_s \in \mathcal{P}_T$.

4. If $\xi_t, \xi_t^1, \xi_t^2 \in \mathcal{M}_T$, then

$$E \int_0^t \xi_s dw_s = 0, \quad \forall t \in [0,T],$$

$$E\left(\int_0^t \xi_\lambda dw_\lambda | \mathcal{F}_s\right) = \int_0^s \xi_\lambda dw_\lambda, \quad s \le t \tag{1.22}$$

i.e. $\left(\int_0^t \xi_\lambda dw_\lambda, \ \mathcal{F}_t\right)$ is a martingale, and

$$E\left(\int_0^t \xi_s^1 dw_s\right)\left(\int_0^t \xi_s^2 dw_s\right) = E \int_0^t \xi_s^1 \xi_s^2 ds. \tag{1.23}$$

These properties can easily be verified for simple processes and then for more general processes by passing to the limit.

The stochastic integral can be defined with respect to a semi-martingale which is much more general than the Wiener process. For this the reader is referred to [LSh], [GSk] and [IW].

1.7 Stochastic Differential Equations

Let $\{a_t, \mathcal{F}_t\}$, $\{b_t, \mathcal{F}_t\}$ be measurable processes and $|a_t|^{\frac{1}{2}} \in \mathcal{P}_T$, $b_t \in \mathcal{P}_T$.

The process ξ_t defined by

$$\xi_t = \xi_0 + \int_0^t a_s ds + \int_0^t b_s dw_s \tag{1.24}$$

is called an Ito process. If a_t and b_t are adapted to $\mathcal{F}_t^\xi \triangleq \sigma(\xi_\lambda, \lambda \leq t)$, then ξ_t is called a diffusion process. We also say that the process defined by (1.24) has the stochastic differential

$$d\xi_t = a_t dt + b_t dw_t \tag{1.25}$$

Theorem 1.15 *(Ito's Formula). Assume that $\{a_t, \mathcal{F}_t\}$ is an l-dimensional measurable process, $\|a_t\|^{\frac{1}{2}} \in \mathcal{P}_T$, $\{B_t, \mathcal{F}_t\}$ is an $l \times m$ measurable matrix, $\|B_t\| \in \mathcal{P}_T$, $\{w_t, \mathcal{F}_t\}$ is an m-dimensional Wiener process.*

$$d\xi_t = a_t dt + B_t dw_t.$$

Further, assume that the functions $f_t(t,x)$, $f_x(t,x)$ and $f_{xx}(t,x)$ are continuous, where $f_t(t,x) = \dfrac{\partial f(t,x)}{\partial t}$,

$$f_x(t,x) \triangleq \begin{bmatrix} \dfrac{\partial f(t,x)}{\partial x^1} \\ \vdots \\ \dfrac{\partial f(t,x)}{\partial x^l} \end{bmatrix}, \quad f_{xx} \triangleq \begin{bmatrix} \dfrac{\partial^2 f(t,x)}{\partial x^1 \partial x^1} & \cdots & \dfrac{\partial^2 f(t,x)}{\partial x^l \partial x^1} \\ \vdots & \ddots & \vdots \\ \dfrac{\partial^2 f(t,x)}{\partial x^1 \partial x^l} & \cdots & \dfrac{\partial^2 f(t,x)}{\partial x^l \partial x^l} \end{bmatrix},$$

$$x = [x^1 \cdots x^l]^\tau.$$

Then

$$\begin{aligned} df(t,\xi_t) &= \left[f_t(t,\xi_t) + f_x^\tau(t,\xi_t) a_t + \frac{1}{2} tr f_{xx}(t,\xi_t) B_t B_t^\tau \right] dt \\ &\quad + f_x^\tau(t,\xi_t) B_t dw_t. \end{aligned} \tag{1.26}$$

Comparing (1.26) with the deterministic calculation, we find that in (1.26) there is an additional term which equals $\frac{1}{2} f''_{xx}(t,\xi_t) b_t^2$ for the one-dimensional ξ_t defined by (1.25). This term appears because, roughly speaking, $(dw_t)^2$ is of order dt rather than $(dt)^2$.

Example 1.13. For $f(t,x) = x^\tau C_t x$ the Ito's formula (1.26) gives us

$$\begin{aligned} d\xi_t^\tau C_t \xi_t &= \left[\xi_t^\tau \dot{C}_t \xi_t + \xi_t^\tau (C_t + C_t^\tau) a_t + tr C_t B_t B_t^\tau \right] dt \\ &\quad + \xi_t^\tau C_t B_t dw_t + \xi_t^\tau C_t^\tau B_t dw_t. \end{aligned} \tag{1.27}$$

Let $(C_T, \mathcal{B}_T)$ be the measurable space of continuous functions x defined on $[0,T]$ and let $\mathcal{B}_t = \sigma[x : x_s\ s \leq t]$ be the sub-σ-algebra of $\mathcal{B}_T$.

Definition. Let $a_t(x)$ and $b_t(x)$ be $\mathcal{B}_t$-measurable for any $t \in [0,T]$ and $x \in \mathcal{B}_T$. The $\mathcal{F}_t$-adapted process $\{\xi_t,\ \mathcal{F}_t\}$ is called the strong solution of the stochastic differential equation

$$d\xi_t = a_t(\xi)dt + b_t(\xi)dw_t \tag{1.28}$$

with $\mathcal{F}_0$-measurable initial value η if $|a_t(\xi)|^{\frac{1}{2}} \in \mathcal{P}_T$, $b_t(\xi) \in \mathcal{P}_T$ and for any $t \in [0,T]$

$$\xi_t = \eta + \int_0^t a_s(\xi)ds + \int_0^t b_s(\xi)dw_s.$$

For detailed discussions on existence and uniqueness of the strong solution, we refer the reader to [LSh], [GSk] and [IW].

We now consider the linear stochastic differential equation

$$d\xi_t = A_t\xi_t dt + b_t dt + D_t dw_t, \tag{1.29}$$

where $\{w_t,\ \mathcal{F}_t\}$ is an m-dimensional Wiener process, $\{A_t,\ \mathcal{F}_t\}$ $\{b_t,\ \mathcal{F}_t\}$ and $\{D_t,\ \mathcal{F}_t\}$ are of compatible dimensions and $\|b_t\| \in \mathcal{P}_T$, $\|D_t\| \in \mathcal{P}_T$, $\|A_t\|^{\frac{1}{2}} \in \mathcal{P}_T$.

Let ϕ_{ts} be the fundamental solution matrix of the linear ordinary differential equation, i.e.

$$\frac{d}{dt}\phi_{ts} = A_t\phi_{ts}, \quad \phi_{ss} = I, \forall t > s \quad a.s. \tag{1.30}$$

Then it is easy to verify that the strong solution of (1.29) is given by

$$\xi_t = \phi_{t0}\xi_0 + \int_0^t \phi_{ts}b_s ds + \phi_{t0}\int_0^t \phi_{s0}^{-1}D_s dw_s. \tag{1.31}$$

When A_t, b_t and D_t are deterministic, then ξ_t expressed by (1.31) is Gaussian.

CHAPTER 2

Limit Theorems on Martingales

The martingale method is a powerful tool for analyzing recursive identification algorithms and stochastic adaptive control systems. In this chapter, we first present classical convergence theorems for martingales [Do], [Cho], [St], [N], [LSh]. Then we estimate the sums of martingale difference sequence with single– and double–indexed random weights [St], [LW1], [CG3], [GHH], [HG]. These results are the bases of analysis for the least–square–based estimates for coefficients and orders which will be considered in later chapters.

2.1 Martingale Convergence Theorems

Definition 2.1 *Let $\{\mathcal{F}_n\}$ be a nondecreasing family of σ–algebras. A random variable valued in the set $\{0, 1, 2, \ldots, \infty\}$ is called a Markov time (with respect to $\{\mathcal{F}_n\}$) if*

$$\{\tau = n\} \in \mathcal{F}_n \quad \textit{for any } n \geq 0.$$

If, in addition, $P(\tau < \infty) = 1$, then τ is called a stopping time.

Lemma 2.1. *Let $\{\xi_n, \mathcal{F}_n\}$ be a stochastic sequence, τ a Markov time and B a Borel set. Then after τ the first time τ_B hitting the set B*

$$\tau_B = \begin{cases} \inf\{n : \tau < n, \xi_n \in B\} \\ \infty, \quad \textit{if } \xi_n \notin B \textit{ for } \forall n > \tau \end{cases}$$

is a Markov time.

Proof. For the proof we need only to note that

$$\{\tau_B = n\} = \bigcup_{i=0}^{n-1} \{\{\tau = i\} \cap \{\xi_{i+1} \notin B, \cdots, \xi_{n-1} \notin B, \xi_n \in B\}\} \in \mathcal{F}_n, \ \forall n \geq 0.$$

Q.E.D.

We now define the number of up–crossing of the interval (a, b) by a submartingale $\{\xi_n, \mathcal{F}_n\}$, $n = 1, \ldots, N$.

$$\tau_0 = 0$$

$$\tau_1 = \begin{cases} \min\{0 < n < N : \xi_n \leq a\} \\ N, \text{ if } \{0 < n < N : \xi_n \leq a\} = \emptyset \end{cases}$$

$$\tau_2 = \begin{cases} \min\{\tau_1 < n \leq N : \xi_n \geq b\} \\ N, \text{ if } \{\tau_1 < n \leq N : \xi_n \geq b\} = \emptyset \end{cases}$$

$$\cdots$$

$$\tau_{2m-1} = \begin{cases} \min\{\tau_{2m-2} < n \leq N : \xi_n \leq a\} \\ N, \text{ if } \{\tau_{2m-2} < n \leq N : \xi_n \leq a\} = \emptyset \end{cases}$$

$$\tau_{2m} = \begin{cases} \min\{\tau_{2m-1} < n \leq N : \xi_n \geq b\} \\ N, \text{ if } \{\tau_{2m-1} < n \leq N : \xi_n \geq b\} = \emptyset. \end{cases}$$

The largest m for which $\xi_{\tau_{2m}} \geq b$ is called the number of up–crossing of the interval (a, b) by the process $\{\xi_n, \mathcal{F}_n\}$ and is denoted by $\beta(a, b)$.

By Lemma 2.1 we see $\{\tau_1 = n\} \in \mathcal{F}_n$, $\forall n < N$ and $\{\tau_1 = N\} = \{\bigcup_{i=0}^{N-1} \{\tau_1 = i\}\}^c \in \mathcal{F}_N$. Hence τ_1 is a Markov time.

Assume τ_i is a Markov time. Then again by Lemma 2.1

$$\{\tau_{i+1} = n\} \in \mathcal{F}_n \quad \forall n < N$$

and $\{\tau_{i+1} = N\} = \{\bigcup_{j=0}^{N-1} \{\tau_{i+1} = j\}\}^c \in \mathcal{F}_N$.

This shows that all τ_i are Markov times, $i = 0, 1, \ldots, 2m$.

Theorem 2.1 *(Doob). The following inequalities are valid for a submartingale $\{\xi_n, \mathcal{F}_n\}$, $n = 1, \ldots, N$*

$$E\beta(a, b) \leq \frac{E(\xi_N - a)^+}{b - a} \leq \frac{E\xi_N^+ + |a|}{b - a}, \tag{2.1}$$

where $\xi^+ = \begin{cases} \xi, & \text{if } \xi > 0 \\ 0, & \text{if } \xi \leq 0. \end{cases}$

Proof. We note that $\beta(a,b)$ equals the number $\beta(0,b-a)$ of up-crossing of the interval $(0,b-a)$ by the submartingale $\{\xi_n-a,\mathcal{F}_n\}$ or by $\{(\xi_n-a)^+,\mathcal{F}_n\}$.

Since for $n \geq m$

$$\begin{aligned}(\xi_m - a)^+ \leq \; & [E(\xi_n - a|\mathcal{F}_m)]^+ \\ = \; & \{E[(\xi_n - a)^+ - (\xi_n - a)^-]|\mathcal{F}_m\}^+ \leq E\left[(\xi_n - a)^+|\mathcal{F}_m\right],\end{aligned}$$

$\{(\xi_n - a)^+, \mathcal{F}_n\}$ is a submartingale. Then, without loss of generality, for (2.1) it suffices to prove that for a nonnegative submartingale $\{\xi_n,\mathcal{F}_n\}$ we have

$$E\beta(0,b) \leq \frac{E\xi_N}{b}. \tag{2.2}$$

Define $\xi_0 = 0$,

$$\eta_i = \begin{cases} 0, & \text{if } \tau_{m-1} < i \leq \tau_m \text{ and } m \text{ is odd}, \\ 1, & \text{if } \tau_{m-1} < i \leq \tau_m \text{ and } m \text{ is even}. \end{cases} \tag{2.3}$$

If m is even, then from τ_{m-1} to τ_m the trajectory of ξ_n once up-crosses the interval $(0,b)$. Hence we have

$$\begin{aligned}&\sum_{i=\tau_{m-1}+1}^{\tau_m} \eta_i(\xi_i - \xi_{i-1}) = \sum_{i=\tau_{m-1}+1}^{\tau_m} (\xi_i - \xi_{i-1}) \\ &= \xi_{\tau_m} - \xi_{\tau_{m-1}} \geq \xi_{\tau_m} \geq b\end{aligned}$$

and

$$\sum_{i=1}^{N} \eta_i(\xi_i - \xi_{i-1}) \geq b\beta(0,b). \tag{2.4}$$

Further, the set $\{\eta_i = 1\}$ is $\mathcal{F}_{i-1}$-measurable, because τ_i is Markov time, $i = 1,2,\ldots$ and

$$\{\eta_i = 1\} = \bigcup_{k\geq 1} \{\{\tau_{2k-1} < i\} \cap \{\tau_{2k} < i\}^c\}.$$

Taking expectation from both sides of (2.4) leads to

$$\begin{aligned}bE\beta(0,b) \; &\leq \; E\sum_{i=1}^{N} \eta_i(\xi_i - \xi_{i-1}) \\ &= \sum_{i=1}^{N} \int_{\{\eta_i=1\}} (\xi_i - \xi_{i-1})\, dP\,. \end{aligned} \tag{2.5}$$

Using (1.1) and the fact that $\{\eta_i = 1\}$ is $\mathcal{F}_{i-1}$- measurable we then have

$$\begin{aligned}
bE\beta(0,b) &\leq \sum_{i=1}^{N} \int_{\{\eta_i=1\}} E[(\xi_i - \xi_{i-1})|\mathcal{F}_{i-1}]\, dP \\
&\leq \sum_{i=1}^{N} \int_{\{\eta_i=1\}} [E(\xi_i|\mathcal{F}_{i-1}) - \xi_{i-1}]\, dP \\
&\leq \sum_{i=1}^{N} \int_{\Omega} [E(\xi_i|\mathcal{F}_{i-1}) - \xi_{i-1}]\, dP \\
&= E\xi_N
\end{aligned}$$

where the last inequality holds because $\{\xi_i, \mathcal{F}_i\}$ is a submartingale and hence the integrand is nonnegative.

Thus (2.2) and hence (2.1) is proved. Q.E.D.

Theorem 2.2 *(Doob). Let $\{\xi_n, \mathcal{F}_n\}$ be a submartingale with*

$$\sup_n E\xi_n^+ < \infty \quad a.s.$$

Then there is a random variable ξ such that

$$\lim_{n\to\infty} \xi_n = \xi \quad a.s. \text{ and } E|\xi| < \infty.$$

Proof. Set

$$\limsup_{n\to\infty} \xi_n \triangleq \xi^* \quad \liminf_{n\to\infty} \xi_n \triangleq \xi_*.$$

Assume that

$$P(\xi^* > \xi_*) > 0. \tag{2.6}$$

Then

$$\{\xi^* > \xi_*\} = \bigcup_{a<b} \{\xi^* > b > a > \xi_*\}$$

where a and b run over all rational numbers.

By (2.6) there exist rational numbers $a < b$ such that

$$P(\xi^* > b > a > \xi_*) > 0\,. \tag{2.7}$$

Let $\beta_N(a,b)$ be the number of up-crossing of the interval (a,b) by $\{\xi_n, \mathcal{F}_n\}$, $n \leq N$.

By Lemma 2.1 we have

$$E\beta_N(a,b) \leq \frac{E\xi_N^+ + |a|}{b-a}\,. \tag{2.8}$$

Setting
$$\beta_\infty(a,b) = \lim_{N\to\infty} \beta_N(a,b)$$
and using Theorem 1.1 from (2.8) we have
$$E\beta_\infty(a,b) = \lim_{N\to\infty} E\beta_N(a,b) \le \frac{\sup_N E\xi_N^+ + |a|}{b-a} < \infty.$$

It is clear that (2.7) means $P(\beta_\infty(a,b) = \infty) > 0$. This contradicts $E\beta_\infty(a,b) < \infty$. Hence
$$P(\xi^* = \xi_*) = 1,$$
or ξ_n converges to a limit ξ which may be infinite for some ω.

By Lemma 1.2 we have
$$E\xi^+ = E\liminf_{n\to\infty}\xi_n^+ \le \liminf_{n\to\infty} E\xi_n^+ \le \sup_n E\xi_n^+ < \infty$$
and
$$\begin{aligned} E\xi^- &= E\liminf_{n\to\infty}\xi_n^- \le \liminf_{n\to\infty} E\xi_n^- \le \sup_n E\xi_n^- \\ &= \sup_n (E\xi_n^+ - E\xi_n) \le \sup_n (E\xi_n^+ - E\xi_1) < \infty \end{aligned}$$
since $E\xi_1 \le E\xi_n$, $n \ge 1$ for submartingales. Hence $E|\xi| < \infty$. Q.E.D.

Corollary 2.1. *If $\{\xi_n, \mathcal{F}_n\}$ is a nonnegative supermartingale or nonpositive submartingale, then*
$$\lim_{n\to\infty} \xi_n = \xi \ a.s. \quad and \quad E|\xi| < \infty.$$

Proof. If $\{\xi_n, \mathcal{F}_n\}$ is a nonpositive submartingale, then $\xi_n^+ = 0$ and the assertions directly follow from the theorem. If $\{\xi_n, \mathcal{F}_n\}$ is a nonnegative super-martingale, then $\{-\xi_n \mathcal{F}_n\}$ is a nonpositive submartingale. Q.E.D.

Corollary 2.2. *If $\{\xi_n, \mathcal{F}_n\}$ is a martingale and* $\sup_n E|\xi_n| < \infty$, *then* $\lim_{n\to\infty} \xi_n = \xi$ *a.s. and* $E|\xi| < \infty$.

Proof. For the martingale $\{\xi_n, \mathcal{F}_n\}$, $E\xi_n = E\xi_1$, $\forall n$. Hence we have
$$\begin{aligned} E|\xi_n| &= E\xi_n^+ + E\xi_n^- = 2E\xi_n^+ - E\xi_n \\ &= 2E\xi_n^+ - E\xi_1 . \end{aligned}$$
Then by assumption we see that
$$\begin{aligned} \sup_n E\xi_n^+ &= \tfrac{1}{2}\sup_n (E|\xi_n| - E\xi_1) \\ &= \tfrac{1}{2}\sup_n E|\xi_n| - \frac{1}{2}E\xi_1 < \infty . \end{aligned}$$

The assertions then follow from the theorem because a martingale is also a submartingale.

2.2 Local Convergence Theorems

In the last section we have established the a.s. convergence for martingales. However, a martingale may converge not on the whole space Ω but on a subset of Ω. This section indicates the set where a martingale converges.

Lemma 2.2. *Let $\{\xi_n, \mathcal{F}_n\}$ be a sequence of random vectors, $\xi_n \in R^m$ and let G be a Borel set in R^m. Then the first exit time τ from G defined by*

$$\tau = \begin{cases} \min\{n : \xi_n \notin G\} \\ \infty, \quad \text{if } \xi_n \in G \quad \forall n \end{cases}$$

is a Markov time.

Proof. Noticing

$$\{\tau = n\} = \{\xi_0 \in G, \xi_1 \in G, \ldots, \xi_{n-1} \in G, \xi_n \notin G\} \in \mathcal{F}_n$$

we see that τ is a Markov time. Q.E.D.

Lemma 2.3. *Let $\{\xi_n, \mathcal{F}_n\}$ be a martingale (supermartingale, submartingale) and τ a Markov time. Then the process $\{\xi_{\tau\wedge n}, \mathcal{F}_n\}$ stopped at τ is again a martingale (supermartingale, submartingale), where $\tau \wedge k \triangleq$* $\min(\tau, k)$.

Proof. From the expression

$$\xi_\tau I_{\{\tau \le k-1\}} = \xi_0 I_{\{\tau=0\}} + \cdots + \xi_{k-1} I_{\{\tau=k-1\}}$$

we find that $\xi_\tau I_{\{\tau\le k-1\}}$ is $\mathcal{F}_{k-1}$-measurable.

If $\{\xi_n, \mathcal{F}_n\}$ is a martingale, then we have

$$\begin{aligned} E(\xi_{\tau\wedge k}|\mathcal{F}_{k-1}) &= E\{(\xi_\tau I_{\{\tau\le k-1\}} + \xi_k I_{\{\tau>k-1\}})|\mathcal{F}_{k-1}\} \\ &= \xi_\tau I_{\{\tau\le k-1\}} + E(\xi_k I_{\{\tau\le k-1\}^c}|\mathcal{F}_{k-1}) \\ &= \xi_\tau I_{\{\tau\le k-1\}} + I_{\{\tau>k-1\}} E(\xi_k|\mathcal{F}_{k-1}) \\ &= \xi_\tau I_{\{\tau\le k-1\}} + I_{\{\tau>k-1\}}\xi_{k-1} \\ &= \xi_{\tau\wedge(k-1)}. \end{aligned}$$

This shows that $\xi_{\tau\wedge(k-1)}$ is $\mathcal{F}_{k-1}$-measurable, and $\{\xi_{\tau\wedge k}, \mathcal{F}_k\}$ is a martingale. For the case of supermartingale or submartingale the proof is completely the same. Q.E.D.

Definition 2.2 *An adapted sequence of random vectors $\{x_n, \mathcal{F}_n\}$ is called a martingale difference sequence (m.d.s.) if $E(x_n|\mathcal{F}_{n-1}) = 0$, $\forall n \ge 1$.*

It is clear that $\{\xi_n, \mathcal{F}_n\}$ is a martingale if $\xi_n = \sum_{i=1}^{n} x_i$ and $\{x_n, \mathcal{F}_n\}$ is a m.d.s.

Theorem 2.3. *Let $\{x_n, \mathcal{F}_n\}$ be a one-dimensional m.d.s.. Then as $n \to \infty$ $\xi_n \triangleq \sum_{i=1}^{n} x_i$ converges on*

$$A \triangleq \left\{ \sum_{i=1}^{\infty} E(x_i^2 | \mathcal{F}_{i-1}) < \infty \right\}.$$

Proof. Since $\sum_{i=1}^{n+1} E(x_i^2|\mathcal{F}_{i-1})$ is $\mathcal{F}_n$–measurable, by Lemma 2.2 the first exit time

$$\tau_M \triangleq \begin{cases} \min\{n : n \geq 1, \sum_{i=1}^{n+1} E(x_i^2 \,|\, \mathcal{F}_{i-1}) > M\} \\ \infty, \text{ if } \sum_{i=1}^{\infty} E(x_i^2|\mathcal{F}_{i-1}) \leq M \end{cases}$$

is a Markov time and by Lemma 2.3 $\xi_{n \wedge \tau_M}$ is a martingale, where M is a constant.

Noticing that

$$\xi_{n\wedge\tau_M} = \sum_{i=1}^{n} x_i I_{\{i \leq \tau_M\}}$$

and that $\{i \leq \tau_M\} = \{\tau_M < i\}^c = \{\tau_M \leq i-1\}^c$ is $\mathcal{F}_{i-1}$–measurable, we find

$$\begin{aligned} \left(E|\xi_{n\wedge\tau_M}|\right)^2 \;\leq\; & E\xi^2_{n\wedge\tau_M} = E\sum_{i=1}^{n} x_i^2 I_{\{i\leq\tau_M\}} \\ = \;& E\Big\{\sum_{i=1}^{n} E(x_i^2 I_{\{i\leq\tau_M\}}|\mathcal{F}_{i-1})\Big\} \\ = \;& E\Big\{\sum_{i=1}^{n} I_{\{i\leq\tau_M\}} E(x_i^2|\mathcal{F}_{i-1})\Big\} \\ = \;& E\Big\{\sum_{i=1}^{n\wedge\tau_M} E(x_i^2|\mathcal{F}_{i-1}\Big\} \leq M\,. \end{aligned}$$

By Corollary 2.2, $\xi_{n\wedge\tau_M}$ converges as $n \to \infty$. It is clear that $\xi_{n\wedge\tau_M} = \xi_n$ on the set $\{\tau_M = \infty\}$. Hence as $n \to \infty$ ξ_n pathwisely converges on

$\{\tau_M = \infty\}$. Noticing that M is arbitrary, we find that ξ_n converges to $\bigcup_{M=1}^{\infty} \{\tau_M = \infty\}$ which obviously equals A. Thus the proof is complete.

Q.E.D.

Theorem 2.4. *Let $\{x_n, \mathcal{F}_n\}$ be a m.d.s. If $E(\sup_n x_n)^+ < \infty$, then $\xi_n (\triangleq \sum_{i=1}^{n} x_i)$ converges on*

$$A_1 \triangleq \{\sup_n \xi_n < \infty\}.$$

If $E(\inf_n x_n)^- < \infty$, then ξ_n converges on

$$A_2 \triangleq \{\inf_n \xi_n > -\infty\}.$$

Proof. It suffices to prove the first assertion, because the second one is reduced to the first one by replacing x_n by $-x_n$.

Set

$$\tau_M = \begin{cases} \min\{n : \xi_n > M, \quad n \geq 1\} \\ \infty, \text{if } \xi_n < M, \forall n. \end{cases}$$

By Lemmas 2.2 and 2.3, $(\xi_{n\wedge\tau_M}, \mathcal{F}_n)$ is a martingale. It is easy to see that

$$\xi_{n\wedge\tau_M} = \begin{cases} \leq M, \quad if \quad \mathrm{n} < \tau_M \\ = \xi_{\tau_M - 1} + x_{\tau_M} \leq M + \sup_n x_n, \quad \text{if } n \geq \tau_M. \end{cases}$$

Hence we have

$$\sup_n E(\xi_{n\wedge\tau_M})^+ \leq E[\sup_n (\xi_{n\wedge\tau_M})^+]$$
$$\leq E[M + (\sup_n x_n)^+] < \infty.$$

By Theorem 2.2 $\xi_{n\wedge\tau_M}$ converges a.s. as $n \to \infty$.

Paying attention to the fact that

$$\xi_{n\wedge\tau_M} = \xi_n \quad \text{on} \quad \{\tau_M = \infty\}$$

we conclude that ξ_n converges on $\{\tau_M = \infty\}$ and consequently on

$$\bigcup_{M=1}^{\infty} \{\tau_M = \infty\} = A_1 \text{ as } n \to \infty.$$

Q.E.D.

Theorem 2.5 *(Borel-Cantelli-Levy). Let $\{B_i\}$ be a sequence of events, $B_i \in \mathcal{F}_i$. Then $\sum_{i=1}^{\infty} I_{B_i} < \infty$ if and only if $\sum_{i=1}^{\infty} P(B_i|\mathcal{F}_{i-1}) < \infty$, or equivalently,*

$$\bigcap_{n=1}^{\infty} \bigcup_{i=n}^{\infty} B_i = \Big\{\sum_{i=1}^{\infty} P(B_i|\mathcal{F}_{i-1}) = \infty\Big\}. \tag{2.9}$$

Proof. Set

$$U_n = \sum_{i=1}^{n} \{I_{B_i} - E(I_{B_i}|\mathcal{F}_{i-1})\}. \tag{2.10}$$

Clearly, $\{U_n, \mathcal{F}_n\}$ is a martingale and $(I_{B_i} - E(I_{B_i}|\mathcal{F}_{i-1}), \mathcal{F}_i)$ is a m.d.s. Notice that

$$\left|I_{B_i} - E(I_{B_i}|\mathcal{F}_{i-1})\right| \le 1.$$

Then by Theorem 2.4 U_n converges on

$$\{\sup_n U_n < \infty\} \bigcup \{\inf_n U_n > -\infty\}.$$

If $\sum_{i=1}^{\infty} I_{B_i} < \infty$, then from (2.10) we find that

$$\sup_n U_n < \infty.$$

Hence U_n converges. From (2.10) this implies that

$$\sum_{i=1}^{\infty} P(B_i|\mathcal{F}_{i-1}) < \infty.$$

Conversely, if $\sum_{i=1}^{\infty} P(B_i|\mathcal{F}_{i-1}) < \infty$, then from (2.10) we see that $\inf_n U_n > -\infty$, but $\{\inf_n U_n > -\infty\}$ is contained in the set where U_n converges. From the convergence of U_n again by (2.10) it follows that $\sum_{i=1}^{\infty} I_{B_i} < \infty$.

Q.E.D.

Corollary 2.3 *(Borel-Cantelli lemma). Let $\{B_i\}$ be a sequence of events. If $\sum_{i=1}^{\infty} P(B_i) < \infty$, then the probability of the fact that B_i occur infinitely often is 0, i.e.*

$$P(\bigcap_{n=1}^{\infty} \bigcup_{i=n}^{\infty} B_i) = 0. \tag{2.11}$$

If B_i are mutually independent and $\sum_{i=1}^{\infty} P(B_i) = \infty$, then

$$P(\bigcap_{n=1}^{\infty} \bigcup_{i=n}^{\infty} B_i) = 1. \tag{2.12}$$

Proof. Denote by $\mathcal{F}_n$ the σ-algebra generated by $\{B_1, \dots, B_n\}$. Then $\{\mathcal{F}_n\}$ is a nondecreasing family of σ–algebras.

If $\sum_{i=1}^{\infty} P(B_i) < \infty$, then

$$\infty > \sum_{i=1}^{\infty} P(B_i) = E\left\{\sum_{i=1}^{\infty} E(I_{B_i}|\mathcal{F}_{i-1})\right\}.$$

Hence

$$\sum_{i=1}^{\infty} E(I_{B_i}|\mathcal{F}_{i-1}) < \infty \quad a.s.$$

and (2.11) follows from (2.9).

When B_i are independent, then

$$\sum_{i=1}^{\infty} P(B_i|\mathcal{F}_{i-1}) = \sum_{i=1}^{\infty} P(B_i).$$

Consequently, $\sum_{i=1}^{\infty} P(B_i) = \infty$ implies

$$\sum_{i=1}^{\infty} P(B_i|\mathcal{F}_{i-1}) = \infty.$$

Then by Theorem 2.4, (2.12) follows. Q.E.D.

Corollary 2.4. *Let $\{y_i, \mathcal{F}_i\}$ be an adapted sequence of random variables, $\{b_i\}$ a sequence of positive numbers. Then*

$$\left\{\sum_{i=1}^{\infty} y_i \quad converges\right\} \cap A = \left\{\sum_{i=1}^{\infty} y_i I_{[|y_i| \le b_i]} \quad converges\right\} \cap A,$$

where

$$A \triangleq \left\{\sum_{i=1}^{\infty} P(|y_i| > b_i|\mathcal{F}_{i-1}) < \infty\right\}.$$

Proof. Set

$$B_i = \{|y_i| > b_i\}.$$

Then by Theorem 2.5 we have

$$A = \left\{\sum_{i=1}^{\infty} I_{B_i} < \infty\right\}$$

which means that A is the set where events B_i occur only finite times. Hence on A, $\sum_{i=1}^{\infty} y_i$ converges if and only if $\sum_{i=1}^{\infty} y_i I_{\{|y_i| \le b_i\}}$ converges. Q.E.D.

Theorem 2.6 *(Three Series Criterion). For any adapted sequence of random variables* $\{x_n, \mathcal{F}_n\}$, $\xi_n \triangleq \sum_{i=1}^{n} x_i$ *converges on the set denoted by* S *on which the following three series converge:*

$$\sum_{i=1}^{\infty} P(|x_i| \geq c|\mathcal{F}_{i-1}), \tag{2.13}$$

$$\sum_{i=1}^{\infty} E\big[x_i I_{\{|x_i| \leq c\}}|\mathcal{F}_{i-1}\big], \tag{2.14}$$

and

$$\left\{\sum_{i=1}^{\infty} E\big[x_i^2 I_{\{|x_i| \leq c\}}|\mathcal{F}_{i-1}\big] - \big(E\big[x_i I_{\{|x_i| \leq c\}}|\mathcal{F}_{i-1}\big]\big)^2\right\} \tag{2.15}$$

where c *is a positive constant.*

Proof. By Corollary 2.4 we have

$$\left\{\sum_{i=1}^{\infty} x_i \quad converges\right\} \cap A = \left\{\sum_{i=1}^{\infty} x_i I_{\{|x_i| \leq c\}} \quad converges\right\} \cap A,$$

where

$$A \triangleq \left\{\sum_{i=1}^{\infty} P(|x_i| \geq c|\mathcal{F}_{i-1}) < \infty\right\}.$$

Hence we know

$$\left\{\sum_{i=1}^{\infty} x_i \quad converges\right\} \cap S = \left\{\sum_{i=1}^{\infty} x_i I_{\{|x_i| \leq c\}} \quad converges\right\} \cap S,$$

since $S \subset A$.

Noticing that

$$\sum_{i=1}^{\infty} E\big[x_i I_{\{|x_i| \leq c\}}|\mathcal{F}_{i-1}\big]$$

converges on S, we then have

$$\left\{\sum_{i=1}^{\infty} x_i I_{\{|x_i| \leq c\}} \quad converges\right\} \cap S = \left\{\sum_{i=1}^{\infty} y_i \quad converges\right\} \cap S$$

and hence

$$\left\{\sum_{i=1}^{\infty} x_i \quad converges\right\} \cap S = \left\{\sum_{i=1}^{\infty} y_i \quad converges\right\} \cap S \tag{2.16}$$

where

$$y_i = x_i I_{\{|x_i| \le c\}} - E\big[x_i I_{\{|x_i|\le c\}} | \mathcal{F}_{i-1}\big].$$

Obviously, $(y_i, \mathcal{F}_i)$ is a m.d.s. and we have that

$$E(y_i^2|\mathcal{F}_{i-1}) = E(x_i^2 I_{\{|x_i|\le c\}}|\mathcal{F}_{i-1}) - E^2(x_i I_{\{|x_i|\le c\}}|\mathcal{F}_{i-1})$$

and

$$S \subset \left\{\sum_{i=1}^{\infty} E(y_i^2|\mathcal{F}_{i-1}) < \infty\right\}.$$

Hence by Theorem 2.3 $\sum_{i=1}^{\infty} y_i$ converges on S, or

$$\left\{\sum_{i=1}^{\infty} y_i \quad converges\right\} \cap S = S.$$

Then from this and (2.16) it follows that

$$\left\{\sum_{i=1}^{\infty} x_i \quad converges\right\} \cap S = S,$$

or

$$\left\{\sum_{i=1}^{\infty} x_i \quad converges\right\} \supset S. \qquad \text{Q.E.D.}$$

Theorem 2.7 *(Chow). Let $\{x_i, \mathcal{F}_i\}$ be a m.d.s.. Then as $n \to \infty$, $\xi_n = \sum_{i=1}^{n} x_i$ converges on A where*

$$A = \left\{\sum_{i=1}^{\infty} E\big[|x_i|^p|\mathcal{F}_{i-1}\big] < \infty\right\}, \qquad 1 \le p \le 2.$$

Proof. By Theorem 2.6 it suffices to prove that $A \subset S$, where S is defined in Theorem 2.6. In other words, we need only to verify that the three series (2.13)-(2.15) converge on A.

For convergence of (2.13) it is sufficient to note

$$\begin{aligned} P(|x_i| \ge c|\mathcal{F}_{i-1}) &= E(I_{\{|x_i|\ge c\}}|\mathcal{F}_{i-1}) \\ &\le \frac{1}{c^p} E\big[|x_i|^p I_{\{|x_i|\ge c\}}|\mathcal{F}_{i-1}\big] \\ &\le \frac{1}{c^p} E\big[|x_i|^p|\mathcal{F}_{i-1}\big]. \end{aligned}$$

Further, since $E(x_i|\mathcal{F}_{i-1}) = 0$ we see that

$$\frac{1}{c}\sum_{i=1}^{\infty}\left|E\left[x_i I_{\{|x_i|\leq c\}}|\mathcal{F}_{i-1}\right]\right| = \frac{1}{c}\sum_{i=1}^{\infty}\left|E\left[x_i I_{\{|x_i|>c\}}|\mathcal{F}_{i-1}\right]\right|$$

$$\leq \sum_{i=1}^{\infty} E\left[\frac{|x_i|}{c} I_{\{|x_i|>c\}}|\mathcal{F}_{i-1}\right] \leq \frac{1}{c^p}\sum_{i=1}^{\infty} E\left[|x_i|^p|\mathcal{F}_{i-1}\right].$$

Hence the series (2.14) also converges on A.

Finally, the series (2.15) converges on A because

$$\frac{1}{c^2}\sum_{i=1}^{\infty} E\left[x_i^2 I_{\{|x_i|\leq c\}}|\mathcal{F}_{i-1}\right] \leq \frac{1}{c^p}\sum_{i=1}^{\infty} E\left[|x_i|^p|\mathcal{F}_{i-1}\right]$$

and by Theorem 1.6

$$\left(E\left[x_i I_{\{|x_i|\leq c\}}|\mathcal{F}_{i-1}\right]\right)^2 \leq E\left[x_i^2 I_{\{x_i\leq c\}}|\mathcal{F}_{i-1}\right].$$

Thus, we have verified that (2.13)–(2.15) converge on A. Q.E.D.

Corollary 2.5. *Theorem 2.6 is also valid for* $0 < p < 1$.

Proof. Set

$$y_i = |x_i|^p - E\left[|x_i|^p|\mathcal{F}_{i-1}\right].$$

Then we have that

$$\sum_{i=1}^{\infty} E\left[|y_i||\mathcal{F}_{i-1}\right] \leq \sum_{i=1}^{\infty} E\left[\left(|x_i|^p + E\left(|x_i|^p|\mathcal{F}_{i-1}\right)\right)|\mathcal{F}_{i-1}\right]$$

$$\leq 2\sum_{i=1}^{\infty} E\left(|x_i|^p|\mathcal{F}_{i-1}\right) < \infty \quad \text{on} \quad A.$$

Applying Theorem 2.7 with $p = 1$ to the m.d.s. $\{y_i, \mathcal{F}_i\}$ we conclude that $\sum_{i=1}^{\infty} y_i$ converges on A, that is,

$$\left\{\sum_{i=1}^{\infty}\left(|x_i|^p - E\left(|x_i|^p|\mathcal{F}_{i-1}\right)\right) \quad \text{converges}\right\} \supset \left\{\sum_{i=1}^{\infty} E\left(|x_i|^p|\mathcal{F}_{i-1}\right) < \infty\right\},$$

which is equivalent to

$$\left\{\sum_{i=1}^{\infty}|x_i|^p \quad \textit{converges}\right\} \supset \left\{\sum_{i=1}^{\infty} E\left(|x_i|^p|\mathcal{F}_{i-1}\right) < \infty\right\}. \tag{2.17}$$

It is easy to see that for $0 < p < 1$ the convergence of $\sum_{i=1}^{\infty} |x_i|^p$ implies the convergence of $\sum_{i=1}^{\infty} |x_i|$ since $|x_i| < 1$ for sufficiently large i. Hence from (2.17) we have

$$\left\{\sum_{i=1}^{\infty} x_i \quad converges\right\} \supset \left\{\sum_{i=1}^{\infty} E(|x_i|^p|\mathcal{F}_{i-1}) < \infty\right\}. \qquad \text{Q.E.D.}$$

Corollary 2.6. *Let $\{x_i, \mathcal{F}_i\}$ be a m.d.s.. If either $\sup_i E[|x_i|^p|\mathcal{F}_{i-1}] < \infty$ a.s. or $\sup_i E|x_i|^p < \infty$ for some $p \in [1,2]$, then as $n \to \infty$ for any $q > 1$*

$$\frac{1}{n^{q/p}} \sum_{i=1}^{n} x_i \to 0 \quad a.s.$$

Proof. It is easy to verify that

$$\sum_{i=1}^{\infty} E\left[\left|\frac{x_i}{i^{q/p}}\right|^p |\mathcal{F}_{i-1}\right] < \infty \quad a.s.$$

Hence applying Theorem 2.7 to the martingale $\sum_{i=1}^{n} \frac{x_i}{i^{q/p}}$, we know that $\sum_{i=1}^{\infty} \frac{x_i}{i^{q/p}}$ converges almost surely. So the desired result follows directly from the next lemma, known as the Kronecker lemma.

Lemma 2.4 *(Kronecker Lemma). If $\{b_i\}$, $i = 1, 2, \ldots$, is a sequence of positive numbers nondecreasingly diverging to infinity and if for a sequence of matrices $\{M_i\}$*

$$\sum_{i=1}^{\infty} \frac{1}{b_i} M_i < \infty,$$

then $\sum_{i=1}^{n} M_i = o(b_n)$.

Proof. Set $N_0 = 0$, $N_n = \sum_{i=1}^{n} \frac{1}{b_i} M_i$, $b_0 = 0$.

By the condition of the lemma

$$N_n \longrightarrow N < \infty, \quad as\ n \to \infty,$$

i.e. for an arbitrarily given $\varepsilon > 0$ there is n_ε so that

$$\|N_i - N\| < \varepsilon \quad for \quad i \geq n_\varepsilon.$$

Then we have

$$\begin{aligned}\|\frac{1}{b_n}\sum_{i=1}^{n} M_i\| &= \|\frac{1}{b_n}\sum_{i=1}^{n} b_i(N_i - N_{i-1})\| = \|N_n + \frac{1}{b_n}\sum_{i=2}^{n}(b_{i-1} - b_i)N_{i-1}\| \\ &= \|N_n - \frac{b_n - b_1}{b_n}N + \frac{1}{b_n}\sum_{i=2}^{n}(b_{i-1} - b_i)(N_{i-1} - N)\| \\ &\le \|N_n - N\| + \frac{b_1}{b_n}\|N\| + \frac{1}{b_n}\sum_{i=2}^{n_\varepsilon}(b_{i-1} - b_i)\|N_{i-1} - N\| + \varepsilon \\ &\xrightarrow[\substack{n \to \infty \\ \varepsilon \to 0}]{} 0\,.\end{aligned}$$

Q.E.D.

2.3 Estimation for Weighted Sums of a Martingale Difference Sequence

The classical limit results such as the central limit theorem and the laws of iterated logarithm have been extended to a more general martingale setting. For the central limit theorem there is the following result [Br]. Let $\{x_n, \mathcal{F}_n\}$ be a m.d.s. with $Ex_i^2 < \infty$.

If

$$\left(\sum_{i=1}^{n} Ex_i^2\right)^{-1} \sum_{i=1}^{n} E(x_i^2|\mathcal{F}_{i-1}) \xrightarrow{P} 1$$

and for all $\varepsilon > 0$

$$\left(\sum_{i=1}^{n} Ex_i^2\right)^{-1} \sum_{i=1}^{n} E(x_i^2 | I_{\{|x_i| \ge \varepsilon (\sum_{i=1}^{n} Ex_i^2)^{\frac{1}{2}}\}}) \longrightarrow 0$$

(Lindeberg Condition)

then

$$\lim_{n\to\infty} P\left(\left(\sum_{i=1}^{n} Ex_i^2\right)^{\frac{-1}{2}} \sum_{i=1}^{n} x_i \le x\right) = (2\pi)^{\frac{-1}{2}} \int_{-\infty}^{x} e^{\frac{-t^2}{2}}\, dt. \tag{2.18}$$

For the law of the iterated logarithm there are various versions described below. If $\{x_n, \mathcal{F}_n\}$ is a stationary ergodic m.d.s. with $Ex_n^2 = \sigma^2$, then [St]

$$\limsup_{n\to\infty}(2\sigma^2 n \log\log n)^{\frac{-1}{2}} \sum_{i=1}^{n} x_i = +1 \qquad a.s., \tag{2.19}$$

$$\liminf_{n\to\infty}(2\sigma^2 n\log\log n)^{\frac{-1}{2}}\sum_{i=1}^{n}x_i=-1 \qquad a.s. \tag{2.20}$$

If $\{x_n, \mathcal{F}_n\}$ is an uniformly bounded m.d.s., then on the set $\{s_n\to\infty\}$

$$\limsup_{n\to\infty}(2s_n\log\log s_n)^{\frac{-1}{2}}\sum_{i=1}^{n}x_i=+1 \qquad a.s., \tag{2.21}$$

$$\liminf_{n\to\infty}(2s_n\log\log s_n)^{\frac{-1}{2}}\sum_{i=1}^{n}x_i=-1 \qquad a.s., \tag{2.22}$$

where

$$s_n^2=\sum_{i=1}^{n}E(x_i^2|\mathcal{F}_{i-1}).$$

For a general m.d.s. (2.21) and (2.22) hold with s_n replaced by $\sum_{i=1}^{n}x_i^2$, if

$$\left(\sum_{i=1}^{n}Ex_i^2\right)^{-1}\sum_{i=1}^{n}E(x_i^2)\longrightarrow x>0 \quad a.s. \quad as \quad n\to\infty \tag{2.23}$$

$$\sum_{i=1}^{\infty}\left(\sum_{j=1}^{i}Ex_j^2\right)^{\frac{-1}{2}}E(|x_i|I_{\{|x_i|>\varepsilon(\sum_{j=1}^{i}Ex_j^2)^{\frac{1}{2}}\}})<\infty \tag{2.24}$$

for any $\varepsilon>0$ and

$$\sum_{i=1}^{\infty}\left(\sum_{j=1}^{i}Ex_j^2\right)^{-2}E(x_i^4 I_{\{|x_i|\le\delta(\sum_{j=1}^{i}Ex_j^2)^{1/2}\}})<\infty \tag{2.25}$$

for any $\delta>0$ [HH].

For analyzing stochastic adaptive control systems we need to know the behavior of partial sums of a martingale difference sequence with weights. In other words, the asymptotic properties like those given by (2.20) and (2.22) are crucial for us. However, (2.20) and (2.22) hold only under additional requirements imposed on $\{x_i, \mathcal{F}_i\}$, which, in general, are not satisfied by adaptive control systems; that is, we actually need to know the convergence rate of

$$\sum_{i=0}^{n}M_iX_{i+1}$$

where $\{X_n, \mathcal{F}_n\}$ is a matrix m.d.s. and M_n is an $\mathcal{F}_n$-measurable random matrix in the case where neither the boundedness of $\|M_iX_{i+1}\|$ nor the finiteness of $E\|M_iX_{i+1}\|$ takes place.

In the sequel for a sequence of matrices $\{M_k\}$ and a nondecreasing sequence of positive real numbers $\{b_k\}$, by $M_n = O(b_n)$ we mean

$$\limsup_{k\longrightarrow\infty} \|M_k\|/b_k < \infty$$

and by $M_k = o(b_k)$

$$\lim_{k\to\infty} \|M_k\|/b_k = 0.$$

In particular, by $M_k = O(\varepsilon)$ we mean that there is a constant $C \geq 0$ such that

$$\|M_k\| \leq C\varepsilon, \quad \forall k \geq 0.$$

Theorem 2.8. *Let $\{X_n, \mathcal{F}_n\}$ be a matrix m.d.s. and $\{M_n, \mathcal{F}_n\}$ an adapted sequence of random matrices $\|M_n\| < \infty$ a.s., $\forall n \geq 0$. If*

$$\sup_n E[\|X_{n+1}\|^\alpha|\mathcal{F}_n] < \infty \quad a.s.$$

for some $\alpha \in (0, 2]$, then as $n \to \infty$

$$\sum_{i=0}^{n} M_i X_{i+1} = O\left(s_n(\alpha)\log^{\frac{1}{\alpha}+\eta}(s_n^\alpha(\alpha)+e)\right) \; a.s., \; \forall \eta > 0, \tag{2.26}$$

where

$$s_n(\alpha) = \left(\sum_{i=0}^{n}\|M_i\|^\alpha\right)^{\frac{1}{\alpha}}.$$

Proof. Without loss of generality assume $M_0 \neq 0$. Let

$$\sigma = \sup_{n\geq 0} E[\|X_{n+1}\|^\alpha|\mathcal{F}_n].$$

Noticing Property 5 of the conditional expectation we have

$$\begin{aligned}
&\sum_{i=1}^{\infty} E\left[\left\|\left(s_i(\alpha)\log^{\frac{1}{\alpha}+\eta}(s_i^\alpha(\alpha)+e)\right)^{-1} M_i X_{i+1}\right\|^\alpha |\mathcal{F}_i\right] \\
\leq\ & \sigma\sum_{i=1}^{\infty}\left(s_i^\alpha(\alpha)\log^{1+\eta\alpha}(s_i^\alpha(\alpha)+e)\right)^{-1}\|M_i\|^\alpha \\
=\ & \sigma\sum_{i=1}^{\infty}\left(s_i^\alpha(\alpha)\log^{1+\eta\alpha}(s_i^\alpha(\alpha)+e)\right)^{-1}\int_{s_{i-1}^\alpha(\alpha)}^{s_i^\alpha(\alpha)} dx \\
\leq\ & \sigma\sum_{i=1}^{\infty}\int_{s_{i-1}^\alpha(\alpha)}^{s_i^\alpha(\alpha)}\frac{dx}{x\log^{1+\eta\alpha}(x+e)} \\
\leq\ & \sigma\int_{s_0^\alpha(\alpha)}^{\infty}\frac{dx}{x\log^{1+\eta\alpha}(x+e)} < \infty.
\end{aligned}$$

By Theorem 2.7 and Corollary 2.5 we know that

$$\sum_{i=1}^{\infty} a_i^{-1} M_i X_{i+1}$$

converges a.s., where for simplicity we have set

$$a_i = s_i(\alpha) \log^{\frac{1}{\alpha}+\eta}(s_i^{\alpha}(\alpha) + e).$$

Clearly, a_i is nondecreasing. Then, the sequence

$$y_n = \sum_{i=0}^{n} a_i^{-1} M_i X_{i+1}$$

is bounded in n, say, $\|y_n\| \leq \zeta, \quad \forall n$, where ζ may depend on ω.

Finally, we have

$$\begin{aligned} \|\sum_{i=1}^{n} M_i X_{i+1}\| &= \|\sum_{i=1}^{n} a_i(y_i - y_{i-1})\| \\ &= \|a_n y_n - a_0 y_0 - \sum_{i=1}^{n}(a_i - a_{i-1})y_{i-1}\| \\ &\leq O(a_n) + \sum_{i=1}^{n}(a_i - a_{i-1})\|y_{i-1}\| \\ &= O(a_n) + O(\sum_{i=1}^{n}(a_i - a_{i-1})) = O(a_n). \end{aligned}$$

This is what we want to prove. Q.E.D.

Remark 2.1.

(i) These type of results can be found in [St], [LW1], [CG3] and [Gu1]. One of the differences between Theorem 2.8 and Theorem 3.3.10 in [St] is that the requirement $S_n(\alpha) \xrightarrow[n \to \infty]{} \infty$ is not needed here.

(ii) In the one-dimensional case if $\alpha = 2$ then by Property 5 of the conditional expectation

$$E(M_i^2 X_{i+1}^2 | \mathcal{F}_i) = M_i^2 E(X_{i+1}^2 | \mathcal{F}_i)$$

and $s_n(2) = s_n$ if $E(X_{i+1}^2 | \mathcal{F}_i) = 1$ where s_n is the same as that in (2.21), (2.22). We thus find that the estimate (2.26) is not as sharp as those given by (2.21) and (2.22) but the conditions needed here are much more general.

(iii) If in Theorem 2.8 α is greater than 2, then a better result is obtainable (see, e.g. [We]).

2.4 Estimation for Double Array Martingales

Among the vast number of martingale limit theorems appearing in the literature, only a few results on limit behaviors of double array martingales are available (e.g. [St], [HH]). Moreover these results, due to various constraints, are difficult to be used directly in the analysis of stochastic adaptive algorithms. In particular, in the order estimation problems of Chapter 5, when upper bounds for system orders are not available, a crucial step in analyzing estimation algorithms is to study asymptotic behaviors of sequences of the form

$$\max_{1\le k\le h_n}\max_{1\le i\le n}\Big|\sum_{j=0}^{i} f_j(k)w_{j+1}\Big|$$

where $\{h_n\}$ is a nondecreasing sequence of integers. It is usually the case that $\{w_j\}$ is a m.d.s., and the double indexed weights $f_j(k)$ are random rather than deterministic and may contain signals derived from feedback. The aim of this section is to establish upper bounds for such sequences.

We first present a lemma.

Lemma 2.5 *[St]. Let $\{U_n,\mathcal{F}_n,n\ge 1\}$ be a supermartingale with $EU_1=0$. Let $U_0=0$ and $x_i=U_i-U_{i-1}$ for $i\ge 1$. Suppose $x_i\le c$ a.s. for some $0\le c<\infty$ and all $i\ge 1$. Fix $\lambda>0$ such that $\lambda c\le 1$. Define*

$$T_n=\exp(\lambda U_n)\exp[-(\lambda^2/2)(1+\frac{\lambda c}{2})\sum_{i=1}^{n}E(x_i^2|\mathcal{F}_{i-1})],\quad T_o=1\quad n\ge 1.$$

Then $\{T_n,\mathcal{F}_n,n\ge 0\}$ is a nonnegative supermartingale.

Proof. Noticing $0\le\lambda c\le 1$ and taking $n=2$ and 5 in the Taylor's expansion

$$e^x=\sum_{i=0}^{n-1}\frac{x^i}{i!}+\frac{x^n}{n!}e^{\zeta},$$

$\zeta\in[0,x]$ or $\zeta\in[x,0]$, respectively for the cases $x_i\le 0$ and $x_i\ge 0$ we directly verify the following inequality:

$$\exp(\lambda x_j)\le 1+\lambda x_j+x_j^2(\lambda^2/2)(1+\lambda c/2).$$

Consequently

$$\begin{aligned}E[\exp(\lambda x_j)|\mathcal{F}_{j-1}]\le&\ 1+(\lambda^2/2)(1+\lambda c/2)E[x_j^2|\mathcal{F}_{j-1}]\\ \le&\ \exp[(\lambda^2/2)(1+\lambda c/2)E(x_j^2|\mathcal{F}_{j-1})]\quad a.s.\end{aligned}$$

Hence for any $n \geq 2$,

$$E[T_n|\mathcal{F}_{n-1}] = \exp(\lambda \sum_{i=1}^{n-1} x_i) \exp\{-(\lambda^2/2)(1+\lambda c/2) \sum_{i=1}^{n} E(x_i^2|\mathcal{F}_{i-1})\} \cdot E[\exp(\lambda x_n)|\mathcal{F}_{n-1}] \leq T_{n-1} \quad a.s.$$

Q.E.D.

Corollary 2.7. *Under the assumptions of Lemma 2.5,*

$$P(\sup_{n\geq 0} T_n > \alpha) \leq \alpha^{-1}, \quad \forall \alpha > 0.$$

Proof. For any $\alpha > 0$, let us define a Markov time τ as follows

$$\tau = \begin{cases} \min\{n \geq 0 : T_n > \alpha\} \\ \infty, \quad \text{if } T_n \leq \alpha, \forall n. \end{cases}$$

Then by Lemma 2.3, $\{T_{\tau\wedge n}, n \geq 0\}$ is also a nonnegative supermartingale with the first element equal to one. Hence for each $n \geq 1$

$$1 \geq ET_{\tau\wedge n} = E\{T_\tau I_{\{\tau\leq n\}} + T_n I_{\{\tau>n\}}\}$$

$$\geq ET_\tau I_{\{\tau\leq n\}} \geq \alpha P(\tau \leq n).$$

Letting $n \to \infty$, we get the desired result:

$$1 \geq \alpha P(\tau < \infty) = \alpha P(\sup_{n\geq 0} T_n > \alpha).$$

Theorem 2.9 *[GHH],[HG]. Let $\{w_t, \mathcal{F}_t\}$ be an m-dimensional martingale difference sequence satisfying*

$$\|w_t\| = o(\varphi(t)) \quad a.s. \tag{2.27}$$

where $\varphi(x)$ is a positive, deterministic and nondecreasing function that satisfies

$$\sup_k \varphi(e^{k+1})/\varphi(e^k) < \infty.$$

Suppose that $f_t(k)$, $t, k = 1, 2, \ldots$, is an $\mathcal{F}_t$-measurable, $p \times m$-dimensional random matrix satisfying

$$\|f_t(k)\| \leq A < \infty \quad a.s. \quad \text{for all } t, k \text{ and some constant } A. \tag{2.28}$$

Then for $h_n = O([\log n]^\alpha)$, $\alpha > 0$ the following properties hold as $n \to \infty$,

$$(i). \quad \max_{1\leq k\leq h_n} \max_{1\leq i\leq n} \|\sum_{j=1}^{i} f_j(k) w_{j+1}\| = O(\max_{1\leq k\leq h_n} \sum_{j=1}^{n} \|f_j(k)\|^2) + o(\varphi(n) \log\log n) \; a.s. \tag{2.29}$$

provided that

$$\sup_j E(\|w_{j+1}\|^2|\mathcal{F}_j) < \infty \quad a.s. \tag{2.30}$$

(*ii*).

$$\max_{1\le k\le h_n}\max_{1\le i\le n}\|\sum_{j=1}^{i} f_j(k)w_{j+1}\|$$

$$= O(\max_{1\le k\le h_n}\sum_{j=1}^{n}\|f_j(k)\|) + o(\varphi(n)\log\log n)\, a.s. \tag{2.31}$$

provided that

$$\sup_j E(\|w_{j+1}\||\mathcal{F}_j) < \infty \quad a.s. \tag{2.32}$$

Proof. (i) We need only to consider the case of scalar variables and $A = 1$. For any $\varepsilon > 0$, let us set

$$\tilde{w}_j = w_j I_{\{|w_j|\le\varepsilon\varphi(j)\}}, \quad \bar{w}_j = \tilde{w}_j - E(\tilde{w}_j|\mathcal{F}_{j-1}).$$

Then

$$|\sum_{j=1}^{i} f_j(k)w_{j+1}| \le |\sum_{j=1}^{i} w_{j+1}I_{\{|w_{j+1}|>\varepsilon\varphi(j+1)\}}f_j(k)|$$

$$+|\sum_{j=1}^{i} E(\tilde{w}_{j+1}|\mathcal{F}_j)f_j(k)| + |\sum_{j=1}^{i} f_j(k)\bar{w}_{j+1}|. \tag{2.33}$$

We have from (2.27) and (2.28) that

$$\max_{1\le k\le h_n}\max_{1\le i\le n}\|\sum_{j=1}^{i} f_j(k)w_{j+1}I_{\{|w_{j+1}|>\varepsilon\varphi(j+1)\}}\|$$

$$\le \sum_{j=1}^{n}|w_{j+1}|I_{\{|w_{j+1}|>\varepsilon\varphi(j+1)\}}$$

$$\le o(\varphi(n+1))\sum_{j=1}^{\infty} I_{\{|w_{j+1}|>\varepsilon\varphi(j+1)\}}$$

$$= o(\varphi(n+1)) \quad a.s. \tag{2.34}$$

Also, letting $a_n = \max_{1\le k\le h_n}\{\sum_{j=1}^{n}|f_j(k)|^2\}^{1/2}$, we have under (2.28),

$$\begin{aligned}
&\leq \{\sum_{j=1}^{n}[E(w_{j+1}I_{\{|w_{j+1}|\leq\varepsilon\varphi(j+1)\}}|\mathcal{F}_j)]^2\}^{1/2}a_n \\
&= \{\sum_{j=1}^{n}[E(w_{j+1}I_{\{|w_{j+1}|>\varepsilon\varphi(j+1)\}}|\mathcal{F}_j)]^2\}^{1/2}a_n \\
&\leq \{\sup_j E(|w_{j+1}|^2|\mathcal{F}_j)\}\{\sum_{j=1}^{n}P(|w_{j+1}|>\varepsilon\varphi(j+1)|\mathcal{F}_j)\}^{1/2}a_n \\
&= O(a_n), \qquad (2.35)
\end{aligned}$$

where the last relation is derived by using (2.27) and Theorem 2.5.

Then, to prove (2.29) we need only to consider the last term on the right-hand side of (2.33).
Set

$$\begin{aligned}
S_i(k) &= \sum_{j=1}^{i}\bar{w}_{j+1}f_j(k),\ 1\leq i\leq n,\ S_0(k)=0; \\
d(x) &= 2\varepsilon\varphi(x+1),\ \lambda(x)=\frac{1}{d(x)}; \\
T_i(k,t) &= exp\{\lambda(e^t)S_i(k) \\
&\quad -\frac{3\lambda^2(e^t)}{4}\sum_{j=1}^{i}E(\bar{w}_{j+1}^2|\mathcal{F}_j)[f_j(k)]^2\},\quad 1\leq i\leq e^t; \\
T_0(k,t) &= 0.
\end{aligned}$$

It is easy to see that

$$|\bar{w}_{j+1}f_j(k)|\leq d(e^t)$$

for any $0\leq j\leq e^t$, $t>0$, $k\geq 0$. Hence by taking $c=d(e^t)$, $\lambda=\lambda(e^t)$ in Lemma 2.5, we know that for any fixed k and t, $\{T_i(k,t),\ 0\leq i\leq e^t\}$ is a supermartingale. Further, from the properties of $\varphi(x)$ we have

$$\lambda(i)=[2\varepsilon\phi(i+1)]^{-1}\geq[2\varepsilon\phi(e^t+1)]^{-1}=\lambda(e^t),\ i\leq e^t.$$

Then

$$\begin{aligned}
&S_i(k)-\frac{3}{4}\lambda(i)\sum_{j=1}^{i}E(\bar{w}_{j+1}^2|\mathcal{F}_j)f_j^2(k) \\
\leq\ &S_i(k)-\frac{3\lambda(e^t)}{4}\sum_{j=1}^{i}E(\bar{w}_{j+1}^2|\mathcal{F}_j)f_j^2(k),\ i<e^t.
\end{aligned}$$

So for any constant $b > 0$ we see that

$$\begin{aligned}
&P\left(\max_{1\le k\le bt^\alpha}\max_{1\le i\le e^t}[S_i(k) - \frac{3\lambda(i)}{4}\sum_{j=1}^{i} E(\bar{w}_{j+1}^2|\mathcal{F}_j)f_j^2(k)]\right.\\
&\qquad\left.\ge (2+\alpha)d(e^t)\log\log e^t\right)\\
\le\ & \sum_{k=1}^{[bt^\alpha]} P\left(\max_{1\le i\le e^t}[S_i(k) - \frac{3\lambda(e^t)}{4}\sum_{j=1}^{i} E(\bar{w}_{j+1}^2|\mathcal{F}_j)f_j^2(k)]\right.\\
&\qquad\left.\ge (2+\alpha)d(e^t)\log\log e^t\right)\\
=\ & \sum_{k=1}^{[bt^\alpha]} P(\max_{1\le i\le e^t} T_i(k,t) \ge \exp[(2+\alpha)logt])\\
\le\ & \sum_{k=1}^{[bt^\alpha]} t^{-(2+\alpha)} = O(t^{-2}), \quad \text{(by Corollary 2.7)}.
\end{aligned}$$

Then, according to Corollary 2.3, we have

$$\limsup_{t\to\infty}\max_{1\le k\le bt^\alpha}\max_{1\le i\le e^t}\frac{[S_i(k) - \frac{3\lambda(i)}{4}\sum_{j=1}^{i} E(\bar{w}_{j+1}2|\mathcal{F}_j)f_j^2(k)]}{\varphi(e^t+1)\log t} \le 2(2+\alpha)\epsilon \text{ a.s.}$$

Since $h_n = O([logn]^\alpha)$, there exits $b > 0$ so that $h_n \le b(\log n)^\alpha$ or $h_n \le bt^\alpha$ for $n \le e^t$. We then have from (2.9) that for $n \in [e^{t-1}, e^t]$

$$\begin{aligned}
&\max_{1\le k\le h_n}\max_{1\le i\le n}\frac{[S_i(k) - \frac{3\lambda(i)}{4}\sum_{j=1}^{i} E(\bar{w}_{j+1}^2|\mathcal{F}_j)f_j^2(k)]}{\varphi(n)loglogn}\\
\le\ & \max_{1\le k\le h_n}\max_{1\le i\le n}\frac{[S_i(k) - \frac{3\lambda(i)}{4}\sum_{j=1}^{i} E(\bar{w}_{j+1}^2|\mathcal{F}_j)f_j^2(k)]}{\varphi(e^{t-1})logloge^{t-1}}\\
\le\ & \frac{\varphi(e^{t+1})}{\varphi(e^t)}\frac{\varphi(e^t)}{\varphi(e^{t-1})}\frac{\log t}{\log(t-1)}\\
&\cdot\frac{\max_{1\le k\le bt^\alpha}\max_{1\le i\le e^t}[S_i(k) - \frac{3\lambda(i)}{4}\sum_{j=1}^{i} E(\bar{w}_{j+1}^2|\mathcal{F}_j)f_j^2(k)]}{\varphi(e^t+1)\log t}\\
=\ & O(\epsilon) \quad \text{a.s., as } t\to\infty.
\end{aligned}$$

Thus, we know

$$\max_{1\le k\le h_n}\max_{1\le i\le n}[S_i(k)-\frac{3\lambda(i)}{4}\sum_{j=1}^{i}E(\bar{w}_{j+1}^2|\mathcal{F}_j)f_j^2(k)]=O(\epsilon(\varphi(n)\log\log n)\text{ a.s.}$$

and by noting that $|\bar{w}_{j+1}|\le d(j)=1/\lambda(j)$ we find

$$\begin{aligned}&\max_{1\le k\le h_n}\max_{1\le i\le n}S_i(k)\\ \le\ &\max_{1\le k\le h_n}\frac{3\lambda(n)}{4}\sum_{j=1}^{n}E(\bar{w}_{j+1}^2|\mathcal{F}_j)f_j^2(k)+O(\epsilon\varphi(n)\log\log n)\\ \le\ &\max_{1\le k\le h_n}\frac{3\lambda(n)}{4}\sum_{j=1}^{n}E(|\bar{w}_{j+1}|\lambda(j)^{-1}|\mathcal{F}_j)f_j^2(k)+O(\epsilon\varphi(n)\log\log n)\\ \le\ &\sup_j E(|w_{j+1}||\mathcal{F}_j)\max_{1\le k\le h_n}\sum_{j=1}^{n}f_j^2(k)\\ &+O(\epsilon\varphi(n)\log\log n)\,.\end{aligned}\tag{2.36}$$

A similar result holds also for $\{-S_i(k)\}$. Then the desired result (2.29) follows from (2.33)–(2.36) and the arbitrariness of ϵ.

(ii) Note that the condition (2.30) is crucial only in the proof of (2.35). When (2.30) is relaxed to (2.32), (2.35) can be replaced by

$$\begin{aligned}&\max_{1\le k\le h_n}\max_{1\le i\le n}|\sum_{j=1}^{i}E(\tilde{w}_{j+1}|\mathcal{F}_j)f_j(k)|\\ \le\ &\sup_j E(|w_{j+1}||\mathcal{F}_j)\max_{1\le k\le h_n}\sum_{j=1}^{n}|f_j(k)|\,.\end{aligned}\tag{2.37}$$

Note also that since $|f_j(k)|\le 1$, $\max_{1\le k\le h_n}\sum_{j=1}^{n}|f_j(k)|^2\le\max_{1\le k\le h_n}\sum_{j=1}^{n}|f_j(k)|$. Then (2.31) follows from (2.33), (2.34), (2.37) and (2.36). Q.E.D.

We remark that if h_n is only assumed to satisfy $h_n=O(n^\alpha)$, $\alpha>0$, then (2.29) and (2.31) with "$\log\log n$" replaced by "$\log n$" still hold.

In the next theorem we remove the boundedness restriction (2.28) on $|f_i(k)|$ and obtain an estimate similar to (2.29). It is worth noting that the estimates given by (2.29) and by the next theorem, in general, are incompatible. For example, if $\varphi(n)$ is bounded and $a_n=\sqrt{n}$, then the next theorem gives an estimate even better than (2.29) [GHH], [HG].

Theorem 2.10 *[HG]. Under the conditions of Theorem 2.9 except (2.28) if (2.30) holds, then*

$$\max_{1\le k\le h_n}\max_{1\le i\le n}\left\|\sum_{j=1}^{i} f_j(k)w_{j+1}\right\| = O(a_n\log a_n)+o(a_n\varphi(n)\log\log n)\ a.s. \tag{2.38}$$

where $a_i = \max_{1\le k\le h_n} g_i(k)$, $g_i(k) = [\sum_{j=1}^{i}\|f_j(k)\|^2+1]^{\frac{1}{2}}$, $g_0(k)=1$.

Proof. Let $x_i(k) = f_j(k)/g_j(k)$, $1\le j\le n$. Then $\|x_j(k)\|\le 1$. So we have from Theorem 2.9 that

$$\max_{1\le k\le h_n}\max_{1\le i\le n}\left\|\sum_{j=1}^{i} x_j(k)w_{j+1}\right\| = O(\max_{1\le k\le h_n}\sum_{j=1}^{n}\|x_j(k)\|^2)+o(\varphi(n)\log\log n)\ \text{a.s.} \tag{2.39}$$

Note that

$$\sum_{j=1}^{n}\|x_j\|^2 = \sum_{j=1}^{n}\frac{g_j^2-g_{j-1}^2}{g_j^2} \le \sum_{j=1}^{n}\int_{g_{j-1}^2}^{g_j^2}\frac{dx}{x} = 2\log g_n, \tag{2.40}$$

and

$$\begin{aligned}\sum_{j=1}^{i} f_j w_{j+1} &= \sum_{j=1}^{i} g_j x_j w_{j+1}\\ &= -\sum_{j=1}^{i}\sum_{t=j}^{i-1}(g_{t+1}-g_t)x_j w_{j+1} + g_i\sum_{j=1}^{i} x_j w_{j+1}\\ &= -\sum_{t=1}^{i-1}(g_{t+1}-g_t)\sum_{j=1}^{t} x_j w_{j+1} + g_i\sum_{j=1}^{i} x_j w_{j+1},\end{aligned}$$

where the dependence on k is suppressed for simplicity of writing.

So we have

$$\max_{1\le k\le h_n}\max_{1\le i\le n}\left\|\sum_{j=1}^{i} f_j(k)w_{j+1}\right\| \le \max_{1\le k\le h_n}\max_{1\le i\le n}\left\|\sum_{j=1}^{i} x_j(k)w_{j+1}\right\|$$

$$\max_{1\le k\le h_n}\max_{1\le i\le n}\{\sum_{t=1}^{i-1}[g_{t+1}(k)-g_t(k)]+g_i(k)\}$$

$$\le\ 2a_n\max_{1\le k\le h_n}\max_{1\le i\le n}\|\sum_{j=1}^{i}x_j(k)w_{j+1}\|. \qquad (2.41)$$

Thus, (2.38) follows from (2.39)–(2.41). Q.E.D.

Remark 2.2. In contrast to the estimate for sums of m.d.s with single-indexed weights (Theorem 2.8), here in Theorem 2.10 the second term on the right-hand side of (2.38), i.e. $o(a_n\varphi(n)loglogn)$, cannot be removed in general. A simple example for this is

$$f_i(k)=\begin{cases}1, & if \quad i=k,\\ 0, & if \quad i\ne k\end{cases}$$

with $h_n = logn$ and with $\{w_j\}$ being any unbounded independent and identically distributed random sequence satisfying conditions in Theorem 2.10, because in this case a_n is bounded.

CHAPTER 3

Filtering and Control for Linear Systems

The main concepts of control theory for linear systems with known parameters are needed in order to work on problems arising from the control system with unknown parameters which is the subject of the present book. Aiming at those who are not in the field of control theory, this chapter provides the necessary concepts and results of control theory which will be used in later chapters.

3.1 Controllability and Observability

Controllability and observability are basic concepts in control theory [Kal], [AM1], [KFA], [Wo1], [Won]. Roughly speaking, they are about whether or not the system output can be made to behave in a desirable way by choosing an appropriate input and the system state can be determined by observing its output.

Consider the linear system with time-varying parameters:

$$\dot{x}_t = A_t x_t + B_t u_t, \tag{3.1}$$

$$y_t = C_t x_t, \quad t \in [t_0, T), \quad 0 < T \leq \infty \tag{3.2}$$

where x_t is n-dimensional state, u_t r-dimensional input (control) and y_t m-dimensional output.

Assume $\|A_t\|$ is integrable on $[t_0, T)$ and $\|B_t\|$ and $\|C_t\|$ are square-integrable on $[t_0, T)$. The set $\mathcal{U}$ of admissible controls consists of square-integrable functions on $[t_0, T)$. If the system is time invariant then (3.1) (3.2) are simply denoted by (A, B, C).

Definition. The system (3.1) is said to be controllable at t_0 if for any initial value $x_{t_0} = x_0$ there exist an admissible control u_t and a time t_f such that $x_{t_f} = 0$. If for any $t_1 \in [t_0, T)$ the system (3.1) (3.2) is controllable at t_1 then the system is said to be completely controllable.

Let ϕ_{ts} be the fundamental solution matrix for (3.1), i.e.

$$\frac{d}{dt}\phi_{ts} = A_t\phi_{ts}, \qquad \phi_{ss} = I.$$

It can be shown that the system (3.1) is controllable at t_0 if and only if there is a t_f such that the following matrix is of full rank:

$$\int_{t_0}^{t_f} \phi_{t_f s} B_s B_s^\tau \phi_{t_f s}^\tau ds > 0. \tag{3.3}$$

If A_t and B_t are constant matrices denoted by A and B, respectively, then controllability at a fixed time is equivalent to the complete controllability. In this case condition (3.3) is reduced to the fact that the matrix

$$Q \triangleq [B \quad AB \quad A^2B \ ... \ A^{n-1}B] \tag{3.4}$$

is of full rank.

Definition. The system (3.1) (3.2) is said to be observable at $t_1 \in [t_0, T)$ if there is a time $t_f \in [t_0, T)$, $t_f > t_1$ such that any initial value $x_{t_1} = x_0$ can be determined on the basis of observation y_t on $[t_1, t_f]$. If the system is observable at any $t_1 \in [t_0, T)$, then the systems is called completely observable.

There exists duality between controllability and observability. To be specific, we have the following criterion.

The system (3.1) (3.2) is observable at t_0 if and only if there is a time t_f such that

$$\int_{t_0}^{t_f} \phi_{st_0}^\tau C_s^\tau C_s \phi_{st_0} ds > 0. \tag{3.5}$$

Further, for the linear time invariant system condition (3.5) is equivalent to that the matrix

$$\Gamma \triangleq [C^\tau \quad A^\tau C^\tau \quad A^{2\tau} C^\tau \ ... \ (A^{n-1})^\tau C^\tau] \tag{3.6}$$

is of full rank.

Controllability and observability are the important concepts of control theory. For example, they are necessary and sufficient for the arbitrary pole assignment of a time invariant system by output feedback. To be precise, if $u_t = Ky_t$, then the closed loop system is

$$\dot{x}_t = (A - BKC)x_t$$

and, in order that there be a feedback gain K making $A - BKC$ have prescribed zeros, the necessary and sufficient conditions are controllability and observability of the system (A, B, C).

For discrete-time systems there are similar concepts and results on controllability and observability. The system now is given by

$$x_{k+1} = A_{k+1,k}x_k + B_k u_k \tag{3.7}$$

$$y_k = C_k x_k. \tag{3.8}$$

For the time invariant case the system is simply denoted by (A, B, C).

The definitions of controllability and observability for the system (3.7) (3.8) are the same as before. The only thing we need to change is the replacement of the initial t_0 and the end time t_f by integers k_0 and k_f, respectively.

The system (3.7) and (3.8) is controllable on $[k_0, k_f]$ if and only if the matrix

$$[B_{k_f-1} \quad A_{k_f\, k_f-1}B_{k_f-2} \;\dots\; A_{k_f\, k_0+1}B_{k_0}] \tag{3.9}$$

is of full rank where

$$A_{i\,j} \triangleq \prod_{k=j}^{i-1} A_{k+1\,k}, \qquad A_{ii} = I.$$

For the time invariant case, the matrix (3.9) takes the form:

$$Q_{n-1} \triangleq [B \quad AB \;\dots\; A^{n-1}B].$$

Similarly, the system (3.7) and (3.8) is observable on $[k_0, k_f]$ if and only if the matrix

$$\begin{bmatrix} C_{k_0} \\ C_{k_0+1}A_{k_0+1\,k_0} \\ \vdots \\ C_{k_f}A_{k_f\,k_0} \end{bmatrix}$$

is of full rank.

For the time invariant system this matrix is expressed as

$$\begin{bmatrix} C \\ CA \\ \vdots \\ CA^{n-1} \end{bmatrix}. \tag{3.10}$$

There are many monographs and textbooks on linear control systems theory, for example [Wol], [Won], [AM1], [Ki] etc.

3.2 Kalman Filtering for Systems with Random Coefficients

Consider the stochastic system

$$x_{k+1} = A_k x_k + B_k u_k + D_k w_{k+1}, \qquad k \geq 0 \tag{3.11}$$

$$y_k = C_{k-1} x_k + G_{k-1} v_{k-1} + F_{k-1} w_k, \qquad k \geq 1 \tag{3.12}$$

where $x_k \in \mathbb{R}^n$, $y_k \in \mathbb{R}^m$, $u_k \in \mathbb{R}^r$, $w_k \in \mathbb{R}^l$, $v_k \in \mathbb{R}^s$.

The Kalman filtering is a recursive algorithm estimating the state x_k on the basis of past observation $\{y_0, ..., y_k\}$ [Ka2], [AM1], [Da], [LSh] etc. The previous results either require the system matrices A_k, B_k, D_k, C_k, G_k and F_k to be deterministic or require the finiteness of the second moments of $B_k u_k$, $G_k v_k$, D_k and F_k, and a.s. uniform boundedness of $\|A_k\|$ and $\|C_k\|$ in the case they are random [LSh], [Che3]. These restrictive conditions are not assumed here. Roughly speaking, both the random coefficients and the controls are only assumed to be finite a.s. The Kalman filtering for systems with coefficients under relaxed conditions is important for its application to system identification and adaptive control [CKS].

The precise conditions imposed on the system are as follows.

1. The l-dimensional driven noise $\{w_k\}$ is independent and identically distributed (iid) and $w_k \in N(0, I)$.

2. The entries of A_k, B_k, u_k, D_k, C_k, G_k, v_k and F_k are all a.s. finite and measurable with respect to $\sigma(y_0, ..., y_k)$, the σ-algebra generated by $(y_0, ..., y_k)$.

3. $\begin{bmatrix} x_0 \\ y_0 \end{bmatrix}$ is independent of $\{w_k\}$ and x_0 is conditionally Gaussian given y_0 with conditional mean $\hat{x}_0$ and conditional covariance P_0.

We recall (see Section 1.4) that x is called conditionally Gaussian given y if there exist a y-measurable random vector $\hat{x}$ and a y-measurable random matrix $P = P^\tau \geq 0$ a.s. such that the conditional characteristic function of x given y is expressed by

$$E\left(e^{i\lambda^\tau x} | y\right) = exp\left(i\lambda^\tau \hat{x} - \frac{1}{2}\lambda^\tau P \lambda\right) \quad a.s. \tag{3.13}$$

for every constant vector λ. In this case

$$\hat{x} = E(x|y), \qquad P = E[(x-\hat{x})(x-\hat{x})^\tau | y] \quad a.s. \tag{3.14}$$

Lemma 3.1. *If* $\begin{bmatrix} x \\ y \end{bmatrix}$ *is conditionally Gaussian given z, then x and y are conditionally independent given z if and only if*

$$P_{xy|z} \triangleq E[(x - E(x|z))(y - E(y|z))^{\tau}|z] = 0 \quad a.s. \tag{3.15}$$

Proof. By Theorem 1.10 it suffices to show that

$$E(exp(i\lambda^{\tau}x + i\mu^{\tau}y)|z) = E(exp(i\lambda^{\tau}x)|z)E(exp(i\mu^{\tau}y)|z) \ a.s. \tag{3.16}$$

if and only if (3.15) holds.

Set $\gamma^{\tau} = [\lambda^{\tau} \quad \mu^{\tau}]$. Since $w \triangleq \begin{bmatrix} x \\ y \end{bmatrix}$ is conditionally Gaussian given z we have

$$\begin{aligned} & E[exp(i\gamma^{\tau}w)|z] \\ = \ & exp[i\gamma^{\tau}E(w|z)]exp\left\{-\frac{1}{2}\gamma^{\tau}\begin{bmatrix} P_{xx|z} & P_{xy|z} \\ P_{yx|z} & P_{yy|z} \end{bmatrix}\gamma\right\} \end{aligned} \tag{3.17}$$

where $P_{xx|z} = E[(x - E(x|z))(x - E(x|z))^{\tau}]$.

Noting that under the condition of the lemma $x = [I \quad 0]\begin{bmatrix} x \\ y \end{bmatrix}$ and $y = [0 \quad I]\begin{bmatrix} x \\ y \end{bmatrix}$ are also conditionally Gaussian given z we find that the right-hand sides of (3.16) and (3.17) are equal if and only if $P_{xy|z} = 0$.

Q.E.D.

For the later discussion we need to introduce the pseudo-inverse A^+ of an $n \times m$ matrix A.

Let the rank of A is $s \le n \wedge m (\equiv min(n, m))$. Then there are an $n \times s$ matrix B and an $s \times m$ matrix C such that

$$A = BC. \tag{3.18}$$

Clearly, B is of column-full-rank and C is of row-full-rank.

The matrix defined by

$$A^+ = C^{\tau}(CC^{\tau})^{-1}(B^{\tau}B)^{-1}B^{\tau} \tag{3.19}$$

is called the pseudo-inverse of A and it is uniquely defined. To see this let A have a representation other than (3.18)

$$A = \overline{B}\,\overline{C} \tag{3.20}$$

with $\overline{B}$ and $\overline{C}$ being $n \times s$ and $s \times m$ matrices respectively.

From (3.18) and (3.20) we have

$$B = AC^{\tau}(CC^{\tau})^{-1} = \overline{B}\,\overline{C}C^{\tau}(CC^{\tau})^{-1} = \overline{B}U \tag{3.21}$$

where $U = \overline{C}C^{\tau}(CC^{\tau})^{-1}$ is an $s \times s$ nonsingular matrix.

Then by (3.18), (3.20) and (3.21) we see $\overline{B}\,\overline{C} = A = BC = \overline{B}UC$, which implies, by multiplying $(\overline{B}^{\tau}\overline{B})^{-1}\overline{B}^{\tau}$ from the left of both sides,

$$\overline{C} = UC. \tag{3.22}$$

Hence, from (3.21) and (3.22) we have

$$\begin{aligned} &\overline{C}^{\tau}(\overline{C}\,\overline{C}^{\tau})^{-1}(\overline{B}^{\tau}\overline{B})^{-1}\overline{B}^{\tau} \\ = \;& C^{\tau}U^{\tau}(UCC^{\tau}U^{\tau})^{-1}(U^{-\tau}B^{\tau}BU^{-1})^{-1}U^{\tau}B^{\tau} \\ = \;& C^{\tau}(CC^{\tau})^{-1}B^{\tau}. \end{aligned}$$

This verifies that A^+ is unique.

From the definition of (3.19) it is easy to see the following relationships:

$$AA^+A = A, \quad A^+AA^+ = A^+, \quad (AA^+)^{\tau} = AA^+, \quad (A^+A)^{\tau} = A^+A; \tag{3.23}$$

$$(A^{\tau})^+ = (A^+)^{\tau}, \quad (A^+)^+ = A; \tag{3.24}$$

$$(A^{\tau}A)^+ = A^+(A^{\tau})^+ = A^+(A^+)^{\tau}; \tag{3.25}$$

$$A^+ = (A^{\tau}A)^+A^{\tau} = A^{\tau}(AA^{\tau})^+; \tag{3.26}$$

$$A^+AA^{\tau} = A^{\tau}AA^+ = A^{\tau}. \tag{3.27}$$

Lemma 3.2. *Let* $\begin{bmatrix} x \\ y \end{bmatrix}$ *be conditionally Gaussian given* z *with conditional covariance*

$$\begin{bmatrix} P_{xx|z} & P_{xy|z} \\ P_{yx|z} & P_{yy|z} \end{bmatrix} \quad a.s.$$

Then

1. *Given* (y, z), x *is conditionally Gaussian with conditional mean*

$$E(x|z,y) = E(x|z) + P_{x,y|z}P^+_{y,y|z}(y - E(y|z)) \tag{3.28}$$

and conditional covariance

$$P_{xx|zy} = P_{xx|z} - P_{xy|z}P^+_{yy|z}P_{yx|z}. \tag{3.29}$$

2. *Given z, $x - E(x|z,y)$ is conditionally Gaussian and independent of y.*

Proof. We first note that a z-measurable version of $P^+_{yy|z}$ can be chosen since $P_{yy|z}$ is z-measurable.

Define

$$w \triangleq x - E(x|z) - P_{xy|z}P^+_{yy|z}(y - E(y|z)). \tag{3.30}$$

Clearly, $\begin{bmatrix} w \\ y \end{bmatrix}$ is a linear combination of x and y with coefficients being z-measurable. Then by the fact mentioned in Section 1.4 $\begin{bmatrix} w \\ y \end{bmatrix}$ is also conditionally Gaussian given z.

Computing the conditional covariance of w and y given z shows that

$$\begin{aligned} & E[(w - E(w|z))(y - E(y|z))^T|z] \\ = \; & E[w(y - E(y|z))^T|] = P_{xy|z} - P_{xy|z}P^+_{yy|z}P_{yy|z}. \end{aligned} \tag{3.31}$$

We now show that the right-hand side of (3.31) is zero. Note that

$$(I - P^+_{yy|z}P_{yy|z})(y - E(y|z)) = 0 \quad a.s. \tag{3.32}$$

since

$$\begin{aligned} & E\left\{(I - P^+_{yy|z}P_{yy|z})(y - E(y|z))(y - E(y|z))^T(I - P^+_{yy|z}P_{yy|z})\,|z\right\} \\ = \; & (I - P^+_{yy|z}P_{yy|z})P_{yy|z}(I - P^+_{yy|z}P_{yy|z}) = 0. \end{aligned}$$

From (3.32) we have

$$\begin{aligned} 0 = \; & E\left\{(x - E(x|z))\left[(I - P^+_{yy|z}P_{yy|z})(y - E(y|z))\right]^T |z\right\} \\ = \; & P_{xy|z} - P_{xy|z}P^+_{yy|z}P_{yy|z}. \end{aligned}$$

This, together with (3.31), implies that w and y are conditionally independent given z by Lemma 3.1. Thus by Theorem 1.10 we have

$$E(w|z,y) = E(w|z) \quad a.s.$$

But from (3.30) a straightforward computation leads to $E(w|z) = 0$. Hence we have

$$E(w|z,y) = 0$$

and from this and (3.30) we obtain (3.28).

Further, by (3.28) we see

$$\begin{aligned} P_{xx|z,y} = &\ E[x - E(x|z,y))(x - E(x|z,y))^\tau|z] \\ = &\ P_{xx|z} + P_{xy|z}P^+_{yy|z}P_{yy|z}P^+_{yy|z}P_{xy|z} \\ &-2R_{xy|z}P^+_{yy|z}P_{yx|z} = P_{xx|z} - R_{xy|z}P^+_{yy|z}P_{yx|z} \end{aligned}$$

which verifies (3.29).

From (3.30) we find

$$w = x - E(x|y,z).$$

However, we have already shown that given z, w is conditionally Gaussian and independent of y. This is the second conclusion of the lemma.

To prove the first assertion we compute the characteristic function

$$\begin{aligned} &E(exp(i\lambda^\tau x)|z,y) \\ = &\ exp(i\lambda^\tau E(x|z,y))E(exp(i\lambda^\tau(x - E(x|z,y))|z,y). \end{aligned} \tag{3.33}$$

By Theorem 1.10, from the second conclusion it follows that

$$\begin{aligned} &E(exp(i\lambda^\tau(x - E(x|z,y))|z,y) = E(exp(i\lambda^\tau(x - E(x|z,y))|z) \\ = &\ exp\left\{-\frac{1}{2}\lambda^\tau E[(x - E(x|z,y))(x - E(x|z,y))^\tau|z]\lambda\right\} \\ = &\ exp\left\{-\frac{1}{2}\lambda^\tau(P_{xx|z} - P_{xy|z}P^+_{yy|z}P_{yx|z})\lambda\right\}. \end{aligned} \tag{3.34}$$

Combining (3.33) and (3.34) shows that $E(exp(i\lambda^\tau x)|z,y)$ is a conditionally Gaussian characteristic function. This proves the first assertion of the lemma.

Corollary 3.1. *If z is a constant in the lemma, then the conditional Gaussianity of $\begin{bmatrix} x \\ y \end{bmatrix}$ given z is equivalent to Gaussianity of $\begin{bmatrix} x \\ y \end{bmatrix}$. In this case the lemma concludes that*

$$E(x|y) = Ex + P_{xy}P^+_{yy}(y - Ey) \tag{3.28'}$$

$$P_{xx|y} = P_{xx} - P_{xy}P^+_{yy}P_{yx}, \tag{3.29'}$$

i.e. in the Gaussian case the conditional expectation $E(x|y)$ is a linear function of y.

Remark 3.1. Formula (3.28) is the basic relationship leading to the Kalman filtering [Ka2]. From the next lemma we shall see that $E(x|z)$ is the minimum variance estimate of x based on z. Hence, (3.28) means that the minimum variance estimate of x based on z and y is obtained from the minimum variance estimate of x based on z by updating the information contained in y but not in z.

Lemma 3.3. *Let x, y be random vectors with $E\|x\|^2 < \infty$. Then $E(x|y)$ is the minimum variance estimate of x based on y in the sense that*

$$E(x - E(x|y))(x - E(x|y))^{\tau} = \min_{z \in \mathcal{F}_y} E(x - z)(x - z)^{\tau}$$

where $\mathcal{F}_y$ denotes the totality of all y-measurable random vectors with dimension compatible with x.

Proof. By Properties 4 and 5 of the conditional expectation we have

$$\begin{aligned} & E(x - E(x|y))(E(x|y) - z)^{\tau} \\ = \; & E\{E[(x - E(x|y))(E(x|y) - z)^{\tau} |y]\} \\ = \; & E\{E[(x - E(x|y))|y](E(x|y) - z)^{\tau}\} = 0, \end{aligned}$$

and hence

$$\begin{aligned} & E[x - z][x - z]^{\tau} \\ = \; & E[x - E(x|y) + E(x|y) - z][x - E(x|y) + E(x|y) - z]^{\tau} \\ = \; & E(x - E(x|y))(x - E(x|y))^{\tau} + E[(E(x|y) - z)(E(x|y) - z)^{\tau}] \\ \geq \; & E(x - E(x|y))(x - E(x|y))^{\tau}. \end{aligned}$$

Q.E.D.

Corollary 3.2. *If $\begin{bmatrix} x \\ y \end{bmatrix}$ is Gaussian, then by (3.28′) and Lemma 3.3 the minimum variance estimate for x based on y equals the linear unbiased minimum variance estimate $\hat{x}$ for x based on y, where $E(x - \hat{x})(x - \hat{x})^{\tau} = \min_{c,\, C}(x - c - Cy)(x - c - Cy)^{\tau}$.*

Theorem 3.1 *(Kalman filter). The state x_k is conditionally Gaussian given $[y_0, y_1, ..., y_k]$, with conditional mean $\hat{x}$ and conditional covariance P_k which are given recursively by the following system of equations,*

$$\begin{aligned} \hat{x}_{k+1} &= A_k \hat{x}_k + B_k u_k \\ &\quad + K_k[y_{k+1} - C_k A_k \hat{x}_k - C_k B_k u_k - G_k v_k], && (3.35) \\ P_{k+1} &= R_k - K_k[C_k R_k + F_k D_k^{\tau}] && (3.36) \end{aligned}$$

with initial values $\hat{x}_0$ and P_0, where

$$\begin{aligned} K_k &= [R_k C_k^{\tau} + D_k F_k^{\tau}] \\ &\quad \cdot [C_k R_k C_k^{\tau} + F_k F_k^{\tau} + C_k D_k F_k^{\tau} + F_k D_k^{\tau} C_k^{\tau}]^{+} && (3.37) \\ R_k &= A_k P_k A_k^{\tau} + D_k D_k^{\tau}, && (3.38) \end{aligned}$$

which is the prediction error covariance matrix.

Proof. Denote

$$Y_k = [y_0^T \;\; y_1^T \; \dots \; y_k^T]^T.$$

We first show that given Y_k, both x_k and $\begin{bmatrix} x_{k+1} \\ y_{k+1} \end{bmatrix}$ are conditionally Gaussian.

Notice that $\begin{bmatrix} x_0 \\ w_1 \end{bmatrix}$ is conditionally Gaussian with finite conditional mean and covariance, since

$$\begin{aligned}
& E[exp(i\lambda^T x_0 + i\mu^T w_1)|y_0] \\
= \; & E\left\{exp(i\lambda^T x_0)E[exp(i\mu^T w_1)|x_0, y_0]|y_0\right\} \\
= \; & E[exp(i\mu^T w_1)]E[exp(i\lambda^T x_0)|y_0] \\
= \; & exp(i\lambda^T \hat{x}_0 - \frac{1}{2}\lambda^T P_0 \lambda - \frac{1}{2}\mu^T \mu)
\end{aligned}$$

by Conditions 1 and 3 of the theorem.

Further, we see that $\begin{bmatrix} x_1 \\ y_1 \end{bmatrix}$ is a linear function of $\begin{bmatrix} x_0 \\ w_1 \end{bmatrix}$ with y_0-measurable and a.s. finite coefficients:

$$\begin{bmatrix} x_1 \\ y_1 \end{bmatrix} = \begin{bmatrix} A_0 & D_0 \\ C_0 A_0 & C_0 D_0 + F_0 \end{bmatrix} \begin{bmatrix} x_0 \\ w_1 \end{bmatrix} + \begin{bmatrix} B_0 u_0 \\ C_0 B_0 u_0 + G_0 v_0 \end{bmatrix}.$$

Hence both $\begin{bmatrix} x_1 \\ y_1 \end{bmatrix}$ and x_0 are conditionally Gaussian given y_0 and the conditional means and conditional covariances are finite a.s.

We now assume that given Y_{k-1} both x_{k-1} and $\begin{bmatrix} x_k \\ y_k \end{bmatrix}$ are conditionally Gaussian with a.s. finite conditional means and conditional covariances. By Lemma 3.2 (1), given Y_k, x_k is conditionally Gaussian with finite conditional mean and conditional covariance. Then given Y_k, $\begin{bmatrix} x_k \\ w_{k+1} \end{bmatrix}$ is conditionally Gaussian with a.s. finite conditional mean and conditional variance because the conditional characteristic function

$$\begin{aligned}
& E\left[exp(i\lambda^T x_k + i\mu^T w_{k+1})|Y_k\right] \\
= \; & E\left[exp(i\lambda^T x_k) \cdot E[exp(i\mu^T w_{k+1})|x_k, Y_k]|Y_k\right] \\
= \; & E[exp(i\mu^T w_{k+1})]E[exp(i\lambda^T x_k)|Y_k]
\end{aligned}$$

is Gaussian.

Finally, $\begin{bmatrix} x_{k+1} \\ y_{k+1} \end{bmatrix}$ can be expressed as a linear function of $\begin{bmatrix} x_k \\ w_{k+1} \end{bmatrix}$:

$$\begin{bmatrix} x_{k+1} \\ y_{k+1} \end{bmatrix} = \begin{bmatrix} A_k & D_k \\ C_k A_k & C_k D_k + F_k \end{bmatrix} \begin{bmatrix} x_k \\ w_{k+1} \end{bmatrix} + \begin{bmatrix} B_k u_k \\ C_k B_k u_k + G_k v_k \end{bmatrix}$$

where the coefficients are Y_k-measurable and a.s. finite. Hence, given Y_k, $\begin{bmatrix} x_{k+1} \\ y_{k+1} \end{bmatrix}$ is conditionally Gaussian with a.s. finite conditional mean and conditional covariance, $k = 0, 1, 2, \ldots$.

Using Lemma 3.2, by identifying z with Y_k, y with y_{k+1}, and x with x_{k+1} we obtain

$$\begin{aligned} &E[x_{k+1}|Y_{k+1}] \\ = \;& E[x_{k+1}|Y_k] + P_{xy|z}P^+_{yy|z}[y_{k+1} - E(y_{k+1}|Y_k)], \quad a.s., \end{aligned} \tag{3.39}$$

where

$$P_{xy|z} = E\left[(x_{k+1} - E(x_{k+1}|Y_k))(y_{k+1} - E(y_{k+1}|Y_k))^\tau |Y_k\right], \tag{3.40}$$

$$P_{yy|z} = E\left[(y_{k+1} - E(y_{k+1}|Y_k))(y_{k+1} - E(y_{k+1}|Y_k))^\tau |Y_k\right]. \tag{3.41}$$

Substituting the following expressions

$$E[x_{k+1}|Y_k] = A_k E(x_k|Y_k) + B_k u_k \qquad a.s. \tag{3.42}$$

$$E[y_{k+1}|Y_k] = C_k E(x_{k+1}|Y_k) + G_k v_k \qquad a.s. \tag{3.43}$$

into (3.40) we find that

$$\begin{aligned} P_{xy|z} &= E\{[A_k(x_k - E(x_k|Y_k)) + D_k w_{k+1}] \\ &\quad \times [C_k A_k(x_k - E(x_k|Y_k)) + (F_k + C_k D_k)w_{k+1}]^\tau |Y_k\} \\ &= A_k P_k A_k^\tau C_k^\tau + D_k(F_k + C_k D_k)^\tau = R_k C_k^\tau + D_k F_k^\tau, \end{aligned} \tag{3.44}$$

since by independence $E[w_{k+1}|Y_k, x_k] = E[w_{k+1}] = 0$ and the cross-term is zero:

$$\begin{aligned} &E\{D_k w_{k+1}[C_k A_k(x_k - E(x_k|Y_k))]^\tau |Y_k\} \\ = \;& E\{D_k E[w_{k+1}|Y_k, x_k][C_k A_k(x_k - E(x_k|Y_k))]^\tau |Y_k\} \\ = \;& 0 \quad a.s. \end{aligned}$$

A completely similar computation for (3.41) leads to

$$\begin{aligned} P_{yy|z} &= C_k A_k P_k A_k^\tau C_k^\tau + (F_k + C_k D_k)(F_k + C_k D_k)^\tau \\ &= C_k R_k C_k^\tau + F_k F_k^\tau + C_k D_k F_k^\tau + F_k D_k^\tau C_k^\tau. \end{aligned} \tag{3.45}$$

Putting (3.40)-(3.45) into (3.39) we obtain (3.35). It remains to show (3.36).

By (3.29), (3.44) and (3.45) we have

$$\begin{aligned}
P_{k+1} &= E\left[(x_{k+1} - E(x_{k+1}|Y_{k+1}))(x_{k+1} - E(x_{k+1}|Y_{k+1}))^{\tau}|Y_{k+1}\right] \\
&= E\left[(x_{k+1} - E(x_{k+1}|Y_k))(x_{k+1} - E(x_{k+1}|Y_k))^{\tau}|Y_k\right] \\
&\quad -(R_k C_k^{\tau} + D_k P_k^{\tau})[C_k P_k C_k^{\tau} + F_k F_k^{\tau} \\
&\qquad +C_k D_k F_k^{\tau} + F_k D_k^{\tau} C_k^{\tau}]^{+}(C_k R_k + F_k D_k^{\tau}) \\
&= E\{[A_k(x_k - E(x_k|Y_k)) + D_k w_{k+1}] \\
&\quad \times[A_k(x_k - E(x_k|Y_k)) + D_k w_{k+1}]^{\tau}|Y_k\} \\
&\quad -K_k(C_k R_k + F_k D_k^{\tau}) \\
&= A_k P_k A_k^{\tau} + D_k D_k^{\tau} - K_k(C_k R_k + F_k D_k^{\tau}) \\
&= R_k - K_k(C_k R_k + F_k D_k^{\tau})
\end{aligned}$$

which verifies (3.36).

Finally, from (3.42) it is seen that

$$R_k = E\{[x_{k+1} - E(x_{k+1}|Y_k)][x_{k+1} - E(x_{k+1}|Y_k)]^{\tau}|Y_k\}.$$

Q.E.D.

Theorem 3.2. *Suppose that 1) in system (3.11) (3.12), A_k, B_k, D_k, C_k, G_k and F_k, $k \geq 0$ are deterministic; 2) both u_k and v_k are linear functions of Y_k; 3) $Ew_k = 0$, $Ew_i w_j^{\tau} = \delta_{ij}$, $E\begin{bmatrix} x_0 \\ y_0 \end{bmatrix} w_k^{\tau} = 0$, $k \geq 1$ ($\{w_k\}$ are not necessarily to be iid, independent of $\begin{bmatrix} x_0 \\ y_0 \end{bmatrix}$ and Gaussian).*

Denote by $\hat{x}_k$ and P_k the linear unbiased minimum variance estimate for x_k based on Y_k and the estimation error covariance respectively. Then $\hat{x}_k$ and P_k are recursively given also by (3.35)-(3.38).

Proof. If $\{w_k\}$ and $\begin{bmatrix} x_0 \\ y_0 \end{bmatrix}$ are Gaussian, then $\{x_i, y_i, w_{i+1}, 0 \leq i \leq k\}$ are Gaussian for any k and (3.35)-(3.38) are the recursive estimation formulas for $E(x_k|Y_k)$ and its estimation error covariance. Hence, by Corollary 3.2, (3.35)-(3.38) are also satisfied by the linear unbiased minimum variance estimates, which, however, do not depend upon distributions of $\{w_k\}$ and $\begin{bmatrix} x_0 \\ y_0 \end{bmatrix}$. Thus the Gaussian assumption for $\{w_k\}$ and $\begin{bmatrix} x_0 \\ y_0 \end{bmatrix}$ is without loss of generality. Q.E.D.

3.3 Discrete-Time Riccati Equations

Let us consider (3.36)-(3.38) in the case $F_k D_k^{\tau} \equiv 0$, i.e. the system noise $D_k w_{k+1}$ and the measurement noise $F_k w_{k+1}$ are uncorrelated. Then the prediction error covariance matrix R_k satisfies the following discrete-time Riccati equation:

$$\begin{aligned} R_{k+1} &= A_{k+1}P_{k+1}A_{k+1}^{\tau} + D_{k+1}D_{k+1}^{\tau} \\ &= A_{k+1}R_kA_{k+1}^{\tau} + D_{k+1}D_{k+1}^{\tau} \\ &\quad -A_{k+1}R_kC_k^{\tau}(C_kR_kC_k^{\tau} + F_kF_k^{\tau})^{+}C_kR_kA_{k+1}^{\tau}. \end{aligned} \tag{3.46}$$

Further, if the matrices appearing in (3.46) are time invariant, then (3.46) turns to

$$R_{k+1} = AR_kA^{\tau} + DD^{\tau} - AR_kC^{\tau}(CR_kC^{\tau} + FF^{\tau})^{+}CR_kA^{\tau}. \tag{3.47}$$

which is the equation satisfied by the prediction error covariance matrix for the system

$$x_{k+1} = Ax_k + Dw_{k+1}, \tag{3.48}$$

$$y_k = Cx_k + Fw_k, \quad x_k \in \mathbb{R}^n, \quad y_k \in \mathbb{R}^m, \quad w_k \in \mathbb{R}^l. \tag{3.49}$$

It is clear that if R_k tends to a limit R as $k \to \infty$, then R must satisfy the algebraic Riccati equation

$$R = ARA^{\tau} + DD^{\tau} - ARC^{\tau}(CRC^{\tau} + FF^{\tau})^{+}CRA^{\tau} \tag{3.50}$$

which plays an important role in control problems with quadratic costs.

We now give conditions guaranteeing convergence of R_k to a limit as $k \to \infty$.

Theorem 3.3. *Assume that 1) (A, D, C) is controllable and observable, i.e. both*

$$Q_{n-1} \triangleq [D \quad AD \quad \dots \quad A^{n-1}D]$$

and

$$G_{n-1} \triangleq [C^{\tau} \quad A^{\tau}C^{\tau} \quad \dots \quad (A^{n-1})^{\tau}C^{\tau}]$$

are of full rank where A is an $n \times n$-matrix and D and C are of compatible dimensions; 2) $FF^{\tau} > 0$. Then R_k defined by (3.47) as $k \to \infty$ uniformly in $\{R_0\colon \|R_0\| \le c\}$ converges to a finite limit R, where c is an arbitrarily fixed constant. Further, the limit R is independent of initial value R_0 and is the unique solution of the equation (3.50) in the class of nonnegative definite matrices. Moreover, as FF^{τ} increases R also increases.

Proof. We prove the theorem in four steps. First we show that the solution (denoted by R_k^0) of (3.47) with initial value equal to zero is positive definite $R_k^0 > 0$ for $k \ge n-1$. Secondly we prove that R_k is bounded for any initial value. Then, we complete the proof of the main part of the theorem by showing that R_k^0 is nondecreasing and the limit is independent of the initial value. Finally, we show that the limit R is nondecreasing with respect to FF^τ.

By Corollaries 3.1 and 3.2, without loss of generality, we may assume that x_0 and $\{w_k\}$ are mutually independent with Gaussian distributions.

Step 1.
Set

$$w^k = [w_0^\tau \quad w_1^\tau \quad \dots \quad w_k^\tau]^\tau, \tag{3.51}$$

$$y^k = [y_0^\tau \quad y_1^\tau \quad \dots \quad y_k^\tau]^\tau, \tag{3.52}$$

$$\phi_k = \begin{bmatrix} 0 & 0 & 0 & \cdots & 0 \\ 0 & CD & 0 & \cdots & 0 \\ 0 & CAD & CD & \cdots & 0 \\ \vdots & \vdots & \vdots & \ddots & \vdots \\ 0 & CA^{k-1}D & CA^{k-2}D & \cdots & CD \end{bmatrix},$$

$$\psi_k = \begin{bmatrix} F & 0 & \cdots & 0 \\ 0 & F & \vdots & 0 \\ \vdots & \vdots & \ddots & \vdots \\ 0 & 0 & \cdots & F \end{bmatrix}_{l(k+1)\times m(k+1)},$$

$$H_k = \phi_k + \psi_k, \quad L_k = [\underbrace{0}_{l} \quad A^{k-1}D \quad A^{k-2}D \quad \dots \quad D].$$

It is easy to verify that from (3.48) and (3.49) we have

$$y^k = G_k x_0 + H_k w^k, \tag{3.53}$$

$$x_k = A^k x_0 + L_k w^k. \tag{3.54}$$

Since $DF^\tau = 0$, $FF^\tau > 0$, we have $H_k H_k^\tau > 0$. Let $x_0 = 0$. Then

$$y^k = H_k w^k, \quad x_k = L_k w^k \tag{3.55}$$

and we have

$$P_{y^k y^k} \triangleq E[(y^k - Ey^k)(y^k - Ey^k)^\tau] = H_k H_k^\tau > 0 \tag{3.56}$$

and

$$P_{x_k x_k} \triangleq E[(x_k - Ex_k)(x_k - Ex_k)^\tau] = L_k L_k^\tau > 0 \tag{3.57}$$

for $k \geq n$ since Q_{n-1} and G_{n-1} are of full rank.

By Corollaries 3.1 and 3.2 it follows that

$$R_{k+1}^0 = P_{x_{k+1}x_{k+1}} - P_{x_{k+1}y^k} P_{y^k y^k}^{-1} P_{y^k x_{k+1}}, \tag{3.58}$$

$$P_{y^k y^k} = P_{y^k | x_{k+1}} + P_{y^k x_{k+1}} P_{x_{k+1}x_{k+1}}^{-1} P_{x_{k+1}y^k} \tag{3.59}$$

where

$$P_{y^k x_{k+1}} = E(y^k - Ey^k)(x_{k+1} - Ex_{k+1})^\tau. \tag{3.60}$$

From (3.55) we have

$$E\left(y^k|x_{k+1}\right) = E\left((\phi_k + \psi_k)w^k|x_{k+1}\right) = \phi_k E(w^k|x_{k+1})$$

since by the property $FD^\tau = 0$,

$$E(\psi_k w^k x_{k+1}^\tau) = \psi_k E w^k (w^{k+1})^\tau L_{k+1}^\tau$$

$$= \psi_k [I \quad 0] L_{k+1}^\tau = 0$$

which implies that

$$E\left(\psi_k w^k|x_{k+1}\right) = \left(E\psi_k w^k x_{k+1}^\tau\right) P^+_{x_{k+1}x_{k+1}} x_{k+1} = 0.$$

Then it is clear that

$$y^k - E(y^k|x_{k+1}) = \phi_k(w^k - E(w^k|x_{k+1})) + \psi_k w^k.$$

Noticing that $\phi_k \psi_k^\tau = 0$ we find that

$$\begin{aligned} P_{y^k|x_{k+1}} &= E[y^k - E(y^k|x_{k+1})][y^k - E(y^k|x_{k+1})]^\tau \\ &= \phi_k E[w^k - E(w^k|x_{k+1})][w^k - E(w^k|x_{k+1})]^\tau \phi_k^\tau + \psi_k \psi_k^\tau \\ &\geq \psi_k \psi_k^\tau > 0. \end{aligned} \tag{3.61}$$

We now apply the matrix inverse identity

$$(C^{-1} + C^{-1}B^\tau A^{-1}BC^{-1})^{-1} = C - B^\tau(A + BC^{-1}B^\tau)^{-1}B \tag{3.62}$$

which is verified as follows. From the trivial identity

$$0 = -(A + BC^{-1}B^\tau)^{-1} + A^{-1} - A^{-1}BC^{-1}B^\tau(A + BC^{-1}B^\tau)^{-1}$$

we derive

$$I = I \quad -C^{-1}B^\tau(A + BC^{-1}B^\tau)^{-1}B + C^{-1}B^\tau A^{-1}B$$

$$-C^{-1}B^\tau A^{-1}BC^{-1}B^\tau(A + BC^{-1}B^\tau)^{-1}B$$

or

$$\begin{aligned} I = &\ I - C^{-1}B^\tau(A + BC^{-1}B^\tau)^{-1}B \\ &+ C^{-1}B^\tau A^{-1}B(I - C^{-1}B^\tau(A + BC^{-1}B^\tau)^{-1}B) \\ = &\ (I + C^{-1}B^\tau A^{-1}B)(I - C^{-1}B^\tau(A + BC^{-1}B^\tau)^{-1}B) \end{aligned}$$

which is equivalent to (3.62). It is worth noting that (3.62) holds whenever C and A are invertable.

In (3.58) and (3.59) identifying $P_{x_{k+1}x_{k+1}}$ to C, $P_{y^k x_{k+1}}$ to B, $P_{y^k|x_{k+1}}$ to A and using (3.62) we get

$$\begin{aligned} R^0_{k+1} &= \left[P^{-1}_{x_{k+1}x_{k+1}} + P^{-1}_{x_{k+1}x_{k+1}} P_{x_{k+1}y^k} P^{-1}_{y^k|x_{k+1}} P_{y^k x_{k+1}} P^{-1}_{x_{k+1}x_{k+1}}\right]^{-1} \\ &> 0 \end{aligned} \tag{3.63}$$

for $k \geq n-1$ because Q_{n-1} is of full rank.

Step 2.

Let R'_k be the solution of (3.47) with initial value R'_0 which is arbitrarily chosen. We now show that R'_k is bounded.

Consider the following control problem with quadratic cost

$$\xi_{k+1} = A^T \xi_k + C^T u_k, \tag{3.64}$$

$$J_N(u) = \xi_N^T R'_0 \xi_N + \sum_{k=0}^{N-1} (\xi_k^T D D^T \xi_k + u_k^T F F_k^T u_k) \tag{3.65}$$

where the initial state ξ_0 is chosen to equal an arbitrary vector ξ.

Paying attention to (3.47) (3.64) from the following chain of equalities

$$\begin{aligned} J_N(u) &= \xi^T R'_N \xi \\ &+ \sum_{k=0}^{N-1} \{(\xi_{k+1}^T R'_{N-k-1} \xi_{k+1} - \xi_k^T R'_{N-k} \xi_k) + \xi_k^T D D^T \xi_k + u_k^T F F_k^T u_k\} \\ &= \xi^T R'_N \xi + \sum_{k=0}^{N-1} \{(A^T \xi_k + C^T u_k)^T R'_{N-k-1} (A^T \xi_k + C^T u_k) \\ &\quad -\xi_k^T \left[A R'_{N-k-1} A^T + D D^T \right. \\ &\quad \left. -A R'_{N-k-1} C^T (C R'_{N-k-1} C^T + F F^T)^{-1} C R'_{N-k-1} A^T\right] \xi_k \\ &\quad + \xi_k^T D D^T \xi_k + u_k^T F F_k^T u_k\} \\ &= \xi^T R'_N \xi + \sum_{k=0}^{N-1} (u_k + L_k \xi_k)^T (C R'_{N-k-1} C^T + F F^T)(u_k + L_k \xi_k) \end{aligned} \tag{3.66}$$

we conclude that the optimal control minimizing $J_N(u)$ is

$$u_k^0 = -L_k \xi_k \tag{3.67}$$

and

$$J_N(u^0) = \xi^T R'_N \xi \tag{3.68}$$

where

$$L_k = (CR'_{N-k-1}C^\tau + FF^\tau)^{-1}CR'_{N-k-1}A^\tau. \tag{3.69}$$

We now take a special control for (3.64)

$$u_k = \tilde{u}_k = \begin{cases} -CA^{n-k-1}(G^\tau_{n-1}G_{n-1})^{-1}(A^\tau)^n\xi_0, & k \le n-1 \\ 0, & k \ge n, \end{cases}$$

which leads to

$$\begin{aligned} \xi_n = & \quad (A^\tau)^n\xi_0 + \sum_{k=0}^{n-1}(A^\tau)^{n-k-1}C^\tau\tilde{u}_k \\ = & \quad \left\{ I - \left[\sum_{k=0}^{n-1}(A^\tau)^{n-k-1}C^\tau CA^{n-k-1}\right](G^\tau_{n-1}G_{n-1})^{-1}\right\}(A^\tau)^n\xi_0 \\ = & \quad 0 \end{aligned}$$

and

$$\xi_k = 0 \quad \text{for} \quad k \ge n.$$

Hence, we see that

$$\sup_{N\ge n} J_N(\tilde{u}) < \infty.$$

However, u^0 is optimal, we then conclude that

$$\sup_{N\ge n} J_N(u^0) \le \sup_{N\ge n} J_N(\tilde{u}) < \infty$$

and by (3.68)

$$\sup_{N\ge n} \xi^\tau R'_N\xi < \infty, \quad \text{for any } n\text{-dimensional } \xi.$$

Hence, R_k defined by (3.47) is bounded for any initial value R_0.

Step 3.

If $R'_0 = 0$ in (3.65), then we rewrite $J_N(u)$ as $J^0_N(u)$.

To emphasize the dependence upon the end time we denote by $u^0_k(N)$ the optimal control u^0_k defined by (3.67) for the cost $J^0_N(u)$. Then it is easy to see that

$$\xi^\tau R^0_N\xi = J^0_N(u^0(N)) \le J^0_N(u^0(N+1)) \le J^0_{N+1}(u^0(N+1)) = \xi^\tau R^0_{N+1}\xi.$$

Hence $\xi^\tau R^0_N\xi$ tends to a finite limit. Since R^0_N is symmetric by choosing appropriate ξ one can easily be convinced of convergence of R^0_N to a finite limit as $N \to \infty$.

Hence, from (3.63) we conclude that

$$R_k^0 \uparrow R > 0. \tag{3.70}$$

Take

$$\bar{u}_k = -(CRC^\tau + FF^\tau)^{-1} CRA^\tau \xi_k \tag{3.71}$$

for (3.64). Then the closed-loop system (3.64) becomes

$$\xi_{k+1} = \left[A^\tau - C^\tau (CRC^\tau + FF^\tau)^{-1} CRA^\tau\right] \xi_k. \tag{3.72}$$

We now show that (3.72) is a stable system. Using (3.50) we have

$$\begin{aligned}
&\xi_{k+1}^\tau R \xi_{k+1} \\
=\;& \xi_k^\tau \{ARA^\tau - 2ARC^\tau (CRC^\tau + FF^\tau)^{-1} CRA^\tau \\
&+ ARC^\tau (CRC^\tau + FF^\tau)^{-1} CRC^\tau (CRC^\tau + FF^\tau)^{-1} CRA^\tau\} \xi_k \\
=\;& \xi_k^\tau \{ARA^\tau - ARC^\tau (CRC^\tau + FF^\tau)^{-1} CRA^\tau \\
&- ARC^\tau (CRC^\tau + FF^\tau)^{-1} FF^\tau (CRC^\tau + FF^\tau)^{-1} CRA^\tau\} \xi_k \\
=\;& \xi_k^\tau R \xi_k - \xi_k^\tau DD^\tau \xi_k - \bar{u}_k^\tau FF^\tau \bar{u}_k.
\end{aligned}$$

Summing up both sides of this expression we find that

$$\begin{aligned}
\xi_N^\tau R \xi_N &= \xi_0^\tau R \xi_0 - \sum_{k=0}^{N-1} (\xi_k^\tau DD^\tau \xi_k + \bar{u}_k^\tau FF^\tau \bar{u}_k) \\
&= \xi_0^\tau R \xi_0 - J_N^0(\bar{u})
\end{aligned} \tag{3.73}$$

and by the optimality of $u^0(N)$ for J_N^0

$$0 \le \xi_N^\tau R \xi_N \le \xi_0^\tau R \xi_0 - J_N^0(u^0(N)) = \xi_0^\tau R \xi_0 - \xi_0^\tau R_N^0 \xi_0 \xrightarrow[N \to \infty]{} 0.$$

This means that $\xi_N \xrightarrow[N \to \infty]{} 0$ for an arbitrary initial value ξ_0 because $R > 0$. Thus system (3.72) is stable.

Further, from (3.73) and (3.65) we see that

$$\begin{aligned}
J_N(\bar{u}) &= J_N^0(\bar{u}) + \xi_N^\tau R_0' \xi_N \\
&= \xi_0^\tau R \xi_0 + \xi_N^\tau (R_0' - R) \xi_N \xrightarrow[N \to \infty]{} \xi_0^\tau R \xi_0.
\end{aligned} \tag{3.74}$$

Obviously, this convergence is uniform with respect to $\{R_0' \colon \|R_0'\| \le c\}$ for any fixed $c > 0$, where we have used the fact that $\xi_N \xrightarrow[N \to \infty]{} 0$ when $\{\bar{u}_k\}$ is applied.

Finally, applying (3.70) and (3.74) to the following expression:

$$\xi_0^\tau R_N^0 \xi_0 = J_N^0(u^0(N)) \leq J_N(u^0) = \xi_0^\tau R_N' \xi_0 \leq J_N^0(u^0) \leq J_N(\bar{u})$$

we see that both of its left-hand side and right-hand side uniformly with respect to $\{R_0': \|R_0'\| \leq c\}$ converge to the same limit $\xi_0^\tau R \xi_0$. Hence the middle term $\xi_0^\tau R_N' \xi_0$ also converges to $\xi_0^\tau R \xi_0$ uniformly in R_0' for $\|R_0'\| \leq c$. By the arbitrariness of ξ_0 we conclude that the convergence

$$R_N' \xrightarrow[N \to \infty]{} R$$

is uniform with respect to $\{R_0': \|R_0'\| \leq c\}$ for any fixed $c > 0$. In other words, starting from any initial value R_0' the solution of (3.47) tends to the same limit R. From this we can also assert the uniqueness of solution for (3.50) in the class of nonnegative definite matrices. To see this let R' be a solution of (3.50) other than R. We have just proved that with initial value R' the recursive algorithm (3.47) provides a solution tending to R. On the other hand, starting with R', (3.47) gives the same R' at any iteration because R' is the solution of (3.50). Hence, $R = R'$.

Step 4.

We now show that R is nondecreasing with respect to FF^τ. Let $F_1F_1^\tau \geq F_2F_2^\tau > 0$. We denote by $R_k(1)$, $R_k(2)$ the solutions of (3.47) with FF^τ replaced by $F_1F_1^\tau$ and $F_2F_2^\tau$ respectively, and denote their limits by $R(1)$ and $R(2)$ respectively.

Take $R_0(1) = R_0(2) > 0$. We prove that $R_k(1) \geq R_k(2) > 0$, $\forall k \geq 0$. Clearly, this is true for $k = 0$. Assume $R_k(1) \geq R_k(2) > 0$ for $0 \leq k \leq n$. From (3.47) by (3.62) it follows that

$$\begin{aligned}
& R_{n+1}(1) \\
= \;& DD^\tau + A[R_n(1) - R_n(1)C^\tau(CR_n(1)C^\tau + F_1F_1^\tau)^{-1}CR_n(1)]A^\tau \\
= \;& DD^\tau + A[R_n^{-1}(1) + C^\tau(F_1F_1^\tau)^{-1}C]^{-1}A^\tau \\
\geq \;& DD^\tau + A[R_n^{-1}(2) + C^\tau(F_1F_1^\tau)^{-1}C]^{-1}A^\tau = R_{n+1}(2) > 0.
\end{aligned}$$

Tending $n \to \infty$ we derive the required inequality

$$R(1) \geq R(2).$$

This completes the proof of the theorem. Q.E.D.

Remark 3.2. From (3.37) it is easy to see that, under the conditions of the theorem, the Kalman filtering gain K_k defined by (3.37) also tends to a steady gain

$$K_k \xrightarrow[k \to \infty]{} K \triangleq RC^\tau(CRC^\tau + FF^\tau)^{-1},$$

which is frequently used for simplifying calculation in the Kalman filtering.

For adaptive control systems coefficients A and C are unavailable. In the best case we can consistently estimate them. So it is important to know whether the solution R of (3.50) can adaptively be estimated. More precisely, the question arising here is as follows. Let A_k and C_k be the estimates at time k for A and C respectively. Recursively define

$$\begin{aligned} R_k &= A_k R_{k-1} A_k^{\tau} + DD^{\tau} \\ &\quad - A_k R_{k-1} C_k^{\tau} (C_k R_{k-1} C_k^{\tau} + FF^{\tau})^{-1} C_k R_{k-1} A_k^{\tau}, \end{aligned} \tag{3.75}$$

where $FF^{\tau} > 0$ and $R_0 \geq 0$ is arbitrary. We want to answer whether or not $R_n \xrightarrow[n \to \infty]{} R$ where R is the solution of (3.50).

Theorem 3.4. *Suppose that* $FF^{\tau} > 0$.

$$A_k \xrightarrow[k \to \infty]{} A, \qquad C_k \xrightarrow[k \to \infty]{} C,$$

(A,C) is observable and $\det A \neq 0$. *Then* R_k *defined by (3.75) is bounded in* k. *If, in addition, (A, D) is controllable, then* R_k *converges to the solution* R *of (3.50).*

Proof. Consider the following stochastic system:

$$x_{k+1} = A_{k+1} x_k + D w_{k+1}^1, \tag{3.76}$$

$$y_k = C_{k+1} x_k + F w_k^2, \tag{3.77}$$

where $w_k \triangleq [w_k^{1\tau} \quad w_k^{2\tau}]^{\tau} \in N(0, I)$ and is mutually independent and independent of $[x_0^{\tau} \quad y_0^{\tau}]^{\tau}$.

Clearly, (3.76) and (3.77) can be written in the form of (3.11) and (3.12). For this we need only to note that

$$D w_k^1 = D^1 w_k, \qquad F w_k^2 = F^1 w_k,$$

where $D^1 = [D \quad 0]$, $F^1 = [0 \quad F]$ and obviously, $D^1 F^1 = 0$. Then by (3.36)-(3.38) R_n defined by (3.75) is the prediction error covariance matrix for system (3.76) and (3.77).

From (3.38) we see that for boundedness of R_k it suffices to show that P_k is bounded since $A_k \xrightarrow[k \to \infty]{} A$ and D_k in (3.38) is a constant matrix in the present case.

We use the previous notation

$$Y_k = [y_0^{\tau} \cdots y_k^{\tau}]^{\tau}.$$

Then

$$P_k = E[x_k - E(x_k | Y_k)][x_k - E(x_k | Y_k)]^{\tau}. \tag{3.78}$$

Observability of (A,C) implies that

$$G_{n-1}G_{n-1}^{\tau} = \sum_{i=0}^{n-1}(A^i)^{\tau}C^{\tau}CA^i > 0$$

and

$$(A^{\tau})^{-(n-1)}G_{n-1}G_{n-1}^{\tau}A^{-(n-1)} = \sum_{i=0}^{n-1}(A^{-i})^{\tau}C^{\tau}CA^{-i} > 0.$$

Hence, there is a constant $\alpha > 0$ such that

$$\sum_{i=0}^{k}(A^{-i})^{\tau}C^{\tau}CA^{-i} \geq \alpha I, \quad \forall k \geq n-1. \tag{3.79}$$

Define

$$\phi(k+1,i) = A_{k+1}\phi(k,i), \quad \phi(i,i) = I, \quad \forall k \geq i. \tag{3.80}$$

Then

$$x_k = \phi(k,i)x_i + \sum_{j=i+1}^{k}\phi(k,j)Dw_j^1, \quad \forall k \geq i. \tag{3.81}$$

Paying attention to (3.79) and the convergence of A_k we know that there is k_0 such that for $\forall k \geq k_0$

$$W(k) = \sum_{i=k-n+1}^{k}\phi^{-\tau}(k,i)C_{i+1}^{\tau}C_{i+1}\phi^{-1}(k,i) \geq \frac{\alpha}{2}I, \tag{3.82}$$

where and hereafter by $\phi^{-\tau}(k,i)$ we mean $(\phi^{\tau}(k,i))^{-1}$.

We now take a special estimate $\bar{x}_k$ for x_k

$$\bar{x}_k = W^{-1}(k)\sum_{i=k-n+1}^{k}\phi^{-\tau}(k,i)C_{i+1}^{\tau}y_i$$

and express its estimation error

$$x_k - \bar{x}_k = W^{-1}(k)\left[W(k)x_k - \sum_{i=k-n+1}^{k}\phi^{-\tau}(k,i)C_{i+1}^{\tau}y_i\right]$$

$$= W^{-1}(k)\left\{\sum_{i=k-n+1}^{k}\phi^{-\tau}(k,i)C_{i+1}^{\tau}C_{i+1}\phi^{-1}(k,i)\right.$$

$$\left[\phi(k,i)x_i + \sum_{j=i+1}^{k}\phi(k,j)Dw_j^1\right]$$

$$- \sum_{i=k-n+1}^{k} \phi^{-\tau}(k,i)C_{i+1}^{\tau}(C_{i+1}x_i + Fw_i^2)\Big\}$$

$$= W^{-1}(k)\Big\{\sum_{i=k-n+1}^{k} \phi^{-\tau}(k,i)C_{i+1}^{\tau}C_{i+1}\phi^{-1}(k,i)\sum_{j=i+1}^{k}\phi(k,j)Dw_j^1$$

$$- \sum_{i=k-n+1}^{k} \phi^{-\tau}(k,i)C_{i+1}^{\tau}Fw_i^2\Big\}.$$

Since $A_i \xrightarrow[i\to\infty]{} A$, $C_i \xrightarrow[i\to\infty]{} C$, by (3.82) we find that there is a constant $\gamma > 0$ such that

$$\|x_k - \bar{x}_k\|^2 \leq \gamma \sum_{i=k-n+1}^{k} (\|w_i^1\|^2 + \|w_i^2\|^2).$$

From this by Lemma 3.3 we then have

$$trP_k = E\|x_k - E(x_k|Y_k)^2\|^2 \leq E\|x_k - \bar{x}_k\|^2$$

$$\leq \gamma n E(\|w_i^1\|^2 + \|w_i^2\|^2) < \infty.$$

Hence P_k is bounded and so is R_k.

We now prove that $R_k \xrightarrow[k\to\infty]{} R$.

For simplicity of notations, denote by $P(A, C, R)$ the right-hand side of (3.50). Then (3.50) and (3.75) can be written as

$$R = P(A, C, R) \tag{3.83}$$

and

$$R_k = P(A_k, C_k, R_{k-1}) \tag{3.84}$$

respectively.

By consistency of A_k and C_k and the boundedness of R_k it follows that

$$P(A, C, R_{k-1}) - P(A_k, C_k, R_{k-1}) \xrightarrow[k\to\infty]{} 0,$$

or

$$\Delta R_k \triangleq P(A, C, R_{k-1}) - R_k \xrightarrow[k\to\infty]{} 0.$$

Hence, for any $\varepsilon > 0$ we can find $N > 0$ such that

$$\|\Delta R_{i+k}\| \leq \varepsilon, \quad \forall k \geq 0, \quad \forall i \geq N. \tag{3.85}$$

Further, we set

$$P_1(\Gamma) \triangleq P(A, C, \Gamma), \qquad P_n(\Gamma) = P_1(P_{n-1}(\Gamma)). \tag{3.86}$$

It is easy to see that for the classes of matrices $\{\Gamma: \|\Gamma\| \le C\}$ and $\{\Delta\Gamma: \|\Delta\Gamma\| \le \varepsilon\}$ there is a constant ζ independent of the particular Γ and $\Delta\Gamma$ such that

$$P_1(\Gamma + \Delta\Gamma) = P_1(\Gamma) + \Delta P_1(\Delta\Gamma), \quad \|\Delta P_1(\Delta\Gamma)\| \le \zeta\varepsilon. \tag{3.87}$$

We now by induction prove that for any $i \ge N$ and $k \ge 1$

$$R_{i+k} = P_k(R_i) + Z_{ik}(\varepsilon) \quad \textit{with} \quad \|Z_{ik}(\varepsilon)\| \le c_k\varepsilon, \tag{3.88}$$

where c_k is a real number independent of $i \ge N$.

By (3.87) we see that (3.88) is true for $k = 1$. Assume (3.88) holds for k.

We have shown that R_i is bounded, say, $\|R_i\| \le c$, $\forall i \ge 0$. By Theorem 3.3, $P_k(R_0)$ uniformly in R_0 for $\{R_0: \|R_0\| \le c\}$ converges to R. Hence $P_k(R_i)$ is bounded in $i \ge 0$ and $k \ge 0$. Then by (3.85) (3.87) it follows that

$$\begin{aligned} R_{i+k+1} = & \; P_1(R_{i+k}) + \Delta R_{i+k+1} \\ = & \; P_1[P_k(R_i) + Z_{ik}(\varepsilon)] + \Delta R_{i+k+1} \\ = & \; P_{k+1}(R_i) + \Delta P_1(Z_{ik}(\varepsilon)) + \Delta R_{i+k+1} \\ = & \; P_{k+1}(R_i) + Z_{ik+1}(\varepsilon), \end{aligned}$$

where

$$\begin{aligned} \|Z_{ik+1}(\varepsilon)\| = & \; \|\Delta P_1(Z_{ik}(\varepsilon)) + \Delta R_{i+k+1}\| \le \zeta \|Z_{ik}(\varepsilon)\| + \Delta \|R_{i+k+1}\| \\ \le & \; \zeta c_k \varepsilon + \varepsilon = c_{k+1}\varepsilon \qquad \textit{with} \quad c_{k+1} = (\zeta c_k + 1). \end{aligned}$$

This proves (3.88), from which we have

$$\|R_{i+k} - R\| \le \|P_k(R_i) - R\| + \|Z_{ik}(\varepsilon)\|, \quad i \ge n. \tag{3.89}$$

Since $\{R_i\}$ is bounded and $P_k(R_0)$ converges to R uniformly in R_0 for $\{R_0: \|R_0\| \le c\}$, for any $\delta > 0$ we can find a sufficiently large k_0 such that

$$\sup_{i \ge N} \|P_k(R_i) - R\| \le \delta, \qquad k \ge k_0.$$

Then from (3.89) we see that

$$\begin{aligned} \sup_{i \ge N} \|R_{i+k_0} - R\| \le & \; \delta + \sup_{i \ge N} \|Z_{ik_0}(\varepsilon)\| \\ \le & \; \delta + c_{k_0}\varepsilon \le 2\delta \quad \textit{as} \quad N \to \infty \end{aligned}$$

since $\varepsilon \to 0$ as $N \to \infty$ by (3.85).

This proves that $R_k \xrightarrow[k \to \infty]{} R$. Q.E.D.

Remark 3.3. If A is stable, then the requirement $\det A \neq 0$ can be removed from Theorem 3.4. In this case, the boundedness of $\{R_k\}$ follows directly from (3.75) and the assumption $A_k \to A$.

3.4 Optimal Control for Quadratic Costs

For stochastic adaptive control we shall mainly be concerned with the ARMAX process which models the stochastic input-output feedback control system in discrete time. We now describe its multidimensional version.

Let A_1, A_2, ..., A_p, C_1, C_2, ..., C_r be $m \times m$ matrices and B_1, B_2, ..., B_q be $m \times l$ matrices. We denote by y_k the m-dimensional output, u_k the l-dimensional control (input) and w_k the m-dimensional driven noise. The ARMAX system is in fact a stochastic difference equation:

$$\begin{aligned} & y_n + A_1 y_{n-1} + \cdots + A_p y_{n-p} \\ = \; & B_1 u_{n-d} + B_2 u_{n-d-1} + \cdots + B_q u_{n-d-q+1} \\ & + w_n + C_1 w_{n-1} + \cdots + C_r w_{n-r}, \quad n \geq 0; \end{aligned}$$

$$y_n = w_n = 0, \; u_n = 0, \; n < 0; \quad p \geq 0, \; r \geq 0, \; d \geq 1. \tag{3.90}$$

which can also be written in a compact form

$$A(z) y_n = B(z) u_{n-d} + C(z) w_n, \tag{3.91}$$

where

$$A(z) = I + A_1 z + \ldots + A_p z^p, \tag{3.92}$$

$$B(z) = B_1 + B_2 z + \ldots + B_q z^{q-1}, \tag{3.93}$$

$$C(z) = I + C_1 z + \ldots + C_r z^r \tag{3.94}$$

and z denotes the shift-back operator: $z y_n = y_{n-1}$.

(p,q,r) are called the system orders, while d is the time-delay.

If there is no explanation, we always assume $\{w_n, \mathcal{F}_n\}$ is a martingale difference sequence with respect to a nondecreasing family of σ-algebras $\{\mathcal{F}_n\}$.

Adaptive control laws to be given in later chapters are inspired by the optimal controls for the system (3.91) with known parameters. Let us consider the quadratic loss function

$$J(u) = \limsup_{n \to \infty} \frac{1}{n} \sum_{i=0}^{n-1} [(y_i - y_i^*)^T Q_1 (y_i - y_i^*) + u_i^T Q_2 u_i], \tag{3.95}$$

where $Q_1 \geq 0$ and $Q_2 > 0$ are the weighting matrices and $\{y_i^*\}$ is a bounded deterministic reference signal. The control $\{u_i\}$ has to be designed to minimize $J(u)$. For deriving the optimal quadratic control it is convenient to represent (3.91) in the state space form.

Set

$$A = \begin{bmatrix} -A_1 & I & 0 & \cdots & 0 \\ -A_2 & 0 & \ddots & \ddots & \vdots \\ \vdots & \vdots & \ddots & \ddots & 0 \\ \vdots & \vdots & \ddots & \ddots & I \\ -A_s & 0 & \cdots & \cdots & 0 \end{bmatrix}, \qquad B = \begin{bmatrix} 0 \\ \vdots \\ 0 \\ B_1 \\ \vdots \\ B_s \end{bmatrix} \left.\vphantom{\begin{matrix}0\\ \vdots\\ 0\end{matrix}}\right\} d-1 \tag{3.96}$$

$$C^T = [I \quad C_1^T \ldots C_{s-1}^T], \qquad H = [\underbrace{I \quad 0 \ldots 0}_{ms}]\}m \tag{3.97}$$

where $s = p \vee (q+d-1) \vee (r+1)$ and $A_i = 0$, $B_j = 0$, $C_k = 0$ for $i > p$, $j > q$, $k > r$ ($p \vee q \triangleq max(p,q)$). Then it is immediately verified that

$$x_{k+1} = Ax_k + Bu_k + Cw_{k+1}, \tag{3.98}$$

$$y_k = Hx_k, \qquad x_0^T = [y_0^T \quad 0 \ldots 0]. \tag{3.99}$$

Define the set U of admissible controls:

$$U = \left\{ u : \sum_{i=0}^{n-1} (\|u_i\|^2 + \|x_i\|^2) = O(n), \ \|x_n\|^2 = o(n) \ a.s. \ u_i \in \mathcal{F}_i, \ i \geq 0 \right\}. \tag{3.100}$$

Theorem 3.5. *Suppose that*
1) $\{w_n, \mathcal{F}_n\}$ *is a martingale difference sequence with*

$$\sup_{n\geq 0} E[\|w_n\|^2 | \mathcal{F}_{n-1}\} < \infty \quad a.s., \tag{3.101}$$

$$\lim_{n\to\infty} \frac{1}{n} \sum_{i=1}^{n} w_i w_i^T = R > 0 \quad a.s. \tag{3.102}$$

2) (A,B,D) is controllable and observable, where D is any matrix satisfying

$$D^T D = H^T Q_1 H. \tag{3.103}$$

Then in the class of nonnegative definite matrices there is a unique $S > 0$ *satisfying*

$$S = A^T SA - A^T SB(Q_2 + B^T SB)^{-1} B^T SA + H^T Q_1 H \tag{3.104}$$

and

$$F \triangleq A - B(Q_2 + B^\tau SB)^{-1}B^\tau SA \tag{3.105}$$

is stable.

The optimal control minimizing $J(u)$ given by (3.95) is

$$u_n^0 = Lx_n + d_n, \qquad \forall n \geq 0 \tag{3.106}$$

where

$$L = -(Q_2 + B^\tau SB)^{-1}B^\tau SA, \tag{3.107}$$

$$d_n = -(Q_2 + B^\tau SB)^{-1}B^\tau b_{n+1}, \tag{3.108}$$

$$b_i = -\sum_{j=0}^{\infty} F^{j\tau} H^\tau Q_1 y_{i+j}^* = F^\tau b_{i+1} - H^\tau Q_1 y_i^* \tag{3.109}$$

and

$$\begin{aligned}\min_{u\in U} J(u) &= J(u^0) \\ &= \limsup_{n\to\infty} \frac{1}{n}\sum_{i=0}^{n-1}[y_i^{*\tau} Q_1 y_i^* - b_{i+1}^\tau B(Q_2 + B^\tau SB)^{-1}B^\tau b_{i+1}] \\ &\quad + trSCRC^\tau \end{aligned} \tag{3.110}$$

Proof. Note that (A, B, D) is controllable and observable if and only if (A^τ, D^τ, B^τ) is controllable and observable. Hence identifying (A^τ, D^τ, B^τ) and Q_2 to (A, B, D) and FF^τ in Theorem 3.3 respectively, we find that (3.104) coincides with (3.50) and the existence and uniqueness of solution for (3.104) follow from Theorem 3.3, while stability of (3.105) is the same as that for (3.72).

We now express

$$J_n(u) = \sum_{i=0}^{n-1}[(y_i - y_i^*)^\tau Q_1(y_i - y_i^*) + u_i^\tau Q_2 u_i] \tag{3.111}$$

in a complete square. For this we first note that by (3.104)

$$\begin{aligned} x_n^\tau S x_n - x_0^\tau S x_0 &= \sum_{i=0}^{n-1}(x_{i+1}^\tau S x_{i+1} - x_i^\tau S x_i) \\ &= \sum_{i=0}^{n-1}[(Ax_i + Bu_i + Cw_{i+1})^\tau S(Ax_i + Bu_i + Cw_{i+1}) \\ &\quad - x_i^\tau (A^\tau SA - A^\tau SB(Q_2 + B^\tau SB)^{-1}B^\tau SA \\ &\quad + H^\tau Q_1 H)x_i], \end{aligned}$$

and hence

$$\begin{aligned}
&\sum_{i=0}^{n-1} x_i^\tau H^\tau Q_1 H x_i \\
= \;& x_0^\tau S x_0 - x_n^\tau S x_n + \sum_{i=0}^{n-1} [x_i^\tau A^\tau SB(Q_2 + B^\tau SB)^{-1} B^\tau SAx_i \\
& +2(Bu_i + Cw_{i+1})^\tau SAx_i \\
& +(Bu_i + Cw_{i+1})^\tau S(Bu_i + Cw_{i+1})]. \qquad (3.112)
\end{aligned}$$

Similarly, we have

$$\begin{aligned}
b_n^\tau x_n - b_0^\tau x_0 &= \sum_{i=0}^{n-1}(b_{i+1}^\tau x_{i+1} - b_i^\tau x_i) \\
&= \sum_{i=0}^{n-1}(b_{i+1}^\tau(Ax_i + Bu_i + Cw_{i+1}) - b_i^\tau x_i]. \qquad (3.113)
\end{aligned}$$

From (3.105) and (3.109) we see

$$\begin{aligned}
& x_i^\tau [A^\tau - A^\tau SB(Q_2 + B^\tau SB)^{-1} B^\tau] b_{i+1} \\
= \;& x_i^\tau b_i + x_i^\tau H^\tau Q_1 y_i^*. \qquad (3.114)
\end{aligned}$$

Using (3.112), (3.113) and (3.114) we obtain

$$\begin{aligned}
J_n(u) =& \sum_{i=0}^{n-1}(x_i^\tau H^\tau Q_1 H x_i - 2y_i^{*\tau} Q_1 H x_i + y_i^{*\tau} Q_1 y_i^* + u_i^\tau Q_2 u_i) \\
= \;& x_0^\tau S x_0 - x_n^\tau S x_n + \sum_{i=0}^{n-1}\{[x_i^\tau A^\tau SB(Q_2 + B^\tau SB)^{-1} B^\tau SAx_i \\
& +2(Bu_i + Cw_{i+1})^\tau SAx_i \\
& +(Bu_i + Cw_{i+1})^\tau S(Bu_i + Cw_{i+1})] \\
& +2b_0^\tau x_0 - 2b_n^\tau x_n + \sum_{i=0}^{n-1} 2\{[b_{i+1}^\tau(Ax_i + Bu_i + Cw_{i+1}) - b_i^\tau x_i] \\
& -2y_i^{*\tau} Q_1 H x_i + y_i^{*\tau} Q_1 y_i^* + u_i^\tau Q_2 u_i\} \\
= \;& x_0^\tau S x_0 - x_n^\tau S x_n + 2b_0^\tau x_0 - 2b_n^\tau x_n \\
& + \sum_{i=0}^{n-1}\{x_i^\tau A^\tau SB(Q_2 + B^\tau SB)^{-1} B^\tau SAx_i \\
& +2(Bu_i + Cw_{i+1})^\tau SAx_i \\
& +(Bu_i + Cw_{i+1})^\tau S(Bu_i + Cw_{i+1}) + 2b_{i+1}^\tau(Bu_i + Cw_{i+1})
\end{aligned}$$

$$
\begin{aligned}
&+2x_i^\tau A^\tau SB(Q_2+B^\tau SB)^{-1}B^\tau b_{i+1}\\
&+y_i^{*\tau}Q_1y_i^*+u_i^\tau Q_2u_i\}\\
=\;& x_0^\tau Sx_0-x_n^\tau Sx_n+2b_0^\tau x_0-2b_n^\tau x_n+\sum_{i=0}^{n-1}w_{i+1}^\tau C^\tau SCw_{i+1}\\
&+\sum_{i=0}^{n-1}[y_i^{*\tau}Q_1y_i^*-b_{i+1}^\tau B(Q_2+B^\tau SB)^{-1}B^\tau b_{i+1}]\\
&+2\sum_{i=0}^{n-1}[(Ax_i+Bu_i)^\tau S+b_{i+1}^\tau]Cw_{i+1}\\
&+\sum_{i=0}^{n-1}[u_i+(Q_2+B^\tau SB)^{-1}B^\tau(SAx_i+b_{i+1})]^\tau\\
&\cdot(Q_2+B^\tau SB)[u_i+(Q_2+B^\tau SB)^{-1}B^\tau(SAx_i+b_{i+1})]. \qquad (3.115)
\end{aligned}
$$

From this the optimal control (3.106) follows immediately. It remains to prove (3.110). From (3.106) and (3.115) we have

$$
\begin{aligned}
J(u^0) =\;& \limsup_{n\to\infty}\frac{1}{n}\{x_0^\tau Sx_0-x_n^\tau Sx_n+2b_0^\tau x_0-2b_n^\tau x_n\\
&+\sum_{i=0}^{n-1}w_{i+1}^\tau C^\tau SCw_{i+1}\\
&+\sum_{i=0}^{n-1}[y_i^{*\tau}Q_1y_i^*-b_{i+1}^\tau B(Q_2+B^\tau SB)^{-1}B^\tau b_{i+1}]\\
&+2\sum_{i=0}^{n-1}[(Ax_i+Bu_i)^\tau SC+b_{i+1}^\tau C]w_{i+1}\}. \qquad (3.116)
\end{aligned}
$$

Since $u\in U$ with U defined by (3.100), by Theorem 2.8 we see that

$$
\begin{aligned}
&\frac{1}{n}\sum_{i=0}^{n-1}(Ax_i+Bu_i)^\tau SCw_{i+1}\\
=\;& O\left(\frac{1}{n}\left(\sum_{i=0}^{n-1}\|(Ax_i+Bu_i)\|^2\right)^{\frac{1}{2}}\left(\log\sum_{i=0}^{n-1}\|(Ax_i+Bu_i)\|^2\right)^{\frac{1}{2}+\delta}\right)\\
=\;& O\left(\frac{1}{n}\cdot\sqrt{n}(\log n)^{\frac{1}{2}+\delta}\right)\xrightarrow[n\to\infty]{}0. \qquad (3.117)
\end{aligned}
$$

As shown for (3.72) the matrix F defined by (3.105) is stable. Then from (3.109) it is seen that the boundedness of $\{y_i^*\}$ implies the boundedness of

$\{b_i\}$. Again by use of Theorem 2.8 we find that

$$\frac{1}{n}\sum_{i=0}^{n-1} b_{i+1}^{\tau} C w_{i+1} \xrightarrow[n\to\infty]{} 0. \tag{3.118}$$

Finally, paying attention to (3.102) and $x_n = o(n)$ from (3.117) (3.118) we conclude (3.110) from (3.116). Q.E.D.

Remark 3.4. If $y_i^* \equiv 0$ then $J(u)$ given by (3.95) is a standard quadratic loss function. In this case $b_n \equiv 0$, $d_n \equiv 0$ and the optimal desired control is $u_n^0 = Lx_n$ with L given by (3.107) and the minimum of the loss function is

$$J(u^0) = trSCRC^{\tau}. \tag{3.119}$$

3.5 Optimal Tracking

In this section we design a feedback control u_n at time n such that y_{n+d} approaches y_{n+d}^* as close as possible.

We first prove a lemma.

Lemma 3.4. *The Diophantine equation*

$$(detC(z))I = F(z)(AdjC(z))A(z) + z^d G(z) \tag{3.120}$$

has a unique solution with respect to $F(z)$ *and* $G(z)$ *with*

$$F(z) = I + F_1 z + \dots + F_{d-1} z^{d-1}. \tag{3.121}$$

Proof. Express the polynomials $detC(z)$ and $(AdjC(z))A(z)$ explicitly

$$detC(z) = 1 + c_1 z + \dots + c_{mr} z^{mr}, \tag{3.122}$$

$$(AdjC(z))A(z) = I + H_1 z + \dots + H_{(m-1)r+p} z^{(m-1)r+p}, \tag{3.123}$$

where by setting $H_i = 0$, for $i \in ((m-1)r + p,\ d-1]$ and $c_i = 0$, for $i \in (mr,\ d-1]$. Then we recursively define F_i, $i = 0, \dots, d-1$:

$$F_0 = I$$

$$F_i = c_i I - \sum_{j=1}^{i} F_{i-j} H_j, \qquad i = 1, \dots, d-1.$$

It is easy to see that $[F_0 \quad F_1 \quad \ldots \quad F_{d-1}]$ satisfies the following matrix equation:

$$[I \quad c_1 I \ldots c_{d-1} I] = [F_0 \; F_1 \ldots F_{d-1}] \begin{bmatrix} I & H_1 & H_2 & \cdots & \cdots & H_{d-1} \\ 0 & I & H_1 & \ddots & \cdots & H_{d-2} \\ 0 & 0 & I & \ddots & \ddots & \vdots \\ \vdots & \vdots & \vdots & \vdots & \ddots & H_1 \\ 0 & \cdots & \cdots & \cdots & 0 & I \end{bmatrix}$$

which implies that all the coefficients of z^i, $i = 1, \ldots, d-1$ in

$$(detC(z))I - F(z)(AdjC(z))A(z)$$

are zero. Therefore, there exists a polynomial matrix $G(z)$ such that

$$(detC(z))I - F(z)(AdjC(z))A(z) = z^d G(z).$$

We now prove the uniqueness.

Let $(F_1(z), G_1(z))$ and $(F_2(z), G_2(z))$ be two solutions of (3.12) with $F_1(z)$ and $F_2(z)$ of degree less than d. Then we have

$$z^d(G_1(z) - G_2(z)) = (F_2(z) - F_1(z))(\mathrm{Adj}C(z))A(z),$$

which means that $F_2(z) - F_1(z)$, being of degree less than d must divide by z^d. The contradiction shows the uniqueness. Q.E.D.

Theorem 3.6. *Assume that (3.101) and (3.102) hold. $\{y_n^*\}$ is a given reference signal independent of $\{w_n\}$ and y_{n+d}^* is $\mathcal{F}_n$-measurable. Then for any $\mathcal{F}_n$-measurable control u_n (thus feedback controls are included.) the tracking error has the lower bound:*

$$\limsup_{n\to\infty} \frac{1}{n} \sum_{i=0}^{n} \|y_i - y_i^*\|^2 \geq tr \sum_{j=0}^{d-1} F_j R F_j^\tau . \tag{3.124}$$

Further, if u_n can be solved from

$$(detC(z))^{-1}[G(z)y_n + F(z)(AdjC(z))B(z)u_{n+d}] = y_{n+d}^*, \tag{3.125}$$

then u_n is optimal and it leads to

$$\lim_{n\to\infty} \frac{1}{n} \sum_{i=0}^{n} (y_i - y_i^*)(y_i - y_i^*)^\tau = \sum_{j=0}^{d-1} F_j R F_j^\tau . \tag{3.126}$$

Proof. Using (3.120) and (3.91) we have

$$\begin{aligned}
&(detC(z))(y_n - F(z)w_n)\\
=\ & G(z)y_{n-d} - (detC(z))F(z)w_n + F(z)(AdjC(z))A(z)y_n\\
=\ & G(z)y_{n-d} - (detC(z))F(z)w_n\\
&+F(z)(AdjC(z))B(z)u_{n-d} + (detC(z))F(z)w_n\\
=\ & G(z)y_{n-d} + F(z)(AdjC(z))B(z)u_n
\end{aligned} \tag{3.127}$$

and

$$y_n - y_n^* = F(z)w_n + f_{n-d}, \tag{3.128}$$

where

$$f_{n-d} = (detC(z))^{-1}[G(z)y_{n-d} + F(z)(AdjC(z))B(z)u_n] - y_n^*$$

is $\mathcal{F}_{n-d}$-measurable and is uncorrelated with $F(z)w_n$.

By Theorem 2.8 we know that for $j = 0, 1, \ldots, d-1$,

$$\left\| \sum_{i=1}^{n} f_{i-d}^{T} F_j w_{i-j} \right\| = O\left(\left(\sum_{i=1}^{n} \|f_{i-d}\|^2 \right)^{\alpha} \right), \ \alpha \in (\frac{1}{2}, 1),$$

and hence

$$\left\| \sum_{i=0}^{n} f_{i-d}^{T} F(z) w_i \right\| = O\left(\left(\sum_{i=0}^{n} \|f_i\|^2 \right)^{\alpha} \right), \qquad \alpha \in (\frac{1}{2}, 1).$$

Then as $n \to \infty$

$$\sum_{i=0}^{n} \|f_{i-d}\|^2 + \sum_{i=1}^{n} f_{i-d}^{T} F_j w_{i-j}$$

is either bounded a.s. or positive. Hence we see that as $n \to \infty$

$$\begin{aligned}
&\limsup_{n\to\infty} \frac{1}{n} \sum_{i=0}^{n} \|y_i - y_i^*\|^2\\
=\ & \limsup_{n\to\infty} \frac{1}{n} \left\{ \sum_{i=0}^{n} \|f_{i-d}\|^2 + O\left(\left(\sum_{i=0}^{n} \|f_i\|^2 \right)^{\alpha} \right) + \sum_{i=1}^{n} \|F(z)w_i\|^2 \right\}\\
\geq\ & \limsup_{n\to\infty} \frac{1}{n} \sum_{i=1}^{n} \|F(z)w_i\|^2\\
=\ & \limsup_{n\to\infty} tr \frac{1}{n} \sum_{i=1}^{n} (F(z)w_i)(F(z)w_i)^{T}\\
=\ & tr \sum_{j=0}^{d-1} F_j R F_j^{T}
\end{aligned}$$

where the last equality is derived by Theorem 2.8.

If u_n is solved from (3.125), then $f_i \equiv 0$ and from (3.128) and Theorem 2.8 we obtain (3.126). Q.E.D.

Corollary 3.3. *In the unit delay case* $d = 1$, $F(z) = I$,

$$G(z) = z^{-1}[detC(z)I - (AdjC(z))A(z)]$$

and (3.125) is reduced to

$$(detC(z))^{-1}\{[detC(z)I - (AdjC(z))A(z)]y_{n+1} + (AdjC(z))B(z)u_{n+1}\}$$
$$= y^*_{n+1}$$

or

$$(I - C^{-1}(z)A(z))y_{n+1} + C^{-1}B(z)u_{n+1} = y^*_{n+1}$$

or

$$(C(z) - A(z))y_{n+1} + B(z)u_{n+1} = C(z)y^*_{n+1}.$$

This equation can be written as

$$-A_1y_n - A_2y_{n-1} - \dots - A_py_{n-p+1} + B_1u_n + \dots + B_qu_{n-q+1}$$
$$+C_1(y_n - y^*_n) + C_2(y_{n-1} - y^*_{n-1}) + \dots + C_r(y_{n-r+1} - y^*_{n-r+1})$$
$$= y^*_{n+1}$$

or

$$\theta^\tau \varphi_n = y^*_{n+1}, \tag{3.129}$$

where

$$\theta^\tau = [-A_1 \quad \dots - A_p \quad B_1 \quad \dots \quad B_q \quad C_1 \quad \dots \quad C_r] \tag{3.130}$$

$$\varphi^\tau_n = [y^\tau_n \quad \dots \quad y^\tau_{n-p+1} \quad u^\tau_n \quad \dots \quad u^\tau_{n-q+1} \quad (y_n - y^*_n)^\tau \quad \dots \quad (y_{n-r+1} - y^*_{n-r+1})^\tau]. \tag{3.131}$$

It is clear that (3.129) can be solved with respect to u_n if $m \geq r$ and B_1 is of full rank. To be specific, in this case the optimal control is

$$u_n = (B^\tau_1 B_1)^{-1}B^\tau_1[A_1y_n + A_2y_{n-1} + \dots + A_py_{n-p+1}$$
$$-B_2u_{n-2} - \dots - B_qu_{n-q+1}$$
$$-C_1(y_n - y^*_n) - C_2(y_{n-1} - y^*_{n-1}) - \dots$$
$$-C_r(y_{n-r+1} - y^*_{n-r+1}) + y^*_{n+1}].$$

3.6 Model Reference Control

Consider a reference model

$$A^0(z)y_n = B^0(z)u_n^*, \tag{3.132}$$

where $A^0(z)$ and $B^0(z)$ are given matrix polynomials of orders $\bar{p}$ and $\bar{q}$ respectively, and u_n^* is $\mathcal{F}_n$-measurable external input.

The model reference control [La] consists in designing the $\mathcal{F}_n$-measurable control u_n in order to make the system (3.91) as close to (3.132) as possible.

We rewrite (3.91) in the form

$$A^0(z)y_n = B^0(z)u_n^* + \varepsilon_n \tag{3.133}$$

with

$$\varepsilon_n = (A^0(z) - A(z))y_n + B(z)u_n - B^0(z)u_n^* + C(z)w_n \tag{3.134}$$

and express

$$A^0(z)F(z) = M(z) + z^d N(z) \tag{3.135}$$

where

$$M(z) = M_0 + M_1 z + \quad \ldots \quad + M_{d-1}z^{d-1}. \tag{3.136}$$

Theorem 3.7. *Let (3.101) and (3.102) hold. Then the deviation ε_n from the reference model has the lower bound*

$$\limsup_{n\to\infty} \frac{1}{n}\sum_{i=0}^{n} \|\varepsilon_i\|^2 \geq tr \sum_{j=0}^{d-1} M_j R M_j^T.$$

Further, if u_n can be solved from $g_n = 0$, where

$$\begin{aligned} g_n \triangleq\ & N(z)w_n - B^0(z)u_{n+d}^* + (detC(z))^{-1}A^0(z)[G(z)y_n \\ & + F(z)(AdjC(z))B(z)u_{n+d}] \end{aligned} \tag{3.137}$$

then u_n is optimal and it leads to

$$\lim_{n\to\infty} \frac{1}{n}\sum_{i=0}^{n} \varepsilon_i\varepsilon_i^T = \sum_{j=0}^{d-1} M_j R M_j^T. \tag{3.138}$$

Proof. From (3.127) we see that

$$y_n = F(z)w_n + (detC(z))^{-1}[G(z)y_{n-d} + F(z)(AdjC(z))B(z)u_n].$$

Putting expressions (3.135) and (3.136) into (3.134) we find

$$\begin{aligned}\varepsilon_n = \; & A^0(z)F(z)w_n + (detC(z))^{-1}A^0(z) \\ & \times[G(z)y_{n-d} + F(z)(AdjC(z))B(z)u_n] - B^0(z)u_n^* \\ = \; & M(z)w_n + N(z)w_{n-d} - B^0(z)u_n^* \\ & +(detC(z))^{-1}A^0(z)[G(z)y_{n-d} + F(z)(AdjC(z))B(z)u_n] \\ = \; & M(z)w_n + g_{n-d}\end{aligned}$$

with g_{n-d} defined by (3.137).

It is obvious that $M(z)w_n$ and g_{n-d} are uncorrelated. By the argument similar to that used for (3.128) we have

$$\limsup_{n\to\infty} \frac{1}{n}\sum_{i=0}^{n} \|\varepsilon_i\|^2 \geq \limsup_{n\to\infty} \frac{1}{n}\sum_{i=0}^{n} \|M(z)w_n\|^2 = tr\sum_{j=0}^{d-1} M_j R M_j^T.$$

Next, we note that by (3.91) w_n can be expressed via $\{y_i,\ u_{i-d},\ w_{i-1},\ i \leq n\}$:

$$w_n = A(z)y_n - B(z)u_{n-d} - (C(z) - I)w_n$$

and hence via $\{y_i,\ u_{i-d},\ i \leq n\}$ by using initial values

$$y_i = w_i = 0, \quad u_i = 0, \qquad i < 0.$$

Then u_n can be defined from (3.137) via $\{y_i,\ u_{i-1},\ y_{i+d}^*,\ i \leq n\}$ if $m \geq l$ and $A^0(0)B_1$ is of full rank. In this case $g_n \equiv 0$ and (3.138) follows from Theorem 2.8. Q.E.D.

Remark 3.5. If $A^0(z) = B^0(z) = I$, then the model reference control problem is reduced to the pure tracking. In this case $M(z) \equiv F(z)$, $N(z) \equiv 0$ and (3.125) is identical to $g_n = 0$.

3.7 Control for CARIMA Models

In many application problems the system noise may have a constant bias and for modelling it the moving average process $C(z)w_n$ gives an unsatisfactory result. For this reason a so-called CARIMA model is considered [CMT], where the system noise is modelled by $e_n = (1 - z)^{-1}C(z)w_n$.

We now discuss a more general case. To be specific, the system is described by

$$A(z)y_n = B(z)u_{n-d} + e_n, \tag{3.139}$$

$$D(z)e_n = C(z)w_n, \quad y_i = w_i = 0, \quad u_i = 0, \quad i < 0, \tag{3.140}$$

where $A(z)$, $B(z)$ and $C(z)$ are given by (3.92)-(3.94), while $D(z)$ is a scale polynomial and possibly unstable

$$D(z) = 1 + \alpha_1 z + \quad \dots \quad + \alpha_s z^s. \tag{3.141}$$

Obviously, if $D(z) = 1$ then (3.139) turns to (3.91) .

Theorem 3.8. *Assume that (3.101) and (3.102) are satisfied. Consider the following loss function*

$$\limsup_{n\to\infty} \frac{1}{n} \sum_{i=0}^{n} \|A^0(z)y_i - B^0(z)y_i^* + Q(z)D(z)u_{i-d}\|^2, \tag{3.142}$$

where $A^0(z)$, $B^0(z)$ and $Q(z)$ are given matrix polynomials, y_i^* is an $\mathcal{F}_{i-d}$-measurable reference signal, $A^0(0)B_1 + Q(0)$ is nondegenerate, and $detC(z)$ is stable, then the $\mathcal{F}_{n-d}$-measurable optimal control can be uniquely defined from

$$\begin{aligned} & [A^0(z)F(z)(AdjC(z))B(z) + (detC(z))Q(z) \\ & -N(z)(AdjC(z))B(z)z^d]D(z)u_n \\ = \quad & (detC(z))B^0(z)y_n^* \\ & -[D(z)N(z)(AdjC(z))A(z) + A^0(z)G(z)]y_{n-d} \end{aligned} \tag{3.143}$$

and the minimal value of the cost is

$$\begin{aligned} & \limsup_{n\to\infty} \frac{1}{n} \sum_{i=0}^{n} \|A^0(z)y_i - B^0(z)y_i^* + Q(z)D(z)u_{i-d}\|^2 \\ & = tr \sum_{j=0}^{d-1} M_j R M_j^T, \end{aligned} \tag{3.144}$$

where $F(z)$, $G(z)$ and $M(z)$ and $N(z)$ are given by (3.120) and (3.135), respectively.

Proof. First of all, if $A^0(0)B_1 + Q(0)$ is nonsingular, then (3.143) is clearly solvable with respect to u_{n-d} and the solution is $\mathcal{F}_{n-d}$-measurable.

By Lemma 3.4 there exist matrix polynomials

$$F(z) = I + F_1 z + \dots + F_{d-1}z^{d-1} \tag{3.145}$$

and $G(z)$ such that

$$(\det C(z))I = F(z)(\mathrm{Adj}C(z))A(z)D(z) + z^d G(z),$$

in which $A(z)D(z)$ plays the same role as $A(z)$ in (3.120).

By using (3.145) and (3.135) we have

$$\begin{aligned} & (detC(z))(y_n - F(z)w_n) \\ = \ & G(z)y_{n-d} - (detC(z))F(z)w_n + F(z)(AdjC(z))A(z)D(z)y_n \\ = \ & G(z)y_{n-d} + F(z)(AdjC(z))B(z)D(z)u_n, \end{aligned} \tag{3.146}$$

therefore,

$$\begin{aligned} & (detC(z))(A^0(z)y_n - B^0(z)u_n^* + Q(z)D(z)u_n) \\ = \ & A^0(z)[(detC(z))F(z)w_n \\ & +G(z)y_{n-d} + F(z)(AdjC(z))B(z)D(z)u_n] \\ & +(detC(z))[Q(z)D(z)u_{n-d} - B^0(z)u_n^*] \\ = \ & (detC(z))M(z)w_n + (detC(z))N(z)w_{n-d} \\ & +A^0(z)[G(z)y_{n-d} + F(z)(AdjC(z))B(z)D(z)u_n] \\ & +(detC(z))[Q(z)D(z)u_n - B^0(z)u_n^*]. \end{aligned} \tag{3.147}$$

By Theorem 2.8 and the argument used in Theorems 3.6 and 3.7, from (3.147) it is easy to see that the lower bound of the cost (3.142) is $tr \sum_{i=0}^{d-1} M_i R M_i^T$.

Further, by (3.143) the right-hand side of (3.147) equals

$$\begin{aligned} & (detC(z))M(z)w_n + (detC(z))N(z)w_{n-d} \\ & -D(z)N(z)(AdjC(z))A(z)y_{n-d} \\ & +z^d N(z)(AdjC(z))B(z)D(z)u_n. \end{aligned} \tag{3.148}$$

However, noting $detC(z)I = (AdjC(z))C(z)$ we see

$$\begin{aligned} & (detC(z))N(z)w_{n-d} - D(z)N(z)(AdjC(z))A(z)y_{n-d} \\ & +z^d N(z)(AdjC(z))B(z)D(z)u_n \\ = \ & N(z)(AdjC(z))[C(z)w_{n-d} \\ & -D(z)A(z)y_{n-d} + z^d D(z)B(z)u_{n-d}] = 0. \end{aligned} \tag{3.149}$$

Then from (3.147)-(3.149) we find that under control defined from (3.143)

$$A^0(z)y_n - B^0(z)y_n^* + Q(z)D(z)u_n = M(z)w_n \tag{3.150}$$

which leads to (3.144) by Theorem 2.8. Q.E.D.

Remark 3.6. If $Q(z) \equiv 0$, then the equation (3.143) defining the optimal control coincides with $g_n = 0$ with g_n defined by (3.137) where w_n should be replaced by $C^{-1}(z)[A(z)y_n - B(z)u_n]$.

Remark 3.7. If consider $\overline{y}_n \triangleq D(z)y_n$, $\overline{u}_n \triangleq D(z)u_n$ as the new output and input of the system, then (3.139) is reduced to (3.91). However, for this new system in the loss function (3.142) $A^0(z)y_i$ should be written as $\frac{A^0(z)}{D(z)}\overline{y}_i$. Therefore Theorem 3.8 cannot be proved by a direct application of Theorem 3.7.

CHAPTER 4

Coefficient Estimation for ARMAX Models

The system considered in this chapter is the same as that described by (3.91)-(3.94), for which it is assumed that (p, q, r) are the known upper bounds for system orders and $d = 1$ is the lower bound for time-delay. The assumption $d = 1$ is not a constraint, because in the case of $d > 1$ we may regard B_i as zero for $1 \leq i < d$.

We shall estimate the unknown matrix coefficient

$$\theta^{\tau} = [-A_1 \quad \dots \quad -A_p \quad B_1 \quad \dots \quad B_q \quad C_1 \quad \dots \quad C_r] \tag{4.1}$$

based on the input-output data $\{u_i,\ y_i\}$. Since the feature of adaptive control is the on-line adjustment of the control law, our attention will mainly be paid to recursive estimation algorithms.

We first describe various algorithms and then analyze their convergence.

4.1 Estimation Algorithms

We start with the simplest case, where the noise is uncorrelated, i.e. temporarily assume $r = 0$. Then the system (3.91) can be written as

$$y_{n+1} = \theta^{\tau} \varphi_n + w_{n+1}, \tag{4.2}$$

where by setting

$$\varphi_n^{\tau} = [y_n^{\tau} \quad \cdots \quad y_{n-p+1}^{\tau} \quad u_n^{\tau} \quad \cdots \quad u_{n-q+1}^{\tau}] \tag{4.3}$$

with $y_i = 0,\ u_i = 0,\ i < 0$ in mind. φ_n is called the stochastic regressor.

Further, introducing the following notations

$$Y_{n+1} = \begin{bmatrix} y_1^\tau \\ \vdots \\ y_{n+1}^\tau \end{bmatrix}, \quad \Phi_n = \begin{bmatrix} \varphi_0^\tau \\ \vdots \\ \varphi_n^\tau \end{bmatrix}, \quad \Sigma_n = \begin{bmatrix} w_1^\tau \\ \vdots \\ w_{n+1}^\tau \end{bmatrix} \tag{4.4}$$

and using equation (4.2) we obtain

$$Y_{n+1} = \Phi_n \theta + \Sigma_n. \tag{4.5}$$

The problem is to estimate θ on the basis of Y_{n+1} and Φ_n. Expression (4.5) is very similar to the linear model considered in the mathematical statistics. The only difference consists in that here Φ_n is a random matrix while in statistics Φ_n is deterministic.

The intuitive and commonly used method for estimating θ is the least squares (LS) method according to which the estimate θ_{n+1} for θ is chosen from

$$\begin{aligned} (Y_{n+1} - \Phi_n \theta_{n+1})^\tau (Y_{n+1} - \Phi_n \theta_{n+1}) \\ = \min_\theta (Y_{n+1} - \Phi_n \theta)^\tau (Y_{n+1} - \Phi_n \theta). \end{aligned} \tag{4.6}$$

By (3.26) and (3.27) we note that

$$\Phi_n^\tau \Phi_n (\Phi_n^\tau \Phi_n)^+ \Phi_n^\tau = \Phi_n^\tau \Phi_n \Phi_n^+ = \Phi_n^\tau.$$

Then it is immediate to check that

$$\begin{aligned} (Y_{n+1} - \Phi_n \theta)^\tau (Y_{n+1} - \Phi_n \theta) \\ = [\theta - (\Phi_n^\tau \Phi_n)^+ \Phi_n^\tau Y_{n+1}]^\tau \Phi_n^\tau \Phi_n [\theta - (\Phi_n^\tau \Phi_n)^+ \Phi_n^\tau Y_{n+1}] \\ + Y_{n+1}^\tau [I - \Phi_n (\Phi_n^\tau \Phi_n)^+ \Phi_n^\tau] Y_{n+1} \end{aligned} \tag{4.7}$$

whose last term is free of θ. Consequently, the minimum for the right-hand side of (4.7) is achieved at

$$\theta_{n+1} = (\Phi_n^\tau \Phi_n)^+ \Phi_n^\tau Y_{n+1} \tag{4.8}$$

which is the LS estimate for θ.

Expressing Φ_n and Y_{n+1} with the help of (4.4) we find that

$$\begin{aligned} \theta_{n+1} &= \left(\sum_{i=0}^n \varphi_i \varphi_i^\tau \right)^+ \sum_{i=0}^n \varphi_i y_{i+1}^\tau \\ &= P_{n+1} \sum_{i=0}^n \varphi_i y_{i+1}^\tau \end{aligned} \tag{4.9}$$

where by definition

$$P_{n+1} = \left(\sum_{i=0}^{n} \varphi_i \varphi_i^T \right)^+ . \tag{4.10}$$

Using the matrix inverse identity (3.62) we can write (4.9) in the recursive form. Identifying C, B and A in (3.62) to P_n, $\varphi_n^T P_n$ and 1 respectively we find that for invertible P_n

$$\begin{aligned} P_{n+1} &= (P_n^{-1} + \varphi_n \varphi_n^T)^{-1} \\ &= (C^{-1} + C^{-1} B^T A^{-1} B C^{-1})^{-1} \\ &= P_n - P_n \varphi_n (1 + \varphi_n^T P_n P_n^{-1} P_n \varphi_n)^{-1} \varphi_n^T P_n \\ &= P_n - a_n P_n \varphi_n \varphi_n^T P_n, \end{aligned} \tag{4.11}$$

where

$$a_n = (1 + \varphi_n^T P_n \varphi_n)^{-1}. \tag{4.12}$$

Further, from (4.9) and (4.11) it follows that

$$\begin{aligned} \theta_{n+1} = {} & (P_n - a_n P_n \varphi_n \varphi_n^T P_n) \left(\sum_{i=0}^{n-1} \varphi_i y_{i+1}^T + \varphi_n y_{n+1}^T \right) \\ = {} & \theta_n - a_n P_n \varphi_n \varphi_n^T \theta_n + P_n \varphi_n y_{n+1}^T \\ & - a_n P_n \varphi_n \varphi_n^T P_n \varphi_n y_{n+1}^T \\ = {} & \theta_n - a_n P_n \varphi_n \varphi_n^T \theta_n + P_n \varphi_n (1 - a_n \varphi_n^T P_n \varphi_n) y_{n+1}^T \\ = {} & \theta_n - a_n P_n \varphi_n \varphi_n^T \theta_n + a_n P_n \varphi_n y_{n+1}^T \\ = {} & \theta_n + a_n P_n \varphi_n (y_{n+1}^T - \varphi_n^T \theta_n). \end{aligned}$$

Thus, we have obtained the recursive algorithm for LS estimate:

$$\theta_{n+1} = \theta_n + a_n P_n \varphi_n (y_{n+1}^T - \varphi_n^T \theta_n), \tag{4.13}$$

$$P_{n+1} = P_n - a_n P_n \varphi_n \varphi_n^T P_n, \quad a_n = (1 + \varphi_n^T P_n \varphi_n)^{-1}. \tag{4.14}$$

The real computation of (4.13) and (4.14) requires the initial values θ_0 and P_0.

We set

$$P_0 = \alpha_0 I, \qquad \frac{1}{e} > \alpha_0 > 0$$

and take θ_0 arbitrary. For such a selection of θ_0 and P_0 we have

$$P_n = \left(\sum_{i=0}^{n-1} \varphi_i \varphi_i^T + \frac{1}{\alpha_0} I \right)^{-1}, \quad P_n^{-1} \geq \frac{1}{\alpha_0} I > eI \tag{4.15}$$

$$\theta_n = P_n \sum_{i=0}^{n-1} \varphi_i y_{i+1}^T + P_n P_0^{-1} \theta_0. \tag{4.16}$$

We now return to the $r \geq 0$ case. Replacing φ_n by

$$\begin{aligned} \varphi_n^0 \triangleq [y_n^T \; \cdots \; y_{n-p+1}^T \; u_n^T \; \cdots \; u_{n-q+1}^T \\ w_n^T \; \cdots \; w_{n-r+1}^T]^T \end{aligned} \tag{4.17}$$

in (4.2)-(4.16) we obtain a LS recursive estimation algorithm for θ defined by (4.1). However, the driven noise w_n is unavailable; hence φ_n^0 cannot be used for the real computation. We have to use an estimate $\hat{w}_n$ for w_n. According to (4.2)

$$\widehat{w}_{n+1} \triangleq y_{n+1} - \theta_{n+1}^T \varphi_n \quad \text{and} \quad \widehat{w}'_{n+1} \triangleq y_{n+1} - \theta_n^T \varphi_n$$

may serve as the *a posteriori* and *a priori* estimate for w_{n+1}, respectively in the case $r = 0$.

This motivates us to apply the following stochastic regressors replacing φ_n^0 in the general $r \geq 0$ case:

$$\begin{aligned} \varphi_n = [y_n^T \; \cdots \; y_{n-p+1}^T \; u_n^T \; \cdots \; u_{n-q+1}^T \\ \widehat{w}_n^T \; \cdots \; \widehat{w}_{n-r+1}^T]^T, \end{aligned} \tag{4.18}$$

$$\begin{aligned} \psi_n = [y_n^T \; \cdots \; y_{n-p+1}^T \; u_n^T \; \cdots \; u_{n-q+1}^T \\ \widehat{w}_n'^T \; \cdots \; \widehat{w}_{n-r+1}'^T]^T, \end{aligned} \tag{4.19}$$

with

$$\widehat{w}_n \triangleq y_n - \theta_n^T \varphi_{n-1} \tag{4.20}$$

$$\widehat{w}'_n \triangleq y_n - \theta_{n-1}^T \psi_{n-1}. \tag{4.21}$$

The algorithm (4.13) and (4.14) with φ_n defined by (4.18) and (4.20) can be used to estimate A_i, B_j and C_k, $i = 1, \ldots, p$, $j = 1, \ldots, q$, $k = 1, \ldots, r$ and is called the extended least squares (ELS) algorithm.

In the subsequent sections it will be shown that the condition number of P_n plays an important role in convergence of θ_n to the true parameter. Therefore, there have been designed many modified least squares algorithms with the attempt to improve the behavior of P_n [SG], [Che2]. The other direction of modification of ELS algorithm is to simplify the algorithm to reduce the computation load. One of the simplest algorithms is the stochastic gradient (SG) algorithm [GRC2] which is described as follows:

$$\theta_{n+1} = \theta_n + \frac{a\psi_n}{r_n}(y_{n+1}^T - \psi_n^T \theta_n), \quad 0 < a \leq 1; \tag{4.22}$$

$$r_n = 1 + \sum_{i=0}^{n} \|\psi_i\|^2 \tag{4.23}$$

where ψ_n is defined by (4.19) (4.21) with θ_n generated from (4.22).

4.2 Convergence of ELS Without the PE Condition

Denote by $\lambda_{max}(n)$ and $\lambda_{min}(n)$ the largest and smallest eigenvalue of $P_{n+1}^{-1} = \sum_{i=0}^{n} \varphi_i \varphi_i^{\tau} + \frac{1}{\alpha_0} I$, respectively, and by $\mu_{max}(n)$ and $\mu_{min}(n)$ the largest and smallest eigenvalue of $\sum_{i=0}^{n} \psi_i \psi_i^{\tau} + \frac{1}{h} I$, respectively, where $h = mp + lq + mr$. The purpose of taking such a rather complicated h is to have a simple relationship: $tr\left(\sum_{i=0}^{n} \psi_i \psi_i^{\tau} + \frac{1}{h} I\right) = r_n$, where r_n is defined by (4.23).

The persistent excitation (PE) condition means that the condition number

$$\limsup_{n\to\infty} \frac{\lambda_{max}(n)}{\lambda_{min}(n)} < \infty \tag{4.24}$$

for ELS or

$$\limsup_{n\to\infty} \frac{\mu_{max}(n)}{\mu_{min}(n)} < \infty \tag{4.25}$$

for SG is bounded. This is the conventional condition guaranteeing strong consistency of θ_n, i.e. convergence of θ_n to θ [Lj1], [Mo1], [Mo2], [So1].

However, PE condition is not necessary for consistency of θ_n for both LS and SG algorithms even when the conditional variance of the driven noise unboundedly grows up [Che1], [Che3]. We intend to show what is the limit for divergence rates of the condition number in order for the estimate θ_n to remain strongly consistent. On this issue basic references are [CG1]-[CG5], [GH], [LW1] and [LW2].

In order to treat the correlated noise we need the following concept.

Definition. A matrix $H(z)$ of rational functions with real coefficients is called strictly positive-real (SPR), if $H(z)$ has no poles in $|z| \le 1$ and

$$H(e^{i\lambda}) + H^{\tau}(e^{-i\lambda}) > 0, \qquad \forall \lambda \in [0,\ 2\pi]. \tag{4.26}$$

Lemma 4.1. *Let $H(z)$ be a matrix of rational functions with real coefficients.*

1. *$H(z)$ is SPR if and only if $H^{-1}(z)$ is SPR;*

2. If $H(z)$ is SPR, then there is a constants $\varepsilon > 0$ such that

$$\sum_{i=0}^{n} u_i^\tau y_i \geq \varepsilon \sum_{i=0}^{n} \left(\|u_i\|^2 + \|y_i\|^2\right), \qquad \forall n \geq 0,$$

where $\{u_i\}$ and $\{y_i\}$ are connected by the transfer function $H(z)$:

$$y_n = H(z)u_n,\ n \geq 0;\ u_n = 0,\ \nu < 0,\ y_n = u_n = 0,\ n < 0.$$

Proof. Assume $H(z)$ is SPR. Since $H^{-1}(e^{i\lambda})$ is the complex conjugate matrix of $H^{-\tau}(e^{-i\lambda})$ we see that

$$\begin{aligned} & H^{-1}(e^{i\lambda}) + H^{-\tau}(e^{-i\lambda}) \\ = \;& H^{-1}(e^{i\lambda})[H(e^{i\lambda}) + H^\tau(e^{-i\lambda})]H^{-\tau}(e^{-i\lambda}) \\ > \;& 0, \qquad \forall \lambda \in [0,\ 2\pi]. \end{aligned} \tag{4.27}$$

We now show that $H(z)$ has no zeros in the closed unit disk. By (4.26) we can find a positive number $a > 0$, such that

$$H(e^{i\lambda}) + H^\tau(e^{-i\lambda}) - a^2 H(e^{i\lambda})H^\tau(e^{-i\lambda}) > 0,\ \forall \lambda \in [0,\ 2\pi] \tag{4.28}$$

or equivalently

$$\left[H(e^{i\lambda}) - \frac{1}{a}I\right]\left[H(e^{i\lambda}) - \frac{1}{a}I\right]^* < \frac{1}{a^2}I, \quad \forall \lambda \in [0,\ 2\pi]. \tag{4.29}$$

Hence, for any complex vector x: $\|x\| = 1$,

$$\left|x^* H(e^{i\lambda})x - \frac{1}{a}\right| \leq \left\|H(e^{i\lambda}) - \frac{1}{a}I\right\| < \frac{1}{a} \tag{4.30}$$

and consequently by the maximum principle for analytic functions,

$$\left|x^* H(z)x - \frac{1}{a}\right| < \frac{1}{a}, \qquad |z| \leq 1,$$

which means that $H(z)$ has no zeros in $|z| \leq 1$,or alternatively, $H^{-1}(z)$ has no poles in $|z| \leq 1$. This and (4.27) mean that $H^{-1}(z)$ is SPR.

The converse is immediate, i.e. the strictly positive realness of $H^{-1}(z)$ implies the same property of $H(z)$ since $(H^{-1}(z))^{-1} = H(z)$.

We now prove the second part of the lemma.

From the definition we know that $D(z) \triangleq H(z) - 2\varepsilon I$ is also SPR if $H(z)$ is SPR and ε is small enough. Since $D(z)$ is analytic in $|z| \leq 1$, we can expand $D(z)$ to the series

$$D(z) = \sum_{k=0}^{\infty} D_k z^k, \qquad \sum_{k=0}^{\infty} \|D_k\| < \infty.$$

Note that

$$\begin{aligned}
0 &\le \frac{1}{2\pi}\int_{-\pi}^{\pi}\left(\sum_{n=0}^{N} u_n e^{-in\lambda}\right)^{\tau}\left[\frac{D(e^{i\lambda})+D^{\tau}(e^{i\lambda})}{2}\right]\left(\sum_{m=0}^{N} u_m e^{-im\lambda}\right)d\lambda\\
&= \frac{1}{2\pi}\int_{-\pi}^{\pi}\left(\sum_{n=0}^{N} u_n e^{-in\lambda}\right)^{\tau} D(e^{i\lambda})\left(\sum_{m=0}^{N} u_m e^{-im\lambda}\right)d\lambda\\
&= \frac{1}{2\pi}\sum_{k=0}^{\infty}\sum_{n=0}^{N}\sum_{m=0}^{N} u_n^{\tau} D_k u_m \int_{-\pi}^{\pi} e^{i(m-n)\lambda} e^{ik\lambda}\, d\lambda\\
&= \sum_{k=0}^{N}\sum_{n=0}^{N}\sum_{\substack{m=0\\ m-n=-k}}^{N} u_n^{\tau} D_k u_m = \sum_{k=0}^{N}\sum_{n=k}^{N} u_n^{\tau} D_k u_{n-k}\\
&= \sum_{n=0}^{N} u_n^{\tau}\sum_{k=0}^{n} D_k u_{n-k} = \sum_{k=0}^{N} u_n^{\tau}(D(z)u_n)\\
&= \sum_{n=0}^{N} u_n^{\tau}(y_n - 2\varepsilon u_n). \qquad (4.31)
\end{aligned}$$

Since $H^{-1}(z)$ is also SPR by the first conclusion of the lemma, we also have

$$\sum_{k=0}^{n} y_i^{\tau}(u_i - 2\varepsilon y_i) \ge 0, \qquad (4.32)$$

where, without loss of generality, the same ε has been used.

Combining (4.31) and (4.32) yields the desired assertion. Q.E.D.

Corollary 4.1. *If $H(z)$ is a polynomial matrix, with real coefficients, then $H(z)$ is SPR if and only if (4.26) or its equivalence*

$$H^{-1}(e^{\lambda i}) + H^{-\tau}(e^{-\lambda i}) > 0, \quad \forall\lambda \in [0, 2\pi]$$

is satisfied and in this case $\det H(z) \ne 0$, $\forall |z| \le 1$.

This is because a polynomial automatically has no poles and (4.26) is sufficient for $H(z)$ to be SPR. The rest follows from the first part of the lemma.

For the SPR property we also refer to [An1], [La], [Ca].

The following theorem is a generalization of [LW1].

Theorem 4.1 *[CG3], [LW2]. Assume the following conditions are fulfilled.*

1. $\{w_n, \mathcal{F}_n\}$ is a martingale difference sequence with

$$\sup_{n\ge 0} E\left[\|w_{n+1}\|^{\beta}|\mathcal{F}_n\right] \triangleq \sigma < \infty \quad a.s., \quad \beta \ge 2. \qquad (4.33)$$

2. $C^{-1}(e^{i\lambda}) + C^{-\tau}(e^{-i\lambda}) - I > 0, \ \forall \lambda \in [0, 2\pi]$.

3. u_n *is* $\mathcal{F}_n$*-measurable.*

Then as $n \to \infty$ *the convergence (or divergence) rate of the estimate produced by the ELS algorithm is expressed by*

$$\|\theta_{n+1} - \theta\|^2 = O\left(\frac{\log \lambda_{max}(n)(\log\log \lambda_{max}(n))^{\delta(\beta-2)}}{\lambda_{min}(n)}\right) \quad a.s., \qquad (4.34)$$

where

$$\delta(x) \triangleq \begin{cases} 0, & x \neq 0 \\ c, & x = 0 \end{cases} \qquad (4.35)$$

with arbitrary $c > 1$.

Proof. Denote by $\tilde{\theta}_n$ the estimation error

$$\tilde{\theta}_n = \theta - \theta_n. \qquad (4.36)$$

Noticing $P_{n+1}^{-1} \geq \lambda_{min}(n+1)I$, we see that

$$\|\tilde{\theta}_{n+1}\|^2 \leq \frac{1}{\lambda_{min}(n)} tr\tilde{\theta}_{n+1}^{\tau} P_{n+1}^{-1} \tilde{\theta}_{n+1}. \qquad (4.37)$$

Therefore, to prove the theorem it suffices to show

$$tr\tilde{\theta}_{n+1}^{\tau} P_{n+1}^{-1} \tilde{\theta}_{n+1} = O\left(\log \lambda_{max}(n)(\log\log \lambda_{max}(n))^{\delta(\beta-2)}\right). \qquad (4.38)$$

Let ξ_{n+1} denote the *a posteriori* estimation error for w_{n+1}:

$$\xi_{n+1} = y_{n+1} - \theta_{n+1}^{\tau}\varphi_n - w_{n+1}. \qquad (4.39)$$

Then $\tilde{\theta}_{n+1}^{\tau}\varphi_n$ is driven by ξ_{n+1} via the transfer function $C(z)$ because

$$\begin{aligned} C(z)\xi_{n+1} &= y_{n+1} + (C(z) - I)(y_{n+1} - \theta_{n+1}^{\tau}\varphi_n) \\ &\quad - \theta_{n+1}^{\tau}\varphi_n - C(z)w_{n+1} \\ &= -(A(z) - I)y_{n+1} + B(z)u_n + (C(z) - I)(y_{n+1} - \theta_{n+1}^{\tau}\varphi_n) - \theta_{n+1}^{\tau}\varphi_n \\ &= \theta^{\tau}\varphi_n - \theta_{n+1}^{\tau}\varphi_n = \tilde{\theta}_{n+1}^{\tau}\varphi_n. \end{aligned} \qquad (4.40)$$

This is equivalent to

$$C^{-1}(z)\tilde{\theta}_{n+1}^{\tau}\varphi_n = \xi_{n+1}.$$

By Condition 2 and Corollary 4.1 we know that $C(z)$ has no poles in $|z| \leq 1$. Hence by definition $C^{-1}(z) - \frac{1}{2}I$ is SPR, and by Lemma 4.1 there are constants $k_0 > 0$ and $k_1 \geq 0$ such that

$$s_n \triangleq \sum_{i=0}^{n} \varphi_i^{\tau}\tilde{\theta}_{i+1}(\xi_{i+1} - \frac{1}{2}(1 + k_0)\tilde{\theta}_{i+1}^{\tau}\varphi_i) \geq 0, \quad \forall n \geq 0. \qquad (4.41)$$

From (4.13) it is easy to see that

$$\begin{aligned}
& y_{n+1}^{\tau} - \varphi_n^{\tau}\theta_{n+1} \\
= \; & y_{n+1}^{\tau} - \varphi_n^{\tau}[\theta_n + a_n P_n \varphi_n (y_{n+1}^{\tau} - \varphi_n^{\tau}\theta_n)] \\
= \; & (1 - a_n \varphi_n^{\tau} P_n \varphi_n)(y_{n+1}^{\tau} - \varphi_n^{\tau}\theta_n) \\
= \; & a_n (y_{n+1}^{\tau} - \varphi_n^{\tau}\theta_n).
\end{aligned} \tag{4.42}$$

Hence by (4.39) and (4.42) we can rewrite (4.13) as

$$\tilde{\theta}_{n+1} = \tilde{\theta}_n - P_n \varphi_n (\xi_{n+1}^{\tau} + w_{n+1}^{\tau}). \tag{4.43}$$

In order to prove (4.38) we expand $tr\tilde{\theta}_{n+1}^{\tau} P_{n+1}^{-1} \tilde{\theta}_{n+1}$ using (4.15) and (4.43)

$$\begin{aligned}
& tr\tilde{\theta}_{k+1}^{\tau} P_{k+1}^{-1} \tilde{\theta}_{k+1} = tr\tilde{\theta}_{k+1}^{\tau} \varphi_k \varphi_k^{\tau} \tilde{\theta}_{k+1} + tr\tilde{\theta}_{k+1}^{\tau} P_k^{-1} \tilde{\theta}_{k+1} \\
= \; & \|\tilde{\theta}_{k+1}^{\tau} \varphi_k\|^2 + tr \left[\tilde{\theta}_k - P_k \varphi_k (\xi_{k+1}^{\tau} + w_{k+1}^{\tau})\right]^{\tau} P_k^{-1} \\
& \left[\tilde{\theta}_k - P_k \varphi_k (\xi_{k+1}^{\tau} + w_{k+1}^{\tau})\right] \\
= \; & \|\tilde{\theta}_{k+1}^{\tau} \varphi_k\|^2 - 2(\xi_{k+1}^{\tau} + w_{k+1}^{\tau}) \tilde{\theta}_k^{\tau} \varphi_k + \varphi_k^{\tau} P_k \varphi_k \|\xi_{k+1} + w_{k+1}\|^2 \\
& + tr\tilde{\theta}_k^{\tau} P_k^{-1} \tilde{\theta}_k \\
= \; & tr\tilde{\theta}_k^{\tau} P_k^{-1} \tilde{\theta}_k - 2(\xi_{k+1}^{\tau} + w_{k+1}^{\tau})[\tilde{\theta}_{k+1} + P_k \varphi_k (\xi_{k+1}^{\tau} + w_{k+1}^{\tau})]^{\tau} \varphi_k \\
& + \varphi_k^{\tau} P_k \varphi_k \|\xi_{k+1} + w_{k+1}\|^2 \\
\le \; & tr\tilde{\theta}_k^{\tau} P_k^{-1} \tilde{\theta}_k + \|\tilde{\theta}_{k+1}^{\tau} \varphi_k\|^2 - 2\xi_{k+1}^{\tau} \tilde{\theta}_{k+1}^{\tau} \varphi_k - 2w_{k+1}^{\tau} \tilde{\theta}_{k+1}^{\tau} \varphi_k \\
= \; & tr\tilde{\theta}_k^{\tau} P_k^{-1} \tilde{\theta}_k - 2[\varphi_k^{\tau} \tilde{\theta}_{k+1} (\xi_{k+1} - \frac{1}{2}(1 + k_0) \tilde{\theta}_{k+1}^{\tau} \varphi_k)] \\
& - k_0 \|\tilde{\theta}_{k+1}^{\tau} \varphi_k\|^2 - 2w_{k+1}^{\tau} \tilde{\theta}_{k+1}^{\tau} \varphi_k.
\end{aligned}$$

Summing up both sides of this expression from 0 to $n+1$ and using (4.41) we derive

$$\begin{aligned}
& tr\tilde{\theta}_{n+1}^{\tau} P_{n+1}^{-1} \tilde{\theta}_{n+1} \\
\le \; & tr\tilde{\theta}_0^{\tau} P_0^{-1} \tilde{\theta}_0 - 2s_n - k_0 \sum_{i=0}^{n} \|\tilde{\theta}_{i+1}^{\tau} \varphi_i\|^2 - 2 \sum_{i=0}^{n} w_{i+1}^{\tau} \tilde{\theta}_{i+1}^{\tau} \varphi_i \\
\le \; & O(1) - k_0 \sum_{i=0}^{n} \|\tilde{\theta}_{i+1}^{\tau} \varphi_i\|^2 - 2 \sum_{i=0}^{n} w_{i+1}^{\tau} \tilde{\theta}_{i+1}^{\tau} \varphi_i
\end{aligned} \tag{4.44}$$

for which the key task is to estimate the last term while the second term is nonpositive and may compensate the similar quantity arising from estimating the last term.

Denote by η_n the *a priori* estimation error for w_{n+1}:

$$\eta \triangleq y_{n+1} - \theta_n^{\tau} \varphi_n - w_{n+1}. \tag{4.45}$$

Obviously, η_n is $\mathcal{F}_n$-measurable and (4.13) can be expressed as

$$\tilde{\theta}_{n+1} = \tilde{\theta}_n - a_n P_n \varphi_n (w_{n+1}^T + \eta_n^T). \tag{4.46}$$

Then the last term of (4.44) can be estimated as follows:

$$\begin{aligned}
&\left|\sum_{i=0}^{n} w_{i+1}^T \tilde{\theta}_{i+1}^T \varphi_i\right| \\
= &\left|\sum_{i=0}^{n} w_{i+1}^T (\tilde{\theta}_i^T - a_i (w_{i+1} + \eta_i) \varphi_i^T P_i) \varphi_i\right| \\
\leq &\sum_{i=0}^{n} a_i \varphi_i^T P_i \varphi_i \|w_{i+1}\|^2 + \left|\sum_{i=0}^{n} w_{i+1}^T (\tilde{\theta}_i^T - a_i \eta_i \varphi_i^T P_i) \varphi_i\right|
\end{aligned}$$

for which Theorem 2.8 is applicable to its second term.

Hence, we have

$$\begin{aligned}
&\left|\sum_{i=0}^{n} w_{i+1}^T \tilde{\theta}_{i+1}^T \varphi_i\right| \\
\leq &\sum_{i=0}^{n} a_i \varphi_i^T P_i \varphi_i \|w_{i+1}\|^2 + O\left(\left[\sum_{i=0}^{n} \|(\tilde{\theta}_i^T - a_i \eta_i \varphi_i^T P_i) \varphi_i\|^2\right]^\alpha\right) \\
= &\sum_{i=0}^{n} a_i \varphi_i^T P_i \varphi_i \|w_{i+1}\|^2 + O\left(\left[\sum_{i=0}^{n} \|(\tilde{\theta}_{i+1}^T + a_i w_{i+1} \varphi_i^T P_i) \varphi_i\|^2\right]^\alpha\right) \\
= &\sum_{i=0}^{n} a_i \varphi_i^T P_i \varphi_i \|w_{i+1}\|^2 + O\left(\left[\sum_{i=0}^{n} \|\tilde{\theta}_{i+1}^T \varphi_i\|^2\right]^\alpha\right) \\
&+ O\left(\left[\sum_{i=0}^{n} (a_i \varphi_i^T P_i \varphi_i)^2 \|w_{i+1}\|^2\right]^\alpha\right)
\end{aligned} \tag{4.47}$$

whenever $\alpha \in (\frac{1}{2}, 1)$.

Combining (4.44) and (4.47) and noticing $\alpha < 1$ we obtain that

$$tr \tilde{\theta}_{n+1}^T P_{n+1}^{-1} \tilde{\theta}_{n+1} \leq O(1) + \left(\sum_{i=0}^{n} (a_i \varphi_i^T P_i \varphi_i)^2 \|w_{i+1}\|^2\right) \quad a.s. \tag{4.48}$$

Thus, the problem is reduced to estimating the last term of (4.48).

Note that the matrix $P_i \varphi_i \varphi_i^T$ has only one nonzero eigenvalue $\varphi_i^T P_i \varphi_i$, so,

$$det(I + P_i \varphi_i \varphi_i^T) = 1 + \varphi_i^T P_i \varphi_i. \tag{4.49}$$

Hence, we have

$$detP_{i+1}^{-1} = det(P_i^{-1} + \varphi_i\varphi_i^T) = detP_i^{-1}det(I + P_i\varphi_i\varphi_i^T)$$

or alternatively,

$$\varphi_i^T P_i \varphi_i = \frac{detP_{i+1}^{-1} - detP_i^{-1}}{detP_i^{-1}}. \tag{4.50}$$

Consequently, using (4.50) and the definition of a_n given by (4.14) we obtain

$$\begin{aligned}\sum_{i=0}^{n} a_i\varphi_i^T P_i\varphi_i &= \sum_{i=0}^{n} \frac{detP_{i+1}^{-1} - detP_i^{-1}}{detP_{i+1}^{-1}} \\ &= \sum_{i=0}^{n} \int_{detP_i^{-1}}^{detP_{i+1}^{-1}} \frac{dx}{detP_{i+1}^{-1}} \le \int_{detP_0^{-1}}^{detP_{n+1}^{-1}} \frac{dx}{x} \\ &= \log\left(detP_{n+1}^{-1}\right) + \alpha_0 \log \alpha_0 \end{aligned} \tag{4.51}$$

since $detP_0^{-1} = \frac{1}{\alpha_0}I$.

We now connect $detP_{n+1}^{-1}$ with $\lambda_{max}(n)$. Noticing $detP_n^{-1} \ge \frac{1}{\alpha_0}I$, $\forall n \ge 0$, we see that

$$\left(\frac{1}{\alpha_0}\right)^{mp+lq+mr-1} \lambda_{max}(n) \le detP_{n+1}^{-1} \le (\lambda_{max}(n))^{mp+lq+mr}$$

and hence

$$\begin{aligned}\log\lambda_{max}(n) - (mp+lq+mr)\log\alpha_0 &\le \log detP_{n+1}^{-1} \\ \le\ (mp+lq+mr)\log\lambda_{max}(n)&\end{aligned} \tag{4.52}$$

where $mp + lq + mr$ is the dimension of φ_n.

By (4.51) and (4.52), we have

$$\sum_{i=0}^{n} a_i\varphi_i^T P_i\varphi_i = O\left(\log\lambda_{\max}(n)\right) \quad \text{a.s.} \tag{4.53}$$

Now, taking $\alpha \in \left[1, \min\left(\frac{\beta}{2}, 2\right)\right]$ and applying Theorem 2.8 with $M_i = a_i\varphi_i^T P_i\varphi_i$, $x_{i+1} = \|w_{i+1}\|^2 - E(\|w_{i+1}\|^2|\mathcal{F}_i)$, we obtain

$$\begin{aligned}&\sum_{i=0}^{n} a_i\varphi_i^T P_i\varphi_i\|w_{i+1}\|^2 \\ =\ &\sum_{i=0}^{n} M_i x_{i+1} + \sum_{i=0}^{n} a_i\varphi_i^T P_i\varphi_i E(\|w_{i+1}\|^2|\mathcal{F}_i)\end{aligned}$$

$$= O\left(\left\{\sum_{i=0}^{n}(M_i)^\alpha\right\}^{\frac{1}{\alpha}} \log^{\frac{1}{\alpha}+\eta}\left(\sum_{i=0}^{n}(M_i)^\alpha + e\right)\right)$$
$$+O\left(\log \lambda_{\max}(n)\right)$$
$$= O\left(\{\log \lambda_{\max}(n)\}^{\frac{1}{\alpha}} \log^{\frac{1}{\alpha}+\eta}\left(\log \lambda_{\max}(n) + e\right)\right)$$
$$+O\left(\log \lambda_{\max}(n)\right) \tag{4.54}$$

for all $\eta > 0$.

If $\beta = 2$, then $\alpha = 1$; while if $\beta > 2$, α can be taken as $\alpha > 1$. Hence by (4.54) we have

$$\sum_{i=0}^{n} a_i \varphi_i^T P_i \varphi_i \|w_{i+1}\|^2$$
$$= O\left(\log \lambda_{\max}(n) \left[\log\log\left(\lambda_{\max}(n)\right)\right]^{\delta(\beta-2)}\right) \tag{4.55}$$

Substituting (4.55) into (4.48) and noting $a_i \varphi_i^T P_i \varphi_i \le 1$ we see that (4.38) is true. Hence the proof is complete. Q.E.D.

Remark 4.1. Let $\lambda_{max}(n)$ be a magnitude of order $O(n^b)$, $b > 0$. Then in order θ_n to be consistent it suffices to require $(\log n)^{1+\varepsilon} = 0(\lambda_{min}(n))$, $\varepsilon > 0$. In this case

$$\|\theta_{n+1} - \theta\|^2 = O\left(\frac{\log n (\log\log n)^{\delta(\beta-2)}}{(\log n)^{1+\varepsilon}}\right) \xrightarrow[n\to\infty]{} 0 \tag{4.56}$$

but the PE condition does not hold because

$$\frac{\lambda_{max}(n)}{\lambda_{min}(n)} = O\left(\frac{n^b}{(\log n)^{1+\varepsilon}}\right) \xrightarrow[n\to\infty]{} \infty.$$

We now express the convergence rate of θ_n in terms of $\lambda^0_{max}(n)$ and $\lambda^0_{min}(n)$ which respectively denote the maximum and minimum eigenvalue of $\sum_{i=0}^{n} \varphi_i^0 \varphi_i^{0T} + \frac{1}{\alpha_0} I$.

Theorem 4.2. *Let conditions of Theorem 4.1 be satisfied. If*

$$\log(\lambda^0_{max}(n))(\log\log(\lambda^0_{max}(n)))^{\delta(\beta-2)} = o\left(\lambda^0_{min}(n)\right) \text{ as } n \to \infty, \tag{4.57}$$

then

$$\|\theta_{n+1} - \theta\|^2 = O\left(\frac{\log(\lambda^0_{max}(n))(\log\log(\lambda^0_{max}(n)))^{\delta(\beta-2)}}{\lambda^0_{min}(n)}\right) \text{ as } n \to \infty, \tag{4.58}$$

where $\delta(x)$ is defined by (4.35).

Proof. Since $C^{-1}(z)$ is strictly positive-real by Corollary 4.1 it must be stable; then from (4.40) it follows that

$$\sum_{i=0}^{n} \|\xi_{i+1}\|^2 = O\left(\sum_{i=0}^{n} \|\tilde{\theta}_{i+1}^{\tau}\varphi_i\|^2\right). \tag{4.59}$$

However, from (4.44) and (4.47) we see

$$\begin{aligned} & tr\tilde{\theta}_{n+1}^{\tau} P_{n+1}^{-1}\tilde{\theta}_{n+1} \\ \leq\ & O(1) - k_0 \sum_{i=0}^{n} \|\tilde{\theta}_{i+1}^{\tau}\varphi_i\|^2 + O\left(\left[\sum_{i=0}^{n} \|\tilde{\theta}_{i+1}^{\tau}\varphi_i\|^2\right]^{\alpha}\right) \\ & + \sum_{i=0}^{n} a_i\varphi_i^{\tau} P_i \varphi_i \|w_{i+1}\|^2 + O\left(\left[\sum_{i=0}^{n} (a_i\varphi_i^{\tau} P_i\varphi_i)^2 \|w_{i+1}\|^2\right]^{\alpha}\right) \end{aligned}$$

which implies that

$$\sum_{i=0}^{n} \|\tilde{\theta}_{i+1}^{\tau}\varphi_i\|^2 = O\left(\sum_{i=0}^{n} a_i\varphi_i^{\tau} P_i\varphi_i \|w_{i+1}\|^2\right) \tag{4.60}$$

since $\alpha \in (\frac{1}{2}, 1)$ and $tr\tilde{\theta}_{n+1}^{\tau} P_{n+1}^{-1}\tilde{\theta}_{n+1} \geq 0$.

Combining (4.59) and (4.60) yields

$$\sum_{i=0}^{n} \|\xi_{i+1}\|^2 = O\left(\sum_{i=0}^{n} a_i\varphi_i^{\tau} P_i\varphi_i \|w_{i+1}\|^2\right) \tag{4.61}$$

and

$$\sum_{i=0}^{n} \|\xi_{i+1}\|^2 = O\left(\log(\lambda_{max}(n))(\log\log(\lambda_{max}(n)))^{\delta(\beta-2)}\right). \tag{4.62}$$

Denote

$$\varphi_n^{\xi} = \varphi_n - \varphi_n^0 = [0 \ \ldots \ 0 \ \xi_n^{\tau} \ \ldots \ \xi_{n-r+1}^{\tau}]^{\tau}. \tag{4.63}$$

Then (4.62) means that

$$\sum_{i=0}^{n} \|\varphi_{i+1}^{\xi}\|^2 = O\left(\log(\lambda_{max}(n))(\log\log(\lambda_{max}(n)))^{\delta(\beta-2)}\right). \tag{4.64}$$

Let x be a unit vector and have the same dimension as φ_n. Then we have

$$\begin{aligned} \sum_{i=0}^{n} (x^{\tau}\varphi_i)^2 &= \sum_{i=0}^{n} (x^{\tau}\varphi_i^{\xi} + x^{\tau}\varphi_i^0)^2 \\ &\leq 2\sum_{i=0}^{n} (x^{\tau}\varphi_i^0)^2 + 2\sum_{i=0}^{n} \|\varphi_i^{\xi}\|^2 \end{aligned} \tag{4.65}$$

and

$$\lambda_{max}(n) \le 2\lambda^0_{max}(n) + O\left(\log(\lambda_{max}(n))(\log\log(\lambda_{max}(n)))^{\delta(\beta-2)}\right),$$

$$\lambda_{max}(n) = O\left(\lambda^0_{max}(n)\right). \tag{4.66}$$

Similar to (4.65) we have

$$\sum_{i=0}^{n}(x^\tau \varphi_i^0)^2 \le 2\sum_{i=0}^{n}(x^\tau \varphi_i)^2 + 2\sum_{i=0}^{n}\|\varphi_i^\xi\|^2$$

and by (4.64), (4.66) and (4.57)

$$\begin{aligned}\lambda^0_{min}(n) &\le 2\lambda_{min}(n) + O\left(\log(\lambda^0_{max}(n))(\log\log(\lambda^0_{max}(n)))^{\delta(\beta-2)}\right)\\ &= 2\lambda_{min}(n) + o(\lambda^0_{min}(n))\end{aligned}$$

which yields

$$\lambda^0_{min}(n) = O\left(\lambda_{min}(n)\right). \tag{4.67}$$

Using (4.66) and (4.67), from (4.34) we derive the desired estimate (4.58). Q.E.D.

Remark 4.2.

(i) The SPR condition plays a crucial role in the analysis. It seems that this condition was first exposed as a convergence condition in [LM] and [Lj2]. Also, the quantity $tr\widetilde{\theta}_n P_n^{-1}\widetilde{\theta}_n$ may be regarded as a stochastic Lyapunov function; earlier references on its recursion are [Mo1] and [So1].

(ii) Theorem 4.1 states that in the case of $\beta > 2$, if $\log \lambda_{\max}(n)/\lambda_{\min}(n)$ tends to zero, then θ_n is strongly consistent. It is worth noting that this condition is the weakest in some sense (see, [LW1]).

4.3 Local Convergence of SG

We now consider the estimate θ_n for θ given by the algorithm (4.22) and (4.23) which is simplified from ELS. Intuitively, it provides a convergence rate not as fast as does ELS. However, the SG algorithm is a useful tool in theory of stochastic adaptive control. For example, the optimality of the SG-based tracker is relatively easy to establish while analyzing the ELS-based tracker is much more difficult.

Let us denote by ζ_{n+1} the *a priori* estimation error given by SG for w_{n+1}

$$\zeta_{n+1} = y_{n+1} - w_{n+1} - \theta_n^\tau \psi_n \tag{4.68}$$

which differs from η_n defined by (4.45) by that the latter is provided by ELS.

Similar to φ_n^ξ given by (4.63) define

$$\varphi_n^\zeta = [0 \ \ldots \ 0 \ \zeta_n^\tau \ \ldots \ \zeta_{n-r+1}^\tau]^\tau.$$

Then

$$\psi_n = \varphi_n^0 + \varphi_n^\zeta. \tag{4.69}$$

Noticing

$$y_{n+1} = \theta^\tau \varphi_n^0 + w_{n+1}$$

we have for θ_n given by (4.22) and (4.23)

$$\begin{aligned}
\theta_{n+1} &= \theta_n + \frac{a\psi_n}{r_n}(\varphi_n^{0\tau}\theta + w_{n+1}^\tau - \psi_n^\tau\theta_n) \\
&= \theta_n + \frac{a\psi_n}{r_n}(\psi_n^\tau\theta - \varphi_n^{\zeta\tau}\theta + w_{n+1}^\tau - \psi_n^\tau\theta_n) \\
&= \theta_n + \frac{a\psi_n}{r_n}(\psi_n^\tau\tilde{\theta}_n - \varphi_n^{\zeta\tau}\theta + w_{n+1}^\tau)
\end{aligned} \tag{4.70}$$

and

$$\tilde{\theta}_{n+1} = \left(I - \frac{a\psi_n\psi_n^\tau}{r_n}\right)\tilde{\theta}_n + \frac{a\psi_n\varphi_n^{\zeta\tau}}{r_n}\theta - \frac{a\psi_n}{r_n}w_{n+1}^\tau, \quad 0 < a \leq 1 \tag{4.71}$$

where $\tilde{\theta}_n$ denotes the estimation error

$$\tilde{\theta}_n = \theta - \theta_n. \tag{4.72}$$

Define $\Phi(n+1,i)$, $n+1 \geq i$, $i = 0, 1, \ldots$ by recursion

$$\Phi(n+1,i) = \left(I - \frac{a\psi_n\psi_n^\tau}{r_n}\right)\Phi(n,i), \ \ \Phi(i,i) = I. \tag{4.73}$$

Then (4.71) yields

$$\begin{aligned}
\tilde{\theta}_{n+1} &= \Phi(n+1,0)\tilde{\theta}_0 + a\sum_{j=0}^{n}\Phi(n+1,j+1)\frac{\psi_j\varphi_j^{\zeta\tau}}{r_j}\theta \\
&\quad -a\sum_{j=0}^{n}\Phi(n+1,j)\frac{\psi_j}{r_j}w_{j+1}^\tau.
\end{aligned} \tag{4.74}$$

We now derive the set where $\tilde{\theta}_n \xrightarrow[n\to\infty]{} 0$. The convergence set is not necessary to be a whole space, so the convergence of θ_n discussed in this section has a local property.

We first prove lemmas.

Lemma 4.2. *Let h denote the dimension of ψ_n, $h \triangleq mp + lq + mr$. Then*

$$\sum_{j=0}^{\infty} \frac{\|\Phi(j,0)\psi_j\|^2}{r_j} \le \frac{h}{a} \tag{4.75}$$

$$\sum_{i=0}^{n-1} \frac{\|\Phi^\tau(n,i+1)\psi_i\|^2}{r_i} \le \frac{h}{a}. \tag{4.76}$$

Proof. By paying attention to (4.73) the estimates (4.75) and (4.76) follow from the chains of equalities and inequalities:

$$\begin{aligned}
h \ge\ & tr\Phi^\tau(n,0)\Phi(n,0) \\
\ge\ & \sum_{j=n}^{\infty} tr\left[\Phi^\tau(j,0)\Phi(j,0) - \Phi^\tau(j+1,0)\Phi(j+1,0)\right] \\
=\ & \sum_{j=n}^{\infty} tr\Phi^\tau(j,0)\left[I - \left(I - \frac{a\psi_j\psi_j^\tau}{r_j}\right)\left(I - \frac{a\psi_j\psi_j^\tau}{r_j}\right)\right]\Phi(j,0) \\
\ge\ & \sum_{j=n}^{\infty} tr\Phi^\tau(j,0)\left[\frac{a\psi_j\psi_j^\tau}{r_j} + \frac{a\psi_j}{r_j}\left(I - \frac{a\|\psi_j\|^2}{r_j}\right)\psi_j^\tau\right]\Phi(j,0) \\
\ge\ & \sum_{j=n}^{\infty} a\frac{\|\Phi(j,0)\psi_j\|^2}{r_j}
\end{aligned}$$

and

$$\begin{aligned}
h \ge\ & tr\Phi(n,n)\Phi^\tau(n,n) \\
\ge\ & \sum_{i=0}^{n-1} tr\left[\Phi(n,i+1)\Phi^\tau(n,i+1) - \Phi(n,i)\Phi^\tau(n,i)\right] \\
=\ & tr\sum_{i=0}^{n-1} \Phi(n,i+1)\left[I - \Phi(i+1,i)\Phi^\tau(i+1,i)\right]\Phi^\tau(n,i+1) \\
=\ & tr\sum_{i=0}^{n-1} \Phi(n,i+1)\left[I - \left(I - \frac{a\psi_i\psi_i^\tau}{r_i}\right)\left(I - \frac{a\psi_i\psi_i^\tau}{r_i}\right)\right]\Phi^\tau(n,i+1) \\
=\ & tr\sum_{i=0}^{n-1} \Phi(n,i+1)\left[\frac{a\psi_i\psi_i^\tau}{r_i} + \frac{a\psi_i}{r_i}\left(I - \frac{a\|\psi_i\|^2}{r_i}\right)\psi_i^\tau\right]\Phi^\tau(n,i+1) \\
\ge\ & \sum_{i=0}^{n-1} a\frac{\|\Phi^\tau(n,i+1)\psi_i\|^2}{r_i}.
\end{aligned}$$

Q.E.D.

We adopt the following conditions:

1. $\{w_n, \mathcal{F}_n\}$ is a martingale difference sequence with

$$\sup_{n\geq 0} E\left[\|w_{n+1}\|^2|\mathcal{F}_n\right] \triangleq \sigma < \infty \quad a.s. \tag{4.77}$$

2. The polynomial $C(z) - \frac{a}{2}I$ is SPR for some $a \in (0, 1]$. (4.78)

3. u_n is $\mathcal{F}_n$-measurable. (4.79)

We note that in (4.77) σ may be random. The following lemma plays a crucial role in the convergence analysis of SG.

Lemma 4.3. *Let (4.77)-(4.79) be held. Then for ξ_{n+1} given by (4.68)*

$$\sum_{n=0}^{\infty} \frac{\|\zeta_{n+1}\|^2}{r_n} < \infty \quad a.s. \tag{4.80}$$

and θ_n is bounded a.s.

Proof. From (4.22) and (4.68) it follows that

$$\theta_{n+1} = \theta_n + \frac{a\psi_n}{r_n}(\zeta_{n+1}^{\tau} + w_{n+1}^{\tau}) \tag{4.81}$$

and by (4.72)

$$\tilde{\theta}_{n+1} = \tilde{\theta}_n - \frac{a\psi_n}{r_n}(\zeta_{n+1}^{\tau} + w_{n+1}^{\tau}). \tag{4.82}$$

Notice

$$\begin{aligned} & C(z)\zeta_n = C(z)(y_n - w_n - \theta_{n-1}^{\tau}\psi_{n-1}) \\ = \; & [(y_n - C(z)w_n) + (C(z) - I)(y_n - \theta_{n-1}^{\tau}\psi_{n-1})] - \theta_{n-1}^{\tau}\psi_{n-1} \\ = \; & \theta^{\tau}\psi_{n-1} - \theta_{n-1}^{\tau}\psi_{n-1} = \tilde{\theta}_{n-1}^{\tau}\psi_{n-1}. \end{aligned} \tag{4.83}$$

Then by Condition (4.78) and Lemma 4.1, there is a constant $k_1 > 0$ such that

$$t_n \triangleq 2a\sum_{i=1}^{n} \zeta_i^{\tau}(\tilde{\theta}_{i-1}^{\tau}\psi_{i-1} - \frac{a(1+k_1)}{2}\zeta_i) \geq 0, \quad s_0 = 0. \tag{4.84}$$

Using (4.82) it is easy to see

$$
\begin{aligned}
& tr\tilde{\theta}_{n+1}^{\tau}\tilde{\theta}_{n+1} + \frac{t_{n+1}}{r_n} \\
= \;& tr\tilde{\theta}_n^{\tau}\tilde{\theta}_n - 2a(\zeta_{n+1}^{\tau} + w_{n+1}^{\tau})\frac{\tilde{\theta}_n^{\tau}\psi_n}{r_n} + \frac{a^2\|\psi_n\|^2}{r_n^2}\|\zeta_{n+1} + w_{n+1}\|^2 + \frac{t_{n+1}}{r_n} \\
= \;& tr\tilde{\theta}_n^{\tau}\tilde{\theta}_n - \frac{2a[\zeta_{n+1}^{\tau}(\tilde{\theta}_n^{\tau}\psi_n - \frac{a}{2}(1+k_1)\zeta_{n+1})]}{r_n} \\
& -\frac{a^2(1+k_1)\|\zeta_{n+1}\|^2}{r_n} - \frac{2aw_{n+1}^{\tau}\tilde{\theta}_n^{\tau}\psi_n}{r_n} + \frac{a^2\|\psi_n\|^2}{r_n^2}\|\zeta_{n+1}\|^2 \\
& +\frac{a^2\|\psi_n\|^2}{r_n^2}\|w_{n+1}\|^2 + \frac{2a^2\|\psi_n\|^2}{r_n^2}\zeta_{n+1}^{\tau}w_{n+1} + \frac{t_{n+1}}{r_n} \\
\le \;& tr\tilde{\theta}_n^{\tau}\tilde{\theta}_n + \frac{t_n}{r_{n-1}} - \frac{a^2k_1\|\zeta_{n+1}\|^2}{r_n} - \frac{2aw_{n+1}^{\tau}\tilde{\theta}_n^{\tau}\psi_n}{r_n} \\
& +\frac{a^2\|\psi_n\|^2}{r_n^2}\|w_{n+1}\|^2 + \frac{2a^2\|\psi_n\|^2}{r_n^2}\zeta_{n+1}^{\tau}w_{n+1}.
\end{aligned}
$$

Summing this inequality from 0 to $n+1$ leads to ($r_{-1} \triangleq 1$)

$$
\begin{aligned}
& tr\tilde{\theta}_{n+1}^{\tau}\tilde{\theta}_{n+1} + \frac{t_{n+1}}{r_n} + \frac{a^2k_1}{2}\sum_{i=0}^{n}\frac{\|\zeta_{i+1}\|^2}{r_i} \\
\le \;& tr\tilde{\theta}_0^{\tau}\tilde{\theta}_0 + \frac{t_0}{r_{-1}} - \frac{a^2k_1}{2}\sum_{i=0}^{n}\frac{\|\zeta_{i+1}\|^2}{r_i} \\
& -2a\sum_{i=0}^{n}\frac{w_{i+1}^{\tau}\tilde{\theta}_i^{\tau}\psi_i}{r_i} + a^2\sum_{i=0}^{n}\frac{\|\psi_i\|^2}{r_i^2}\|w_{i+1}\|^2 \\
& +2a^2\sum_{i=0}^{n}\frac{\|\psi_i\|^2}{r_i^2}\zeta_{i+1}^{\tau}w_{i+1}.
\end{aligned} \tag{4.85}
$$

We now show that the right-hand side of the last inequality is pathwisely bounded in n.

Fix a small constant $\delta \in (0, \frac{1}{2})$. By Theorem 2.8 we have

$$
\begin{aligned}
\left|\sum_{i=0}^{n}\frac{w_{i+1}^{\tau}\tilde{\theta}_i^{\tau}\psi_i}{r_i}\right| &= O\left(\left(\sum_{i=0}^{n}\frac{\|\tilde{\theta}_i^{\tau}\psi_i\|^2}{r_i^2}\right)^{\frac{1}{2}+\delta}\right) \\
&= O\left(\left(\sum_{i=0}^{n}\frac{\|\zeta_{i+1}\|^2}{r_i^2}\right)^{\frac{1}{2}+\delta}\right),
\end{aligned} \tag{4.86}
$$

where the last equality holds because by (4.83)

$$\|\tilde{\theta}_{n-1}^T \psi_{n-1}\| \le \|\zeta_n\| + \|C_1\|\|\zeta_{n-1}\| + \ \dots\ + \|C_r\|\|\zeta_{n-r}\|.$$

Notice that ζ_{i+1} is $\mathcal{F}_i$-measurable. Again, applying Theorem 2.8 to the last term of (4.85) yields

$$\begin{aligned}\sum_{i=0}^{n} \frac{\|\psi_i\|^2}{r_i^2} \zeta_{i+1}^T w_{i+1} &= O\left(\left(\sum_{i=0}^{n} \frac{\|\psi_i\|^4}{r_i^4} \|\zeta_{i+1}\|^2\right)^{\frac{1}{2}+\delta}\right) \\ &= O\left(\left(\sum_{i=0}^{n} \frac{\|\zeta_{i+1}\|^2}{r_i^2}\right)^{\frac{1}{2}+\delta}\right). \qquad (4.87)\end{aligned}$$

Clearly, for each sample with possible exception of a set with probability zero, the right-hand sides of (4.86) and (4.87) are either bounded by a constant possibly depending upon sample or dominated by $\sum_{i=0}^{n} \frac{\|\zeta_{i+1}\|^2}{r_i}$ since $r_i \ge 1$ and $\delta < \frac{1}{2}$. Hence the right-hand side of (4.85) will be pathwisely bounded if we can prove the finiteness of

$$\sum_{i=0}^{\infty} \frac{\|\psi_i\|^2}{r_i^2} \|w_{i+1}\|^2.$$

For this applying Theorem 2.7 to

$$\sum_{i=0}^{\infty} \frac{\|\psi_i\|^2}{r_i^2} (\|w_{i+1}\|^2 - E(\|w_{i+1}\|^2 | \mathcal{F}_i))$$

we find that it is convergent a.s. since

$$\begin{aligned}&\sum_{i=0}^{\infty} E\left[\left(\frac{\|\psi_i\|^2}{r_i^2} |\|w_{i+1}\|^2 - E(\|w_{i+1}\|^2 | \mathcal{F}_i)|\right) | \mathcal{F}_i\right] \\ &\le 2\sigma^2 \sum_{i=0}^{\infty} \frac{\|\psi_i\|^2}{r_i^2} \\ &\le 2\sigma^2 \sum_{i=0}^{\infty} \frac{r_i - r_{i-1}}{r_i r_{i-1}} < \infty.\end{aligned}$$

Hence, we have

$$
\begin{aligned}
&\sum_{i=0}^{\infty} \frac{\|\psi_i\|^2}{r_i^2} \||w_{i+1}\|^2 \\
\leq\ & \sum_{i=0}^{\infty} \frac{\|\psi_i\|^2}{r_i^2} (\|w_{i+1}\|^2 - E(\|w_{i+1}\|^2|\mathcal{F}_i)) + \sum_{i=0}^{\infty} \frac{\|\psi_i\|^2}{r_i^2} E(\||w_{i+1}\|^2|\mathcal{F}_i) \\
\leq\ & \sum_{i=0}^{\infty} \frac{\|\psi_i\|^2}{r_i^2} (\|w_{i+1}\|^2 - E(\|w_{i+1}\|^2|\mathcal{F}_i)) + \sigma^2 \sum_{i=0}^{\infty} \frac{\|\psi_i\|^2}{r_i^2} \\
<\ & \infty, \quad a.s.
\end{aligned}
$$

Thus, the right-hand side of (4.85) is pathwisely bounded in n. This implies all conclusions of the lemma because $t_n \geq 0$ by (4.84). Q.E.D.

Remark 4.3. It can be shown by the supermartingale convergence theorem that $\lim_{n\to\infty} \|\tilde{\theta}_n\|$ exists. Also (4.77) can be considerably relaxed. For details see [CG11].

We recall that there always possibly exists an exceptional set with probability zero when we talk about a relationship between random quantities or sets.

Theorem 4.3 *[Gu1], [CG11]. Assume Conditions (4.77) -(4.79) are satisfied. Then θ_n given by (4.22) and (4.23) with an arbitrary θ_0 converges to θ as $n \to \infty$ on the set*

$$S \triangleq \left\{\omega : \quad \Phi(n,0) \xrightarrow[n\to\infty]{} 0\right\}. \tag{4.88}$$

In the special case $r = 0$ the converse is also true, i.e.

$$\left\{\omega : \Phi(n,0) \xrightarrow[n\to\infty]{} 0\right\} = \left\{\omega : \theta_n \xrightarrow[n\to\infty]{} \theta \text{ for any } \theta_0\right\} \tag{4.89}$$

Proof. Assume $\Phi(n,0) \xrightarrow[n\to\infty]{} 0$. Let us consider each term on the right-hand side of (4.74).

Obviously, for the first term

$$\Phi(n+1,0)\tilde{\theta}_0 \xrightarrow[n\to\infty]{} 0 \quad \text{for any } \theta_0, \tag{4.90}$$

while for the second term we have

$$\left\| \sum_{j=0}^{n} \Phi(n+1, j+1) \frac{\psi_j \varphi_j^{\zeta\tau}}{r_j} \theta \right\|$$

$$\leq \left\| \sum_{j=0}^{N} \Phi(n+1, j+1) \frac{\psi_j \varphi_j^{\zeta\tau}}{r_j} \right\| \|\theta\|$$

$$+ \left(\sum_{j=N+1}^{n} \frac{\|\Phi(n+1, j+1)\psi_j\|^2}{r_j} \right)^{\frac{1}{2}} \left(\sum_{j=N+1}^{n} \frac{\|\varphi_j^{\zeta}\|^2}{r_j} \right)^{\frac{1}{2}} \|\theta\|. \quad (4.91)$$

Similar to (4.49) we know that

$$det\left(I - \frac{a\psi_j \psi_j^\tau}{r_j}\right) = 1 - \frac{a\|\psi\|^2}{r_j} > 0$$

and

$$\Phi(j+1, 0) = \left(I - \frac{a\psi_j \psi_j^\tau}{r_j}\right)\left(I - \frac{a\psi_{j-1}\psi_{j-1}^\tau}{r_{j-1}}\right) \cdots \left(I - \frac{a\psi_0 \psi_0^\tau}{r_0}\right)$$

is invertible. Then for any $j \in \{0, 1, ..., N\}$ with N fixed

$$\Phi(n+1, j+1) = \Phi(n+1, 0)(\Phi(j+1, 0))^{-1} \xrightarrow[n \to \infty]{} 0. \quad (4.92)$$

Hence the first term on the right-hand side of (4.91) converges to zero on S for any fixed N, while for the last term by Lemma 4.2 it is bounded by

$$\sqrt{\frac{h}{a}} \|\theta\| \left(\sum_{j=N+1}^{n} \frac{\|\varphi_j^{\zeta}\|^2}{r_j} \right)^{\frac{1}{2}} \leq \sqrt{\frac{h}{a}} \|\theta\| \left(\sum_{j=N+1}^{\infty} \frac{\|\varphi_j^{\zeta}\|^2}{r_j} \right)^{\frac{1}{2}}$$

which goes to zero as $N \to \infty$.

Thus, from (4.74) we see that to prove $\theta_n \xrightarrow[n \to \infty]{} \theta$ on S it suffices to show that

$$\sum_{j=0}^{n} \Phi(n+1, j+1) \frac{\psi_j}{r_j} w_{j+1}^\tau \xrightarrow[n \to \infty]{} 0 \quad on\ S. \quad (4.93)$$

For this we first show that

$$\sum_{i=n}^{\infty} \frac{\psi_i}{r_i} w_{i+1}^\tau = O(r_n^{-\delta}) \qquad a.s. \quad on\ S \quad (4.94)$$

for $\forall \delta \in [0, \frac{1}{2})$.

Noticing

$$\sum_{i=1}^{\infty} E\left[\left\| \frac{\psi_i w_{i+1}^\tau}{r_i^{1-\delta}} \right\| |\mathcal{F}_i \right] \leq \sigma^2 \sum_{i=1}^{\infty} \frac{\|\psi_i\|^2}{r_i^{2(1-\delta)}}$$

$$= \sigma^2 \sum_{i=1}^{\infty} \frac{\int_{r_{i-1}}^{r_i} dx}{r_i^{2(1-\delta)}} \leq \sigma^2 \int_{r_0}^{\infty} \frac{dx}{x^{2(1-\delta)}} < \infty \quad a.s., \quad (4.95)$$

by Theorem 2.7 we find that

$$\sum_{j=1}^{\infty} \frac{\psi_j w_{j+1}^{\tau}}{r_j^{1-\delta}}$$

converges a.s. This implies that as $n \to \infty$

$$s_n^{\delta} \triangleq \sum_{i=n}^{\infty} \frac{\psi_i w_{i+1}^{\tau}}{r_i^{1-\delta}} = o(1). \tag{4.96}$$

Then (4.94) follows from the following chain of inequalities

$$\begin{aligned}
\left\| r_n^{\delta} \sum_{i=n}^{\infty} \frac{\psi_i w_{i+1}^{\tau}}{r_i} \right\| &= \left\| r_n^{\delta} \sum_{i=n}^{\infty} \frac{\psi_i w_{i+1}^{\tau}}{r_i^{1-\delta}} \times \frac{1}{r_i^{\delta}} \right\| \\
= \quad & \left\| r_n^{\delta} \sum_{i=n}^{\infty} (s_i^{\delta} - s_{i+1}^{\delta}) \frac{1}{r_i^{\delta}} \right\| = \left\| s_n^{\delta} - r_n^{\delta} \sum_{i=n}^{\infty} s_{i+1}^{\delta} \left(\frac{1}{r_i^{\delta}} - \frac{1}{r_{i+1}^{\delta}} \right) \right\| \\
= \quad & o(1) + o(1) \left\| \sum_{i=n}^{\infty} r_n^{\delta} \left(\frac{1}{r_i^{\delta}} - \frac{1}{r_{i+1}^{\delta}} \right) \right\| = o(1).
\end{aligned}$$

Set

$$s \triangleq \sum_{i=0}^{\infty} \frac{\psi_i w_{i+1}^{\tau}}{r_i}, \qquad s_1 \triangleq 0,$$

$$s_n \triangleq \sum_{i=0}^{n} \frac{\psi_i w_{i+1}^{\tau}}{r_i}, \qquad \tilde{s}_n = s - s_n.$$

By (4.94) we see

$$\|\tilde{s}_n\| \le c r_n^{-\delta} \tag{4.97}$$

for some c possibly depending upon the sampling path.

We now prove (4.93) as follows:

$$\begin{aligned}
& \left\| \sum_{j=0}^{n} \Phi(n+1, j+1) \frac{\psi_j}{r_j} w_{j+1}^{\tau} \right\| \\
= \quad & \left\| \sum_{j=0}^{n} \Phi(n+1, j+1)(s_j - s_{j+1}) \right\| \\
= \quad & \left\| s_n - \sum_{j=0}^{n} [\Phi(n+1, j+1) - \Phi(n+1, j)] s_{j-1} \right\|
\end{aligned}$$

$$
\begin{aligned}
&= \left\| s_n - \sum_{j=0}^{n}[\Phi(n+1,j+1) - \Phi(n+1,j)]s \right. \\
&\quad \left. + \sum_{j=0}^{n}[\Phi(n+1,j+1) - \Phi(n+1,j)]\tilde{s}_{j-1} \right\| \\
&= \left\| s_n - s + \Phi(n+1,0)s + \sum_{j=0}^{n}\Phi(n+1,j+1)[I - \Phi(j+1,j)]\tilde{s}_{j-1} \right\| \\
&\le \|\tilde{s}_n\| + \|\Phi(n+1,0)s\| + a\sum_{j=0}^{n}\|\Phi(n+1,j+1)\frac{\psi_j\psi_j^\tau}{r_j}\tilde{s}_{j-1}\| \\
&\le \|\tilde{s}_n\| + \|\Phi(n+1,0)s\| + ca\sum_{j=0}^{n}\frac{\|\Phi(n+1,j+1)\psi_j\|}{r_j^{\frac{1}{2}}}\frac{\|\psi_j\|}{r_j^{\frac{1}{2}+\delta}} \\
&\le \|\tilde{s}_n\| + \|\Phi(n+1,0)s\| + ca\sum_{j=0}^{N}\frac{\|\Phi(n+1,j+1)\psi_j\|}{r_j^{\frac{1}{2}}}\frac{\|\psi_j\|}{r_j^{\frac{1}{2}+\delta}} \\
&\quad + ca\left\{\sum_{j=N+1}^{n}\frac{\|\Phi(n+1,j+1)\psi_j\|^2}{r_j}\sum_{j=N+1}^{n}\frac{\|\psi_j\|^2}{r_j^{1+2\delta}}\right\}^{\frac{1}{2}} \qquad (4.98)
\end{aligned}
$$

for which the second term tends to 0 on S as $n \to \infty$, by (4.92) the third term also goes to 0 on S for any fixed N, while the last term by Lemma 4.2 is bounded by

$$
c\sqrt{ah}\left(\sum_{j=N+1}^{\infty}\frac{\|\psi_j\|^2}{r_j^{1+2\delta}}\right)^{\frac{1}{2}}
$$

which goes to 0 as shown in (4.95).

Thus, from (4.98) and (4.97) we obtain that

$$
\begin{aligned}
&\left\|\sum_{j=0}^{n}\Phi(n+1,j+1)\frac{\psi_j}{r_j}w_{j+1}^\tau\right\| \\
&= \|\tilde{s}_n\| + o(1) \le cr_n^{-\delta} + o(1) \qquad on\ S. \qquad (4.99)
\end{aligned}
$$

Finally, we note that on S

$$
\begin{aligned}
det\Phi(n+1,0) &= det\prod_{i=0}^{n}\Phi(i+1,i) \\
&= \prod_{i=0}^{n}det\left(I - \frac{a\psi_i\psi_i^\tau}{r_i}\right) = \prod_{i=0}^{n}\left(1 - \frac{a\|\psi_i\|^2}{r_i}\right)\xrightarrow[n\to\infty]{} 0. \qquad (4.100)
\end{aligned}
$$

This implies that $r_n \xrightarrow[n\to\infty]{} \infty$ on S, because otherwise, $\sum_{i=0}^{\infty} \frac{\|\psi_i\|^2}{r_i} < \infty$ and this would contradict with (4.100). Hence (4.99) yields (4.93). This completes the proof of the first part of the theorem. In other words, we have shown that

$$\left\{\omega: \quad \Phi(n,0) \xrightarrow[n\to\infty]{} 0\right\} \subset \left\{\omega: \quad \theta_n \xrightarrow[n\to\infty]{} \theta \quad \text{for any } \theta_0\right\}.$$

For the second part of the theorem we need to prove the converse inclusion for the case $r = 0$.

If $r = 0$ we note that

$$\psi_n = [y_n^T \quad \cdots \quad y_{n-p+1}^T \quad u_n^T \quad \cdots \quad u_{n-q+1}^T]^T = \varphi_n^0.$$

Then in (4.70) the term with φ_n^{ζ} disappears and (4.74) is reduced to

$$\tilde{\theta}_{n+1} = \Phi(n+1,0)\tilde{\theta}_0 - a\sum_{j=0}^{n} \Phi(n+1,j)\frac{\psi_j}{r_j} w_{j+1}^T. \tag{4.101}$$

Notice that the last term is independent of θ_0. Consequently, if $\tilde{\theta}_n \xrightarrow[n\to\infty]{} 0$ for any θ_0, then

$$\tilde{\theta}_{n+1}(\tilde{\theta}_0^1) - \tilde{\theta}_{n+1}(\tilde{\theta}_0^2) = \Phi(n+1,0)(\tilde{\theta}_0^1 - \tilde{\theta}_0^2) \xrightarrow[n\to\infty]{} 0 \tag{4.102}$$

for any $\tilde{\theta}_0^1$ and $\tilde{\theta}_0^2$ of compatible dimensions, where we use the notation $\tilde{\theta}_{n+1}(\tilde{\theta}_0^i)$ ($i = 1, 2$) to emphasize the dependence of $\tilde{\theta}_{n+1}$ on the initial value $\tilde{\theta}_0^i$. Obviously, (4.102) is possible only if $\Phi(n,0) \xrightarrow[n\to\infty]{} 0$. This proves (4.89). Q.E.D.

Theorem 4.4. *Under Conditions (4.77)-(4.79) the estimate θ_n given by (4.22) and (4.23) with an arbitrary θ_0 converges to θ as $n \to \infty$ on*

$$S^0 \triangleq \{\omega: \quad \Phi^0(n,0) \xrightarrow[n\to\infty]{} 0\} \tag{4.103}$$

where $\Phi^0(n,0)$ is defined in a similar way as for $\Phi(n,0)$, i.e.,

$$\Phi^0(n+1,0) = \left(I - \frac{a\varphi_n^0\varphi_n^{0T}}{r_n^0}\right)\Phi^0(n,0), \quad \Phi^0(i,i) = I, \tag{4.104}$$

$$r_n^0 = 1 + \sum_{i=0}^{n} \|\varphi_i^0\|^2. \tag{4.105}$$

Proof. By Theorem 4.3 it suffices to show $S^0 \subset S$, where S is defined by (4.88). In fact, we proceed to prove a stronger result, namely,

$$S^0 = S$$

under conditions of the theorem.

Let $\Phi^0(n,0) \xrightarrow[n\to\infty]{} 0$. Then as shown by (4.100) $r_n^0 \xrightarrow[n\to\infty]{} \infty$. We now show $r_n \to \infty$ as $n \to \infty$.

By Lemma 2.4 from the expression (4.68) for φ_n^ζ we find that under the conditions of the theorem

$$\sum_{i=0}^{\infty} \frac{\|\varphi_i^\zeta\|^2}{r_i} < \infty \qquad a.s. \tag{4.106}$$

Assume the converse were true, i.e. $r_n \xrightarrow[n\to\infty]{} c < \infty$. Then from (4.106) it follows that

$$\sum_{i=0}^{\infty} \|\varphi_i^\zeta\|^2 < \infty.$$

Using (4.69) it is easy to see that

$$\begin{aligned} r_n^0 &= 1 + \sum_{i=0}^{n} \|\psi_i - \varphi_i^\zeta\|^2 \\ &= r_n - 2\sum_{i=0}^{n} \psi_i^T \varphi_i^\zeta + \sum_{i=0}^{n} \|\varphi_i^\zeta\|^2 \\ &\le r_n + 2\sqrt{r_n}\left(\sum_{i=0}^{n} \|\varphi_i^\zeta\|^2\right)^{\frac{1}{2}} + \sum_{i=0}^{n} \|\varphi_i^\zeta\|^2 \\ &\le c + 2\sqrt{c}\left(\sum_{i=0}^{\infty} \|\varphi_i^\zeta\|^2\right)^{\frac{1}{2}} + \sum_{i=0}^{\infty} \|\varphi_i^\zeta\|^2 < \infty \end{aligned} \tag{4.107}$$

which contradicts the fact $r_n^0 \xrightarrow[n\to\infty]{} \infty$. Hence r_n diverges to ∞.

Then by the Kronecker lemma (Lemma 2.4) from (4.106) it follows that

$$\frac{1}{r_n} \sum_{i=0}^{n} \|\varphi_n^\zeta\|^2 \xrightarrow[n\to\infty]{} 0. \tag{4.108}$$

Using this we obtain that

$$\frac{r_n^0}{r_n} = \frac{r_n - 2\sum_{i=0}^{n} \psi_i^T \varphi_i^\zeta + \sum_{i=0}^{n} \|\varphi_i^\zeta\|^2}{r_n} \xrightarrow[n\to\infty]{} 1 \qquad on \quad S^0 \tag{4.109}$$

and by Lemma 4.3

$$\sum_{i=0}^{\infty} \frac{\|\zeta_{i+1}\|^2}{r_i^0} < \infty \qquad on \quad S^0. \tag{4.110}$$

Finally, to show $\Phi(n,0) \xrightarrow[n\to\infty]{} 0$ on S^0 we rewrite $\Phi(n+1,0)$ by the following chain of equalities:

$$\begin{aligned}
&\Phi(n+1,0) \\
= \ & \left(I - \frac{a\varphi_n^0\varphi_n^{0\tau}}{r_n^0}\right)\Phi(n,0) + \left(\frac{a\varphi_n^0\varphi_n^{0\tau}}{r_n^0} - \frac{a\psi_n\psi_n^\tau}{r_n}\right)\Phi(n,0) \\
= \ & \Phi^0(n+1,0) + a\sum_{j=0}^{n}\Phi^0(n+1,j+1)\left(\frac{\varphi_j^0\varphi_j^{0\tau}}{r_j^0} - \frac{\psi_j\psi_j^\tau}{r_j}\right)\Phi(j,0) \\
= \ & \Phi^0(n+1,0) \\
& + a\sum_{j=0}^{n}\Phi^0(n+1,j+1)\left[-\frac{\varphi_j^0\varphi_j^{\zeta\tau}}{r_j^0}\Phi(j,0) + \frac{\varphi_j^0\psi_j^\tau}{\sqrt{r_j^0}}\frac{\Phi(j,0)}{\sqrt{r_j^0}}\frac{\sqrt{r_j}}{\sqrt{r_j}}\right] \\
& - a\sum_{j=0}^{n}\Phi^0(n+1,j+1)\left[\frac{\varphi_j^\zeta\psi_j^\tau}{r_j}\Phi(j,0) + \frac{\varphi_j^0\psi_j^\tau}{\sqrt{r_j}}\frac{\Phi(j,0)}{\sqrt{r_j}}\frac{\sqrt{r_j^0}}{\sqrt{r_j^0}}\right] \\
= \ & \Phi^0(n+1,0) + a\sum_{j=0}^{n}\Phi^0(n+1,j+1)\left[\frac{\varphi_j^\zeta\psi_j^\tau}{r_j} - \frac{\varphi_j^0\varphi_j^{\zeta\tau}}{r_j^0}\right]\Phi(j,0) \\
& + a\sum_{j=0}^{n}\frac{\Phi^0(n+1,j+1)\varphi_j^0}{\sqrt{r_j^0}}\left(\frac{\sqrt{r_j}}{\sqrt{r_j^0}} - \frac{\sqrt{r_j^0}}{\sqrt{r_j}}\right)\frac{\psi_j^\tau\Phi(j,0)}{\sqrt{r_j}}.
\end{aligned} \tag{4.111}$$

On the right-hand side of (4.111) the first term tends to 0 on S^0 by assumption, the second term vanishes as $n \to \infty$ because by Lemma 4.2 and (4.106) and (4.110)

$$\begin{aligned}
& \left\|\sum_{j=0}^{n}\Phi^0(n+1,j+1)\frac{\varphi_j^\zeta\psi_j^\tau}{r_j}\Phi(j,0)\right\| \\
\leq \ & \left\|\sum_{j=0}^{N}\Phi^0(n+1,j+1)\frac{\varphi_j^\zeta\psi_j^\tau}{r_j}\Phi(j,0)\right\| \\
& + \sqrt{\frac{h}{a}}\left(\sum_{j=N+1}^{n}\frac{\|\varphi_j^\zeta\|^2}{r_j}\right)^{\frac{1}{2}} \xrightarrow[\substack{n\to\infty \\ N\to\infty}]{} 0 \qquad on \quad S^0
\end{aligned}$$

and

$$\left\| \sum_{j=0}^{n} \Phi^0(n+1,j+1) \frac{\varphi_j^0 \varphi_j^{\zeta\tau}}{r_j} \Phi(j,0) \right\|$$

$$\le \left\| \sum_{j=0}^{N} \Phi^0(n+1,j+1) \frac{\varphi_j^0 \varphi_j^{\zeta\tau}}{r_j^0} \Phi(j,0) \right\|$$

$$+\sqrt{\frac{h}{a}} \left(\sum_{j=N+1}^{n} \frac{\|\varphi_j^\zeta\|^2}{r_j^0} \right)^{\frac{1}{2}} \xrightarrow[\substack{n\to\infty \\ N\to\infty}]{} 0 \quad on \quad S^0.$$

Finally, paying attention to the fact that $\left| \frac{\sqrt{r_j}}{\sqrt{r_j^0}} - \frac{\sqrt{r_j^0}}{\sqrt{r_j}} \right|$ can be made arbitrarily small, say, less than $\varepsilon > 0$ for $j > N$ we estimate the last term of (4.111) by using Lemma 4.2 as follows

$$\left\| a \sum_{j=0}^{n} \frac{\Phi^0(n+1,j+1)\varphi_j^0}{\sqrt{r_j^0}} \left(\frac{\sqrt{r_j}}{\sqrt{r_j^0}} - \frac{\sqrt{r_j^0}}{\sqrt{r_j}} \right) \frac{\psi_j^\tau \Phi(j,0)}{\sqrt{r_j}} \right\|$$

$$\le \left\| a \sum_{j=0}^{n} \frac{\Phi^0(n+1,j+1)\varphi_j^0}{\sqrt{r_j^0}} \left(\frac{\sqrt{r_j}}{\sqrt{r_j^0}} - \frac{\sqrt{r_j^0}}{\sqrt{r_j}} \right) \frac{\psi_j^\tau \Phi(j,0)}{\sqrt{r_j}} \right\| + h\varepsilon$$

$$\xrightarrow[\substack{n\to\infty \\ \varepsilon\to 0}]{} 0 \quad on \quad S^0.$$

Thus, we have shown $\Phi(n,0) \xrightarrow[n\to\infty]{} 0$ on S^0. Conversely, if $\Phi(n,0) \xrightarrow[n\to\infty]{} 0$, then $r_n \xrightarrow[n\to\infty]{} \infty$ and (4.108)-(4.110) hold as before.

Instead of the first line of (4.111) we now have

$$\Phi^0(n+1,0)$$

$$= \left(I - \frac{a\psi_n\psi_n^\tau}{r_n} \right) \Phi^0(n,0) + \left(a\frac{\psi_n\psi_n^\tau}{r_n} - \frac{\varphi_n^0\varphi_n^{0\tau}}{r_n^0} \right) \Phi^0(n,0).$$

The derivation similar to that used for (4.111) leads to that $\Phi^0(n,0) \xrightarrow[n\to\infty]{} 0$.

Hence $S = S^0$. Q.E.D.

4.4   Convergence of SG Without the PE Condition

In the last section we have proved that the SG algorithm produces a strongly consistent estimate θ_n on S or S^0. We now connect $\Phi(n,0) \xrightarrow[n \to \infty]{} 0$ with the condition numbers

$$\frac{\mu_{max}(n)}{\mu_{min}(n)} \qquad and \qquad \frac{\mu_{max}^0(n)}{\mu_{min}^0(n)}$$

where $\mu_{max}(n)$ and $\mu_{min}(n)$ respectively denote the largest and the smallest eigenvalues of $\sum_{i=0}^{n} \psi_i \psi_i^\tau + \frac{1}{h}I$ and $\mu_{max}^0(n)$ and $\mu_{min}^0(n)$ denote the largest and the smallest eigenvalues of $\sum_{i=0}^{n} \varphi_i^0 \varphi_i^{0\tau} + \frac{1}{h}I$, where $h = mp + lq + mr$. As a matter of fact we shall derive a condition guaranteeing $\Phi(n,0) \xrightarrow[n \to \infty]{} 0$ and showing that the PE condition consisting in boundedness of $\frac{\mu_{max}(n)}{\mu_{min}(n)}$ or $\frac{\mu_{max}^0(n)}{\mu_{min}^0(n)}$ is not necessary for strong consistency of θ_n.

Theorem 4.5 *[Gu1], [CG11]. Assume Conditions (4.77)-(4.79) are fulfilled. Then θ_n given by (4.22) and (4.23) with arbitrarily given θ_0 converges to θ as $n \to \infty$ on G which denotes the set where the following conditions are satisfied*

$$(4.112)$$

i) $r_n \xrightarrow[n \to \infty]{} \infty$

ii) There are $\delta \in [0, \frac{1}{4}]$, $N_0 > 0$, $M > 0$ which possibly depend upon the sample path such that

$$\frac{r_n}{r_{n-1}} \le M(\log r_{n-1})^\delta, \qquad \forall n \ge N_0; \qquad (4.113)$$

$$\frac{\mu_{max}(n)}{\mu_{min}(n)} \le M(\log r_n)^{\frac{1}{4}-\delta}, \qquad \forall n \ge N_0; \qquad (4.114)$$

Proof. By Theorem 4.3 it is sufficient to show that

$$G \subset S$$

where S is defined by (4.88).

Let us assume that for some sample path (4.112)-(4.114) hold.

We first explain the idea of the proof, but for this we need to introduce an integer-valued function $m(t)$ of continuous parameter t and clarify its simple properties.

$$m(t) \triangleq max[n: \quad t_n \leq t], \tag{4.115}$$

$$t_n \triangleq \sum_{i=N_0}^{n-1} \frac{\|\psi_i\|^2}{r_i(logr_{i-1})^{1/4}}, \tag{4.116}$$

where by (4.112), without loss of generality, we assume N_0 is sufficiently large so that $logr_{N_0-1} > 0$.

It is worth noting at once that by (4.115) and (4.116)

$$t < t_{m(t)+1} \leq t_{m(t)} + 1 \leq t + 1 \tag{4.117}$$

and by (4.113)

$$\frac{1}{r_n} \geq \frac{1}{Mr_{n-1}(logr_{n-1})^{\delta}} \qquad \forall n \geq N_0.$$

Substituting this in (4.116) yields for $n \geq N_0$

$$\begin{aligned} t_{n+1} \geq & \frac{1}{M}\sum_{i=N_0}^{n} \frac{\|\psi_i\|^2}{r_{i-1}(logr_{i-1})^{\frac{1}{4}+\delta}} \\ \geq & \frac{1}{M}\sum_{i=N_0}^{n} \int_{r_{i-1}}^{r_i} \frac{dx}{x(logx)^{\frac{1}{4}+\delta}} \\ = & \frac{4}{(3-4\delta)M}\left[(\log r_n)^{\frac{3}{4}-\delta} - (\log r_{N_0-1})^{\frac{3}{4}-\delta}\right]. \end{aligned}$$

Assume N is large enough so that

$$logr_n > 3logr_{N_0-1} \qquad \forall n \geq N \geq N_0. \tag{4.118}$$

We then have

$$t_{n+1} \geq \frac{2}{(3-4\delta)M}(\log r_n)^{\frac{3}{4}-\delta} \xrightarrow[n\to\infty]{} \infty \tag{4.119}$$

where the divergence follows from (4.112).

This together with (4.115) implies that

$$m(t) < \infty \quad \forall t \quad \text{and} \quad m(t) \xrightarrow[t\to\infty]{} \infty. \tag{4.120}$$

Notice that $\|\Phi(k,j)\| \leq 1$ for any $k \geq j$. Then by (4.120) for proving $\Phi(n,0) \xrightarrow[n\to\infty]{} 0$ it suffices to show for some $\alpha > 1$

$$\Phi(m(N+k\alpha),0) \xrightarrow[k\to\infty]{} 0. \tag{4.121}$$

We group $\Phi(m(N+k\alpha),0)$ by factors $\Phi(m(N+i\alpha),m(N+(i-1)\alpha))$, and estimate it by

$$\begin{aligned} &\|\Phi(m(N+k\alpha),0)\| \\ \le\ &\|\Phi(m(N+(k_0-1)\alpha),0)\| \\ &\cdot \prod_{i=k_0}^{k} \|\Phi(m(N+i\alpha),m(N+(i-1)\alpha))\|, \end{aligned} \tag{4.122}$$

where k_0 is large enough so that for $k \ge k_0$

$$m(N+k\alpha) > N \ge N_0. \tag{4.123}$$

Clearly, k_0 depends on N.

The crucial step is to prove

$$\|\Phi(m(N+i\alpha),m(N+(i-1)\alpha))\| \le 1-a_i, \quad i \ge k_0 \tag{4.124}$$

with $a_i > 0$ and $\sum_{i=k_0}^{\infty} a_i = \infty$.

If this is done, then from (4.122) we shall have the desired result:

$$\|\Phi(m(N+k\alpha),0)\| \le \prod_{i=k_0}^{k}(1-a_i) \xrightarrow[k\to\infty]{} 0. \tag{4.125}$$

The integer N appeared above will be fixed later on.

We now proceed to prove (4.124) by 3 steps.

Step 1. We establish the upper and lower bounds for $log r_{m(N+k\alpha)-1}$, with $N \ge N_0$, $k > k_0$.

The upper bound directly follows from (4.119) and (4.117)

$$\begin{aligned} \log r_{m(N+k\alpha)-1} &\le \left[\frac{(3-4\delta)M}{2} t_{m(N+k\alpha)}\right]^{\frac{4}{3-4\delta}} \\ &\le \left[\frac{(3-4\delta)M}{2}(N+k\alpha)\right]^{\frac{4}{3-4\delta}} \end{aligned} \tag{4.126}$$

since by (4.123) $m(N+k\alpha) > N \ge N_0$ for $k \ge k_0$ and (4.119) is applicable.

For the lower bound noting from (4.113) that for $\forall n \ge N_0$

$$\log r_n \le log r_{n-1} + log M + \delta log log r_{n-1}$$

and hence

$$\log r_n \le 2 log r_{n-1}, \quad \forall n \ge N \ge N_0 \quad \text{with } N \text{ sufficiently large.} \tag{4.127}$$

From this and (4.116) we find that for $n \geq N$

$$\begin{aligned} t_n \leq & \ 2\sum_{i=N}^{n-1} \frac{\|\psi_i\|^2}{r_i(logr_i)^{\frac{1}{4}}} \\ \leq & \ 2\sum_{i=N}^{n-1} \int_{r_{i-1}}^{r_i} \frac{dt}{t(logt)^{\frac{1}{4}}} = \frac{3}{8}(logr_{n-1})^{\frac{3}{4}} \end{aligned}$$

or

$$logr_{n-1} \geq (\frac{8}{3}t_n)^{4/3}. \tag{4.128}$$

Noticing $k > k_0$, $k-1 \geq k_0$ and $m(N+(k-1)\alpha)) > N \geq N_0$ by (4.128) we have

$$\begin{aligned} logr_{m(N+(k-1)\alpha)} \ \geq & \ (\frac{3}{8}t_{m(N+(k-1)\alpha)+1})^{4/3} \\ \geq & \ (\frac{3}{8}(N+(k-1)\alpha))^{4/3} \end{aligned} \tag{4.129}$$

where for the last inequality we have used the first inequality of (4.117).

Step 2. We now show that for N large enough

$$\sum_{i=m(N+(k-1)\alpha)}^{m(N+k\alpha)-1} \frac{\psi_i\psi_i^T}{r_i} \geq bk^{\frac{4\delta}{3}}I, \qquad \forall k > k_0, \tag{4.130}$$

where

$$b \triangleq \frac{\alpha-1}{Mh}\left(\frac{3\alpha}{8}\right)^{\frac{4\delta}{3}}, \qquad h = mp + lq + mr.$$

For any $k > k_0$ summing by parts we derive

$$\begin{aligned} & \sum_{i=m(N+(k-1)\alpha)}^{m(N+k\alpha)-1} \frac{\psi_i\psi_i^T}{r_i} \\ \geq & \sum_{i=m(N+(k-1)\alpha)}^{m(N+k\alpha)} \frac{1}{r_i}\left(\sum_{j=0}^{i}\psi_j\psi_j^T + \frac{1}{h}I - \sum_{j=0}^{i-1}\psi_j\psi_j^T - \frac{1}{h}I\right) - I \\ \geq & \sum_{i=m(N+(k-1)\alpha)+1}^{m(N+k\alpha)} \left(\frac{1}{h}I + \sum_{j=0}^{i-1}\psi_j\psi_j^T\right)\frac{\|\psi_i\|^2}{r_ir_{i-1}} - 2I \\ \geq & \sum_{i=m(N+(k-1)\alpha)+1}^{m(N+k\alpha)} \left[\frac{\mu_{max}(i-1)}{M(logr_{i-1})^{\frac{1}{4}-\delta}}\right]\frac{\|\psi_i\|^2}{r_ir_{i-1}}I - 2I, \end{aligned} \tag{4.131}$$

where the last inequality follows because by (4.113)

$$\left(\frac{1}{h}I+\sum_{j=0}^{i-1}\psi_j\psi_j^T\right)\geq\mu_{min}(i-1)I\geq\frac{\mu_{max}(i-1)}{M(logr_{i-1})^{\frac{1}{4}-\delta}}I.$$

Note that

$$r_{i-1}=tr\left(\frac{1}{h}I+\sum_{j=0}^{i-1}\psi_j\psi_j^T\right)\leq h\mu_{max}(i-1).$$

We then continue the estimation (4.131) as follows

$$\begin{aligned}
&\sum_{i=m(N+(k-1)\alpha)}^{m(N+k\alpha)-1}\frac{\psi_i\psi_i^T}{r_i}\\
\geq\ &\frac{1}{Mh}\sum_{i=m(N+(k-1)\alpha)+1}^{m(N+k\alpha)}\frac{\|\psi_i\|^2}{r_i(logr_{i-1})^{\frac{1}{4}-\delta}}I-2I\\
\geq\ &\frac{log^{\delta}r_{m(N+(k-1)\alpha)}}{Mh}\sum_{i=m(N+(k-1)\alpha)+1}^{m(N+k\alpha)}\frac{\|\psi_i\|^2}{r_ilogr_{i-1}}I-2I\\
=\ &\frac{log^{\delta}r_{m(N+(k-1)\alpha)}}{Mh}\left(t_{m(N+k\alpha)+1}-t_{m(N+(k-1)\alpha)+1}\right)I-2I \qquad (4.132)
\end{aligned}$$

where the last equality directly follows from (4.116) and $m(N+(k-1)\alpha)+1>N_0$ for $k>k_0$ by (4.123).

By (4.117) we have

$$t_{m(N+k\alpha)+1}>N+k\alpha,\quad t_{m(N+(k-1)\alpha)+1}<N+(k-1)\alpha+1.$$

Substituting these estimates in (4.132) leads to

$$\begin{aligned}
&\sum_{i=m(N+(k-1)\alpha)}^{m(N+k\alpha)-1}\frac{\psi_i\psi_i^T}{r_i}\\
\geq\ &\frac{(log^{\delta}r_{m(N+(k-1)\alpha)})(\alpha-1)}{Mh}I-2I\\
\geq\ &\left\{\frac{\alpha-1}{Mh}\left(\frac{3}{8}(N+(k-1)\alpha)\right)^{\frac{4\delta}{3}}-2\right\}I
\end{aligned}$$

where the estimate (4.129) is applied.

Hence we have the desired result

$$\sum_{i=m(N+(k-1)\alpha)}^{m(N+k\alpha)-1} \frac{\psi_i\psi_i^T}{r_i}$$

$$\geq \frac{\alpha-1}{Mh}\left(\frac{3}{8}k\alpha\right)^{\frac{46}{3}}\left[\left(1+\frac{N-\alpha}{k\alpha}\right)^{\frac{46}{3}} - \frac{2Mh}{\alpha-1}\left(\frac{3}{8}k\alpha\right)^{-\frac{46}{3}}\right]I$$

$$\geq bk^{\frac{46}{3}}I$$

because the quantity in the square bracket is greater than or equal to

$$1+\left(\frac{N-\alpha}{k\alpha}\right)^{\frac{46}{3}} - \frac{2Mh}{\alpha-1}\left(\frac{3}{8}k\alpha\right)^{-\frac{46}{3}}$$

$$\geq 1+(k\alpha)^{-\frac{46}{3}}\left[(N-\alpha)^{\frac{46}{3}} - \frac{2Mh}{\alpha-1}\left(\frac{8}{3}\right)^{\frac{46}{3}}\right] > 1$$

if N is sufficiently large so that

$$(N-\alpha)^{\frac{46}{3}} - \frac{2Mh}{\alpha-1}\left(\frac{8}{3}\right)^{\frac{46}{3}} > 0. \qquad (4.133)$$

Assume that N is fixed and is chosen such that (4.118), (4.127) and (4.133) are fulfilled, and for this fixed N, k_0 is selected to satisfy (4.123).

Step 3. We now prove (4.124).

Let ρ_k be the largest eigenvalue of

$$\Phi^T(m(N+k\alpha), m(N+(k-1)\alpha))\Phi(m(N+k\alpha), m(N+(k-1)\alpha))$$

and let $x_{m(N+(k-1)\alpha)}$ be the corresponding unit eigenvector.

For $i \in [m(N+(k-1)\alpha), m(N+k\alpha)-1]$ recursively define x_i

$$x_{i+1} = \left(I - \frac{a\psi_i\psi_i^T}{r_i}\right)x_i. \qquad (4.134)$$

Then we have

$$x^T_{m(N+k\alpha)}x_{m(N+k\alpha)}$$

$$= x^T_{m(N+(k-1)\alpha)}\Phi^T(m(N+k\alpha), m(N+(k-1)\alpha))$$

$$\times\Phi(m(N+k\alpha), m(N+(k-1)\alpha))x_{m(N+(k-1)\alpha)}$$

$$= x^T_{m(N+(k-1)\alpha)}\rho_k x_{m(N+(k-1)\alpha)} = \rho_k.$$

From (4.134) it is easy to see

$$x_{i+1}^T x_{i+1} \le x_i^T x_i - a x_i^T \frac{\psi_i \psi_i^T}{r_i} x_i.$$

Summing both sides of this inequality yields

$$\begin{aligned} & a \sum_{i=m(N+(k-1)\alpha)}^{m(N+k\alpha)-1} \frac{\|\psi_i^T x_i\|^2}{r_i} \\ \le\ & \|x_{m(N+(k-1)\alpha)}\|^2 - \|x_{m(N+k\alpha)}\|^2 = 1 - \rho_k. \end{aligned} \tag{4.135}$$

From (4.134) we also have for $i \in [m(N+(k-1)\alpha), m(N+k\alpha)-1]$,

$$\begin{aligned} & \|x_i - x_{m(N+(k-1)\alpha)}\| \\ = \ & a \| \sum_{j=m(N+(k-1)\alpha)}^{i-1} \frac{\psi_j \psi_j^T}{r_j} x_j \| \\ \le \ & a (log r_{m(N+k\alpha)-1})^{1/8} \sum_{i=m(N+(k-1)\alpha)}^{m(N+k\alpha)-1} \frac{\|\psi_i\|}{r_i^{1/2} (log r_{i-1})^{1/8}} \frac{\|\psi_i^T x_i\|}{r_i^{1/2}} \\ \le \ & \sqrt{a} (log r_{m(N+k\alpha)-1})^{1/8} \sqrt{1+\alpha} \sqrt{1-\rho_k}, \end{aligned} \tag{4.136}$$

where the last inequality is obtained by using the Schwarz inequality, (4.115), (4.116) and (4.135).

From (4.130) we see that

$$\begin{aligned} b k^{\frac{4\delta}{3}} \le \ & x_{m(N+(k-1)\alpha)}^T \\ & \cdot \sum_{i=m(N+(k-1)\alpha)}^{m(N+k\alpha)-1} \frac{\psi_i \psi_i^T}{r_i} (x_{m(N+(k-1)\alpha)} - x_i + x_i) \\ \le \ & (log r_{m(N+k\alpha)-1})^{1/4} \\ & \cdot \sum_{i=m(N+(k-1)\alpha)}^{m(N+k\alpha)-1} \frac{\|\psi_i\|^2}{r_i (log r_{i-1})^{1/4}} \|x_{m(N+(k-1)\alpha)} - x_i\| \\ & + (log r_{m(N+k\alpha)-1})^{1/8} \sum_{i=m(N+(k-1)\alpha)}^{m(N+k\alpha)-1} \frac{\|\psi_i\|}{r_i^{1/2} (log r_{i-1})^{1/8}} \frac{\|\psi_i^T x_i\|}{r_i^{1/2}}. \end{aligned}$$

Applying (4.135) and (4.136) to the last expression we find

$$
\begin{aligned}
bk^{\frac{4\delta}{3}} \le\ & \sqrt{a}(\log r_{m(N+k\alpha)-1})^{3/8} \\
& \sum_{i=m(N+(k-1)\alpha)}^{m(N+k\alpha)-1} \frac{\|\psi_i\|^2}{r_i(\log r_{i-1})^{1/4}}\sqrt{1+\alpha}\sqrt{1-\rho_k} \\
& +(\log r_{m(N+k\alpha)-1})^{1/8} \\
& \left(\sum_{i=m(N+(k-1)\alpha)}^{m(N+k\alpha)-1} \frac{\|\psi_i\|^2}{r_i(\log r_{i-1})^{1/4}}\right)^{1/2} \frac{1}{\sqrt{a}}\sqrt{1-\rho_k} \\
\le\ & (\log r_{m(N+k\alpha)-1})^{3/8}(1+\alpha)^{3/2}\sqrt{1-\rho_k} \\
& +(\log r_{m(N+k\alpha)-1})^{1/8}\frac{1}{\sqrt{a}}\sqrt{1+\alpha}\sqrt{1-\rho_k} \\
\le\ & \frac{2}{\sqrt{a}}(\log r_{m(N+k\alpha)-1})^{3/8}(1+\alpha)^{3/2}\sqrt{1-\rho_k}
\end{aligned}
$$

where as seen from (4.129), without loss of generality, we have assumed

$$\log r_{m(N+k\alpha)-1} > 1.$$

Using (4.126) we then obtain

$$bk^{\frac{4\delta}{3}} \le \frac{2}{\sqrt{a}}\left[\frac{(3-4\delta)M}{2}(N+k\alpha)\right]^{\frac{4}{3-4\delta}\frac{3}{8}}(1+\alpha)^{3/2}\sqrt{1-\rho_k}.$$

Since N is fixed, we can select k_0 large enough such that not only (4.123) holds but also $k_0\alpha > N$. Then we have that for $k > k_0$

$$bk^{\frac{4\delta}{3}} \le \frac{2}{\sqrt{a}}[(3-4\delta)Mk\alpha]^{\frac{3}{2(3-4\delta)}}\sqrt{1-\rho_k}(1+\alpha)^{\frac{3}{2}}.$$

From this a simple calculation leads to

$$\rho_k \le 1-2a_k, \qquad a_k = \frac{c}{2k^{\lambda}},$$

where

$$c = \frac{ab^2}{4(1+\alpha)^3[(3-4\delta)M\alpha]^{\frac{3}{3-4\delta}}}$$

$$\lambda = \frac{3}{3-4\delta} - \frac{8\delta}{3}.$$

It is easy to see that $\frac{4}{5} < \lambda \le 1$ for $\delta \in [0, \frac{1}{4})$. Hence $\sum_{k=k_0}^{\infty} a_k = \infty$.

In conclusion, we have

$$\begin{aligned} & \|\Phi(m(N+k\alpha), m(N+(k-1)\alpha))\| \\ \le \; & \sqrt{\rho_k} \le (1-2a_k)^{1/2} \le 1 - a_k \quad k > k_0 \end{aligned} \tag{4.137}$$

which verifies (4.124), (4.125) and completes the proof of the theorem.

Q.E.D.

Remark 4.4. In the theorems of this section the conditions guaranteeing convergence of θ_n to θ are essentially imposed on the behaviors of φ_n and ψ_n, which involve the estimate θ_n itself in the case where the system noise is correlated, i.e. $r > 0$. Then, in this case these conditions become difficult to verify. We shall return to this issue in Chapter 5.

Remark 4.5. The proof of Theorem 4.5 is purely algebraic. In fact, we have proved an interesting fact: If ψ_n, r_n, $\Phi(n,i)$ are connected by (4.23) and (4.73), then conditions (4.112)-(4.114) imply $\Phi(n,0) \xrightarrow[n\to\infty]{} 0$ whatever the vector ψ_n is. For example, if (4.112)-(4.114) hold with ψ_n, r_n, $\mu_{max}(n)$ and $\mu_{min}(n)$ replaced by φ_n^0, r_n^0, $\mu_{max}^0(n)$ and $\mu_{min}^0(n)$ respectively, then $\Phi^0(n,0) \xrightarrow[n\to\infty]{} 0$ and by Theorem 4.4 $\theta_n \xrightarrow[n\to\infty]{} \theta$.

We note that Condition (4.114) means that the PE condition may not be satisfied if $r_n \xrightarrow[n\to\infty]{} \infty$. More precisely, we allow the condition number to grow at a rate not faster than $(log r_n)^{1/4}$ in order to guarantee $\Phi(n,0) \xrightarrow[n\to\infty]{} 0$. It is natural to ask if we can increase the divergence rate from $(log r_n)^{1/4}$ to, say, $(log r_n)^{1+\delta}$, with $\delta > 0$ and still have $\Phi(n,0) \xrightarrow[n\to\infty]{} 0$. The following example shows that this is not true in general.

Example 4.1 [CG1]. Let $\{w_n, \mathcal{F}_n\}$ be an m-dimensional martingale difference sequence with

$$\sup_i E(\|w_i\|^2 | \mathcal{F}_{i-1}) < \infty,$$

$$\lim_{n\to\infty} \frac{1}{n} \sum_{i=1}^{n} w_i w_i^{\tau} = r > 0$$

and let $\{v_n\}$ be a sequence of m-dimensional iid random vectors independent of $\{w_n\}$ with $Ev_n = 0$, $Ev_n v_n^{\tau} = I$, $\|v_n\| \le v$ where v is a constant.

Define

$$\psi_n^{\tau} = [y_n^{\tau} \quad u_n^{\tau}],$$

where

$$y_n = A_1 y_{n-1} + w_n, \qquad \text{with } A_1 \text{ stable}, \tag{4.138}$$

$$u_n = \frac{v_n}{log^{(1+\delta)/2}(n+1)}. \tag{4.139}$$

We now show that

$$\frac{\mu_{max}(n)}{\mu_{min}(n)} \leq c\,log^{1+\delta}(n+1) \quad a.s. \quad \forall n \geq 1 \tag{4.140}$$

but

$$\Phi(n,0) \not\to 0 \quad a.s. \tag{4.141}$$

We first prove that

$$\lim_{n\to\infty} \frac{1}{n+1}(log(n+1))^{1+\delta}\mu_{min}(n) > 0. \tag{4.142}$$

If (4.142) were not true, then there would exist $\{\alpha_{n_k}\}$ and $\{\beta_{n_k}\}$ such that

$$\|\alpha_{n_k}\|^2 + \|\beta_{n_k}\|^2 = 1 \tag{4.143}$$

and

$$\frac{1}{n_k+1} log^{1+\delta}(n_k+1) \sum_{i=1}^{n_k} (\alpha_{n_k}^T y_i + \beta_{n_k}^T u_i)^2 \xrightarrow[k\to\infty]{} 0. \tag{4.144}$$

By (4.143) and Theorem 2.8 we know

$$\begin{aligned} \left|\sum_{i=1}^{n_k} \alpha_{n_k}^T y_i \beta_{n_k}^T u_i\right| &= \left|\alpha_{n_k}^T \left(\sum_{i=1}^{n_k} y_i u_i^T\right) \beta_{n_k}\right| \\ &\leq \left|\sum_{i=1}^{n_k} y_i u_i^T\right| = O\left(\left(\sum_{i=1}^{n_k} \|y_i\|^2\right)^{\beta}\right) \end{aligned} \tag{4.145}$$

where $1 > \beta > \frac{1}{2}$. By stability assumption of A_1 there exist constants $\rho \in (0,1)$ and $c_1 > 0$ such that

$$\|A_1^k\| \leq c_1 \rho^k, \qquad \forall k \geq 0. \tag{4.146}$$

Then from (4.138) we have

$$y_n = A_1^n y_0 + \sum_{i=1}^{n} A_1^{n-i} w_i, \tag{4.147}$$

and hence by (4.146) and the Schwarz inequality

$$\sum_{i=1}^{n} \|y_i\|^2 \leq \sum_{i=1}^{n} \left[2\|y_0\|^2 c_1^2 \rho^{2i} + 2c_1^2 \left(\sum_{j=1}^{i} \rho^{i-j} \|w_j\| \right)^2 \right]$$

$$\leq \quad 2c_1^2\|y_0\|^2\sum_{i=1}^{n}\rho^{2i} + 2c_1^2\sum_{i=1}^{n}\sum_{k=1}^{i}\rho^{i-k}\sum_{j=1}^{i}\rho^{i-j}\|w_j\|^2$$

$$\leq \quad c_2 + c_3\sum_{j=1}^{n}\|w_j\|^2 = O(n) \qquad (4.148)$$

where c_2 and c_3 denote constants.

Estimates (4.145) and (4.148) imply that

$$\left|\sum_{i=1}^{n_k}\alpha_{n_k}^{\tau}y_i\beta_{n_k}^{\tau}u_i\right| \leq c_4 n_k^{\beta} \qquad (4.149)$$

where $c_4 > 0$ possibly depends upon ω.

From (4.144) and (4.149), it follows that

$$\frac{1}{n_k+1}log^{1+\delta}(n_k+1)\left[\alpha_{n_k}^{\tau}\sum_{i=1}^{n_k}y_iy_i^{\tau}\alpha_{n_k} + \beta_{n_k}^{\tau}\sum_{i=1}^{n_k}u_iu_i^{\tau}\beta_{n_k}\right] \xrightarrow[k\to\infty]{} 0. \qquad (4.150)$$

By Theorem 2.7 and the boundedness of v_n we know that the martingale

$$\sum_{i=1}^{n}\frac{1}{i^{\gamma}}\left[u_iu_i^{\tau} - \frac{1}{log^{1+\delta}(i+1)}I\right]$$

converges a.s. for any $\gamma > \frac{1}{2}$.

Then by Lemma 2.4 it follows that

$$\frac{1}{n^{\gamma}}\sum_{i=1}^{n}\left[u_iu_i^{\tau} - \frac{1}{log^{1+\delta}(i+1)}I\right] \xrightarrow[n\to\infty]{} 0 \quad a.s. \qquad (4.151)$$

Noticing that

$$\sum_{i=1}^{n}\frac{1}{log^{1+\delta}(i+1)}$$

$$= \sum_{i=1}^{n}\int_{i}^{i+1}\frac{dx}{log^{1+\delta}(i+1)} \geq \int_{1}^{n}\frac{dx}{log^{1+\delta}(x+1)}$$

$$= \frac{(n+1)}{log^{1+\delta}(n+1)} - \frac{2}{log^{1+\delta}2} + (1+\delta)\int_{1}^{n}\frac{dx}{log^{2+\delta}(x+1)}, \qquad (4.152)$$

we then from (4.151) and (4.152) find that

$$\begin{aligned}
&\liminf_{n\to\infty} \lambda_{min}\left(\frac{log^{1+\delta}(n+1)}{n}\sum_{i=1}^{n} u_i u_i^T\right) \\
= &\liminf_{n\to\infty}\left[\lambda_{min}\left(\frac{log^{1+\delta}(n+1)}{n}\sum_{i=1}^{n}\left(u_i u_i^T - \frac{1}{log^{1+\delta}(i+1)}I\right)\right)\right. \\
&\left. + \frac{log^{1+\delta}(n+1)}{n+1}\sum_{i=1}^{n}\frac{1}{log^{1+\delta}(i+1)}I\right] \geq 1
\end{aligned}$$

where $\lambda_{min}(A)$ denotes the minimum eigenvalue of a matrix A.

This together with (4.143) and (4.150) implies that

$$\beta_{n_k} \xrightarrow[k\to\infty]{} 0, \qquad \alpha_{n_k} \xrightarrow[k\to\infty]{} 1. \tag{4.153}$$

Further, we note that by (4.148) and Theorem 2.8

$$\left\|\sum_{i=1}^{n} A_1 y_{i-1} w_i^T\right\| = O\left(\left(\sum_{i=1}^{n}\|y_{i-1}\|^2\right)^{\beta}\right) = O(n^{\beta}), \quad \forall \beta \in (\frac{1}{2}, 1)$$

and by (4.138)

$$y_i y_i^T = A_1 y_{i-1} y_{i-1}^T A_1^T + A_1 y_{i-1} w_i^T + w_i y_{i-1}^T A_1^T + w_i w_i^T.$$

Hence, we have

$$\begin{aligned}
&\liminf_{n\to\infty} \lambda_{min}\left(\frac{1}{n}\sum_{i=1}^{n} y_i y_i^T\right) \\
= &\liminf_{n\to\infty} \lambda_{min}\left(\frac{1}{n}\sum_{i=1}^{n}(A_1 y_{i-1} y_{i-1}^T A_1^T + w_i w_i^T)\right) \\
\geq &\liminf_{n\to\infty} \lambda_{min}\left(\frac{1}{n}\sum_{i=1}^{n} w_i w_i^T\right) \\
= &\lambda_{min}(R) > 0.
\end{aligned} \tag{4.154}$$

From this and (4.153) we find that for sufficiently large k

$$\begin{aligned}
&\frac{1}{n_k+1} log^{1+\delta}(n_k+1)\alpha_{n_k}^T \sum_{i=1}^{n_k} y_i y_i^T \alpha_{n_k}^T \\
\geq &\frac{1}{2}\lambda_{min}(R)\|\alpha_{n_k}\|^2 log^{1+\delta}(n_k+1) \xrightarrow[k\to\infty]{} \infty,
\end{aligned}$$

which contradicts (4.150). Thus (4.142) is verified.

By boundedness of u_n and (4.148) we know that

$$\mu_{max}(n) = O\left(\sum_{i=1}^{n} \|\psi_i\|^2\right) = O\left(\sum_{i=1}^{n}(\|y_i\|^2 + \|u_i\|^2)\right) = O(n).$$

This incorporating (4.142) shows that for ψ_n defined by (4.138) and (4.139) Condition (4.140) is really satisfied.

It remains to prove (4.141).

Assume the converse were true, i.e.

$$\Phi(n,0) \xrightarrow[n \to \infty]{} 0. \tag{4.155}$$

Given an arbitrary $2m \times m$ matrix θ_0 we recursively define

$$\theta_{n+1} = \theta_n + \frac{a\psi_n}{r_n}(y_{n+1}^{\tau} - \psi_n^{\tau}\theta_n), \qquad r_n = 1 + \sum_{i=1}^{n} \|\psi_i\|^2. \tag{4.156}$$

Under the converse assumption (4.155) we now show that θ_n defined by (4.156) converges to

$$\theta^x = [-A_1 \quad x]^{\tau} \tag{4.157}$$

with x being an arbitrary $m \times m$ matrix. This paradox will prove impossibility of (4.155).

Set

$$\psi_n^{u\tau} = [\underbrace{0}_{m} \quad u_n^{\tau}].$$

Then by (4.138) y_{n+1} can be written as

$$y_{n+1} = \theta^{x\tau}(\psi_n - \psi_n^u) + w_{n+1}.$$

Hence, (4.156) becomes

$$\theta_{n+1} = \theta_n + \frac{a\psi_n}{r_n}(\psi_n^{\tau}\theta^x - \psi_n^{u\tau}\theta^x - \psi_n^{\tau}\theta_n + w_{n+1}^{\tau})$$

or

$$\tilde{\theta}_{n+1} = \tilde{\theta}_n + \frac{a\psi_n}{r_n}(\psi_n^{\tau}\tilde{\theta}_n - \psi_n^{u\tau}\theta^x + w_{n+1}^{\tau})$$

for estimation errors

$$\tilde{\theta}_n \triangleq \theta^x - \theta_n.$$

Similar to (4.74) we find

$$\tilde{\theta}_{n+1} = \Phi(n+1,0)\tilde{\theta}_0 + a\sum_{j=0}^{n} \Phi(n+1,j+1)\frac{\psi_j\psi_j^{u\tau}}{r_j}\theta^x$$
$$+ a\sum_{j=0}^{n} \Phi(n+1,j+1)\frac{\psi_j w_{j+1}^{\tau}}{r_j}. \tag{4.158}$$

We now prove that all three terms on the right-hand side of (4.158) tend to 0 as $n \to \infty$, if (4.155) is true. Under (4.155), obviously the first term goes to 0, and as proved in Theorem 4.3 (see (4.99)) the last term of (4.158) also vanishes as $n \to \infty$.

Finally, for the second term on the right-hand side of (4.158), by Lemma 4.2 we have

$$\begin{aligned}
&\left\| \sum_{j=0}^{n} \Phi(n+1, j+1) \frac{\psi_j \psi_j^{u\tau}}{r_j} \theta^x \right\| \\
\leq &\left\| \sum_{j=0}^{N} \Phi(n+1, j+1) \frac{\psi_j \psi_j^{u\tau}}{r_j} \theta^x \right\| + \left\| \sum_{j=N+1}^{n} \Phi(n+1, j+1) \frac{\psi_j \psi_j^{u\tau}}{r_j} \theta^x \right\| \\
\leq &\left\| \sum_{j=0}^{N} \Phi(n+1, j+1) \frac{\psi_j \psi_j^{u\tau}}{r_j} \theta^x \right\| \\
&+ \left(\sum_{j=N+1}^{n} \frac{\|\Phi(n+1, j+1)\psi_j\|^2}{r_j} \right)^{\frac{1}{2}} \left(\sum_{j=N+1}^{n} \frac{\|\psi_j^u\|^2}{r_j} \right)^{\frac{1}{2}} \|\theta^x\| \\
\leq &\left\| \sum_{j=0}^{N} \Phi(n+1, j+1) \frac{\psi_j \psi_j^{u\tau}}{r_j} \theta^x \right\| \\
&+ \sqrt{\frac{2m}{a}} \|\theta^x\| \left(\sum_{j=N+1}^{n} \frac{v^2}{r_j \log^{1+\delta}(j+1)} \right)^{\frac{1}{2}} \xrightarrow[\substack{n \to \infty \\ N \to \infty}]{} 0,
\end{aligned}$$

because (4.155) means that

$$r_n \geq \sum_{i=1}^{n} \|y_i\|^2 \geq \frac{n}{2} \lambda_{min}(R)$$

for sufficiently large n and

$$\sum_{j=1}^{\infty} \frac{v^2}{r_j \log^{1+\delta}(j+1)}$$

is a convergent series.

Thus, we have shown the convergence $\theta_n \xrightarrow[n \to \infty]{} \theta^x$. However, this is impossible because θ_n is free of x and x is arbitrary. Hence (4.155) is false.

Q.E.D.

4.5 Convergence Rate of SG

In this section we are restricted to the case where the noise in system (3.81)-(3.84) is uncorrelated, i.e. $C(z) \equiv I$. We are planning to establish the convergence rate of θ_n given by (4.22) and (4.23) with $a = 1$ in terms of r_n [CG11].

In the present case the estimation error given by (4.71) turns to

$$\tilde{\theta}_{n+1} = \Phi(n+1,0)\tilde{\theta}_0 - \sum_{j=0}^{n} \Phi(n+1,j+1)\frac{\psi_j w_{j+1}^\tau}{r_j}. \tag{4.159}$$

So it is natural to express $\tilde{\theta}_n$ via $||\Phi(n,0)||$ first, then to estimate $||\Phi(n,0)||$ by r_n.

Lemma 4.4. *The following estimates take place:*

1). $\frac{1}{r_n} \leq ||\Phi(n+1,0)||^h$, $\forall n \geq 1$ *(h is the dimension of ψ_n).* (4.160)

2). $||\Phi(n,k+1)|| \leq ||\Phi(n,0)||r_k$, $\forall n \geq k \geq 0$. (4.161)

3). $\displaystyle\sum_{j=n+1}^{\infty} \frac{||\psi_j||^2}{r_j^{1+\delta}} \leq \frac{1}{\delta} r_n^{-\delta}$, $\forall n \geq 0$, $\forall \delta > 0$. (4.162)

Proof. When $a = 1$ from (4.100) it follows that

$$det\Phi(n+1,0) = \prod_{i=0}^{n}\left(1 - \frac{||\psi_j||^2}{r_j}\right) = \prod_{i=1}^{n}\frac{r_{i-1}}{r_i}\frac{1}{r_0} = \frac{1}{r_n}$$

and

$$\frac{1}{r_n^2} = det[\Phi(n+1,0)\Phi^\tau(n+1,0)] \leq ||\Phi(n+1,0)||^{2h}$$

which proves 1).

From the following chain of inequalities and equalities we immediately derive 2):

$$\begin{aligned}\|\Phi(n,k+1)\| &\le \|\Phi(n,0)\|\cdot\|\Phi(k+1,0)\|^{-1}\\ &\le \|\Phi(n,0)\|\prod_{i=1}^{k+1}\|\Phi^{-1}(i,i-1)\|\\ &= \|\Phi(n,0)\|\prod_{i=1}^{k+1}\left\|\left(I-\frac{\psi_{i-1}\psi_{i-1}^{\tau}}{r_{i-1}}\right)^{-1}\right\|\\ &= \|\Phi(n,0)\|\prod_{i=2}^{k+1}\frac{r_{i-1}}{r_{i-2}}r_0=\|\Phi(n,0)\|r_k.\end{aligned}$$

Finally, the following calculation leads to 3):

$$\begin{aligned}\sum_{j=n+1}^{\infty}\frac{\|\psi_j\|^2}{r_j^{1+\delta}} &= \sum_{j=n+1}^{\infty}\int_{r_{j-1}}^{r_j}\frac{dx}{r_j^{1+\delta}}\\ &\le \sum_{j=n+1}^{\infty}\int_{r_{j-1}}^{r_j}\frac{dx}{x^{1+\delta}}\le\int_{r_n}^{\infty}\frac{dx}{x^{1+\delta}}=\frac{1}{\delta}r_n^{-\delta}.\end{aligned}$$

Q.E.D.

Lemma 4.5. *Assume (4.77) and (4.79) hold. Then*

$$\|\tilde{\theta}_n\|=O(\|\Phi(n,0)\|^{\delta/(1+\delta)}),\qquad \forall\delta\in(0,\frac{1}{2}).$$

Proof. Set

$$\alpha(t)=max\{k:\ r_k\le t\},\qquad t\ge 0;$$

$$\lambda(n)=\alpha(\|\Phi(n,0)\|^{-1/(1+\delta)}),\qquad n\ge 0.$$

By Lemma 4.4 we have

$$\|\Phi(n+1,\lambda(n+1)+1)\|\le\|\Phi(n,0)\|r_{\lambda(n+1)}\le\|\Phi(n+1,0)\|^{\frac{\delta}{1+\delta}}. \quad (4.163)$$

We recall that in (4.98) we have obtained ($a=1$)

$$\begin{aligned}&\left\|\sum_{j=0}^{n}\Phi(n+1,j+1)\frac{\psi_j w_{j+1}^{\tau}}{r_j}\right\|\\ &\le \|\tilde{s}_n\|+\|\Phi(n+1,0)s\|+c\sum_{j=0}^{n}\|\Phi(n+1,j+1)\|\frac{\|\psi_j\|^2}{r_j^{1+\delta}}. \quad (4.164)\end{aligned}$$

Applying estimates (4.97), (4.160), (4.161) and (4.163) to (4.164) leads to

$$\left\|\sum_{j=0}^{n}\Phi(n+1,j+1)\frac{\psi_j w_{j+1}^{\tau}}{r_j}\right\|$$

$$
\begin{aligned}
&\le cr_n^{-\delta} + \|\Phi(n+1,0)s\| \\
&\quad + c\sum_{j=0}^{\lambda(n)} \|\Phi(n+1,\lambda(n+1)+1)\|\|\Phi(\lambda(n+1)+1,j+1)\|\frac{\|\psi_j\|^2}{r_j^{1+\delta}} \\
&\quad + \sum_{j=\lambda(n+1)+1}^{n} \|\Phi(n+1,j+1)\|\frac{\|\psi_j\|^2}{r_j^{1+\delta}} \\
&\le c\|\Phi(n+1,0)\|^{h\delta} + \|\Phi(n+1,0)s\| \\
&\quad + c\|\Phi(n,0)\|^{\delta/(1+\delta)}\sum_{j=0}^{\infty}\frac{\|\psi_j\|^2}{r_j^{1+\delta}} + \frac{\|\psi_{\lambda(n+1)+1}\|^2}{r_{\lambda(n+1)+1}^{1+\delta}} + \frac{1}{\delta}r_{\lambda(n+1)+1}^{-\delta} \\
&= O(\|\Phi(n,0)\|^{\delta/(1+\delta)}), \qquad (4.165)
\end{aligned}
$$

where for the last equality we have used the following estimate

$$r_{\lambda(n+1)+1} = r_{\alpha(\|\Phi(n+1,0)\|^{-1/(1+\delta)})+1} > \|\Phi(n+1,0)\|^{-1/(1+\delta)}.$$

Combining (4.159) and (4.165) yields the desired result. Q.E.D.

Theorem 4.6. *Suppose that (4.77) and (4.79) hold and that* $r_n \xrightarrow[n\to\infty]{} \infty$ *and* $\sup_n \frac{r_n}{r_{n-1}} < \zeta$, *where* ζ *possibly depends upon* ω.

Then as $n \to \infty$

1) $\|\tilde{\theta}_n\| = O(r_n^{-\delta_1})$ *a.s. with* $\delta_1 > 0$ *if* $\mu_{max}(n)/\mu_{min}(n) \le \gamma < \infty$, (4.166)
where $\mu_{max}(n)$ *and* $\mu_{min}(n)$ *denote the largest and smallest eigenvalue of* $\sum_{i=0}^{n}\psi_i\psi_i^\tau + \frac{1}{h}I$ *respectively.*

2) $\|\tilde{\theta}_n\| = O((logr_n)^{-\delta_2})$ *a.s. with* $\delta_2 > 0$ *if* $\mu_{max}(n)/\mu_{min}(n) \le M(logr_n)^{1/4}$, $\forall n \ge N_0$, (4.167)
where γ, M *and* N_0 *are all positive and possibly depend on* ω.

Proof. 1). Similar to (4.115) and (4.116) define

$$n(t) = max\{k : \quad \tau_k \le t\}, \qquad t \ge 0,$$

where

$$\tau_k = \sum_{i=0}^{k-1}\frac{\|\psi_i\|^2}{r_i}, \qquad t_0 = 0.$$

Let λ_k be the biggest eigenvalue of

$$\Phi^\tau(n(N+k\alpha), n(N+(k-1)\alpha))\Phi(n(N+k\alpha), n(N+(k-1)\alpha))$$

and let $x_{n(N+(k-1)\alpha)}$ be the corresponding unit eigenvector.

For $i \in [n(N+(k-1)\alpha), n(N+k\alpha)-1]$ define x_i by (4.134) with $a=1$, then the estimate similar to (4.136) yields

$$\begin{aligned}
\|x_i - x_{n(N+(k-1)\alpha)}\| &= \|\sum_{j=n(N+(k-1)\alpha)}^{i-1} \frac{\|\psi_j\psi_j^T}{r_j} x_j\| \\
&\le \left(\sum_{j=n(N+(k-1)\alpha)}^{n(N+k\alpha)-1} \frac{\|\psi_j\|^2}{r_j}\right)^{\frac{1}{2}} \left(\sum_{j=n(N+(k-1)\alpha)}^{n(N+k\alpha)-1} \frac{\|\psi_j^T x_j\|^2}{r_j}\right)^{\frac{1}{2}} \\
&\le \sqrt{1+\alpha}\sqrt{1-\lambda_k}. \qquad (4.168)
\end{aligned}$$

Using (4.166) we now show that there are α, β, $N \in (0,\infty)$ such that

$$\sum_{j=n(t)}^{n(t+\alpha)-1} \frac{\psi_j\psi_j^T}{r_j} \ge \beta I, \qquad \forall t \ge N. \qquad (4.169)$$

As a matter of fact, summing by parts and applying (4.166) yield

$$\begin{aligned}
&\sum_{j=n(t)}^{n(t+\alpha)-1} \frac{\psi_j\psi_j^T}{r_j} \\
\ge\ & \sum_{i=n(t)}^{n(t+\alpha)} \frac{1}{r_i}\left(\sum_{j=0}^{i}\psi_j\psi_j^T + \frac{1}{h}I - \sum_{j=0}^{i-1}\psi_j\psi_j^T - \frac{1}{h}I\right) - I \\
\ge\ & \sum_{i=n(t)+1}^{n(t+\alpha)} \left(\sum_{j=0}^{i-1}\psi_j\psi_j^T + \frac{1}{h}I\right)\frac{\|\psi_i\|^2}{r_{i-1}r_i} - 2I \\
\ge\ & \sum_{i=n(t)+1}^{n(t+\alpha)} \frac{1}{h\mu_{max}(i-1)}\left(\sum_{j=0}^{i-1}\psi_j\psi_j^T + \frac{1}{h}I\right)\frac{\|\psi_i\|^2}{r_i} - 2I \\
\ge\ & \frac{1}{h\gamma}\sum_{i=n(t)+1}^{n(t+\alpha)} \frac{\|\psi_i\|^2}{r_i} I - 2I \\
\ge\ & \left(\frac{\alpha-1}{h\gamma} - 2\right) I
\end{aligned}$$

which verifies (4.169).

By.(4.168) and (4.169) we then have

$$\begin{aligned}
\beta \leq\ & x^T_{n(N+(k-1)\alpha)} \sum_{j=n(N+(k-1)\alpha)}^{n(N+k\alpha)-1} \frac{\psi_j \psi_j^T}{r_j} x_{n(N+(k-1)\alpha)} \\
\leq\ & \sum_{j=n(N+(k-1)\alpha)}^{n(N+k\alpha)-1} \frac{\|\psi_j\|^2}{r_j} \|x_{n(N+(k-1)\alpha)} - x_j\| \\
& + \sum_{j=n(N+(k-1)\alpha)}^{n(N+k\alpha)-1} \frac{\|\psi_j\| \|\psi_j^T x_j\|}{r_j} \\
\leq\ & (1+\alpha)\sqrt{1+\alpha}\sqrt{1-\lambda_k} + \sqrt{1+\alpha}\sqrt{1-\lambda_k} \\
=\ & \sqrt{1+\alpha}(\alpha+2)\sqrt{1-\lambda_k}
\end{aligned}$$

which implies

$$\lambda_k \leq 1 - \frac{\beta^2}{(\alpha+1)(\alpha+2)^2} \tag{4.170}$$

and

$$\begin{aligned}
& \|\Phi(n(N+k\alpha), n(N+(k-1)\alpha))\| \\
\leq\ & \left(1 - \frac{\beta^2}{(\alpha+1)(\alpha+2)^2}\right)^{\frac{1}{2}} \triangleq \rho < 1,
\end{aligned} \tag{4.171}$$

$$\begin{aligned}
& \|\Phi(n(N+k\alpha), 0)\| \\
\leq\ & \|\Phi(n(N), 0)\| \prod_{i=1}^{k} \|\Phi(n(N+i\alpha), n(N+(i-1)\alpha))\| \leq \rho^k.
\end{aligned} \tag{4.172}$$

Since $r_k \xrightarrow[k\to\infty]{} \infty$, from the definition of τ_k it follows that

$$\begin{aligned}
\tau_k &= \sum_{i=0}^{k-1} \int_{r_{i-1}}^{r_i} \frac{dx}{r_i} \geq \frac{1}{\zeta} \sum_{i=0}^{k-1} \int_{r_{i-1}}^{r_i} \frac{dx}{r_{i-1}} \\
&\geq \frac{1}{\zeta} \sum_{i=0}^{k-1} \int_{r_{i-1}}^{r_i} \frac{dx}{x} = \frac{1}{\zeta} \log r_{k-1} \xrightarrow[k\to\infty]{} \infty,
\end{aligned} \tag{4.173}$$

and hence $n(t) < \infty$, $\forall t \geq 0$, where by setting $r_{-1} = 1$.

Therefore, for any $n \geq n(N+\alpha)$ there exists a positive constant k such that

$$n(N+k\alpha) \leq n \leq n(N+(k+1)\alpha). \tag{4.174}$$

By the monotonicity of t_n we have

$$\tau_n \leq \tau_{n(N+(k+1)\alpha)} \leq N + (k+1)\alpha$$

and then

$$k \geq \frac{\tau_n - N - \alpha}{\alpha}. \tag{4.175}$$

By (4.172)-(4.175) it follows that

$$\|\Phi(n,0)\| \leq \|\Phi(n(N+k\alpha),0)\| \leq \rho^k$$
$$\leq \rho^{\frac{\tau_n - N - \alpha}{\alpha}} \leq \left(\frac{1}{\rho}\right)^{\frac{N+\alpha}{\alpha}} \cdot \rho^{\frac{\tau_n}{\alpha}} \leq \rho^{-\frac{N+\alpha}{\alpha}} \cdot \rho^{\frac{\log r_{n-1}}{\alpha\zeta}} = c_0 r_{n-1}^{-\alpha_1} \tag{4.176}$$

where $c_0 = \rho^{-\frac{N+\alpha}{\alpha}}$, $\alpha_1 = \dfrac{|\log \rho|}{\alpha\zeta} > 0$.

From (4.176) and Lemma 4.5 we conclude that the first part of the theorem is true for $\delta_1 = \frac{\alpha_1 \delta}{1+\delta}$, $\forall \delta \in (0, \frac{1}{2})$.

2). We recall (4.137) for the case $\lambda = 1$

$$\|\Phi(m(N+k\alpha), m(N+(k-1)\alpha))\| \leq 1 - a_k, \qquad k > k_0,$$

where $a_k = \frac{c}{2k}$.

Noticing $\sum_{i=1}^{k} \frac{1}{i} \geq log(k+1)$ and $1 \cdot x \leq e^{-x}$, $\forall x \geq 0$ we then have

$$\|\Phi(m(N+k\alpha),0)\| \leq \|\Phi(m(N+k_0\alpha),0)\| \prod_{i=k_0+1}^{k} \left(1 - \frac{c}{2i}\right)$$
$$\leq exp\left\{-\frac{c}{2} \sum_{i=k_0+1}^{k} \frac{1}{i}\right\} \leq exp\left\{\frac{c}{2}\sum_{i=1}^{k_0} \frac{1}{i}\right\} exp[-\frac{c}{2} log(k+1)]$$
$$\leq c_1 (k+1)^{-\frac{c}{2}} \tag{4.177}$$

where $c_1 = exp\left\{\frac{c}{2} \sum_{i=1}^{k_0} \frac{1}{i}\right\}$.

From (4.119) we know that $t_n \xrightarrow[n\to\infty]{} \infty$, and hence $m(t) < \infty$, $\forall t \geq 0$. Then for any $n \geq m(N+\alpha)$ there exists some $k \geq 1$ such that

$$m(N+k\alpha) \leq n \leq m(N+(k+1)\alpha).$$

Then

$$t_n \leq t_{m(N+(k+1)\alpha)} \leq N + (k+1)\alpha$$

and therefore

$$k \geq \frac{t_n - N - \alpha}{\alpha}. \tag{4.178}$$

Further, from (4.116) for t_n we have its lower bound

$$\begin{aligned} t_n &= \sum_{i=N_0}^{n-1} \frac{\|\psi_i\|^2}{r_i(log r_{i-1})^{1/4}} \geq \frac{1}{\zeta} \sum_{i=N_0}^{n-1} \frac{\|\psi_i\|^2}{r_{i-1}(\log r_{i-1})^{1/4}} \\ &\geq \frac{1}{\zeta} \int_{r_{N_0-1}}^{r_{n-1}} \frac{dx}{x(log x)^{1/4}} = \frac{4}{3\zeta}\left[log^{\frac{3}{4}} r_{n-1} - log^{\frac{3}{4}} r_{N_0-1}\right]. \end{aligned} \tag{4.179}$$

Finally, from (4.177)-(4.178) we derive as $n \to \infty$

$$\begin{aligned} \|\Phi(m(N+k\alpha),0)\| &\leq c_1 \left(\frac{t_n - N}{\alpha}\right)^{-\frac{c}{2}} \\ &\leq c_1 \alpha^{\frac{c}{2}} \left\{\frac{4}{3\zeta}\left[log^{\frac{3}{4}} r_{n-1} - log^{\frac{3}{4}} r_{N_0-1}\right] - N\right\}^{-\frac{c}{2}} \\ &= O\left(\{log r_{n-1}\}^{-\delta_2}\right) = O\left(\{log r_n\}^{-\delta_2}\right) \end{aligned} \tag{4.180}$$

with some positive δ_2.

From (4.180) and Lemma 4.5 the second conclusion of the theorem follows immediately. Q.E.D.

4.6 Removing the SPR Condition By An Overparameterization Technique

Up to now, we have exclusively used the SPR assumption on the noise model when an ARMAX model is identified. In particular, when the ELS algorithm is applied, we assumed that

$$C^{-1}(e^{i\lambda}) + C^{-\tau}(e^{-i\lambda}) - I > 0, \quad \forall \lambda \in [0, 2\pi]. \tag{4.181}$$

Qualitatively, this condition means that the system noise $\{C(z)w_n\}$ is not too "colored". Also, it implies the stability of the polynomial matrix $C(z)$ as shown in corollary 4.1.

Let us now rearrange Inequality (4.181) as

$$\begin{aligned} I &> I - C(e^{i\lambda}) - C^{\tau}(e^{-i\lambda}) + C(e^{i\lambda})C^{\tau}(e^{-i\lambda}) \\ &= [I - C(e^{i\lambda})][I - C^{\tau}(e^{-i\lambda})]. \end{aligned}$$

Then integrating both sides from 0 to 2π, we have

$$\|[C_1, \ldots, C_r]\| < 1.$$

which in the scalar case means

$$\sum_{i=1}^{r} C_i^2 < 1.$$

So, it is immediately seen that a stable polynomial may not satisfy the SPR condition (4.181). The following $C(z)$ may serve as an example of such a polynomial:

$$C(z) = (1 - 0.3z)(1 - 0.4z)(1 - 0.5z)$$

The SPR condition (4.181) though is very restrictive, there are examples [LSo] showing that the ELS algorithm does not converge if it fails. This suggests we reexamine the construction of the ELS algorithm. We find that the possible reason why the standard ELS algorithm requires a SPR condition for its convergence is that the estimate $\widehat{w}_n$ for w_n is generated by the ELS itself.

Based on this observation, for system (3.91)-(3.94) we now present an identification scheme for whose convergence no SPR condition is imposed on the noise model. This scheme is carried out by two steps: In the first step, estimates for the noise process are formed by using an overparameterized model. In the second step the parameter is estimated by an ELS algorithm with regressors including entries being the noise estimates obtained in the first step. More precisely, the algorithm is described as follows:

Step 1. The noise estimate $\{\widehat{w}_i\}$ is defined by

$$\widehat{w}_i = y_i - \alpha_i^T \psi_{i-1}, \quad i \geq 0 \tag{4.182}$$

with α_i and ψ_i given by

$$\alpha_{i+1} = \alpha_i + b_i P_i \psi_i [y_{i+1} - \psi_i^T \alpha_i] \tag{4.183}$$

$$P_{i+1} = P_i - b_i P_i \psi_i \psi_i^T P_i, \quad b_i = (1 + \psi_i^T P_i \psi_i)^{-1}, \tag{4.184}$$

$$\psi_i = [y_i^T \; ... \; y_{i-\overline{p}+1}^T \quad u_i^T \; ... \; u_{i-\overline{q}+1}^T \quad \widehat{w}_i^T \; ... \; \widehat{w}_{i-\overline{r}+1}^T]^T \tag{4.185}$$

where $\overline{p} > p$, $\overline{q} > q$, $\overline{r} > r$ are suitably large integers and the initial values $\alpha_0 = 0$. $P_0 = \beta I$, $\beta > 0$.

Step 2. The estimate θ_n for the parameter θ given by (4.1) is defined by an extended least squares method with regressors formed by using the noise estimates obtained in Step 1:

$$\theta_n = \left[\sum_{i=0}^{n-1} \varphi_i \varphi_i^T + \beta I\right]^{-1} \sum_{i=0}^{n-1} \varphi_i y_{i+1}^T, \tag{4.186}$$

$$\varphi_i = [y_i^T \; ... \; y_{i-p+1}^T \quad u_i^T \; ... \; u_{i-q+1}^T \quad \widehat{w}_i^T \; ... \; \widehat{w}_{i-r+1}^T]^T. \tag{4.187}$$

We now turn to the convergence aspect of the algorithm (4.186) (4.187). We still use $\lambda_{\max}(n)$ and $\lambda_{\min}(n)$ to denote the maximum and minimum eigenvalues of $\sum_{i=0}^{n} \varphi_i \varphi_i^T + \beta I$, respectively.

Theorem 4.7. *Assume the following conditions are fulfilled for the ARMAX model (3.91)-(3.94):*

1. $\{w_n, \mathcal{F}_n\}$ *is a martingale difference sequence with*

$$\sup_n E[\|w_{n+1}\|^{2+\delta} | \mathcal{F}_n] < \infty, \quad a.s. \text{ for some } \delta \geq 0.$$

2. $\det C(z) \neq 0$, $\forall z$: $|z| \leq 1$, *and* $C^{-1}(z)$ *has the following expansion*

$$C^{-1}(z) = \sum_{i=0}^{\infty} D_i z^i, \quad \forall z : |z| \leq 1$$

with $\|D_i\| \leq M\lambda^i$, $(i \geq 0)$, *for some* $M > 0$ *and* $\lambda \in (0, 1)$.

3. u_n *is* $\mathcal{F}_n$*-measurable.*

Then as $n \to \infty$ *the estimation error produced by the algorithm (4.182)-(4.187) is expressed by*

$$\|\theta_{n+1} - \theta\|^2 = O\left(\frac{\log \lambda_{\max}(n)(\log\log \lambda_{\max}(n))^{\delta(\beta-2)}}{\lambda_{\min}(n)}\right) \ a.s.,$$

provided that in (4.185), $\overline{p} = pk$, $\overline{q} = qk$ *and* $\overline{r} = rk$ *with* k *being any positive integer satisfying*

$$k > \frac{\log\left[\|C(z)\|_\infty^{-1} M^{-1}(1-\lambda)\right]}{|\log \lambda|} - 1,$$

where $\delta(x)$ *is given by (4.35)*

$$\|C(z)\|_\infty = \max_{|z|=1} \lambda_{\max}[C(z)C^T(z^{-1})]$$

and M, λ *are given in Condition 2.*

Proof. Let us consider the overparameterized model

$$D(z)A(z)y_n = D(z)B(z)u_n + D(z)C(z)w_n \tag{4.188}$$

where $D(z)$ is the polynomial matrix

$$D(z) = \sum_{i=0}^{k} D_i z^i,$$

with D_i defined in Condition 2. We now show that $[D(z)C(z)]^{-1} - \frac{1}{2}I$ is SPR. For this it is sufficient (and also necessary) to prove that

$$\|D(z)C(z) - I\|_\infty < 1. \tag{4.189}$$

This can be done by observing that

$$\|D(z)C(z) - I\|_\infty = \left\| \left[C^{-1} - \sum_{i=k+1}^{\infty} D_i z_i \right] C(z) - I \right\|_\infty$$

$$= \left\| \sum_{i=k+1}^{\infty} D_i z^i C(z) \right\|_\infty$$

$$\leq \sum_{i=k+1}^{\infty} \|D_i\| \|C(z)\|_\infty \leq \|C(z)\|_\infty M \sum_{i=k+1}^{\infty} \lambda^i$$

$$\leq \|C(z)\|_\infty M \frac{\lambda^{k+1}}{1-\lambda} < 1.$$

Hence $[D(z)C(z)]^{-1} - \frac{1}{2}I$ is SPR. Then applying the ELS algorithm (4.182)-(4.185) to model (4.188), by (4.62) we have the following estimate:

$$\begin{aligned} \sum_{i=0}^{n} \|\widehat{w}_{i+1} - w_{i+1}\|^2 &= O\left(\log t_n (\log\log t_n)^{\delta(\beta-2)}\right) \\ &= O\left(\log \lambda_{\max}(n) \left(\log\log \lambda_{\max}(n)\right)^{\delta(\beta-2)}\right), \end{aligned} \tag{4.190}$$

where $t_n = 1 + \sum_{i=1}^{n} \|\psi_i\|^2$ with ψ_i defined by (4.185).

This is an important property of the noise estimates produced by the first step of the algorithm. We now turn to the analysis of the second step.

Note that with φ_n^0 defined by (4.17),

$$y_{i+1} = \theta^\tau \varphi_i^0 + w_{i+1} = \theta^\tau \varphi_i + \theta^\tau (\varphi_i^0 - \varphi_i) + w_{i+1}. \tag{4.191}$$

Substituting this into (4.186), we see that

$$\begin{aligned} \theta_n &= \left(\sum_{i=0}^{n-1} \varphi_i \varphi_i^\tau + \beta I \right)^{-1} \\ &\quad \left[\sum_{i=0}^{n-1} \varphi_i \varphi_i^\tau \theta + \sum_{i=0}^{n-1} \varphi_i (\varphi_i^0 - \varphi_i)^\tau \theta + \sum_{i=0}^{n-1} \varphi_i w_{i+1}^\tau \right] \\ &= \theta - \left(\sum_{i=0}^{n-1} \varphi_i \varphi_i^\tau + \beta I \right)^{-1} \beta\theta \end{aligned}$$

$$+\left(\sum_{i=0}^{n-1}\varphi_i\varphi_i^T+\beta I\right)^{-1}\sum_{i=0}^{n-1}\varphi_i(\varphi_i^0-\varphi_i)^T\theta$$
$$+\left(\sum_{i=0}^{n-1}\varphi_i\varphi_i^T+\beta I\right)^{-1}\sum_{i=0}^{n-1}\varphi_i w_{i+1}^T. \tag{4.192}$$

The key issue is to estimate the last two terms on the right-hand side of (4.192). Setting

$$S_n=\left(\sum_{i=0}^{n-1}\varphi_i\varphi_i^T+\beta I\right)^{-1},$$

we then have

$$\begin{aligned}
&\left\|S_n\sum_{i=0}^{n-1}\varphi_i(\varphi_i^0-\varphi_i)^T\right\|^2\\
\leq\ & \|S_n^{\frac{1}{2}}\|^2\left\|\sum_{i=0}^{n-1}S_n^{\frac{1}{2}}\varphi_i(\varphi_i^0-\varphi_i)^T\right\|^2\\
\leq\ & \lambda_{\min}^{-1}(n)\cdot\left\|\sum_{i=0}^{n-1}S_n^{\frac{1}{2}}\varphi_i\varphi_i^T S_n^{\frac{1}{2}}\right\|\cdot\sum_{i=0}^{n-1}\|\varphi_i^0-\varphi_i\|^2\\
\leq\ & O\left(\frac{\log\lambda_{\max}(n)}{\lambda_{\min}(n)}(\log\log\lambda_{\max}(n))^{\delta(\beta-2)}\right)\ \text{a.s.,}
\end{aligned}\tag{4.193}$$

where Property (4.190) has been used.

Finally, for estimating the last term of (4.192), introduce the following linear model

$$z_{i+1}=\alpha^T\varphi_i+w_{i+1}.$$

By Theorem 4.1, the recursive least squares estimate α_n of α satisfies

$$\begin{aligned}
&\left\|\left(\sum_{i=0}^{n-1}\varphi_i\varphi_i^T+\beta I\right)^{-1}\sum_{i=0}^{n-1}\varphi_i w_{i+1}^T\right\|^2\\
=\ & \|\alpha_n-\alpha\|^2+O\left(\frac{1}{\lambda_{\min}(n)}\right)\\
=\ & O\left(\frac{\log\lambda_{\max}(n)}{\lambda_{\min}(n)}(\log\log\lambda_{\max}(n))^{\delta(\beta-2)}\right)\ \text{a.s.}
\end{aligned}\tag{4.194}$$

Substituting (4.193) and (4.194) into (4.192) we get the desired result. This completes the proof of the theorem. Q.E.D.

As in Theorem 4.2, we can translate the excitation condition on φ_i to φ_i^0. Let us keep the notations of Theorem 4.2. Then we have

Theorem 4.8. *Let Conditions of Theorem 4.7 be satisfied. If*

$$\log \lambda_{\max}^0(n) = o(\lambda_{\min}^0(n)), \quad a.s.,$$

then as $n \to \infty$

$$\|\theta_{n+1} - \theta\|^2 = O\left(\frac{\log \lambda_{\max}^0(n)}{\lambda_{\min}^0(n)} \left(\log\log \lambda_{\max}^0(n)\right)^{\delta(\beta-2)}\right) \quad a.s.$$

The proof is exactly the same as that for Theorem 4.2.

To conclude this section we mention that the on-line spectral factorization method may also be used to identify a special class of ARMAX models without requiring the SPR condition (see, e.g. [Mo3] and [GMo]).

4.7 Removing the SPR Condition By Using Increasing Lag Least Squares

In the last section, we have presented an estimation algorithm without requiring any SPR condition for its convergence. However, we need the *a priori* information about the polynomial $C(z)$ in constructing the noise estimates. When such information is not available *a priori*, a natural idea [Du2] is to let the length of the regressor increase with the data size n. Let $\{h_n\}$ be a sequence of nondecreasing positive integers. Introduce the following regression vector for any $n \geq 1$:

$$\psi_i(h_n) = [y_i^\tau \; \dots \; y_{i-h_n+1}^\tau \quad u_i^\tau \; \dots \; u_{i-h_n+1}^\tau]^\tau, \quad 0 \leq i \leq n. \tag{4.195}$$

For any $n \geq 1$, define the noise estimates $\{\widehat{w}_i(n),\ 1 \leq i \leq n\}$ for $\{w_i,\ 1 \leq i \leq n\}$as follows:

$$\widehat{w}_i(n) = y_i - \alpha_i^\tau(n)\psi_{i-1}(h_n), \quad 1 \leq i \leq n \tag{4.196}$$

$$\alpha_{i+1}(n) = \alpha_i(n) + b_i(n)P_i(n)\psi_i(h_n)[y_{i+1} - \psi_i^\tau(h_n)\alpha_i(n)] \tag{4.197}$$

$$P_{i+1}(n) = P_i(n) - b_i(n)P_i(n)\psi_i(h_n)\psi_i^\tau(h_n)P_i(n), \tag{4.198}$$

$$b_i(n) = (1 + \psi_i^\tau(h_n)P_i(n)\psi_i(h_n))^{-1}, \tag{4.199}$$

where the initial values $\alpha_0(n) = 0$ and $P_0(n) = \beta I$, $\beta > 0$.

To analyze properties of $\{\widehat{w}_i(n),\ 1 \leq i \leq n\}$ we need the following lemma, the proof of which is essentially based on the double array martingale theorems developed in Section 2.4.

Lemma 4.6 *[HG], [GHH]. Suppose that* $\{w_i, \mathcal{F}_i\}$ *is an m-dimensional martingale difference sequence satisfying*

$$\sup_i E(\|w_{i+1}\|^2 | \mathcal{F}_i) < \infty, \quad \|w_i\| = o(d(i)) \ a.s. \tag{4.200}$$

where $\{d(i)\}$ is a positive, deterministic and nondecreasing sequence and satisfies $\sup_i d(e^{i+1})/d(e^i) < \infty$.

Let $\psi_j(k) = [x_{j1}^\tau, ..., x_{jk}^\tau]^\tau$ with x_{jt} being $\mathcal{F}_j$-measurable, $t = 1, 2, ...,$ and let $h_n = O([\log n]^\alpha)$, $\alpha > 0$. Set

$$M_i(k) = \sum_{j=1}^{i} \psi_j(k)\psi_j^\tau(k) + \beta I, \quad \beta > 0$$

$$S_i(k) = \sum_{j=1}^{i} \psi_j(k) w_{j+1}^\tau, \quad S_0(k) = 0$$

$$V_i(k) = \|[M_i(k)]^{-\frac{1}{2}} S_i(k)\|^2,$$

$$U_i(k) = \sum_{j=1}^{i} \|\psi_j^\tau(k)[M_j(k)]^{-1} S_j(k)\|^2, \ 1 \le i \le n, \ 1 \le k \le h_n.$$

Then, as $n \to \infty$

$$\max_{1\le k\le h_n} V_n(k) = O(\delta_n) + o\left([d(n)\log\log n]^2\right) \ a.s.$$

$$\max_{1\le k\le h_n} U_n(k) = O(\delta_n) + o\left([d(n)\log\log n]^2\right) \ a.s.$$

where $\delta_n = h_n \log^+ \lambda_{\max}(M_n(h_n))$ and

$$\log^+ x = \begin{cases} \log x, & \text{if } x \ge 1 \\ 0, & \text{if } 0 < x < 1. \end{cases}$$

Proof. Set

$$c_i(k) = \left[1 + \psi_i^\tau(k) M_{i-1}^{-1}(k)\psi_i(k)\right]^{-1}.$$

By the matrix inverse formula (3.62) we know that

$$[M_i(k)]^{-1} = [M_{i-1}(k)]^{-1} - c_i(k)[M_{i-1}(k)]^{-1}\psi_i(k)\psi_i^\tau(k)[M_{i-1}(k)]^{-1}.$$

Consequently we have

$$\begin{aligned} & tr\left\{S_i^\tau(k)[M_i(k)]^{-1}S_i(k)\right\} = tr\left\{S_{i-1}^\tau(k)[M_{i-1}(k)]^{-1}S_{i-1}(k)\right\} \\ & +2c_i(k)\psi_i^\tau(k)[M_{i-1}(k)]^{-1}S_{i-1}(k)w_{i+1} \\ & -c_i(k)\|\psi_i^\tau(k)[M_{i-1}(k)]^{-1}S_{i-1}(k)\|^2 \\ & +c_i(k)\psi_i^\tau(k)[M_{i-1}(k)]^{-1}\psi_i(k)\|w_{i+1}\|^2 \qquad (4.201) \\ \le \ & tr\left\{S_{i-1}(k)[M_{i-1}(k)]^{-1}S_{i-1}(k)\right\} \\ & +2\psi_i^\tau(k)[M_i(k)]^{-1}S_{i-1}(k)w_{i+1} \\ & -\|\psi_i^\tau(k)[M_i(k)]^{-1}S_{i-1}(k)\|^2 \\ & +\psi_i^\tau(k)[M_i(k)]^{-1}\psi_i(k)\|w_{i+1}\|^2. \qquad (4.202) \end{aligned}$$

For any fixed k, summing up from 1 to n we derive

$$\begin{aligned} & V_n(k) + \sum_{i=1}^{n} \|\psi_i^\tau(k)[M_i(k)]^{-1} S_{i-1}(k)\|^2 \\ \leq & \sum_{i=1}^{n} \psi_i^\tau(k)[M_i(k)]^{-1}\psi_i(k)\|w_{i+1}\|^2 \\ & +2\sum_{i=1}^{n} \psi_i^\tau(k)[M_i(k)]^{-1} S_{i-1}(k) w_{i+1}. \end{aligned} \tag{4.203}$$

Similar to (4.50), it is easy to see that

$$\psi_i^\tau(k)[M_i(k)]^{-1}\psi_i(k) = \frac{\det[M_i(k)] - \det[M_{i-1}(k)]}{\det[M_i(k)]} \leq 1,$$

and hence

$$\begin{aligned} & \max_{1\leq k\leq h_n} \sum_{i=1}^{n} \psi_i^\tau(k)[M_i(k)]^{-1}\psi_i(k) \\ \leq & \max_{1\leq k\leq h_n} \int_{\det[M_0(k)]}^{\det[M_n(k)]} x^{-1}\,dx = O(\delta_n) \quad a.s. \end{aligned} \tag{4.204}$$

Thus from Theorem 2.9(ii), (4.200) and (4.204) it follows that

$$\begin{aligned} & \max_{1\leq k\leq h_n} \sum_{i=1}^{n} \psi_i^\tau(k)[M_i(k)]^{-1}\psi_i(k)\|w_{i+1}\|^2 \\ \leq & \max_{1\leq k\leq h_n} \left\{ \sum_{i=1}^{n} \psi_i^\tau(k)[M_i(k)]^{-1}\psi_i(k)[\|w_{i+1}\|^2 - E(\|w_{i+1}\|^2|\mathcal{F}_i)] \right. \\ & \left. + \sum_{i=1}^{n} \psi_i^\tau(k)[M_i(k)]^{-1}\psi_i(k) E(\|w_{i+1}\|^2|\mathcal{F}_i) \right\} \\ = & \; o(d^2(n)\log\log n) + O\left(\max_{1\leq k\leq h_n} \sum_{i=1}^{n} \psi_i^\tau(k)[M_i(k)]^{-1}\psi_i(k) \right) \\ = & \; o(d^2(n)\log\log n) + O(\delta_n) \quad a.s. \end{aligned} \tag{4.205}$$

Let

$$a_n = \left\{ \max_{1\leq k\leq h_n} \sum_{i=1}^{n} \|\psi_i^\tau(k)[M_i(k)]^{-1} S_{i-1}(k)\|^2 \right\}^{\frac{1}{2}}.$$

Then by (4.203), Theorem 2.10 and (4.205) we obtain that

$$\begin{aligned} & \max_{1\leq k\leq h_n} V_n(k) + [a_n]^2 \\ = & \; o(d^2(n)\log\log n) + O(\delta_n) + o(a_n d(n)\log\log n) + O(a_n \log n) \\ \leq & \; O(\delta_n) + o([a_n]^2) + o([d(n)\log\log n]^2) \quad a.s. \end{aligned}$$

From this it is easy to conclude the first assertion of the lemma and

$$[a_n]^2 = O(\delta_n) + o([d(n)\log\log n]^2) \quad a.s. \tag{4.206}$$

Finally, the second assertion follows from (4.205), (4.206) and the following inequality

$$\max_{1\le k\le h_n} U_n(k) \le 2[a_n]^2 + 2\max_{1\le k\le h_n}\sum_{i=1}^{n}\psi_i^T(k)[M_i(k)]^{-1}\psi_i(k)\|w_{i+1}\|^2.$$

This completes the proof of the lemma. Q.E.D.

From the proof of Theorem 4.6 we see that the following is also true.

Corollary 4.2. *If in Lemma 4.6 $\psi_j(n)$ is assumed to be any h_n-dimensional and $\mathcal{F}_j$-measurable vector sequence, then*

$$V_n(n) = O(\delta_n) + o([d(n)\log\log n]^2) \quad a.s.$$

Lemma 4.7. *Let $\{w_i\}$ and $\{\psi_i(n)\}$ be any m- and h_n-dimensional random sequences respectively. Then the following estimates take place:*

(i) $$\sum_{i=0}^{k}\psi_i^T(n)[M_i(n)]^{-1}\psi_i(n) \le h_n \log^+ \lambda_{\max}(M_k(n)) \ a.s., \forall k, \forall n.$$

(ii) $$V_n(n) \le \sum_{j=1}^{n}\|w_{j+1}\|^2$$

where the notations are the same as in Lemma 4.6.

Proof. The assertion (i) directly follows from (4.204). To prove the second assertion, applying the inequality $2ab \le a^2 + b^2$ to the second term on the right-hand side of (4.201), we see that

$$\begin{aligned} &2c_i(k)\psi_i^T(k)[M_{i-1}(k)]^{-1}S_{i-1}(k)w_{i+1} \\ \le\ & c_i(k)\|\psi_i^T(k)[M_{i-1}(k)]^{-1}S_{i-1}(k)\|^2 + c_i(k)\|w_{i+1}\|^2. \end{aligned}$$

Substituting this inequality into (4.201) and summing up from 1 to n, we see that (ii) is true. Q.E.D.

We are now in a position to study properties of the noise estimates $\{\widehat{w}_i(n), \le i \le n\}$ generated by (4.196)-(4.199). The following result is based on [GH] and [HG].

Theorem 4.9. *For ARMAX model (3.91)-(3.94), assume that*

(i). $\{w_i, \mathcal{F}_i\}$ *satisfies Conditions in Lemma 4.6, and*

$$\liminf_{n\to\infty} \frac{1}{n} \sum_{n=1}^{n} \|w_i\|^2 \neq 0 \ a.s.$$

(ii). $\det C(z) \neq 0$, $\forall z$: $|z| \leq 1$.

(iii). u_i *is* $\mathcal{F}_i$*-measurable, and there is a constant* $b > 0$ *such that*

$$\sum_{i=0}^{n-1} (\|y_i\|^2 + \|u_i\|^2) = O(n^b) \quad a.s.$$

If h_n *in (4.195) is chosen as* $h_n = O(\{\log n\}^\alpha)$, $\alpha > 1$, *and* $\log n = o(h_n)$, *then as* $n \to \infty$ *the noise estimates* $\{\widehat{w}_i(n), \leq i \leq n\}$ *produced by (4.196)-(4.199) have the following property:*

$$\sum_{i=0}^{n-1} \|\widehat{w}_i(n) - w_i\|^2 = O(h_n \log n) + o(\{d(n) \log\log n\}^2) \ a.s. \tag{4.207}$$

Proof. By Condition (ii), we have the following expansions

$$\begin{cases} C^{-1}(z)A(z) = I + \sum_{i=1}^{\infty} G_i z^i \\ C^{-1}(z)B(z) = \sum_{i=1}^{\infty} H_i z^i. \end{cases} \tag{4.208}$$

Set

$$\alpha(n) = [-G_1 \ \dots \ -G_{h_n} \quad H_1 \ \dots \ H_{h_n}]^\tau, \tag{4.209}$$

$$\varepsilon_i(n) = \sum_{t=h_n+1}^{\infty} [-G_t y_{i-t+1} + H_t u_{i-t+1}] \tag{4.210}$$

Then by (3.91) and (4.208)-(4.210) we see that

$$y_i = \alpha^\tau(n)\psi_{i-1}(n) + \varepsilon_{i-1}(n) + w_i, \tag{4.211}$$

which in conjunction with (4.196) yields

$$\widehat{w}_i(n) - w_i = [\alpha^\tau(n) - \alpha_i^\tau(n)]\psi_{i-1}(n) + \varepsilon_{i-1}(n), \tag{4.212}$$

where and hereafter in this proof for simplicity of notations we write $\psi_{i-1}(n)$ for $\psi_{i-1}(h_n)$.

Noting that for any fixed n, (4.197), (4.198) give the standard least-squares recursion, we have

$$\alpha_i(n) = \left\{ \sum_{j=0}^{i-1} \psi_j(n)\psi_j^\tau(n) + \beta I \right\}^{-1} \left\{ \sum_{j=0}^{i-1} \psi_j(n) y_{j+1}^\tau \right\}.$$

Substituting (4.211) into this leads to

$$\begin{aligned}\alpha_i(n)-\alpha(n) &= \left\{\sum_{j=0}^{i-1}\psi_j(n)\psi_j^T(n)+\beta I\right\}^{-1} \\ &\quad \left\{\sum_{j=0}^{i-1}\psi_j(n)[w_{j+1}^T+\varepsilon_j^T(n)]-\beta\alpha(n)\right\}.\end{aligned}$$

Consequently, by noting

$$\left\|\psi_{i-1}^T(n)\left\{\sum_{j=0}^{i-1}\psi_j(n)\psi_j^T(n)+\beta I\right\}^{-\frac{1}{2}}\right\|\leq 1,$$

we know that

$$\begin{aligned}&\|\psi_{i-1}^T(n)[\alpha_i(n)-\alpha(n)]\|^2 \\ \leq\ & 3\left\|\psi_{i-1}^T(n)\left\{\sum_{j=0}^{i-1}\psi_j(n)\psi_j^T(n)+\beta I\right\}^{-1}\sum_{j=0}^{i-1}\psi_j(n)w_{j+1}^T\right\|^2 \\ &+3\left\|\psi_{i-1}^T(n)\left\{\sum_{j=0}^{i-1}\psi_j(n)\psi_j^T(n)+\beta I\right\}^{-1}\sum_{j=0}^{i-1}\psi_j(n)\varepsilon_j^T(n)\right\|^2 \\ &+3\left\|\psi_{i-1}^T(n)\left\{\sum_{j=0}^{i-1}\psi_j(n)\psi_j^T(n)+\beta I\right\}^{-1}\beta\alpha(n)\right\|^2 \\ \leq\ & 3\left\|\psi_{i-1}^T(n)\left\{\sum_{j=0}^{i-1}\psi_j(n)\psi_j^T(n)+\beta I\right\}^{-1}\sum_{j=0}^{i-1}\psi_j(n)w_{j+1}^T\right\|^2 \\ &+3\left\|\left\{\sum_{j=0}^{i-1}\psi_j(n)\psi_j^T(n)+\beta I\right\}^{-\frac{1}{2}}\sum_{j=0}^{i-1}\psi_j(n)\varepsilon_j^T(n)\right\|^2 \\ &+O\left(\left\|\psi_{i-1}^T(n)\left\{\sum_{j=0}^{i-1}\psi_j(n)\psi_j^T(n)+\beta I\right\}^{\frac{1}{2}}\right\|^2\right).\end{aligned}$$

Summing up from 1 to n, and applying Lemmas 4.6 and 4.7, we get

$$\begin{aligned}
&\sum_{i=1}^{n} \|\psi_{i-1}^{T}(n)[\alpha_i(n) - \alpha(n)]\|^2 \\
&= O\left(h_n \log^+ \lambda_{\max}\left\{ \sum_{j=0}^{n-1} \psi_j(n)\psi_j^T(n) \right\} \right) \\
&+ o([d(n)\log\log n]^2) + O\left(\sum_{i=1}^{n}\sum_{j=0}^{i-1} \|\varepsilon_j(n)\|^2 \right). \qquad (4.213)
\end{aligned}$$

By Condition (ii), (4.208)-(4.210), and the Schwarz inequality it follows that:

$$\begin{aligned}
\sum_{i=0}^{n-1}\|\varepsilon_i(n)\|^2 &\le 2\sum_{i=0}^{n-1}\left\{ \sum_{j=h_n+1}^{\infty}\|G_j\| \sum_{j=h_n+1}^{\infty}\|G_j\|\|y_{i-j+1}\|^2 \right. \\
&\quad \left. + \sum_{j=h_n+1}^{\infty}\|H_j\| \sum_{j=h_n+1}^{\infty}\|H_j\|\|u_{i-j+1}\|^2 \right\} \\
&\le 2\left(\sum_{j=h_n+1}^{\infty}\|G_j\| \right)^2 \sum_{i=0}^{n-1}\|y_i\|^2 \\
&\quad +2\left(\sum_{j=h_n+1}^{\infty}\|H_j\| \right)^2 \sum_{i=0}^{n-1}\|u_i\|^2 \\
&= O\left(\sum_{i=0}^{n-1}[\|y_i\|^2 + \|u_i\|^2]\cdot \exp\{-\lambda h_n\} \right), \qquad (4.214)
\end{aligned}$$

for some $\lambda > 0$.

Similarly, by (3.91), Condition (i), (4.208) and the Schwarz inequality it is easy to verify that

$$\begin{aligned}
0 &\ne \liminf_{n\to\infty}\frac{1}{n}\sum_{i=0}^{n-1}\|w_i\|^2 \\
&\le \liminf_{n\to\infty}\frac{1}{n}\sum_{i=0}^{n-1}\left\{ \sum_{j=0}^{\infty}(\|G_j y_{i-j}\| + \|H_j u_{i-j}\|) \right\}^2 \\
&\le 2\left\{ \sum_{j=0}^{\infty}(\|G_j\| + \|H_j\|) \right\}^2 \liminf_{n\to\infty}\frac{1}{n}\sum_{j=0}^{n-1}(\|y_j\|^2 + \|u_j\|^2)
\end{aligned}$$

$$= O\left(\liminf_{n\to\infty}\frac{1}{n}\sum_{j=0}^{n-1}(\|y_j\|^2+\|u_j\|^2)\right). \tag{4.215}$$

Then we have

$$\begin{aligned}
&\log^+\lambda_{\max}\left\{\sum_{i=0}^{n-1}\psi_i(n)\psi_i^\tau(n)\right\} \\
&= O\left(\log\left\{h_n\sum_{i=0}^{n-1}(\|y_i\|^2+\|u_i\|^2)\right\}\right) \\
&= O(\log n) \quad a.s.
\end{aligned} \tag{4.216}$$

Finally, substituting (4.214) and (4.216) into (4.213), we see from (4.212) that

$$\begin{aligned}
&\sum_{i=1}^{n}\|\widehat{w}_{i+1}^\tau(n)-w_{i+1}\|^2 \\
&= O\left(\sum_{i=1}^{n}\|\psi_{i-1}^\tau(n)[\alpha_n(n)-\alpha(n)]\|^2\right)+O\left(\sum_{i=1}^{n}\|\varepsilon_{i-1}(n)\|^2\right) \\
&= O(h_n\log n)+o([d(n)\log\log n]^2)+O\left(n^{b+1}\exp\{-\lambda h_n\}\right) \\
&= O(h_n\log n)+o([d(n)\log\log n]^2) \quad a.s.
\end{aligned}$$

This is the desired result (4.207). Q.E.D.

From Theorem 4.9, we see that if the sample path behavior of the noise process $\{w_i\}$ is not "too bad", e.g., $\{w_i\}$ is bounded a.s., or Gaussian and white ($\|w_i\| = O([\log i]^{1/2})$), or has a growth rate of $O(\{\log i\}^{1-\varepsilon})$, ($\varepsilon > 0$), then the second term on the right-hand side of (4.207) is negligible. We now give an example to show that in such cases the result of Theorem 4.9 is the best possible.

Example 4.2. For the ARMAX model (3.91) assume that $\det A(z) \neq 0$, $|z| \leq 1$, $\{w_t\}$ is a Gaussian $N(0,\Sigma)$ white noise (i.i.d) sequence independent of $\{u_t\}$, and $\{u_t\}$ is a Gaussian stationary ARMA process whose spectral density matrix is uniformly positive definite on $[-\pi,\pi]$. If h_n is chosen as in Theorem 4.9, then the noise estimate $\{w_t(n)\}$ satisfies

$$\lim_{n\to\infty}\frac{1}{h_n\log n}\sum_{t=1}^{n}[\widehat{w}_t(n)-w_t][\widehat{w}_t(n)-w_t]^\tau = (m+l)\Sigma \quad a.s.$$

The proof of this is given in [HG]. Instead of giving details of the proof here, we refer the reader to a more general result (Theorem 9.4 and Example 9.3) in Chapter 9.

Having established properties of noise estimates, we now proceed to consider the coefficient estimation problem.

Let $\{\widehat{w}_i(n),\ 1 \le i \le n\}$ be generated from (4.196)-(4.199). Similar to (4.186)-(4.187), define the estimate θ_n for θ as

$$\theta_n = \left(\sum_{i=0}^{n-1} \varphi_i(n)\varphi_i^\tau(n) + \beta I\right)^{-1} \sum_{i=0}^{n-1} \varphi_i(n) y_{i+1}^\tau \tag{4.217}$$

$$\begin{aligned} \varphi_i(n) &= [y_i^\tau \ \dots \ y_{i-p+1}^\tau \quad u_i^\tau \ \dots \ u_{i-q+1}^\tau \\ &\quad \widehat{w}_i^\tau(n) \ \dots \ \widehat{w}_{i-r+1}^\tau(n)]^\tau \quad 0 \le i \le n-1. \end{aligned} \tag{4.218}$$

Theorem 4.10 *[GH]. Suppose that all conditions of Theorem 4.9 are fulfilled. Then the estimation error produced by (4.217) and (4.218) can be expressed as follows:*

$$\|\theta_n - \theta\|^2 = O\left(\frac{h_n \log n}{\lambda_{\min}(n)}\right) + o\left(\frac{[d(n)\log\log n]^2}{\lambda_{\min}(n)}\right) \quad a.s. \tag{4.219}$$

where $\lambda_{\min}(n)$ is defined as

$$\lambda_{\min}(n) \triangleq \lambda_{\min}\left(\sum_{i=1}^{n-1} \varphi_i(n)\varphi_i^\tau(n) + \beta I\right).$$

Proof. The proof is similar to that of Theorem 4.7. Note that in the present case, (4.191) and (4.192) remain valid with φ_i replaced by $\varphi_i(n)$, and the estimation error is bounded by

$$\begin{aligned} \|\theta_n - \theta\| \ \le\ & \left\| \left[\sum_{i=1}^{n-1} \varphi_i(n)\varphi_i^\tau(n) + \beta I\right]^{-1} \sum_{i=1}^{n-1} \varphi_i(n) w_{i+1}^\tau \right\| \\ & + \left\| \left[\sum_{i=1}^{n-1} \varphi_i(n)\varphi_i^\tau(n) + \beta I\right]^{-1} \sum_{i=1}^{n-1} \varphi_i(n)[\varphi_i^0 - \varphi_i(n)]^\tau \theta \right\| \\ & + \left\| \left[\sum_{i=1}^{n-1} \varphi_i(n)\varphi_i^\tau(n) + \beta I\right]^{-1} \beta\theta \right\|. \end{aligned} \tag{4.220}$$

By (4.17), (4.218) and Theorem 4.9, we have

$$\sum_{i=0}^{n-1} \|\varphi_i^0 - \varphi_i(n)\|^2 = O(h_n \log n) + o([d(n)\log\log n]^2). \tag{4.221}$$

By Condition (4.200) and the Markov inequality it is seen that for any $\delta > \frac{1}{2}$,

$$P\left(\|w_{i+1}\| > i^{\delta} | \mathcal{F}_i\right) \leq \frac{E(\|w_{i+1}\|^2 | \mathcal{F}_i)}{i^{2\delta}} = O(i^{-2\delta}).$$

So by Theorem 2.5,

$$\|w_{i+1}\| = O(i^{\delta}) \quad a.s.$$

By taking $d(n) = n^{\delta}$ in (4.221) we obtain

$$\log \sum_{i=0}^{n-1} \|\varphi_i^0 - \varphi_i(n)\|^2 = o(\log n) \quad a.s. \tag{4.222}$$

Using the growth rate assumption on $\{y_i, u_i\}$ from (4.222) we find

$$\begin{aligned} &\log \lambda_{\max} \left\{ \sum_{i=0}^{n-1} \varphi_i(n) \varphi_i^T(n) \right\} \\ &= O\left(\log^+ \left\{ \sum_{i=0}^{n-1} \|\varphi_i^0\|^2 \right\} \right) + O\left(\log^+ \left\{ \sum_{i=0}^{n-1} \|\varphi_i^0 - \varphi_i(n)\|^2 \right\} \right) \\ &= O(\log n) \quad a.s. \end{aligned} \tag{4.223}$$

So, by Corollary 4.2 and (4.223) it is easy to see that

$$\begin{aligned} &\left\| \left[\sum_{i=1}^{n-1} \varphi_i(n) \varphi_i^T(n) + \beta I \right]^{-1} \sum_{i=1}^{n-1} \varphi_i(n) w_{i+1}^T \right\|^2 \\ &\leq \frac{1}{\lambda_{\min}(n)} \left\| \left[\sum_{i=1}^{n-1} \varphi_i(n) \varphi_i^T(n) + \beta I \right]^{-\frac{1}{2}} \sum_{i=1}^{n-1} \varphi_i(n) w_{i+1}^T \right\|^2 \\ &= O\left(\frac{1}{\lambda_{\min}(n)} \{\log n + [d(n) \log\log n]^2\} \right). \end{aligned} \tag{4.224}$$

We now estimate the second term on the right-hand side of (4.220). By Lemma 4.7(ii) and (4.221) it follows that

$$\begin{aligned} &\left\| \left[\sum_{i=1}^{n-1} \varphi_i(n) \varphi_i^T(n) + \beta I \right]^{-1} \sum_{i=1}^{n-1} \varphi_i(n) [\varphi_i^0 - \varphi_i(n)]^{\tau} \theta \right\|^2 \\ &= O\left(\frac{1}{\lambda_{\min}(n)} \sum_{i=1}^{n-1} \|\varphi_i^0 - \varphi_i(n)\| \right) \\ &= O\left(\frac{h_n \log n}{\lambda_{\min}(n)} \right) + o\left(\frac{[d(n) \log\log n]^2}{\lambda_{\min}(n)} \right) \quad a.s. \end{aligned} \tag{4.225}$$

Finally, substituting (4.224) and (4.225) into (4.220) we obtain the desired result (4.219). Q.E.D.

Similar to the proof of Theorem 4.2, we can verify the following result.

Theorem 4.11. *Let all conditions of Theorem 4.10 be satisfied. If* $h_n \log n + [d(n) \log\log n]^2 = o\left(\lambda^0_{\min}(n)\right)$ *a.s., then as* $n \to \infty$

$$\|\theta_n - \theta\|^2 = O\left(\frac{h_n \log n}{\lambda^0_{\min}(n)}\right) + o\left(\frac{[d(n)\log\log n]^2}{\lambda^0_{\min}(n)}\right) \quad a.s.$$

where $\lambda^0_{\min}(n)$ *is defined in Theorem 4.2.*

CHAPTER 5

Stochastic Adaptive Tracking

We continue considering the system described by (3.91)-(3.94), for which (p, q, r) are assumed to be the known upper bounds for system orders and $d = 1$ to be the lower bound for the time-delay of the system. The true orders may be strictly less than p, q and r respectively. The system coefficients written in the matrix form

$$\theta^\tau = [-A_1 \ \dots \ -A_p \quad B_1 \ \dots \ B_q \quad C_1 \ \dots \ C_r] \tag{5.1}$$

are unknown.

Assume again the driven noise $\{w_n, \mathcal{F}_n\}$ is a martingale difference sequence satisfying (3.101) and (3.102).

Let $\{y_n^*\}$ be a bounded reference signal and let y_{n+1}^* be $\mathcal{F}_n$-measurable. At time n the control u_n is designed on the basis of information available at and including time n, i.e. on the basis of $\{y_i,\ i \le n,\ u_j,\ j \le n-1\}$ in order for the system output y_n to follow y_n^*, in other words, to minimize the tracking error

$$\lim_{n\to\infty} \frac{1}{n} \sum_{i=0}^{n} (y_i - y_i^*)(y_i - y_i^*)^\tau \tag{5.2}$$

if it exists.

Corollary 3.3 and (3.125) tell us that the optimal control can be defined from

$$\theta^\tau \varphi_n = y_{n+1}^* \tag{5.3}$$

and the minimal tracking error is R where φ_n is defined by (3.131).

The idea of adaptive control (indirect approach) consists in replacing the unknown parameter θ by its estimate θ_n at time n when one defines u_n.

Thus, the equation defining adaptive control u_n is

$$\theta_n^\tau \psi_n = y_{n+1}^* \tag{5.4}$$

if θ_n is updated in accordance with the SG algorithm (4.19), (4.22) and (4.23), or is

$$\theta_n^\tau \varphi_n = y_{n+1}^* \tag{5.5}$$

if θ_n is calculated by the ELS algorithm (4.13), (4.14) and (4.18).

This chapter mainly analyzes the convergence and optimality of adaptive trackers composed of (3.90), (4.19), (4.22), (4.23) or of (3.90), (4.13), (4.14), (4.18) with u_n respectively defined from (5.4) or (5.5) (or from their modifications). The former is called the SG-based adaptive tracker and the latter the ELS-based adaptive tracker. The last section of this chapter applies an adaptive tracker to solve an adaptive model reference control problem.

The ELS-based adaptive tracker in its special form where the system (3.90) is of single-input single-output with uncorrelated noise, i.e. $C(z) = 1$, $y_n^* \equiv 0$ and B_1 is known, coincides with the self-tuning regulator proposed by Åström–Wittenmark [AW], which has got a great success in applications and has drawn much attention from control theorists in an attempt to establish its convergence.

5.1 SG-Based Adaptive Tracker With $d = 1$

Following the well-known work on adaptive control for deterministic systems [GRC1] the first progress in convergence analysis for stochastic adaptive trackers was made in [GRC2] showing that the SG-based adaptive tracker (with unit delay) is convergent and optimal. Prior to demonstrating this result we give a condition guaranteeing solvability of (5.4) [MC1], [CG4].

Lemma 5.1 *[CG4]. Assume that for System (3.91)-(3.94) $m = l$, and $\{w_n\}$ is a sequence of independent random vectors with continuous distributions. Then (5.4) is solvable with respect to u_n $\forall n \geq 1$ if the initial values u_0 and B_{10} are chosen so that* $\det B_{11} \neq 0$ *(for example, $u_0 = 0$,* $\det B_{10} \neq 0$*), where B_{1n} denotes the estimate for B_1 given by*

$$\theta_n^\tau = [-A_{1n} \ \dots \ -A_{pn} \quad B_{1n} \ \dots \ B_{qn} \quad C_{1n} \ \dots \ C_{rn}]$$

which is calculated according to (4.19), (4.22) and (4.23).

Proof. By (4.22) and (4.68) we have

$$\theta_{n+1} = \theta_n + \frac{a\psi_n}{r_n}(w_{n+1}^\tau + \zeta_{n+1}^\tau) \tag{5.6}$$

and hence

$$B_{1n+1} = B_{1n} + \frac{a}{r_n}(\zeta_{n+1} + w_{n+1})u_n^{\tau}, \tag{5.7}$$

from which we see that $\det B_{11} \neq 0$ if $u_n = 0$ and $\det B_{10} \neq 0$.

For solvability of (5.4) it suffices to show that $\det B_{1n} \neq 0$ a.s. $\forall n \geq 1$. Assume it is true for n, i.e. $P(N) = 0$, where

$$N \triangleq \{\omega : \det B_{1n} = 0\}.$$

Set $D \triangleq \{\omega : \det B_{1n+1} = 0\}$. We need only to prove $P(DN^c) = 0$.

Suppose that the converse were true, i.e. $P(DN^c) > 0$.

From (5.7) we have

$$\det\left(B_{1n} + \frac{a}{r_n}(\zeta_{n+1} + w_{n+1})u_n^{\tau}\right) = 0, \quad \forall \omega \in DN^c. \tag{5.8}$$

Noticing $\det B_{1n} \neq 0$ for $\omega \in DN^c$ we then derive

$$\begin{aligned} 0 &= \det\left(I + \frac{a}{r_n}B_{1n}^{-1}(\zeta_{n+1} + w_{n+1})u_n^{\tau}\right) \\ &= 1 + \frac{a}{r_n}u_n^{\tau}B_{1n}^{-1}(\zeta_{n+1} + w_{n+1}), \quad \forall \omega \in DN^c \end{aligned} \tag{5.9}$$

where the last equality follows by observing that $B_{1n}^{-1}(\zeta_{n+1} + w_{n+1})u_n^{\tau}$ has only one positive nonzero eigenvalue $u_n^{\tau}B_{1n}^{-1}(\zeta_{n+1} + w_{n+1})$.

From (5.8) we then have

$$u_n^{\tau}B_{1n}^{-1} \neq 0, \quad \forall \omega \in DN^c. \tag{5.10}$$

Write $u_n^{\tau}B_{1n}^{-1}$ and w_{n+1} in the component form

$$u_n^{\tau}B_{1n}^{-1} = [\alpha_1 \ \ldots \ \alpha_m], \quad w_{n+1} = [w_{n+1,1} \ \ldots \ w_{n+1,m}]^{\tau}. \tag{5.11}$$

By (5.10) we have some α_i, say α_1, and some $D_1 \subset DN^c$ such that

$$\alpha_1 \neq 0 \quad \forall \omega \in D_1, \quad P(D_1) > 0. \tag{5.12}$$

Setting

$$w \triangleq \begin{cases} \dfrac{1}{\alpha_1}\left[\displaystyle\sum_{i=2}^{m} \alpha_i w_{n+1,i} + r_n + u_n^{\tau}B_{1n}^{-1}\xi_{n+1}\right], & \omega \in D_1 \\ 0, & \omega \in D_1^c \end{cases}$$

from (5.9) we find that

$$P(w_{n+1,1} + w = 0) \geq P(D_1) > 0. \tag{5.13}$$

However, since $w_{n+1,1}$ and w are independent and the distribution function of $w_{n+1,1}$ is continuous, $w_{n+1,1}+w$ must have a continuous distribution. Thus, (5.13) is impossible. The contradiction shows that $P(DN^c)=0$.

Q.E.D.

This lemma gives sufficient conditions for solvability of (5.4). However, in what follows we do not intend to make such a restriction on $\{w_n\}$, instead, we either simply assume that (5.4) is solvable or slightly modify B_{1n} so that u_n is well-defined from (5.4) or (5.5).

Theorem 5.1 *[GRC2]. Suppose that for the system (3.91)-(3.94)*
1) $\{w_n, \mathcal{F}_n\}$ *is a martingale difference sequence with*

$$\sup_{n\geq 1} E\{\|w_n\|^2|\mathcal{F}_{n-1}\}<\infty \quad a.s. \tag{5.14}$$

$$\lim_{n\to\infty}\frac{1}{n}\sum_{i=1}^{n} w_i w_i^{\tau} = R > 0 \quad a.s. \tag{5.15}$$

2) $d=1$, $m=l$ *and zeros of* $\det[z^{-1}B(z)]$ *are outside the closed unit disk (minimum phase condition);*
3) the polynomial $C(z)-\frac{a}{2}I$ *is SPR for some* $a\in(0,1]$*;*
4) u_n *can be defined from (5.4).*
Then the SG-based adaptive tracker is optimal:

$$\lim_{n\to\infty}\frac{1}{n}\sum_{i=1}^{n}(y_i-y_i^*)(y_i-y_i^*)^{\tau}=R \quad a.s. \tag{5.16}$$

Proof. Similar to (3.96) and (3.97) we set

$$B = \begin{bmatrix} -B_1^{-1}B_2 & I & 0 & \cdots & 0 \\ \vdots & 0 & \ddots & \ddots & \vdots \\ \vdots & \vdots & \ddots & \ddots & 0 \\ \vdots & \vdots & \ddots & \ddots & I \\ -B_1^{-1}B_s & 0 & \cdots & \cdots & 0 \end{bmatrix}, \quad A = \begin{bmatrix} B_1^{-1} \\ B_1^{-1}A_1 \\ \vdots \\ B_1^{-1}A_{s-2} \end{bmatrix},$$

$$C = \begin{bmatrix} -B_1^{-1} \\ -B_1^{-1}C_1 \\ \vdots \\ -B_1^{-1}C_{s-2} \end{bmatrix}, \tag{5.17}$$

where $s=q\vee(p+2)\vee(r+2)$ and $A_i=0$, $B_j=0$, $C_k=0$ for $i>p$, $j>q$, $k>r$. The invertibility of B_1 is guaranteed by Assumption 2).

Define the $(s-1)l$-dimensional vector U_n:

$$U_n = BU_{n-1} + Ay_{n+1} + Cw_{n+1}, \tag{5.18}$$
$$U_0^\tau = [u_0^\tau \quad 0 \quad \dots \quad 0].$$

Clearly, the vector composed of the first l components of U_n coincides with u_n. It is easy to verify that $det\, z^{-1}B(z) \neq 0$ for $|z| \leq 1$ if and only if all eigenvalues of B are less than 1 (see, e.g. [Che3, p.48]). Then by Condition 2) there are positive numbers $c > 0$ and $\rho \in (0,1)$ such that

$$\|B^k\| \leq c\rho^k \qquad \forall k \geq 0. \tag{5.19}$$

Consequently, we have

$$\|u_n\| = \left\| [I \quad 0 \quad \dots \quad 0] \left(B^n U_0 + \sum_{i=1}^n B^{n-i}(Ay_{i+1} + Cw_{i+1}) \right) \right\|$$
$$\leq c\|u_0\|\rho^n + \sum_{i=1}^n c\rho^{n-i}(\|A\|\|y_{i+1}\| + \|C\|\|w_{i+1}\|)$$

and by the Schwarz inequality

$$\|u_n\|^2 \leq 2c^2\|u_0\|^2\rho^{2n}$$
$$+4c^2 \sum_{i=1}^n \rho^{n-i} \sum_{i=1}^n \rho^{n-i}(\|A\|^2\|y_{i+1}\|^2 + \|C\|^2\|w_{i+1}\|^2).$$

By (5.15) it follows that

$$\frac{1}{n}\sum_{i=1}^n \|u_i\|^2 \leq O(1) + O\left(\frac{1}{n}\sum_{i=1}^n \|y_{i+1}\|^2\right) \quad a.s. \tag{5.20}$$

Combining (5.4) and (4.68) leads to

$$y_{n+1} = \zeta_{n+1} + w_{n+1} + y_{n+1}^* \tag{5.21}$$

which implies, by (5.15) and the boundedness of $\{y_n^*\}$, that

$$\frac{1}{n}\sum_{i=1}^n \|y_{i+1}\|^2 \leq O(1) + O\left(\frac{1}{n}\sum_{i=1}^n \|\zeta_{i+1}\|^2\right) \quad a.s. \tag{5.22}$$

Notice

$$\sum_{i=1}^n \|y_{i+1}\|^2 = \sum_{i=1}^n \|f_i + w_{i+1}\|^2$$

$$= \sum_{i=1}^n \|f_i\|^2 + 2\sum_{i=1}^n f_i^\tau w_{i+1} + \sum_{i=1}^n \|w_{i+1}\|^2 \tag{5.23}$$

where

$$f_i = -A_1 y_i - \ldots - A_p y_{i-p+1} + B_1 u_i + \ldots + B_q u_{i-q+1} + C_1 w_i + \ldots + C_r w_{i-r+1}$$

which clearly is $\mathcal{F}_i$-measurable.

Then by Theorem 2.8 and (5.15) for sufficiently large n we have

$$\left\| 2\sum_{i=1}^{n} f_i^T w_{i+1} \right\| = O\left(\left(\sum_{i=1}^{n} \|f_i\|^2 \right)^{\frac{1}{2}} log \left(\sum_{i=1}^{n} \|f_i\|^2 + e \right) \right)$$

$$\leq \begin{cases} \frac{1}{2}\sum_{i=1}^{n} \|f_i\|^2, & if \quad \sum_{i=1}^{\infty} \|f_i\|^2 = \infty \\ \frac{1}{2}\sum_{i=1}^{n} \|w_{i+1}\|^2, & if \quad \sum_{i=1}^{\infty} \|f_i\|^2 < \infty. \end{cases}$$

Hence from (5.23) we see

$$\sum_{i=1}^{n} \|y_{i+1}\|^2 \geq \frac{1}{2} \sum_{i=1}^{n} (\|f_i\|^2 + \|w_{i+1}\|^2) \xrightarrow[n \to \infty]{} \infty, \tag{5.24}$$

which yields

$$r_n \xrightarrow[n \to \infty]{} \infty. \tag{5.25}$$

Thus, by Lemmas 2.4 and 4.3 we find

$$\Delta_n \triangleq \frac{1}{r_n} \sum_{i=1}^{n} \|\zeta_{i+1}\|^2 \xrightarrow[n \to \infty]{} 0. \tag{5.26}$$

Noticing (5.4) and (5.20) we obtain

$$\frac{1}{n} \sum_{i=1}^{n} \|\zeta_{i+1}\|^2 = \frac{\Delta_n}{n} \cdot r_n$$

$$\leq \frac{\Delta_n}{n} \left(p \sum_{j=1}^{n} \|y_j\|^2 + q \sum_{j=1}^{n} \|u_j\|^2 + r \sum_{j=1}^{n} \|y_j - y_j^*\|^2 + 1 \right)$$

$$\leq \Delta_n O \left(1 + \frac{1}{n} \sum_{j=1}^{n} \|y_{j+1}\|^2 \right). \tag{5.27}$$

Substituting (5.27) into (5.22) by (5.26) we conclude that

$$\frac{1}{n} \sum_{j=1}^{n} \|y_{j+1}\|^2 = O(1) \quad a.s.$$

which together with (5.27) implies

$$\frac{1}{n}\sum_{i=1}^{n}\|\zeta_{i+1}\|^2 \xrightarrow[n\to\infty]{} 0 \quad a.s. \tag{5.28}$$

Finally, by (5.28) and (5.15) the desired result follows immediately from (5.21). Q.E.D.

5.2 SG-Based Adaptive Tracker With $d \geq 1$

We now consider System (3.91)-(3.94) with multi-delay $d \geq 1$ [CZ1].

In Theorem 3.6 we have derived the minimum tracking error equal to the right-hand side of (3.124). We proceed to design an adaptive control under which this minimum is achieved.

In the last section the adaptive control law is constructed according to the certainty equivalence principle, i.e. the adaptive control at time n is derived in the form of the optimal control for the fictitious system, which is obtained from System (3.91)-(3.94) with unknown parameters replaced by their estimates at time n. This approach is called indirect or explicit. In this section we proceed in a different way. Namely, we do not estimate the unknown parameter of the system but directly estimate the optimal control gain. This method is called direct or implicit.

Let

$$G(z) = G_0 + G_1 z + \ldots + G_{p_1} z^{p_1} \tag{5.29}$$

$$F(z)(\mathrm{Adj}C(z))B(z) = D_0 + D_1 z + \ldots + D_{p_2} z^{p_2} \tag{5.30}$$

and set

$$\theta^{*\tau} = [G_0\ G_1 \ldots G_{p_1} \quad D_0\ D_1 \ldots D_{p_2} \quad c_1 I \ldots c_{mr} I] \tag{5.31}$$

where $G(z)$, $F(z)$ and c_i are given by (3.120) and (3.122).

Replacing the algorithm (4.19), (4.22) and (4.23) which is used to estimate θ defined by (5.1) we now apply the following SG algorithm to estimate θ^*:

$$\theta_n^* = \theta_{n-1}^* + \frac{a}{r_{n-d}^*}\psi_{n-d}^*(y_n^\tau - \psi_{n-d}^{*\tau}\theta_{n-d}^*) \tag{5.32}$$

$$\begin{aligned}\psi_n^* = {} & [y_n^\tau \cdots y_{n-p_1}^\tau \quad u_n^\tau \cdots u_{n-p_2}^\tau \\ & -\psi_{n-1}^{*\tau}\theta_{n-1}^* \cdots -\psi_{n-mr}^{*\tau}\theta_{n-mr}^*]^\tau\end{aligned} \tag{5.33}$$

$$r_n^* = r_{n-1}^* + 2d\|\psi_n^*\|^2, \ n > 0, \ r_n^* = 1, \ n \leq 0 \tag{5.34}$$

with arbitrary initial values θ_0^* and ψ_{-1}^*.

Setting

$$z_n = y_{n+d} - F(z)w_{n+d} - \theta_n^{*\tau}\psi_n^*, \ n > 0, \ z_n = 0, \ n \leq 0 \tag{5.35}$$

from (3.127) we have

$$\begin{aligned}(\det C(z))z_n &= G(z)y_n + F(z)(\mathrm{Adj}C(z))B(z)u_n - (\det C(z))\theta_n^{*\tau}\psi_n^* \\ &= \theta^{*\tau}\psi_n^* - \theta_n^{*\tau}\psi_n^* = \widetilde{\theta}_n^{*\tau}\psi_n^* \qquad (5.36)\end{aligned}$$

where $\widetilde{\theta}_n^* = \theta^* - \theta_n^*$

Lemma 5.2. *If* $\det C(z) - \frac{a}{2}$ *is SPR,* u_n *is* $\mathcal{F}_n$*-measurable and*

$$E[w_{n+1}|\mathcal{F}_n] = 0 \text{ and } \sup_{n\geq 0} E\{\|w_{n+1}\|^2|\mathcal{F}_n\} \leq \sigma^2 < \infty \text{ a.s.},$$

then

$$\sum_{i=0}^{\infty}\frac{\|z_i\|^2}{r_i^*} < \infty \text{ a.s. and } \sup_i \|\widetilde{\theta}_i^*\| < \infty \qquad (5.37)$$

Proof. Since $\det C(z) - \frac{a}{2}$ is SPR there exists $\varepsilon > 0$ so that $\det C(z) - \frac{a(1+\varepsilon)}{2}$ remains to be SPR. Then by Lemma 4.1 we know that

$$\begin{aligned}s_n &\triangleq 2a\sum_{i=0}^{n} z_{i-d}^{\tau}\left[\left(\det C(z) - \frac{a(1+\varepsilon)}{2}\right)z_{i-d}\right] \\ &\geq 0 \quad \forall n \geq 0 \quad \text{a.s.}, \qquad (5.38)\end{aligned}$$

i.e.

$$s_n = s_{n-1} + 2az_{n-d}^{\tau}\widetilde{\theta}_{n-d}^{*\tau}\psi_{n-d}^* - a^2(1+\varepsilon)\|z_{n-d}\|^2 \geq 0. \qquad (5.39)$$

By (5.32) and (5.35) it follows that

$$\widetilde{\theta}_n^* = \widetilde{\theta}_{n-1}^* - \frac{a}{r_{n-d}^*}\psi_{n-d}^*(z_{n-d} + F(z)w_n)^{\tau},$$

which yields

$$\begin{aligned}tr\widetilde{\theta}_n^{*\tau}\widetilde{\theta}_n^* &= tr\widetilde{\theta}_{n-1}^{*\tau}\widetilde{\theta}_{n-1}^* + \frac{a^2\|\psi_{n-d}^*\|^2}{r_{n-d}^{*2}}\|z_{n-d} + F(z)w_n\|^2 \\ &\quad - \frac{2a}{r_{n-d}^*}(z_{n-d} + F(z)w_n)^{\tau}\widetilde{\theta}_{n-1}^{*\tau}\psi_{n-d}^* \\ &= tr\widetilde{\theta}_{n-1}^{*\tau}\widetilde{\theta}_{n-1}^* + \frac{a^2\|\psi_{n-d}^*\|^2}{r_{n-d}^{*2}}\|z_{n-d} + F(z)w_n\|^2 \\ &\quad - \frac{2a}{r_{n-d}^*}(z_{n-d} + F(z)w_n)^{\tau} \\ &\quad \left(\widetilde{\theta}_{n-d}^* - a\sum_{i=1}^{d-1}\frac{\psi_{n-i-d}^*}{r_{n-i-d}^*}(z_{n-i-d} + F(z)w_{n-i})^{\tau}\right)^{\tau}\psi_{n-d}^*\end{aligned}$$

$$
\begin{aligned}
&\leq tr\widetilde{\theta}_{n-1}^{*\tau}\widetilde{\theta}_{n-1}^{*} - \frac{2a}{r_{n-d}^{*}}(z_{n-d}+F(z)w_n)^{\tau}\widetilde{\theta}_{n-d}^{*\tau}\psi_{n-d}^{*} \\
&+\frac{da^2\|\psi_{n-d}^{*}\|^2}{r_{n-d}^{*2}}\|z_{n-d}+F(z)w_n\|^2 \\
&+a^2\sum_{i=1}^{d-1}\frac{\|\psi_{n-i-d}^{*}\|^2}{r_{n-i-d}^{*2}}\|z_{n-i-d}+F(z)w_{n-i}\|^2.
\end{aligned}
\tag{5.40}
$$

Clearly, there is a constant $k_1 > 0$ such that

$$
t_n \triangleq a^2\sum_{i=1}^{n}\left[\frac{\|z_{i-d}\|^2}{2r_{i-d}^{*}} - \sum_{j=1}^{d-1}\frac{\|z_{i-j-d}\|^2}{2dr_{i-j-d}^{*}}\right] + k_1 \geq 0 \quad \forall n \geq 0. \tag{5.41}
$$

Set

$$
v_n = tr\widetilde{\theta}_n^{*\tau}\widetilde{\theta}_n^{*} + t_n + \frac{s_n}{r_{n-d}^{*}}. \tag{5.42}
$$

By (5.34) and (5.39)–(5.41) we have that

$$
\begin{aligned}
v_n \quad &\leq \quad v_{n-1} + \frac{da^2\|\psi_{n-d}^{*}\|^2}{r_{n-d}^{*2}}\|z_{n-d}+F(z)w_n\|^2 \\
&+a^2\sum_{i=1}^{d-1}\frac{\|\psi_{n-i-d}^{*}\|^2}{r_{n-i-d}^{*2}}\|z_{n-i-d}+F(z)w_{n-i}\|^2 \\
&-\frac{2a}{r_{n-d}^{*}}(z_{n-d}+F(z)w_n)^{\tau}\widetilde{\theta}_{n-d}^{*\tau}\psi_{n-d}^{*} \\
&+\frac{2a}{r_{n-d}^{*}}z_{n-d}^{\tau}\widetilde{\theta}_{n-d}^{*\tau}\psi_{n-d}^{*} - \frac{a^2(1+\varepsilon)}{r_{n-d}^{*}}\|z_{n-d}\|^2 \\
&+\frac{a^2\|z_{n-d}\|^2}{2r_{n-d}^{*}} - a^2\sum_{i=1}^{d-1}\frac{\|z_{n-i-d}\|^2}{2dr_{n-i-d}^{*}} \\
&\leq \quad v_{n-1} - \frac{a^2\varepsilon}{r_{n-d}^{*}}\|z_{n-d}\|^2 - \frac{2a}{r_{n-d}^{*}}(F(z)w_n)^{\tau}\widetilde{\theta}_{n-d}^{*\tau}\psi_{n-d}^{*} \\
&+\frac{da^2}{r_{n-d}^{*2}}\|\psi_{n-d}^{*}\|^2(2z_{n-d}^{\tau}F(z)w_n + \|F(z)w_n\|^2) \\
&+a^2\sum_{i=1}^{d-1}\frac{\|\psi_{n-i-d}^{*}\|^2}{r_{n-i-d}^{*2}}(\|F(z)w_{n-i}\|^2 + 2z_{n-i-d}^{\tau}F(z)w_{n-i}).
\end{aligned}
\tag{5.43}
$$

Notice that the order of $F(z)$ is less than or equal to $d-1$. By Theorem 2.8 and (5.36) we see that

$$
\sum_{i=1}^{n}\frac{1}{r_{i-d}^{*}}(F(z)w_i)^{\tau}\widetilde{\theta}_{i-d}^{*\tau}\psi_{i-d}^{*}
$$

$$= O\left(\left(\sum_{i=1}^{n} \frac{\|\widetilde{\theta}_{i-d}^{*\tau}\psi_{i-d}^{*}\|^2}{r_{i-d}^{*2}}\right)^{\frac{1}{2}+\delta}\right)$$

$$= O\left(\left(\sum_{i=1}^{n} \frac{\|z_{i-d}\|^2}{r_{i-d}^{*2}}\right)^{\frac{1}{2}+\delta}\right), \quad \forall\delta > 0 \tag{5.44}$$

$$\sum_{i=1}^{n} \frac{\|\psi_{i-d}^{*}\|^2}{r_{i-d}^{*2}} z_{i-d}^{\tau} F(z) w_i = O\left(\left(\sum_{i=1}^{n} \frac{\|z_{i-d}\|^2}{r_{i-d}^{*2}}\right)^{\frac{1}{2}+\delta}\right), \quad \forall\delta > 0 \tag{5.45}$$

and

$$\sum_{j=1}^{n} \frac{\|\psi_{j-d-i}^{*}\|^2}{r_{j-d-i}^{*2}} \|F(z)w_{j-i}\|^2 = O(1), \quad i = 0,\ 1,\ \ldots,\ d-1, \tag{5.46}$$

since $\sum_{j=1}^{\infty} \frac{\|\psi_{j-d-i}^{*}\|^2}{r_{j-d-i}^{*2}} < \infty$ a.s. and $\|w_{j-i}\|^2 - E(\|w_{j-i}\|^2|\mathcal{F}_{j-i-1})$ is a martingale difference sequence.

Summing up both sides of (5.43) and using (5.44)–(5.46) we find that

$$v_n \leq v_0 - a^2\varepsilon \sum_{i=1}^{n} \frac{\|z_{i-d}\|^2}{r_{i-d}^{*2}} + O\left(\left(\sum_{i=1}^{n} \frac{\|z_{i-d}\|^2}{r_{i-d}^{*2}}\right)^{\frac{1}{2}+\delta}\right)$$

which implies the desired result (5.37). Q.E.D.

Theorem 5.2. *For the system (3.91)-(3.94) assume that*
1) $\{w_n, \mathcal{F}_n\}$ *is a martingale difference sequence with (5.14), (5.15) held;*
2) $d \geq 1$, $m = l$ *and* $\det B(z) \neq 0$, $\forall |z| \leq 1$;
3) $\det C(z) - \frac{a}{2}$ *is SPR for some* $a \in (0, 1]$;
4) u_n *can be defined from*

$$y_{n+d}^{*} = \theta_n^{*\tau}\psi_n^{*}. \tag{5.47}$$

Then the SG-based tracker (3.91)-(3.94), (5.32)-(5.34), (5.47) is optimal, i.e. the tracking error is minimized:

$$\lim_{n\to\infty} \frac{1}{n} \sum_{i=1}^{n} (y_i - y_i^{*})(y_i - y_i^{*})^{\tau} = \sum_{j=0}^{d-1} F_j R F_j^{\tau}. \tag{5.48}$$

Proof. The proof is similar to that for Theorem 5.2. In the present case we define $(s-d)$-dimensional vector U_n by

$$U_n = BU_{n-1} + Ay_{n+d} + Cw_{n+d}, \quad U_0^{\tau} = [u_0^{\tau}\ 0\ \ldots\ 0],$$

where

$$B = \begin{bmatrix} -B_1^{-1}B_2 & I & 0 & \cdots & 0 \\ \vdots & 0 & \ddots & \ddots & \vdots \\ \vdots & \vdots & \ddots & \ddots & 0 \\ \vdots & \vdots & \ddots & \ddots & I \\ -B_1^{-1}B_s & 0 & \cdots & \cdots & 0 \end{bmatrix}, \quad A = \begin{bmatrix} B_1^{-1} \\ B_1^{-1}A_1 \\ \vdots \\ B_1^{-1}A_s \end{bmatrix},$$

$$C = \begin{bmatrix} -B_1^{-1} \\ -B_1^{-1}C_1 \\ \vdots \\ -B_1^{-1}C_s \end{bmatrix}, \quad s = (q-1) \vee (p+1) \vee (r+1), \tag{5.49}$$

$$A_i = 0,\ i > p,\ B_j = 0,\ j > q,\ C_k = 0,\ k > r. \tag{5.50}$$

By Condition 2), similar to (5.20) we have

$$\frac{1}{n}\sum_{i=1}^{n}\|u_i\|^2 \leq O(1) + O\left(\frac{1}{n}\sum_{i=1}^{n}\|y_{i+d}\|^2\right) \quad \text{a.s.} \tag{5.51}$$

From (5.35) and (5.47) it follows that

$$y_{n+d} = z_n + F(z)w_{n+d} + y^*_{n+d} \tag{5.52}$$

and

$$\frac{1}{n}\sum_{i=1}^{n}\|y_{i+d}\|^2 \leq O(1) + O\left(\frac{1}{n}\sum_{i=1}^{n}\|z_i\|^2\right) \quad \text{a.s.} \tag{5.53}$$

which corresponds to (5.22).

In the present case we still have

$$r^*_n \xrightarrow[n\to\infty]{} \infty \quad \text{a.s.} \tag{5.54}$$

and

$$\Delta_n \triangleq \frac{1}{r^*_n}\sum_{i=1}^{n}\|z_i\|^2 \xrightarrow[n\to\infty]{} 0 \quad \text{a.s.} \tag{5.55}$$

Similar to (5.26)-(5.28) we now derive

$$\frac{1}{n}\sum_{i=1}^{n}\|z_i\|^2 \xrightarrow[n\to\infty]{} 0 \quad \text{a.s.} \tag{5.56}$$

From (5.52) we see that

$$\begin{aligned}&\frac{1}{n}\sum_{i=1}^{n}(y_{i+d}-y^*_{i+d})(y_{i+d}-y^*_{i+d})^\tau\\ =\ &\frac{1}{n}\sum_{i=1}^{n}(z_i+F(z)w_{i+d})(z_i+F(z)w_{i+d})^\tau.\end{aligned} \tag{5.57}$$

Applying (5.56) and Theorem 2.8 to (5.57) yields the desired result (5.48). Q.E.D.

Remark 5.1. In the case $d = 1$ the right-hand side of (5.48) reduces to R and (5.48) coincides with (5.16).

5.3 Stability and Optimality of Åström-Wittenmark Self-Tuning Tracker

In the rest part of this chapter only the unit delay ($d = 1$) case is considered.

We note that the key step for establishing convergence of the SG-based tracker is Lemma 4.3 and Lemma 5.2 for $d = 1$ and $d \geq 1$ respectively, which have been proved for general $\mathcal{F}_n$-measurable controls, u_n. However, it is unknown whether or not corresponding results hold for the ELS algorithm. This is why the convergence of the self-tuning regulator proposed by Åström–Wittenmark [AW] had been unclear for a long time [Ku1], [Ku2], if no modification is made on the algorithm.

In this section we prove that the Åström–Wittenmark self-tuning tracker is stable and optimal by showing a result similar to that given in Lemma 4.3 but for adaptive control designed for tracking a bounded signal rather than for general $\mathcal{F}_n$-measurable controls u_n as treated in Lemma 4.3.

We shall use the following conditions.

A1. $\{w_n, \mathcal{F}_n\}$ is a martingale difference sequence with

$$\sup_{n\geq 0} E[\|w_{n+1}\|^\beta|\mathcal{F}_n] < \infty \quad \text{a.s. for some } \beta > 2 \tag{5.58}$$

and

$$\lim_{n\to\infty}\frac{1}{n}\sum_{i=1}^{n}w_i w_i^\tau = R > 0 \quad \text{a.s.} \tag{5.59}$$

A2. $C^{-1}(e^{i\lambda}) + C^{-\tau}(e^{-i\lambda}) - I > 0, \quad \forall\lambda \in [0, 2\pi]$.

A3. $m = l$, $\det(B(z)z^{-1}) \neq 0, \forall z : |z| \leq 1$.

Set

$$s_n = trP^{-1}_{n+1} > e + \sum_{i=0}^{n}\|\varphi_i\|^2 \tag{5.60}$$

where P_n is given by (4.15).

Obviously, $s_n > \lambda_{\max}(n)$, where $\lambda_{\max}(n)$ denotes the maximum eigenvalue of P_{n+1}^{-1}.

By Theorem 4.1 for θ_n given by the ELS algorithm we have

$$\|\theta_{n+1} - \theta\|^2 = O\left(\frac{\log s_n}{\lambda_{\min}(n)}\right) \quad \text{a.s.} \tag{5.61}$$

Let us denote by

$$\widehat{w}_{n+1} = y_{n+1} - \theta_{n+1}^{\tau}\varphi_n \tag{5.62}$$

the *a posteriori* estimate for w_{n+1}. Then by (4.39) and (4.62) it follows that

$$\sum_{i=1}^{n+1} \|\widehat{w}_i - w_i\|^2 = O(\log s_n) \quad \text{a.s.} \tag{5.63}$$

We need one more property of the ELS estimate and formulate it as a lemma.

Lemma 5.3. *Suppose that for System (3.91)-(3.94) $d = 1$ and A1 and A2 hold and u_n is $\mathcal{F}_n$-measurable. Then*

$$\sum_{i=1}^{n} \alpha_i = O(\log s_n) \quad a.s. \tag{5.64}$$

where

$$\alpha_i = \frac{\|\widetilde{\theta}_i^{\tau}\varphi_i\|^2}{1 + \varphi_i^{\tau} P_i \varphi_i} \tag{5.65}$$

$\widetilde{\theta}_i = \theta - \theta_i$, and θ_i is calculated according to (4.13), (4.14) and (4.18), (4.20).

Proof. Setting

$$x_{k+1} = w_{k+1} + \theta^{\tau}(\varphi_k^0 - \varphi_k) \tag{5.66}$$

we have

$$y_{k+1} = \theta^{\tau}\varphi_k^0 + w_{k+1} = \theta^{\tau}\varphi_k + x_{k+1} \tag{5.67}$$

where φ_k^0 is given by (4.17).

Using (4.13) and (4.14) we see

$$\widetilde{\theta}_{k+1} = (I - a_k P_k \varphi_k \varphi_k^{\tau})\widetilde{\theta}_k - a_k P_k \varphi_k x_{k+1}^{\tau} \tag{5.68}$$

and

$$P_{k+1}P_k^{-1} = I - a_k P_k \varphi_k \varphi_k^{\tau}, \quad P_{k+1}^{-1}P_k = I + \varphi_k\varphi_k^{\tau}P_k. \tag{5.69}$$

By (5.68) and (5.69) it is easy to derive the following chain of equalities:

$$
\begin{aligned}
tr\widetilde{\theta}_{k+1}^{\tau} P_{k+1}^{-1} \widetilde{\theta}_{k+1} &= tr[\widetilde{\theta}_k^{\tau}(I - a_k \varphi_k \varphi_k^{\tau} P_k) - a_k x_{k+1} \varphi_k^{\tau} P_k] \\
&\quad \cdot [P_k^{-1} \widetilde{\theta}_k - P_{k+1}^{-1} a_k P_k \varphi_k x_{k+1}^{\tau}] \\
&= tr\widetilde{\theta}_k^{\tau} P_k^{-1} \widetilde{\theta}_k - a_k \|\varphi_k^{\tau} \widetilde{\theta}_k\|^2 - 2 a_k \varphi_k^{\tau} \widetilde{\theta}_k x_{k+1} \\
&\quad + a_k^2 \varphi_k^{\tau} P_k P_{k+1}^{-1} P_k \varphi_k \|x_{k+1}\|^2 \\
&= tr\widetilde{\theta}_k^{\tau} P_k^{-1} \widetilde{\theta}_k - a_k \|\varphi_k^{\tau} \widetilde{\theta}_k\|^2 - 2 a_k \varphi_k^{\tau} \widetilde{\theta}_k x_{k+1} \\
&\quad + a_k \varphi_k^{\tau} P_k \varphi_k \|x_{k+1}\|^2
\end{aligned} \tag{5.70}
$$

By Theorem 2.8, (5.63) and (4.56) it follows that

$$
\begin{aligned}
& 2\sum_{k=1}^{n} a_k \varphi_k^{\tau} \widetilde{\theta}_k x_{k+1} \\
&= 2\sum_{k=1}^{n} a_k \varphi_k^{\tau} \widetilde{\theta}_k w_{k+1} + 2\sum_{k=1}^{n} a_k \varphi_k^{\tau} \widetilde{\theta}_k \theta^{\tau} (\varphi_k^0 - \varphi_k) \\
&= O\left(\left[\sum_{k=1}^{n} a_k \|\varphi_k^{\tau} \widetilde{\theta}_k\|^2\right]^{\frac{1}{2}+\delta}\right) + \delta \sum_{k=1}^{n} a_k \|\varphi_k^{\tau} \widetilde{\theta}_k\|^2 \\
&\quad + \frac{1}{\delta} \sum_{k=1}^{n} \|\theta^{\tau} (\varphi_k^0 - \varphi_k)\|^2 \\
&= 2\delta \sum_{k=1}^{n} a_k \|\varphi_k^{\tau} \widetilde{\theta}_k\|^2 + O(\log s_{n-1}), \quad \forall \delta \in (0, \frac{1}{2})
\end{aligned} \tag{5.71}
$$

for sufficiently large n, and

$$
\begin{aligned}
& \sum_{k=1}^{n} a_k \varphi_k^{\tau} P_k \varphi_k \|x_{k+1}\|^2 \le 2 \sum_{k=1}^{n} a_k \varphi_k^{\tau} P_k \varphi_k [\|w_{k+1}\|^2 + \|\theta^{\tau} (\varphi_k^0 - \varphi_k)\|^2] \\
&= O(\log s_n) + 2\|\theta\|^2 \sum_{k=1}^{n} \|\varphi_k^0 - \varphi_k\|^2 \\
&= O(\log s_n)
\end{aligned} \tag{5.72}
$$

Summing up both side of (5.70) and paying attention to (5.71) and (5.72) we derive (5.64) immediately. Q.E.D.

Remark 5.2. When B_1 is known the unknown parameter θ given by (5.1) will be shortened to

$$
\overline{\theta}^{\tau} = [-A_1 \; \ldots \; -A_p \quad B_2 \; \ldots \; B_q \quad C_1 \; \ldots \; C_r] \tag{5.73}
$$

The ELS estimate for $\overline{\theta}$ is carried out by the following algorithm:

$$\begin{aligned}\overline{\theta}_{n+1} &= \overline{\theta}_n + \overline{a}_n \overline{P}_n \overline{\varphi}_n (y_{n+1} - B_1 u_n - \overline{\theta}_n^{\tau} \overline{\varphi}_n)^{\tau} && (5.74)\\ \overline{P}_{n+1} &= \overline{P}_n - \overline{a}_n \overline{P}_n \overline{\varphi}_n \overline{\varphi}_n^{\tau} \overline{P}_n, \quad \overline{a}_n = (1 + \overline{\varphi}_n^{\tau} \overline{P}_n \overline{\varphi}_n)^{-1}, && (5.75)\\ \overline{\varphi}_n &= [y_n^{\tau} \dots y_{n-p+1}^{\tau} \quad u_{n-1}^{\tau} \dots u_{n-q+1}^{\tau} \quad \overline{w}_n^{\tau} \dots \overline{w}_{n-r+1}^{\tau}]^{\tau}, && (5.76)\\ \overline{w}_n &= y_n - B_1 u_{n-1} - \overline{\theta}_n^{\tau} \overline{\varphi}_{n-1}, \quad \overline{w}_n = 0, \ n < 0 && (5.77)\end{aligned}$$

with arbitrary initial values $\overline{\theta}_0$, $\overline{\varphi}_0 \neq 0$ and $\overline{P}_0 > 0$.

Corresponding to (5.61)-(5.65) there obviously exist analogous relationships:

$$\|\overline{\theta}_{n+1} - \overline{\theta}\|^2 = O\left(\frac{\log \overline{s}_n}{\overline{\lambda}_{\min}(n)}\right) \quad \text{a.s.} \tag{5.78}$$

$$\sum_{i=1}^{n+1} \|\overline{w}_i - w_i\|^2 = O(\log \overline{s}_n) \quad \text{a.s.} \tag{5.79}$$

and

$$\sum_{i=1}^{n} \overline{\alpha}_i = O(\log \overline{s}_n) \quad \text{a.s.}, \tag{5.80}$$

where $\overline{s}_n = tr\overline{P}_{n+1}$, $\overline{\lambda}_{\min}(n)$ denotes the minimum eigenvalue of $\overline{P}_{n+1}^{-1}$ and $\overline{\alpha}_i = \dfrac{\|\tilde{\overline{\theta}}_i^{\tau} \overline{\varphi}_i\|^2}{1 + \overline{\varphi}_i^{\tau} \overline{P}_i \overline{\varphi}_i}$, $\tilde{\overline{\theta}}_i = \overline{\theta} - \overline{\theta}_i$. By (5.24) it is clear that

$$\overline{s}_n \xrightarrow[n \to \infty]{} \infty \quad \text{a.s.}$$

Let $\{y_n^*\}$ be a bounded reference signal and let y_{n+1}^* be $\mathcal{F}_n$-measurable. Motivated by (5.5) we naturally define

$$u_n = B_1^{-1}(y_{n+1}^* - \overline{\theta}_n^{\tau} \overline{\varphi}_n) \tag{5.81}$$

when B_1 is known. It is worth noting that B_1 is nondegenerate if Condition A3 is satisfied.

By the Åström-Wittenmark self-tuning tracker we mean the adaptive control system (3.91)-(3.94) with $\{u_n\}$ given by (5.81) and (5.74)-(5.77).

Theorem 5.3 *[GC1]. If for System (3.91)-(3.94) $d = 1$ and Conditions A1-A3 hold, then the Åström-Wittenmark self-tuning tracker is stable and optimal in the following sense:*

$$\limsup_{n \to \infty} \frac{1}{n} \sum_{i=1}^{n} (\|y_i\|^2 + \|u_i\|^2) < \infty \quad a.s. \tag{5.82}$$

$$\lim_{n \to \infty} \frac{1}{n} \sum_{i=1}^{n} (y_i - y_i^*)(y_i - y_i^*)^{\tau} = R \quad a.s. \tag{5.83}$$

Further, if $\{d_n\}$ is a nondecreasing sequence of positive numbers such that

$$\sup_n \frac{d_{n+1}}{d_n} < c$$

and

$$\|w_n\|^2 = O(d_n) \quad a.s., \tag{5.84}$$

then the following convergence rate holds true:

$$\sum_{i=1}^{n} \|y_i - y_i^* - w_i\|^2 = O(n^{\varepsilon} d_n) \quad a.s. \quad \forall \varepsilon > 0, \tag{5.85}$$

where and hereafter c denotes a constant but may vary from place to place.

Remark 5.3. By (5.58) and the conditional Chebyshev's inequality (Theorem 1.6) we have for any $\delta > 2/\beta$

$$\sum_{n=1}^{\infty} P(\|w_{n+1}\|^2 \geq n^{\delta} | \mathcal{F}_n) = \sum_{n=1}^{\infty} P(\|w_{n+1}\|^{\beta} \geq n^{\frac{\delta\beta}{2}} | \mathcal{F}_n)$$
$$\leq \sum_{n=1}^{\infty} n^{-\frac{\delta\beta}{2}} E(\|w_{n+1}\|^{\beta} | \mathcal{F}_n) < \infty \quad a.s.$$

Hence from Theorem 2.5 it follows that

$$\|w_{n+1}\|^2 = O(n^{\delta}), \quad \forall \delta \in \left(\frac{2}{\beta}, 1\right) \tag{5.86}$$

Consequently, under Condition A1 we can take $d_n = n^{\delta}$ in (5.84), $\forall \delta \in \left(\frac{2}{\beta}, 1\right)$.

Proof of Theorem 5.3.

We prove the theorem by three steps.

Step 1. We show that there exist constants $\lambda \in (0,1)$ and $M > 0$ possibly depending on ω such that

$$\|y_{k+1}\|^2 \leq M \overline{\alpha}_k \overline{\delta}_k \sum_{i=0}^{k} \lambda^{k-i} \|y_i\|^2 + O(d_k \log \overline{s}_k) + O(\log^2 \overline{s}_k) \tag{5.87}$$

where $\overline{\alpha}_i = \dfrac{\|\tilde{\overline{\theta}}_i^T \overline{\varphi}_i\|^2}{1 + \overline{\varphi}_i^T P_i \overline{\varphi}_i}$, and

$$\overline{\delta}_k \triangleq tr(\overline{P}_k - \overline{P}_{k+1}). \tag{5.88}$$

From (5.67) and (5.81) we see

$$\begin{aligned} y_{k+1} &= B_1 u_k + \overline{\theta}^{\tau} \overline{\varphi}_k + x_{k+1} = y^*_{k+1} - \overline{\theta}^{\tau}_k \overline{\varphi}_k + \overline{\theta}^{\tau} \overline{\varphi}_k + x_{k+1} \\ &= \tilde{\overline{\theta}}^{\tau}_k \overline{\varphi}_k + y^*_{k+1} + w_{k+1} + \overline{\theta}^{\tau} (\overline{\varphi}^0_k - \overline{\varphi}_k) \end{aligned} \tag{5.89}$$

where

$$\overline{\varphi}^0_k = [y^{\tau}_k \ \dots \ y^{\tau}_{k-p+1} \quad u^{\tau}_{k-1} \ \dots \ u^{\tau}_{k-q+1} \quad w^{\tau}_k \ \dots \ w^{\tau}_{k-r+1}]^{\tau}. \tag{5.90}$$

By (5.79), (5.84) and the boundedness of $\{y^*_n\}$ from (5.89) it follows that

$$\begin{aligned} \|y_{k+1}\|^2 &\le 2\|\tilde{\overline{\theta}}^{\tau} \overline{\varphi}_k\|^2 + O(\log \overline{s}_k) + O(d_k) \\ &= 2\overline{\alpha}_k[1 + \overline{\varphi}^{\tau}_k \overline{P}_{k+1} \overline{\varphi}_k + \overline{\varphi}^{\tau}_k (\overline{P}_k - \overline{P}_{k+1}) \overline{\varphi}_k] + O(d_k + \log \overline{s}_k) \\ &\le 2\overline{\alpha}_k [2 + \overline{\delta}_k \|\overline{\varphi}_k\|^2] + O(d_k + \log \overline{s}_k) \\ &= 2\overline{\alpha}_k \overline{\delta}_k \|\overline{\varphi}_k\|^2 + O(d_k + \log \overline{s}_k). \end{aligned} \tag{5.91}$$

By Condition A3 from (5.91) we know that there exists a constant $\lambda \in (0,1)$ such that

$$\begin{aligned} \|u_{k-1}\|^2 &= O\left(\sum_{i=0}^{k} \lambda^{k-i} \|y_i\|^2\right) + O\left(\sum_{i=0}^{k} \lambda^{k-i} \|w_i\|^2\right) \\ &= O\left(\sum_{i=0}^{k} \lambda^{k-i} \|y_i\|^2\right) + O(d_k), \end{aligned} \tag{5.92}$$

which incorporating (5.79) yields

$$\begin{aligned} \|\overline{\varphi}_k\|^2 &= \sum_{i=0}^{p-1} \|y_{k-i}\|^2 + \sum_{i=1}^{q-1} \|u_{k-i}\|^2 + \sum_{i=0}^{r-1} \|\overline{w}_{k-i}\|^2 \\ &= O\left(\sum_{i=0}^{k} \lambda^{k-i} \|y_i\|^2\right) + O(d_k) \\ &\quad + O\left(\sum_{i=0}^{k} \lambda^{k-i} (\|\overline{w}_{k-i} - w_{k-i}\|^2 + \|w_{k-i}\|^2)\right) \\ &= O\left(\sum_{i=0}^{k} \lambda^{k-i} \|y_i\|^2\right) + O(d_k + \log \overline{s}_k). \end{aligned} \tag{5.93}$$

Substituting (5.93) into (5.91) and noticing $\overline{\alpha}_k \overline{\delta}_k = O(\log \overline{s}_k)$ we derive (5.87) immediately.

Step 2. We now show that

$$\|\overline{\varphi}_n\|^2 = O(\overline{s}^{\varepsilon}_n d_n) \quad a.s. \quad \forall \varepsilon > 0. \tag{5.94}$$

Set

$$h_n = \sum_{i=0}^{n} \lambda^{n-i} \|y_i\|^2. \tag{5.95}$$

Adding a term $\sum_{i=0}^{k} \lambda^{k+1-i} \|y_i\|^2$ to the both sides of (5.87) we derive

$$h_{n+1} \le (\lambda + M\overline{\alpha}_k \overline{\delta}_k) h_k + O(d_k \log \overline{s}_k) + O(\log^2 \overline{s}_k) \tag{5.96}$$

and hence

$$\begin{aligned} h_{n+1} &= O\left(d_n \log \overline{s}_n + \log^2 \overline{s}_n\right) \\ &+ O\left(\sum_{i=0}^{n-1} \prod_{j=i+1}^{n} (\lambda + M\overline{\alpha}_j \overline{\delta}_j)(d_i \log \overline{s}_i + \log^2 \overline{s}_i)\right) \end{aligned} \tag{5.97}$$

or

$$\begin{aligned} h_{n+1} &= O\left(d_n \log \overline{s}_n + \log^2 \overline{s}_n\right) \\ &+ O\left(\sum_{i=0}^{n-1} \lambda^{n-i} \prod_{j=i+1}^{n} (1 + \lambda^{-1} M\overline{\alpha}_j \overline{\delta}_j)(d_i \log \overline{s}_i + \log^2 \overline{s}_i)\right) \end{aligned}$$

We note that

$$\sum_{j=0}^{\infty} \overline{\delta}_j = \sum_{j=0}^{\infty} [tr\overline{P}_j - tr\overline{P}_{j+1}] \le tr\overline{P}_0 < \infty, \tag{5.98}$$

and hence $\overline{\delta}_j \xrightarrow[j \to \infty]{} 0$.

Then by (5.80) for any $\varepsilon > 0$ there exists i_0 such that

$$\lambda^{-1} M \sum_{j=i}^{n} \overline{\alpha}_j \overline{\delta}_j \le \varepsilon \log \overline{s}_n, \quad \forall n \ge i \ge i_0. \tag{5.99}$$

Using this and the elementary inequality

$$1 + x \le e^x, \quad x \ge 0 \tag{5.100}$$

we have

$$\begin{aligned} &\prod_{j=i}^{n} (1 + \lambda^{-1} M\overline{\alpha}_j \overline{\delta}_j) \le exp\{\lambda^{-1} M \sum_{j=i}^{n} \overline{\alpha}_j \overline{\delta}_j\} \\ \le\ & exp\{\varepsilon \log \overline{s}_n\} = \overline{s}_n^{\varepsilon}, \quad \forall n \ge i \ge i_0. \end{aligned} \tag{5.101}$$

Putting this into (5.98) we obtain

$$h_{n+1} = O\left(\overline{s}_n^{\varepsilon}(d_n \log \overline{s}_n + \log^2 \overline{s}_n)\right) \quad a.s. \quad \forall \varepsilon > 0.$$

which implies

$$h_{n+1} = O(\overline{s}_n^{\varepsilon} d_n) \quad \text{and} \quad \|y_{n+1}\|^2 = O(\overline{s}_n^{\varepsilon} d_n). \tag{5.102}$$

by arbitrariness of ε.

By (5.102) from (5.92) and (5.93) it follows that

$$\|u_n\|^2 = O(\overline{s}_n^{\varepsilon} d_n) \quad \text{and} \quad \|\varphi_n\|^2 = O(\overline{s}_n^{\varepsilon} d_n) \quad \text{a.s.} \quad \forall \varepsilon > 0.$$

Step 3. We now complete the proof of the theorem.

By (5.24) and (5.59) it is easy to see that

$$\liminf_{n \to \infty} \frac{\overline{s}_n}{n} \geq trR > 0 \quad a.s.$$

Then by (5.86) we find that d_n can be taken as

$$d_n = O(\overline{s}_n^{\delta}) \quad \forall \delta \in \left(\frac{2}{\beta}, 1\right).$$

This incorporating (5.94) implies

$$\|\overline{\varphi}_n\|^2 = O(\overline{s}_n^{\delta}) \quad \forall \delta \in \left(\frac{2}{\beta}, 1\right).$$

Then by (5.80) we obtain

$$\begin{aligned}
\sum_{i=0}^{n} \|\overline{\varphi}_i^{\tau} \tilde{\overline{\theta}}_i\|^2 &= \sum_{i=0}^{n} \overline{\alpha}_i (1 + \overline{\varphi}_i^{\tau} \overline{P}_i \overline{\varphi}_i) = O(\log \overline{s}_n) + O(\sum_{i=0}^{n} \overline{\alpha}_i \|\overline{\varphi}_i\|^2) \\
&= O(\log \overline{s}_n) + O(\overline{s}_n^{\delta} \sum_{i=0}^{n} \overline{\alpha}_i) \\
&= O(\overline{s}_n^{\delta} \log \overline{s}_n), \quad \forall \delta \in \left(\frac{2}{\beta}, 1\right).
\end{aligned} \tag{5.103}$$

We now apply (5.103), (5.79), (5.59) and the boundedness of $\{y_n^*\}$ to (5.89) and find that

$$\sum_{k=0}^{n} \|y_{k+1}\|^2 = O(\overline{s}_n^{\delta}) + O(n) + O(\log \overline{s}_n) = O(\overline{s}_n^{\delta}) + O(n) \tag{5.104}$$

and by (5.92)

$$\sum_{k=0}^{n} \|u_{k-1}\|^2 = O(\bar{s}_n^{\delta}) + O(n). \tag{5.105}$$

From (5.59) and (5.79) it is easy to see

$$\sum_{k=0}^{n} \|\overline{w}_i\|^2 = O(\log \bar{s}_n) + O(n). \tag{5.106}$$

Combining (5.104)-(5.106) we conclude that

$$\bar{s}_n = O(\bar{s}_n^{\delta}) + O(n), \quad \forall \delta \in \left(\frac{2}{\beta}, 1\right),$$

which implies $\bar{s}_n = O(n)$. Thus (5.82) has been verified.

From (5.89) we have

$$\begin{aligned} &\sum_{k=0}^{n} (y_{k+1} - y_{k+1}^*)(y_{k+1} - y_{k+1}^*)^{\tau} \\ = \;& \sum_{k=0}^{n} w_{k+1} w_{k+1}^{\tau} + O\left(\left\| \sum_{k=0}^{n} [\tilde{\overline{\theta}}_k^{\tau} \overline{\varphi}_k + \overline{\theta}^{\tau} (\overline{\varphi}_k^0 - \overline{\varphi}_k)] w_{k+1}^{\tau} \right\| \right) \\ & + O\left(\sum_{k=0}^{n} [\|\tilde{\overline{\theta}}_k^{\tau} \overline{\varphi}_k\|^2 + \|\overline{\varphi}_k^0 - \overline{\varphi}_k\|^2] \right), \end{aligned} \tag{5.107}$$

whose last two terms are of the order of $O(\bar{s}_n^{\delta}) = O(n^{\delta})$, $\forall \delta \in \left(\frac{2}{\beta}, 1\right)$ by Theorem 2.8 and (5.79). From (5.107) by (5.59) we immediately derive (5.83).

Using (5.94) we have

$$\begin{aligned} \sum_{i=0}^{n} \|\tilde{\overline{\theta}}_i^{\tau} \overline{\varphi}_i\|^2 &= \sum_{i=0}^{n} \overline{\alpha}_i (1 + \overline{\varphi}_i^{\tau} \overline{P}_i \overline{\varphi}_i) \\ &= O(\log \bar{s}_n) + O(\bar{s}_n^{\varepsilon} d_n \log \bar{s}_n) = O(\bar{s}_n^{\varepsilon} d_n), \quad \forall \varepsilon > 0 \end{aligned} \tag{5.108}$$

and, finally, from (5.89) by (5.79) and (5.108)

$$\begin{aligned} \sum_{i=1}^{n} \|y_i - y_i^* - w_i\|^2 &= O\left(\sum_{i=0}^{n} [\|\tilde{\overline{\theta}}_i^{\tau} \overline{\varphi}_i\|^2 + \|\overline{\varphi}_i^0 - \overline{\varphi}_i\|^2] \right) \\ &= O(\bar{s}_n^{\varepsilon} d_n) \quad a.s., \quad \forall \varepsilon > 0. \end{aligned}$$

Q.E.D.

5.4 Stability and Optimality of ELS-Based Adaptive Trackers

In this section we establish results similar to Theorem 5.3 but for the case where B_1 is unknown. We now use the algorithms (4.13), (4.14), (4.18) and (4.20) for estimating the whole θ including B_1. At time n the estimate θ_n is again written in the block form

$$\theta_n^\tau = [-A_{1n} \ \dots \ -A_{pn} \quad B_{1n} \ \dots \ B_{qn} \quad C_{1n} \ \dots \ C_{rn}].$$

When $\det B_{1n} \neq 0$ the adaptive control u_n that solves (5.5) is expressed by

$$u_n = B_{1n}^{-1}\{y_{n+1}^* + (B_{1n}u_n - \theta_n^\tau \varphi_n)\}. \tag{5.109}$$

Unfortunately, the set $\{\det B_{1n} = 0\}$ may have a positive probability. For the SG-based adaptive tracker we have ruled out this possibility by assuming independence and continuity for distributions of $\{w_n\}$ in Lemma 5.1. Similar results also hold for the ELS algorithm. Instead of making such restrictions we now slightly modify B_{1n} when defining u_n to preclude it from being too small.

To be specific, we replace B_{1n}^{-1} in (5.109) by any $\mathcal{F}_n$-measurable $\widehat{B}_{1n}^{-1}$ satisfying the following conditions (5.110) and (5.111):

$$\widehat{B}_{1n}^\tau \widehat{B}_{1n} \geq \frac{1}{\log^\mu s_{n-1}} I \tag{5.110}$$

$$\|\widehat{B}_{1n} - B_{1n}\|^2 \leq \frac{1}{\log^\nu s_{n-1}} \tag{5.111}$$

where s_n is defined by (5.60), μ is any integer≥ 1 and $\nu > 0$ is a real number.

As a consequence of (5.24) and (5.59) we see that $s_n \xrightarrow[n\to\infty]{} \infty$ a.s. Then (5.110) keeps the minimum eigenvalue of $\widehat{B}_{1n}^\tau \widehat{B}_{1n}$ from zero by a small magnitude $\frac{1}{\log^\mu s_{n-1}}$, while (5.111) means that $\widehat{B}_{1n}$ is asymptotically equivalent to B_{1n}.

Remark 5.4. If it is known that $\|B_{1n}^{-1}\| \leq \log^\mu s_n$, $\forall n \geq 1$ for some integer μ, then $B_{1n}^\tau B_{1n} \geq \frac{1}{\log^{2\mu} s_n} I$ and as $\widehat{B}_{1n}$ we can take B_{1n} itself. In general, $\widehat{B}_{1n}$ satisfying (5.110), (5.111) can be chosen in various ways. For example, if $m = l = 1$, we may simply take

$$\widehat{B}_{1n} \triangleq \begin{cases} B_{1n} & \text{if } |B_{1n}| \geq \dfrac{1}{\log^{\frac{\mu}{2}} s_{n-1}} \\ B_{1n} + \dfrac{1}{\log^{\frac{\mu}{2}} s_{n-1}} \operatorname{Sign}(B_{1n}), & \text{otherwise,} \end{cases} \tag{5.112}$$

where

$$\mathrm{Sign}(x) = \begin{cases} 1 & x \geq 0 \\ -1 & x < 0, \end{cases}$$

and μ is an integer, $\mu \geq 1$ and $\nu = \mu$.

For a multidimensional analogue of (5.112) we need the singular value decomposition of an $m \times m$ matrix B (see e.g. [S], p.318):

$$B = V \begin{bmatrix} \Lambda & 0 \\ 0 & 0 \end{bmatrix} U^\tau \tag{5.113}$$

where $\Lambda = \begin{bmatrix} \lambda_1 & \dots & 0 \\ \vdots & \ddots & \vdots \\ 0 & \dots & \lambda_l \end{bmatrix}$ is a positive definite diagonal matrix with $\lambda_i > 0$ being singular values of B, and V and U are orthogonal matrices. In fact, as U we may take the one diagonalizing $B^\tau B$

$$U^\tau B^\tau B U = \begin{bmatrix} \Lambda^2 & 0 \\ 0 & 0 \end{bmatrix}.$$

With U fixed we then take

$$V = \left[B^{\tau+} U \begin{bmatrix} \Lambda \\ 0 \end{bmatrix} \vdots F \right]$$

where F is an $m \times (m-l)$ matrix such that $F^\tau F = I$ and $F^\tau B^{\tau+} U = 0$.

It is easy to verify that V is an orthogonal matrix and (5.113) is satisfied.

Let us assume that B_{1n} has the singular value decomposition

$$B_{1n} = V_n \begin{bmatrix} \Lambda_n & 0 \\ 0 & 0 \end{bmatrix} U_n^\tau. \tag{5.114}$$

Corresponding to (5.112) in the multidimensional case we set

$$\widehat{B}_{1n} \triangleq \begin{cases} B_{1n} & \text{if } B_{1n}^\tau B_{1n} \geq \frac{1}{\log^\mu s_{n-1}} I \\ B_{1n} + V_n U_n^\tau \frac{1}{\log^{\frac{\mu}{2}} s_{n-1}}, & \text{otherwise,} \end{cases} \tag{5.115}$$

which obviously satisfies (5.110), (5.111), where $\mu \geq 1$ is an integer.

In accordance with (5.109) we define adaptive control u_n as

$$u_n = \widehat{B}_{1n}^{-1} \{ y_{n+1}^* + (B_{1n} u_n - \theta_n^\tau \varphi_n) \}. \tag{5.116}$$

Theorem 5.4 *[GC2]. Assume that for System (3.91)-(3.94) $d = 1$ and Conditions A1-A3 stated in Section 5.3 are satisfied and that the reference signal $\{y_n^*\}$ is a.s. bounded and y_{n+1}^* is $\mathcal{F}_n$-measurable. Then the ELS-based adaptive tracker consisting of (3.91)-(3.94), (4.13), (4.14), (4.18), (4.20) and (5.116) with (5.110), (5.111) satisfied by $\widehat{B}_{1n}$ $\forall n \geq 1$ is stable and optimal in the sense that (5.82) and (5.83) hold. Moreover,*

$$\|y_n\|^2 + \|u_n\|^2 = o(n^\varepsilon d_n) \quad a.s. \quad \forall \varepsilon > 0 \tag{5.117}$$

where $\{d_n\}$ is defined in Theorem 5.3.

Proof. Similar to the previous theorem we divide the proof into three steps.

Step 1. We first show that there exist $c > 0$ and $\lambda \in (0,1)$ such that

$$\|y_{n+1}\|^2 \leq c\eta_n \sum_{i=0}^{n} \lambda^{n-i} \|y_i\|^2 + \xi_n, \tag{5.118}$$

where c possibly depends on ω, $\{\xi_n\}$ is a nondecreasing positive sequence satisfying

$$\xi_k = O\left(d_k \log^{2\mu+2} s_k + \log^{2\mu+3} s_k\right), \tag{5.119}$$

$$\eta_k = (\alpha_k \delta_k \log s_{k-1})^{\mu+1} + \sum_{i=1}^{\mu} (\alpha_k \delta_k)^i + \frac{1}{\log^\nu s_{k-1}}, \tag{5.120}$$

α_k is given by (5.65) and

$$\delta_k = tr(P_k - P_{k+1}). \tag{5.121}$$

By (5.24) and (5.59) it is clear that

$$\liminf_{n \to \infty} \frac{s_n}{n} \geq trR > 0 \quad \text{a.s.} \tag{5.122}$$

Setting

$$\Delta \widehat{B}_{1k} \triangleq \widehat{B}_{1k} - B_{1k} \tag{5.123}$$

then by (5.116) we have

$$y_{k+1}^* = \Delta \widehat{B}_{1k} u_k + \theta_k^\tau \varphi_k \tag{5.124}$$

and

$$\begin{aligned} y_{k+1} &= \theta^\tau \varphi_k + \theta^\tau (\varphi_k^0 - \varphi_k) + w_{k+1} \\ &= \widetilde{\theta}_k^\tau \varphi_k - \Delta \widehat{B}_{1k} u_k + y_{k+1}^* + \theta^\tau (\varphi_k^0 - \varphi_k) + w_{k+1} \end{aligned} \tag{5.125}$$

where φ_k^0 is given by (4.17) and $\widetilde{\theta}_k = \theta - \theta_k$.

By Condition A3 there exists $\lambda \in (0,1)$ such that

$$\|u_k\|^2 = O\left(\sum_{i=0}^{k+1} \lambda^{k-i}\|y_i\|^2\right) + O\left(\sum_{i=0}^{k+1} \lambda^{k-i}\|w_i\|^2\right). \tag{5.126}$$

Noticing (5.124) we have

$$\begin{aligned} B_1 u_k &= B_1 u_k + y^*_{k+1} - \Delta \widehat{B}_{1k} u_k - \theta_k^\tau \varphi_k + \theta^\tau \varphi_k - \theta^\tau \varphi_k \\ &= \widetilde{\theta}_k^\tau \varphi_k - \Delta \widehat{B}_{1k} u_k + y^*_{k+1} + (B_1 u_k - \theta^\tau \varphi_k). \end{aligned} \tag{5.127}$$

By A3, B_1 is nondegenerate. Hence $\|B_1^{-1}\Delta \widehat{B}_{1k}\| < \frac{1}{2}$ for sufficiently large k. Then from (5.127) we derive

$$\|u_k\| \le 2\|B_1^{-1}\|(\|\widetilde{\theta}_k^\tau \varphi_k\| + \|y^*_{k+1}\| + \|B_1 u_k - \theta^\tau \varphi_k\|)$$

$$\|u_k\|^2 \le 12\|B_1^{-1}\|^2(\|\widetilde{\theta}_k^\tau \varphi_k\|^2 + \|y^*_{k+1}\|^2 + \|B_1 u_k - \theta^\tau \varphi_k\|^2). \tag{5.128}$$

Applying (5.126) to $\|B_1 u_k - \theta^\tau \varphi_k\|^2$, which is free of u_k, from (5.128) we find that

$$\begin{aligned} \|u_k\|^2 &\le 12\|B_1^{-1}\|^2\|\widetilde{\theta}_k^\tau \varphi_k\|^2 + O\left(\sum_{i=0}^{k} \lambda^{k-i}\|y_i\|^2\right) \\ &\quad + O(d_k + \log s_k) \end{aligned} \tag{5.129}$$

and

$$\begin{aligned} \|\varphi_k\|^2 &= \|u_k\|^2 + [\|\varphi_k\|^2 - \|u_k\|^2] \\ &\le 12\|B_1^{-1}\|^2\|\widetilde{\theta}_k^\tau \varphi_k\|^2 + O\left(\sum_{i=0}^{k} \lambda^{k-i}\|y_i\|^2\right) \\ &\quad + O(d_k + \log s_k). \end{aligned} \tag{5.130}$$

We have the following estimate

$$\begin{aligned} \|\widetilde{\theta}_k^\tau \varphi_k\|^2 &= \alpha_k(1 + \varphi_k^\tau P_{k+1}\varphi_k + \varphi_k^\tau(P_k - P_{k+1})\varphi_k) \\ &\le \alpha_k \delta_k \|\varphi_k\|^2 + 2\alpha_k = \alpha_k \delta_k \|\varphi_k\|^2 + O(\log s_k) \end{aligned} \tag{5.131}$$

since by (5.64) $\alpha_k = O(\log s_k)$ and by (4.14)

$$\varphi_k^\tau P_{k+1}\varphi_k = \varphi_k^\tau P_k \varphi_k - a_k(\varphi_k^\tau P_k \varphi_k)^2 < 1. \tag{5.132}$$

Combining (5.130) and (5.131) leads to

$$\|\varphi_k\|^2 \le 12\|B_1^{-1}\|^2\alpha_k\delta_k\|\varphi_k\|^2 + O\left(\sum_{i=0}^{k} \lambda^{k-i}\|y_i\|^2\right) + O(d_k + \log s_k)$$

and hence for any integer $s \geq 1$

$$\begin{aligned} \|\varphi_k\|^2 \le\ & (12\|B_1^{-1}\|^2)^s (\alpha_k\delta_k)^s \|\varphi_k\|^2 + \left[\sum_{i=0}^{s-1}(\alpha_k\delta_k)^i\right] \\ & O\left(\sum_{i=0}^{k}\lambda^{k-i}\|y_i\|^2\right) \\ & +O(d_k \log^{s-1} s_k + \log^s s_k) \end{aligned} \tag{5.133}$$

where we have used $\alpha_k = O(\log s_k)$ and the boundedness of δ_k which is easily seen from

$$\sum_{i=0}^{\infty}\delta_i = \sum_{i=0}^{\infty}(trP_i - trP_{i+1}) < \infty. \tag{5.134}$$

Noticing that by (5.63)

$$\sum_{i=0}^{k+1}\|\varphi_i^0 - \varphi_i\|^2 = O(\log s_k) \quad \text{a.s.}$$

and using (5.129) and (5.130) from (5.125) we find for suitably large k

$$\begin{aligned} \|y_{k+1}\|^2 \le\ & 3\|\widetilde{\theta}_k^\tau\varphi_k\|^2 + 3\|\Delta\widehat{B}_{1k}\|^2\|u_k\|^2 + O(d_k + \log s_k) \\ \le\ & (3 + 36\|B_1^{-1}\|^2\|\Delta\widehat{B}_{1k}\|^2)\|\widetilde{\theta}_k^\tau\varphi_k\|^2 \\ & +O\left(\|\Delta\widehat{B}_{1k}\|^2\sum_{i=0}^{k}\lambda^{k-i}\|y_i\|^2\right) + O\left(\|\Delta\widehat{B}_{1k}\|^2(d_k + \log s_k)\right) \\ & +O(d_k + \log s_k) \\ \le\ & 4\alpha_k\delta_k\|\varphi_k\|^2 + O\left(\frac{1}{\log^\nu s_{k-1}}\sum_{i=0}^{k}\lambda^{k-i}\|y_i\|^2\right) \\ & +O(d_k + \log s_k) \end{aligned} \tag{5.135}$$

where for the last inequality we have used (5.131) and $\Delta\widehat{B}_{1k} \xrightarrow[k\to\infty]{} 0$.

Applying the estimate (5.133) to (5.135) yields

$$\begin{aligned} \|y_{k+1}\|^2 \le\ & 4\,(12\|B_1^{-1}\|^2)^s (\alpha_k\delta_k)^{s+1}\|\varphi_k\|^2 \\ & + \left[\frac{1}{\log^\nu s_{k-1}} + \sum_{i=0}^{s}(\alpha_k\delta_k)^i\right] O\left(\sum_{i=0}^{k}\lambda^{k-i}\|y_i\|^2\right) \\ & +O(d_k \log^s s_k + \log^{s+1} s_k), \quad \forall s \geq 1. \end{aligned} \tag{5.136}$$

We now derive the upper bound for $\|\varphi_k\|^2$ in terms of $\|y_i\|^2$, $i \leq k$. By (5.110), (5.61) and the boundedness of $\{y_n^*\}$ from (5.109) we readily obtain

$$\|u_k\|^2 \le O\left(\log^{\mu+1} s_{k-1}\left[\sum_{i=0}^{p-1}\|y_{k-i}\|^2 + \sum_{i=1}^{q-1}\|u_{k-i}\|^2\right.\right.$$

$$+\sum_{i=0}^{r-1}\|\widehat{w}_{k-i}\|^2\Bigg]\Bigg)+O\left(\log^{\mu} s_{k-1}\right). \tag{5.137}$$

Expressing $\|u_{k-i}\|^2$ on the right-hand side of (5.137) by (5.126) we see that

$$\begin{aligned}\|u_k\|^2 &\leq O\left(\log^{\mu+1} s_{k-1}\sum_{i=0}^{k}\lambda^{k-i}\left(\|y_i\|^2+\|\widehat{w}_i\|^2+\|w_i\|^2\right)\right)\\ &\quad +O\left(\log^{\mu} s_{k-1}\right).\end{aligned}$$

By this and (5.63) it is easy to see

$$\begin{aligned}\|\varphi_k\|^2 &= \|u_k\|^2+[\|\varphi_k\|^2-\|u_k\|^2]\\ &= O\left(\log^{\mu+1} s_{k-1}\sum_{i=0}^{k}\lambda^{k-i}\left(\|y_i\|^2+\|\widehat{w}_i\|^2+\|w_i\|^2\right)\right)\\ &\quad +O\left(\log^{\mu} s_{k-1}\right)\\ &= O\left(\log^{\mu+1} s_{k-1}\sum_{i=0}^{k}\lambda^{k-i}\|y_i\|^2\right)\\ &\quad +O\left(d_k\log^{\mu+1} s_{k-1}+\log^{\mu+2} s_{k-1}\right).\end{aligned} \tag{5.138}$$

Substituting this into (5.136) and setting $s=\mu$ we obtain

$$\begin{aligned}\|y_{k+1}\|^2 &\leq O\left(\sum_{i=0}^{k}\lambda^{k-i}\|y_i\|^2\right)\\ &\quad \left[(\alpha_k\delta_k\log s_{k-1})^{\mu+1}+\frac{1}{\log^{\nu} s_{k-1}}+\sum_{i=0}^{\mu}(\alpha_k\delta_k)^i\right]\\ &\quad +O(d_k\log^{2\mu+2} s_k+\log^{2\mu+3} s_k)\end{aligned} \tag{5.139}$$

which proves (5.118).

Step 2. We now show that

$$\|\varphi_k\|^2=O(s_k^{\varepsilon}d_k)\quad \text{a.s.},\quad \forall\varepsilon>0. \tag{5.140}$$

Set

$$h_n=\sum_{i=0}^{n}\lambda^{n-i}\|y_i\|^2.$$

Adding a term $\sum_{i=0}^{n}\lambda^{n+1-i}\|y_i\|^2$ to both sides of (5.118) we have

$$h_{n+1}\leq(\lambda+c\eta_n)h_n+\xi_n$$

and hence

$$
\begin{aligned}
h_{n+1} &\le \xi_n + \sum_{i=0}^{n-1} \prod_{j=i+1}^{n} (\lambda + c\eta_j)\xi_i + \left[\prod_{j=0}^{n}(\lambda + c\eta_j)\right] h_0 \\
&= \xi_n + \sum_{i=0}^{n-1} \left[\prod_{j=i+1}^{n} \left(1 + \frac{c}{\lambda}\eta_j\right)\right] \lambda^{n-i}\xi_i \\
&\quad + \left[\prod_{j=0}^{n} \left(\lambda + \frac{c}{\lambda}\eta_j\right)\right] \lambda^{n+1}\xi_0.
\end{aligned}
\tag{5.141}
$$

For a fixed $\varepsilon > 0$ take $\delta > 0$ sufficiently small and possibly random so that

$$
\delta \sum_{j=0}^{n} \alpha_j \le \varepsilon (\log s_n) \quad \text{a.s.,} \quad \forall n \tag{5.142}
$$

and select i_0 large enough such that

$$
\left(\frac{c}{\lambda}\right)^{\frac{1}{s+1}} \frac{(\mu+1)^2}{\delta} \sum_{j=i_0}^{\infty} \delta_j \le \varepsilon \quad \text{a.s.,} \quad \forall s = 0,\ 1,\ \ldots,\ \mu. \tag{5.143}
$$

Since $s_n \xrightarrow[n \to \infty]{} \infty$ and $\lambda < 1$ we may assume

$$
\sup_{j \ge i_0} \left(1 + \frac{c}{\lambda \log^{\nu} s_{j-1}}\right) < 2 - \lambda.
$$

Then by (5.64), (5.142), (5.143) and the elementary inequalities $1 + xy \le (1+x)(1+y)$, $x,\ y \ge 0$; $1 + x^b \le e^{bx}$, $x \ge 0$, $b \ge 1$ we estimate the product appearing in (5.141) as follows:

$$
\begin{aligned}
&\prod_{j=i}^{n} \left(1 + \frac{c}{\lambda}\eta_j\right) \\
\le\ & \prod_{j=i}^{n} \left[1 + \frac{c}{\lambda}(\alpha_j\delta_j \log s_{j-1})^{\mu+1}\right] \prod_{k=1}^{\mu} \prod_{j=i}^{n} \left(1 + \frac{c}{\lambda}(\alpha_j\delta_j)^k\right) \\
&\cdot \prod_{j=i}^{n} \left(1 + \frac{c}{\lambda \log^{\nu} s_{j-1}}\right) \\
\le\ & \prod_{j=i}^{n} \left[1 + \left(\frac{\delta\alpha_j}{\mu+1}\right)^{\mu+1}\right] \prod_{j=i}^{n} \left[1 + \frac{c}{\lambda}\left(\frac{\mu+1}{\delta}\delta_j \log s_{j-1}\right)^{\mu+1}\right] \\
&\cdot \prod_{k=1}^{\mu} \prod_{j=i}^{n} \left[1 + \left(\frac{\delta}{k}\alpha_j\right)^k\right] \prod_{j=i}^{n} \left[1 + \frac{c}{\lambda}\left(\frac{k}{\delta}\delta_j\right)^k\right] \cdot (2-\lambda)^{n-i+1}
\end{aligned}
$$

$$
\begin{aligned}
&\le \exp\left\{\delta\sum_{j=i}^{n}\alpha_j\right\}\cdot\exp\left\{\left(\frac{c}{\lambda}\right)^{\frac{1}{\mu+1}}\frac{(\mu+1)^2}{\delta}\log s_{n-1}\sum_{j=i}^{n}\delta_j\right\}\\
&\quad\cdot\prod_{k=1}^{\mu}\left[\exp\left\{\delta\sum_{j=i}^{n}\alpha_j\right\}\cdot\exp\left\{\left(\frac{c}{\lambda}\right)^{\frac{1}{k}}\frac{k^2}{\delta}\sum_{j=i}^{n}\delta_j\right\}\right]\cdot(2-\lambda)^{n-i+1}\\
&\le s_n^{\varepsilon}\cdot s_{n-1}^{\varepsilon}\cdot s_n^{\mu\varepsilon}\cdot e^{\mu\varepsilon}(2-\lambda)^{n-i+1}\\
&\le e^{\mu\varepsilon}s_n^{(\mu+2)\varepsilon}(2-\lambda)^{n-i+1},\quad i\ge i_0,\quad \forall\varepsilon>0. \qquad (5.144)
\end{aligned}
$$

Paying attention to (5.119) from (5.141), (5.144) we arrive at that

$$
\begin{aligned}
h_{n+1} &\le O(\xi_n)+O\left(s_n^{(\mu+2)\varepsilon}\sum_{i=1}^{n}(2\lambda-\lambda^2)^{n-i}\xi_{i-1}\right)\\
&\le O\left(s_n^{(\mu+2)\varepsilon}\left[d_n\log^{2\mu+2}s_n+\log^{2\mu+3}s_n\right]\right).
\end{aligned}
$$

By the arbitrariness of ε from this it follows that

$$
h_{n+1}=O(d_n s_n^{\varepsilon}),\quad \forall\varepsilon>0.
$$

From this and (5.138) the desired (5.140) follows immediately.

Step 3. We now complete the proof of the theorem.

From (5.86) and (5.122) we see

$$
d_n=O(s_n^{\delta})\quad \text{a.s.}\quad \forall\delta\in\left(\frac{2}{\beta},1\right)
$$

and from (5.140)

$$
\|\varphi_n\|^2=O(s_n^{\delta})\quad \text{a.s.}\quad \forall\delta\in\left(\frac{2}{\beta},1\right). \qquad (5.145)
$$

Then using (5.64) we have

$$
\begin{aligned}
\sum_{i=0}^{n}\|\tilde{\theta}_i^{T}\varphi_i\|^2 &= \sum_{i=0}^{n}\alpha_i(1+\varphi_i^{T}P_i\varphi_i)\\
&= O(\log s_n)+O\left(\sum_{i=0}^{n}\alpha_i\|\varphi_i\|^2\right)\\
&= O(\log s_n)+O\left(s_n^{\delta}\sum_{i=0}^{n}\alpha_i\right)\\
&= O\left(s_n^{\delta}\right)\quad \text{a.s.}\quad \forall\delta\in\left(\frac{2}{\beta},1\right). \qquad (5.146)
\end{aligned}
$$

By (5.146), (5.59) and (5.63) from (5.125) it is easy to see

$$\begin{aligned}\sum_{i=0}^{n}\|y_{i+1}\|^2 &= O(s_n^\delta)+o\left(\sum_{i=1}^{n}\|u_i\|^2\right)+O(\log s_n)+O(n)\\ &= o(s_n)+O(n),\end{aligned} \tag{5.147}$$

$$\sum_{i=1}^{n}\|\widehat{w}_i\|^2 = 2\sum_{i=1}^{n}[\|\widehat{w}_i-w_i\|^2+\|w_i\|^2] = O(\log s_n)+O(n) \tag{5.148}$$

and from (5.126)

$$\begin{aligned}\sum_{i=0}^{n}\|u_i\|^2 &= O\left(\sum_{i=0}^{n+1}\|y_i\|^2\right)+O\left(\sum_{i=0}^{n+1}\|w_i\|^2\right)\\ &= o(s_n)+O(n).\end{aligned} \tag{5.149}$$

Combining (5.147)-(5.149) leads to

$$s_n = o(s_n)+O(n)$$

or

$$s_n = O(n) \tag{5.150}$$

which implies (5.83).

From (5.125) we know that

$$\begin{aligned}&\frac{1}{n}\sum_{i=1}^{n}(y_i-y_i^*)(y_i-y_i^*)^\tau\\ =\ &\frac{1}{n}\sum_{i=1}^{n}\left[\widetilde{\theta}_{i-1}^\tau\varphi_{i-1}-\Delta\widehat{B}_{1i-1}u_{i-1}+\theta^\tau(\varphi_{i-1}^0-\varphi_{i-1})+w_i\right]\\ &\cdot\left[\widetilde{\theta}_{i-1}^\tau\varphi_{i-1}-\Delta\widehat{B}_{1i-1}u_{i-1}+\theta^\tau(\varphi_{i-1}^0-\varphi_{i-1})+w_i\right]^\tau.\end{aligned} \tag{5.151}$$

Applying (5.59), (5.63), (5.111), (5.146) and (5.150) to (5.151) yields (5.83). Finally (5.117) follows from (5.140) and (5.150). Q.E.D.

Remark 5.5. It is not clear if θ_n in the ELS-based trackers discussed above converges and if the ELS-based tracker (3.91)-(3.94), (4.13), (4.14), (4.18), (4.20), (5.109) is optimal whenever B_{1n} is nondegenerate.

Remark 5.6. Unlike Theorem 5.4, in the previous work [GC3], [LY] for stability and optimality of the ELS-based adaptive tracker a parallel algorithm is usually involved which is used to reduce the magnitude of the system output on the basis of Theorem 5.1.

5.5 Model Reference Adaptive Control

In this section we use the adaptive tracker developed in Section 5.2 to solve the adaptive control problem discussed in Section 3.7 when the system coefficients are unknown.

To be specific, let the system be described by a CARIMA model [Cl], [CMT], [Sc], [CZ2]:

$$\begin{aligned} A(z)y_n &= B(z)u_n + e_n, \\ D(z)e_n &= C(z)w_n, \ y_i = w_i = 0, \ u_i = 0, \ i < 0, \end{aligned} \tag{5.152}$$

where $A(z)$, $B(z)$ and $C(z)$ are given by (3.91)-(3.94) with unknown A_i, $i = 1, ..., p$, B_j, $j = 1, ..., q$, C_k, $k = 1, ..., q$, while $D(z)$ is a known scalar polynomial,

$$D(z) = 1 + \alpha_1 z + \ldots + \alpha_s z^s. \tag{5.153}$$

The loss function is

$$\limsup_{n\to\infty} \frac{1}{n} \sum_{i=0}^{n} \|A^0(z)y_i - B^0(z)y_i^* + Q(z)D(z)u_{i-d}\|^2 \tag{5.154}$$

where $A^0(z)$, $B^0(z)$ and $Q(z)$ are given matrix polynomials, and $\{y_n^*\}$ is a given bounded reference signal with y_n^* being $\mathcal{F}_{i-d}$-measurable.

It is clear that (5.152) is reduced to (3.91) if $D(z) \equiv 1$, and the problem is reduced to the ordinary model reference adaptive control if $Q(z) \equiv 0$.

We recall the relationship (3.147) obtained in the proof of Theorem 3.8:

$$\begin{aligned} &(\det C(z))(\overline{y}_n - M(z)w_n) = (\det C(z))N(z)w_{n-d} \\ &+A^0(z)[G(z)y_{n-d} + F(z)(\mathrm{Adj}C(z))B(z)D(z)u_{n-d}] \\ &(\det C(z))[Q(z)D(z)u_{n-d} - B^0(z)y_n^*] \end{aligned} \tag{5.155}$$

where

$$\overline{y}_n = A^0(z)y_n - B^0(z)y_n^* + Q(z)D(z)u_{n-d}. \tag{5.156}$$

By (5.152) replacing $(\det C(z))N(z)w_{n-d}$ on the right-hand side of (5.155) by the following expression

$$N(z)D(z)(\mathrm{Adj}C(z))(A(z)y_{n-d} - z^d B(z)u_{n-d})$$

leads to

$$\begin{aligned} &(\det C(z))(\overline{y}_n - M(z)w_n) \\ = \ &G_0(z)y_{n-d} + G_1(z)y_n^* + G_2(z)D(z)u_{n-d} \end{aligned} \tag{5.157}$$

where

$$\begin{aligned} G_0(z) &= N(z)D(z)(\mathrm{Adj}C(z))A(z) + A^0(z)G(z), \\ G_1(z) &= -(\det C(z))B^0(z), \\ G_2(z) &= A^0(z)F(z)(\mathrm{Adj}C(z))B(z) + (\det C(z))Q(z) \\ &\quad -N(z)(\mathrm{Adj}C(z))B(z)z^d. \end{aligned}$$

From Theorem 3.8 we know that the optimal control is the one that makes the right-hand side of (5.157) equal to zero. Then the implicit approach of adaptive control suggests we take the control from the following equation

$$\theta_n^{*\tau}\varphi_n^* = 0 \tag{5.158}$$

where θ_n^* is the estimate at time n for θ^* consisting of coefficients of $G_0(z)$, $G_1(z)$, $G_2(z)$ and $\det C(z)$:

$$\begin{aligned} \theta^* = [\ & G_{00}\ G_{01}\ \dots\ G_{0g_0}\quad G_{10}\ G_{11}\ \dots\ G_{1g_1} \\ & G_{20}\ G_{21}\ \dots\ G_{2g_2}\quad c_1 I\ \dots\ c_{mr} I]^\tau \end{aligned} \tag{5.159}$$

and φ_n^* is the corresponding regressor

$$\begin{aligned} \varphi^{*\tau} = & [y_n^\tau\ \dots\ y_{n-g_0}^\tau\quad y_{n+d}^{*\tau}\ \dots\ y_{n+d-g_1}^{*\tau} \\ & D(z)u_n^\tau\ \dots\ D(z)u_{n-g_2}^\tau \\ & -\varphi_{n-1}^{*\tau}\theta_{n-1}^*\ \dots\ -\varphi_{n-mr}^{*\tau}\theta_{n-mr}^*]^\tau \end{aligned} \tag{5.160}$$

Here we have assumed that

$$\begin{aligned} G_0(z) &= G_{00} + G_{01}z + \ \dots\ + G_{0g_0}z^{g_0} \\ G_1(z) &= G_{10} + G_{11}z + \ \dots\ + G_{1g_1}z^{g_1} \\ G_2(z) &= G_{20} + G_{21}z + \ \dots\ + G_{2g_2}z^{g_2} \\ \det C(z) &= 1 + c_1 z + \ \dots\ + c_{mr}z^{mr}, \end{aligned}$$

and that θ_n^* is given by the SG algorithm

$$\theta_n^* = \theta_{n-1}^* + \frac{a}{r_{n-d}^*}\varphi_{n-d}^*(\overline{y}_n^\tau - \varphi_{n-d}^{*\tau}\theta_{n-d}^*), \quad a > 0. \tag{5.161}$$

Thus we are using an SG-based tracker to solve a model reference adaptive control problem for System (5.152) more general than (3.91).

Theorem 5.5 *[CZ2]. Suppose that*
1) Conditions (5.14) and (5.15) hold;
2) $A^0(z)$ *is stable and* $\det C(z) - \frac{a}{2}$ *is SPR;*
3) $m = l$ *and* $(\det A^0(z))B(z) + A(z)(\mathrm{Adj}A^0(z))Q(z)D(z)$ *is stable.*

If adaptive control can be defined from (5.158), then it stabilizes the output and minimizes the loss function (5.154):

$$\limsup_{n\to\infty} \frac{1}{n}\sum_{i=0}^{n}(\|y_i\|^2 + \|D(z)u_i\|^2) < \infty \quad a.s. \tag{5.162}$$

$$\limsup_{n\to\infty} \frac{1}{n}\sum_{i=0}^{n}\|A^0(z)y_i - B^0(z)y_i^* + Q(z)D(z)u_{i-d}\|^2$$

$$= tr\sum_{i=0}^{d-1} M_i R M_i^T \quad a.s. \tag{5.163}$$

Proof. Set

$$z_{i-d} = \overline{y}_i - M(z)w_i - \theta_{i-d}^{*T}\varphi_{i-d}^*, \tag{5.164}$$

$$\widetilde{\theta}_n^* = \theta^* - \theta_n^*. \tag{5.165}$$

By (5.156), (5.157), (5.159) and (5.160) it is obvious that

$$(\det C(z))(\overline{y}_n - M(z)w_n - \theta_{n-d}^{*T}\varphi_{n-d}^*)$$

$$= \theta^{*T}\varphi_{n-d}^* - \theta_{n-d}^{*T}\varphi_{n-d}^* = \widetilde{\theta}_{n-d}^{*T}\varphi_{n-d}^* \tag{5.166}$$

By Lemma 5.2 we have

$$\sum_{i=0}^{\infty}\frac{\|z_i\|^2}{r_i^*} < \infty \quad \text{a.s.} \tag{5.167}$$

and

$$\sup_n \, tr\theta_n^{*T}\theta_n^* < \infty \quad \text{a.s.} \tag{5.168}$$

Notice that (5.152) is equivalent to

$$D(z)A(z)y_n = D(z)B(z)u_n + C(z)w_n. \tag{5.169}$$

Hence y_n can be expressed as

$$y_n = w_n + f_{n-1}$$

where f_{n-1} is $\mathcal{F}_{n-1}$-measurable. Consequently, by Theorem 2.8 and (5.15) we find that

$$\sum_{i=1}^{n}\|y_i\|^2 \to \infty \quad \text{and} \quad r_n^* \xrightarrow[n\to\infty]{} \infty \quad \text{a.s.} \tag{5.170}$$

Applying the Kronecker lemma to (5.167) leads to

$$\sum_{i=1}^{n} \|z_i\|^2 = o(r_n^*) \tag{5.171}$$

Using (5.156) and (5.169) we easily find that

$$\begin{aligned} & A(z)(\mathrm{Adj}A^0(z))D(z)\overline{y}_n \\ = \; & [(\det A^0(z))B(z) + A(z)(\mathrm{Adj}A^0(z))Q(z)D(z)]D(z)u_{n-d} \\ & +(\det A^0(z))C(z)w_n - A(z)(\mathrm{Adj}A^0(z))B^0(z)D(z)y_n^* \end{aligned} \tag{5.172}$$

and by (5.164), (5.158)

$$\begin{aligned} & [(\det A^0(z))B(z) + A(z)(\mathrm{Adj}A^0(z))Q(z)D(z)]D(z)u_{n-d} \\ = \; & A(z)(\mathrm{Adj}A^0(z))D(z)z_n + A(z)(\mathrm{Adj}A^0(z))D(z)M(z)w_n \\ & +(\det A^0(z))C(z)w_n - A(z)(\mathrm{Adj}A^0(z))B^0(z)D(z)y_n^*. \end{aligned} \tag{5.173}$$

Then by Condition 3), (5.15) and the boundedness of $\{y_n^*\}$ we conclude that

$$\sum_{i=1}^{n} \|D(z)u_{i-d}\|^2 = O\left(\sum_{i=1}^{n} \|z_{i-d}\|^2 + n\right)$$

and by stability of $A^0(z)$ and (5.156)

$$\sum_{i=1}^{n} \|y_i\|^2 = O\left(\sum_{i=1}^{n} \|z_{i-d}\|^2 + n\right).$$

Therefore, we have

$$r_n^* = O\left(\sum_{i=1}^{n} \|z_i\|^2 + n\right)$$

which together with (5.171) implies

$$\sum_{i=1}^{n} \|z_i\|^2 = o(n). \tag{5.174}$$

Then (5.162) follows immediately.

Finally, from (5.164) by Theorem 2.8 for any $\eta \in \left(\frac{1}{2}, 1\right)$ we have

$$\begin{aligned} & \limsup_{n\to\infty} \frac{1}{n} \sum_{i=0}^{n} \|\overline{y}_i\|^2 = \limsup_{n\to\infty} \frac{1}{n} \sum_{i=0}^{n} \|M(z)w_i + z_{i-d}\|^2 \\ = \; & \limsup_{n\to\infty} \frac{1}{n} \sum_{i=0}^{n} \|M(z)w_i\|^2 \end{aligned}$$

$$
\begin{aligned}
&+\limsup_{n\to\infty}\left[\frac{1}{n}\sum_{i=0}^{n}\|z_{i-d}\|^2+O\left(\frac{1}{n}\left(\sum_{i=0}^{n}\|z_{i-d}\|^2\right)^{\eta}\right)\right]\\
&= tr\sum_{i=0}^{d-1}M_iRM_i^{\tau}
\end{aligned}
$$

which is the minimum of the loss function (5.154) by Theorem 3.8. The proof is completed. Q.E.D.

CHAPTER 6

Coefficient Estimation in Adaptive Control Systems

In Chapter 4 we discussed the coefficient estimation problem for linear stochastic systems without monitoring, i.e. there is no specific purpose for the system input. In order for the coefficient estimate to be consistent, we have clarified how the stochastic regressors φ_n or φ_n^0 or ψ_n should behave. Unfortunately, as to be shown in this chapter, the required properties of φ_n or φ_n^0 or ψ_n, in general, may not be satisfied for adaptive control systems.

In this chapter it is demonstrated that to make the parameter estimate in an adaptive control system strongly consistent, the system input should be somehow excited either by an external dither or by the control purpose itself. For example, for an adaptive tracking system the sufficient excitation will be provided if the system input is disturbed by an external excitation signal or if the system output is forced to follow a reference signal that is "rich" in a certain sense. However, in general, it is not easy to verify whether or not the system input is sufficiently excited by the control purpose itself, so in order to obtain the consistent parameter estimate while minimizing a loss function, an easier way is to add a diminishing excitation signal to the system input as will be done in the later chapters.

6.1 Necessity of Excitation for Consistency of Estimates

We continue considering the system described by (3.91)-(3.94) with known (p, q, r, d) but with unknown

$$\theta^\tau = [-A_1 \quad \dots \quad -A_p \quad B_1 \quad \dots \quad B_q \quad C_1 \quad \dots \quad C_r]. \tag{6.1}$$

The question raised in this section is whether or not the conditions for strong consistency of parameter estimates derived in Chapter 4 are fulfilled for adaptive control systems. From the following discussion we shall see that the answer , in general, is negative.

Theorem 5.1 shows that the equation (5.4) defines an adaptive control law which minimizes the tracking error (5.2). However, it tells us nothing about the consistency of parameter estimates. As a matter of fact, θ_n does not converge to θ in general. To see this let us assume $y_n^* \equiv 0$. Then $\theta_n^\tau \psi_n \equiv 0$ and from (4.22) it follows that

$$\theta_n^\tau(\theta_{n+1} - \theta_n) = a\theta_n^\tau \frac{\psi_n}{r_n} y_{n+1}^\tau \equiv 0.$$

Hence we have

$$\begin{aligned} \theta_n^\tau \theta_n &= \theta_{n-1}^\tau \theta_{n-1} + (\theta_n - \theta_{n-1})^\tau (\theta_n - \theta_{n-1}) \\ &= \theta_0^\tau \theta_0 + \sum_{i=1}^{n} (\theta_i - \theta_{i-1}^\tau)(\theta_i - \theta_{i-1}) \geq \theta_0^\tau \theta_0, \end{aligned} \tag{6.2}$$

which means that θ_n will never converge to θ if the initial value θ_0 is chosen so that $\theta_0^\tau \theta_0 > \theta^\tau \theta$.

A detailed analysis [BKW] shows that θ_n tends to $\gamma\theta$ where γ is a scalar.

We now prove this for the special case where $m = l = 1$, $r = 0$, $A(z) = 1 + az$, $B(z) = bz$ in (3.91)-(3.94), and the step size $a = 1$ in (4.22).

In the present case we note that $C(z) = I$, $\theta^\tau = [-a \quad b]$, and $\zeta_{n+1} = \widetilde{\theta}_n^\tau \psi_n$. Then from (4.82) we derive that

$$\begin{aligned} \|\widetilde{\theta}_{n+1}\|^2 &= \|\widetilde{\theta}_n\|^2 - \frac{2}{r_n}(\zeta_{n+1} + w_{n+1})\zeta_{n+1} \\ &\quad + \frac{\|\psi_n\|^2}{r_n^2} \|\zeta_{n+1} + w_{n+1}\|^2. \end{aligned} \tag{6.3}$$

Set

$$m_n = \|\widetilde{\theta}_n\|^2 + \sigma^2 E\left(\sum_{i=1}^{\infty} \frac{\|\psi_i\|^2}{r_i^2} | \mathcal{F}_n\right) - \sigma^2 \sum_{i=1}^{n-1} \frac{\|\psi_i\|^2}{r_i^2} \geq 0,$$

where σ^2 denotes $\sup_n E(\|w_{n+1}\|^2 | \mathcal{F}_n)$ which is finite by (5.14).

Noticing that ζ_{n+1} is $\mathcal{F}_n$-measurable from (6.3) we then have

$$\begin{aligned} E(m_{n+1}|\mathcal{F}_n) = & \ \|\widetilde{\theta}_n\|^2 - \frac{2}{r_n}\zeta_{n+1}^2 + \frac{\|\psi_n\|^2}{r_n^2}\zeta_{n+1}^2 + \frac{\|\psi_n\|^2}{r_n^2}E(\|w_{n+1}\|^2|\mathcal{F}_n) \\ & + \sigma^2 E\left(\sum_{i=1}^{\infty}\frac{\|\psi_i\|^2}{r_i^2}|\mathcal{F}_n\right) - \sigma^2\sum_{i=1}^{n}\frac{\|\psi_i\|^2}{r_i^2} \\ \leq & \ \|\widetilde{\theta}_n\|^2 - \frac{\zeta_{n+1}^2}{r_n} + \sigma^2 E\left(\sum_{i=1}^{\infty}\frac{\|\psi_i\|^2}{r_i^2}|\mathcal{F}_n\right) - \sigma^2\sum_{i=1}^{n-1}\frac{\|\psi_i\|^2}{r_i^2} \\ \leq & \ m_n. \end{aligned}$$

Consequently, by Corollary 2.1 m_n converges a.s. as $n \to \infty$, and by Corollary 2.2 $E\left(\sum_{i=1}^{\infty}\frac{\|\psi_i\|^2}{r_i^2}|\mathcal{F}_n\right)$ also converges. Therefore, $\|\theta - \theta_n\|^2$ converges as $n \to \infty$. This implies that $\|\theta_n\|$ is bounded, then from (6.2) it follows that $\|\theta_n\|^2$ also converges as $n \to \infty$.

To explicitly express the convergence of $\|\theta - \theta_n\|^2$ and $\|\theta_n\|^2$ we write

$$\theta_n^\tau = [a_n \quad b_n].$$

Then we have that both $(a - a_n)^2 + (b - b_n)^2$ and $a_n^2 + b_n^2$ converge, and hence both $aa_n + bb_n$ and $\|\theta_n - \gamma\theta\|^2$ converge a.s. as $n \to \infty$ for any γ.

From (5.28) and (5.4) it follows that

$$\frac{1}{n}\sum_{i=1}^{n}\|\widetilde{\theta}_i^\tau\psi_i\|^2 = \frac{1}{n}\sum_{i=1}^{n}\left(a - b\frac{a_i}{b_i}\right)^2 y_i^2 \xrightarrow[n\to\infty]{} 0 \quad a.s. \tag{6.4}$$

where for solvability of (5.4) it is assumed that $b_1 \neq 0$.

We now show that (6.4) implies

$$\liminf_{n\to\infty}\left(a - b\frac{a_n}{b_n}\right) = 0 \quad a.s. \tag{6.5}$$

Otherwise, if $\liminf_{n\to\infty}\left(a - b\frac{a_n}{b_n}\right)^2 \triangleq \eta > 0$ on A with $P(A) > 0$, then by Theorem 5.1

$$\lim_{n\to\infty}\frac{1}{n}\sum_{i=1}^{n}\left(a - b\frac{a_i}{b_i}\right)^2 y_i^2 I_A \geq \lim_{n\to\infty}\eta\frac{1}{n}\sum_{i=1}^{n}y_i^2 I_A > 0 \quad \text{on } A$$

which contradicts (6.4). Hence (6.5) is true, and there is a subsequence $\frac{a_{n_k}}{b_{n_k}}$ such that

$$\frac{a_{n_k}}{b_{n_k}} \longrightarrow \frac{a}{b} \qquad a.s. \tag{6.6}$$

where n_k may depend upon ω.

By convergence of

$$aa_n + bb_n = b_n \left(a\frac{a_n}{b_n} + b \right),$$

its subsequence $b_{n_k}\left(a\frac{a_{n_k}}{b_{n_k}} + b \right)$ also converges; hence by (6.6) b_{n_k} converges as $k \to \infty$. This and (6.6) imply

$$a_{n_k} \longrightarrow \gamma a, \qquad b_{n_k} \longrightarrow \gamma b \quad \text{for some } \gamma. \tag{6.7}$$

Therefore,

$$\|\theta_{n_k} - \gamma\theta\|^2 \to 0$$

and

$$\theta_n \to \gamma\theta$$

because $\|\theta_n - \gamma\theta\|^2$ is convergent.

The idea of proving $\theta_n \xrightarrow[n \to \infty]{} \gamma\theta$ used here is also applicable to the general case [BKW].

6.2 Reference Signal With Decaying Richness

For strong consistency of the parameter estimate we have obtained in Chapter 4 several conditions of decaying excitation, for example, (4.57) for ELS and (4.114) for SG. As discussed in the last section these conditions are not automatically satisfied for adaptive control systems even if the control is optimal. However, if the reference signal in an adaptive tracking system is rich even in a decaying way, then the required excitation will be guaranteed and both the minimality of tracking error and the consistency of parameter estimate will be achieved simultaneously [CG11], [CG2], [KP], [SK].

Before discussing this problem in detail we consider the solvability of (5.4) again. For (5.4) to be solvable with respect to u_n Lemma 5.1 provides a sufficient condition imposed on $\{w_n\}$, which we do not intend to assume now. Instead, we introduce a dither "v_n" to the control u_n and show in the following lemma that such a dither technique makes (5.4) solvable.

We write θ_n in the block form

$$\theta_n^\tau = [-A_{1n} \quad \dots \quad -A_{pn} \quad B_{1n} \quad \dots \quad B_{qn} \quad C_{1n} \quad \dots \quad C_{rn}] \tag{6.8}$$

and write

$$B_{1n} = [\underbrace{B'_{1n}}_{m} \quad \underbrace{B''_{1n}}_{l-m}], \qquad u_n^\tau = [\underbrace{u_n'^\tau}_{m} \quad \underbrace{u_n''^\tau}_{l-m}] \tag{6.9}$$

if $l \geq m$.

Lemma 6.1 *[CG2]. Let $\{v'_n\}$ be a sequence of mutually independent m-dimensional random vectors with independent components with continuous distributions and let $\{v'_n\}$ be independent of both $\{w_n\}$ and $\{y^*_n\}$. If $m \le l$ and the initial value θ_0 is chosen so that $\det B'_{10} \neq 0$, then the following equation, which differs from (5.4) by term $B_{1n}v_n$, is solvable with respect to u_n with $u''_n = 0$:*

$$y^*_{n+1} = \theta^\tau_n \psi_n - B_{1n} v_n \qquad \forall n \ge 0 \tag{6.10}$$

where $v^\tau_n = [v'^\tau_n \quad 0]$ is l-dimensional.

Proof. It suffices to show that $\det B'_{1n} \neq 0$ a.s. $\forall n \ge 0$.
We prove this by induction. Since $\det B'_{10} \neq 0$ by assumption we assume $\det B'_{1i} \neq 0$ a.s. for $\forall i \le n$.

Denoting $\theta^\tau_i \psi_i - B_{1i}u_i$ by $\overline{\theta}^\tau_i \overline{\psi}_i$, from (6.10) we have

$$y^*_{i+1} = \overline{\theta}^\tau_i \overline{\psi}_i + B_{1i} u^s_i, \ \forall i \le n \tag{6.11}$$

where

$$u^s_i = u_i - v_i \qquad \forall i \le n. \tag{6.12}$$

Let us assume $\det B'_{1n} \neq 0$ a.s. Set

$$N \triangleq \{\omega : \ \det B'_{1n} = 0\}, \quad D \triangleq \{\omega : \ \det B'_{1n+1} = 0\}. \tag{6.13}$$

By (4.68) we have

$$a u'^\tau_n (B'_{1n})^{-1} (y_{n+1} - \theta^\tau_n \psi_n) + r_n = 0, \qquad \forall \omega \in DN^c$$

which can be written in the form

$$a u^\tau_n \begin{bmatrix} (B'_{1n})^{-1} \\ 0 \end{bmatrix} \quad (B_1 u_n - B_{1n} u_n + y_{n+1} - B_1 u_n - \theta^\tau_n \psi_n + B_{1n} u_n)$$
$$+ u^\tau_n u_n + r_n - u^\tau_n u_n = 0$$

or

$$u^\tau_n M_n u_n + u^\tau_n f_n + g_n = 0 \qquad \forall \omega \in DN^c \tag{6.14}$$

where

$$M_n = I + a \begin{bmatrix} (B'_{1n})^{-1} \\ 0 \end{bmatrix} (B_1 - B_{1n}),$$

$$f_n = a \begin{bmatrix} (B'_{1n})^{-1} \\ 0 \end{bmatrix} (y_{n+1} - B_1 u_n - \theta^\tau_n \psi_n + B_{1n} u_n),$$

$$g_n = r_n - u^\tau_n u_n.$$

It is obvious, that all M_n, f_n and g_n are free of u_n, and as functions of u_i, they depend upon u_i, $i \le n-1$ only.

From (6.11) we know that

$$u_i^s = \begin{bmatrix} B_{1i}^{-1}(y_{i+1}^* - \overline{\theta}_i^T \overline{\psi}_i) \\ 0 \end{bmatrix} \qquad i \leq n$$

and u_i^s is measurable with respect to $\{y_{j+1}^*, w_j, v_{j-1}, 0 \leq j \leq i\}$ and hence is independent of v_i. Then by (6.12) we can rewrite (6.14) as

$$v_n^T M_n v_n + v_n^T h_n + j_n = 0 \qquad \forall \omega \in DN^c \tag{6.15}$$

where M_n, h_n and j_n are all independent of v_n.

Denoting entries of M_n, v_n and h_n by $m_{ij}(n)$, v_{ni} and h_{ni} respectively, we then have

$$\sum_{i,j=1}^{m} v_{ni} m_{ij}(n) v_{nj} + \sum_{i=1}^{m} v_{ni} h_{ni} + j_n = 0, \ \forall \omega \in DN^c. \tag{6.16}$$

If $m_{ii}(n) \neq 0$ on a subset D_1 of DN^c for some i with $P(D_1) > 0$, then rewriting (6.16) as

$$m_{ii}(n) v_{ni}^2 + v_{ni} h_{ni} + s_n = 0, \qquad \forall \omega \in D_1 \tag{6.17}$$

we know that s_n is independent of v_{ni}.

Therefore, from (6.17) we find

$$(v_{ni} - v_{ni}(1))(v_{ni} - v_{ni}(2)) = 0 \qquad \forall \omega \in D_1 \tag{6.18}$$

where both $v_{ni}(1)$ and $v_{ni}(2)$ are independent of v_{ni}. However, the distribution of v_{ni} is continuous, then $v_{ni} - v_{ni}(1)$ and $v_{ni} - v_{ni}(2)$ also have the continuous distributions and (6.18)is possible only if $P(D_1) = 0$. The contradiction shows that $P(DN^c) = 0$ or $det\ B'_{1\,n+1} \neq 0$. Thus the induction is completed for the case where $m_{ii}(n) \neq 0$ for some i on a subset of DN^c with positive probability.

We now consider the case where $m_{ii}(n) = 0$ on DN^c, $i = 1, \ .. \ m$.

Rewrite (6.16) as

$$\sum_{j>i}^{m} v_{ni}(m_{ij}(n) + m_{ji}(n)) v_{nj} + \sum_{i=1}^{m} v_{ni} h_{ni} + j_n$$
$$= 0, \quad \forall \omega \in DN^c. \tag{6.19}$$

If for some i and j, say, $i = 1$, $j = 2$, $m_{12}(n) + m_{21}(n) \neq 0$ on $D_1 \subset DN^c$ with $P(D_1) > 0$, then from (6.19) it follows that

$$v_{n1} p_n + q_n = 0 \qquad \forall \omega \in D_1 \tag{6.20}$$

where

$$q_n = \sum_{j>i>2}^{m} v_{ni}(m_{ij}(n) + m_{ji}(n))v_{nj} + \sum_{i=2}^{m} v_{ni}h_{ni} + j_n,$$

$$p_n = \sum_{j>2}^{m}(m_{1j}(n) + m_{j1}(n))v_{nj} + (m_{12}(n) + m_{21}(n))v_{n2} + h_{n1}.$$

Since $m_{12}(n) + m_{21}(n) \neq 0$ on D_1 and v_{n2} is independent of the remaining entries in p_n, by the argument used above we conclude $p_n \neq 0$ on D_1. Then (6.20) implies

$$v_{n1} + \frac{q_n}{p_n} = 0 \qquad \forall \omega \in D_1. \tag{6.21}$$

By the same argument we repeatedly used above from (6.21) it follows that $P(D_1) = 0$. The contradiction means $det\, B'_{1n+1} \neq 0$ a.s.

Finally, if $m_{ii}(n) = 0$, $m_{ij}(n) + m_{ji}(n) = 0$, $\forall i\, j = 1, ..., m$. Then from (6.14) we find

$$u_n^\tau f_n + g_n = 0 \qquad \forall \omega \in DN^c. \tag{6.22}$$

Notice $g_n \geq 1$. Then at least one component of f_n differs from zero on a subset D_1 of DN^c with $P(D_1) > 0$.

Without loss of generality assume the first element f_{n1} of f_n is not equal to zero on D_1. Then (6.22) implies that

$$v_{n1} + \frac{1}{f_{n1}}(u_n^\tau f_n - v_{n1}f_{n1} + g_n) = 0 \qquad \forall \omega \in D_1. \tag{6.23}$$

Obviously, v_{n1} is independent of the second term in (6.23). Again, the repeatedly used argument leads to that $P(D_1) = 0$. This in turn implies $det\, B'_{1n+1} \neq 0$ a.s. Q.E.D.

In the sequel $\lambda_{min}(A)$ denotes the minimum eigenvalue of a matrix A.

Let $F(z)$ and $G(z)$ be polynomial matrices of dimension $m \times m$ and $m \times l$ respectively:

$$F(z) = I + F_1 z + ... + F_p z^p, \tag{6.24}$$

$$G(z) = G_1 + G_2 z + ... + G_q z^{q-1}, \tag{6.25}$$

and let $\{f_i\}$ be a sequence of m-dimensional vectors with $f_i = 0$, $\forall i < 0$. Define two increased lag vectors f'_n and f''_n as follows:

$$f'_n = [f_n^\tau \quad f_{n-1}^\tau \quad \cdots \quad f_{n-p-l(q-1)}^\tau]^\tau, \tag{6.26}$$

$$\begin{aligned} f''_n =\ & [(detG_1^+ G(z))f_n^\tau \quad \cdots \quad (detG_1^+ G(z))f_{n-p+1}^\tau \\ & ((AdjG_1^+ G(z))G_1^+ F(z)f_{n+1})^\tau \quad \cdots \\ & ((AdjG_1^+ G(z))G_1^+ F(z)f_{n-q+2})^\tau]^\tau, \end{aligned} \tag{6.27}$$

The following lemma establishes the relationship between $\lambda_{min}\left(\sum_{i=1}^{n} f_i' f_i'^{\tau}\right)$ and $\lambda_{min}\left(\sum_{i=1}^{n} f_i'' f_i''^{\tau}\right)$ as $n \to \infty$.

Lemma 6.2. *Suppose that $G_1^+ F(z)$ and $G_1^+ G(z)$ are left-coprime and $G_1^+ G_q$ is of full rank (This implicitly requires $m \geq l$). Then*

$$\liminf_{n\to\infty} \gamma_n \lambda_{min}\left(\sum_{i=1}^{n} f_i' f_i'^{\tau}\right) \neq 0 \tag{6.28}$$

implies

$$\liminf_{n\to\infty} \gamma_n \lambda_{min}\left(\sum_{i=1}^{n} f_i'' f_i''^{\tau}\right) \neq 0 \tag{6.29}$$

where $\{\gamma_n\}$ is a sequence of nonnegative real numbers.

Proof. Suppose that the converse of (6.29) holds. Then there is a sequence of $mp + lq$-dimensional vectors

$$\alpha_{n_k} \triangleq [\alpha_{n_k 1}^{\tau} \quad \dots \quad \alpha_{n_k p+q}^{\tau}]^{\tau}$$

with $\|\alpha_{n_k}\| = 1$ and $\alpha_{n_k i} \in I\!R^m$, $\alpha_{n_k j} \in I\!R^l$, $i = 1, \dots, p$, $j = p+1, \dots, p+q$ such that

$$\gamma_{n_k} \alpha_{n_k}^{\tau} \left(\sum_{i=1}^{n_k} f_i'' f_i''^{\tau}\right) \alpha_{n_k} \xrightarrow[k\to\infty]{} 0. \tag{6.30}$$

Set

$$\begin{aligned} H_{n_k}(z) \triangleq\ & \alpha_{n_k 1}^{\tau}(det G_1^+ G(z))z + \dots + \alpha_{n_k p}^{\tau}(det G_1^+ G(z))z^p \\ & + \alpha_{n_k p+1}^{\tau}(Adj G_1^+ G(z)) G_1^+ F(z) + \dots \\ & + \alpha_{n_k p+q}^{\tau}(Adj G_1^+ G(z)) G_1^+ F(z) z^{q-1} \\ \triangleq\ & \sum_{i=0}^{p+l(q-1)} h_i^{\tau}(n_k) z^i \end{aligned} \tag{6.31}$$

where $h_i(n_k) \in I\!R^m$.

Write

$$h(n_k) \triangleq [h_0^{\tau}(n_k) \quad \dots \quad h_{p+l(q-1)}^{\tau}(n_k)]^{\tau}. \tag{6.32}$$

Then we can rewrite (6.30) as

$$\gamma_{n_k} h^{\tau}(n_k) \left(\sum_{i=1}^{n_k} f_i' f_i'^{\tau}\right) h(n_k) \xrightarrow[k\to\infty]{} 0 \tag{6.33}$$

from which by (6.28) we conclude that

$$h(n_k) \xrightarrow[k \to \infty]{} 0,$$

and consequently

$$H_{n_k}(z) \xrightarrow[k \to \infty]{} 0. \tag{6.34}$$

Let

$$[\alpha^{\tau}(1) \quad ... \quad \alpha^{\tau}(p+q)]^{\tau}$$

be the limit of a convergent subsequence of α_{n_k}. Then (6.34) yields

$$\begin{aligned}
&\alpha^{\tau}(1)(detG_1^{+}G(z))z + ... + \alpha^{\tau}(p)(detG_1^{+}G(z))z^{p} \\
&+\alpha^{\tau}(p+1)(AdjG_1^{+}G(z))G_1^{+}F(z) + ... \\
&+\alpha^{\tau}(p+q)(AdjG_1^{+}G(z))G_1^{+}F(z)z^{q-1} = 0.
\end{aligned} \tag{6.35}$$

Setting $z = 0$ in (6.35) we find that $\alpha(p+1) = 0$, and (6.35) becomes

$$\begin{aligned}
&\sum_{i=1}^{p} \alpha^{\tau}(i)(detG_1^{+}G(z))z^{i-1} \\
= &-\sum_{i=2}^{q} \alpha^{\tau}(p+i)(AdjG_1^{+}G(z))G_1^{+}F(z)z^{i-2}.
\end{aligned} \tag{6.36}$$

By the left-coprimeness condition there are polynomial matrices $M(z)$ and $N(z)$ such that

$$G_1^{+}G(z)M(z) + G_1^{+}F(z)N(z) = I.$$

Using this we have

$$\begin{aligned}
&\sum_{i=2}^{q} \alpha^{\tau}(p+i)z^{i-2}AdjG_1^{+}G(z) \\
= &\sum_{i=2}^{q} \alpha^{\tau}(p+i)z^{i-2}[AdjG_1^{+}G(z)][G_1^{+}G(z)M(z) + G_1^{+}F(z)N(z)] \\
= &\ detG_1^{+}G(z)\left[\sum_{i=2}^{q} \alpha^{\tau}(p+i)z^{i-2}M(z) - \sum_{i=1}^{p} \alpha^{\tau}(i)z^{i-1}N(z)\right]
\end{aligned} \tag{6.37}$$

where for the last equality we have applied (6.36).

The highest order of z on the left-hand side of (6.37) is less than or equal to $l(q-1)-1$ while the lowest order of the right-hand side of (6.37) is greater than or equal to

$$l(q-1) = deg(detG_1^{+}G(z))$$

because $G_1^+ G_q$ is of full rank. Hence $\alpha(p+i) = 0$, $i = 2, ..., q$, and from (6.36) we see $\alpha(i) = 0$, $i = 1, ..., p$.

On the other hand, $\|\alpha_{n_k}\| = 1, \forall k \geq 1$. Hence the limit of any convergent subsequence of $\{\alpha_{n_k}\}$ must have norm equal to 1. The contradiction proves (6.29). Q.E.D.

Theorem 6.1 *[CG2]. Assume that Conditions 1)-4) of Theorem 5.1 hold, B_1 and B_q are of full rank, and $A(z)$ and $z^{-1}B(z)$ are left-coprime. Further, assume that $\{w_n\}$ are mutually independent with mutually independent components having continuous distributions and that the reference signal $\{y_i^*\}$ is bounded and independent of $\{w_n\}$ and y_i^* is $\mathcal{F}_{i-1}$-measurable with $y_i^* = 0$, $i < 0$. If the initial value is selected so that $\det B_{10} \neq 0$, then u_n can be solved from (5.4) and*

$$\limsup_{n\to\infty} \frac{1}{n}\sum_{i=1}^{n}(\|y_i\|^2 + \|u_i\|^2) < \infty \quad a.s. \tag{6.38}$$

$$\lim_{n\to\infty} \frac{1}{n}\sum_{i=1}^{n}(y_i - y_i^*)(y_i - y_i^*)^T = R \quad a.s. \tag{6.39}$$

If, in addition, $\{y_i^*\}$ satisfies the decaying richness condition

$$\liminf_{n\to\infty} \frac{log^{1/4}n}{n}\lambda_{min}\left(\sum_{i=1}^{n} y_i'^* y_i'^{*T}\right) \neq 0 \tag{6.40}$$

with

$$y_i'^* \triangleq [y_i^{*T} \quad \cdots \quad y_{i-p-l(q-1)}^{*T}]^T, \tag{6.41}$$

then $\theta_n \xrightarrow[n\to\infty]{} \theta$ a.s.

Proof. By Lemma 6.1 u_n can be solved from 5.4. Then by Theorem 5.1, the assertions (6.38) and (6.39) follow immediately. It remains to prove $\theta_n \xrightarrow[n\to\infty]{} \theta$ a.s.

From (3.91) we have

$$u_n = H_1(z)y_{n+1} - H_2(z)w_{n+1}. \tag{6.42}$$

where

$$H_1(z) = zB^{-1}(z)A(z), \qquad H_2(z) = zB^{-1}(z)C(z). \tag{6.43}$$

Set

$$H_3(z) = H_1(z) - H_2(z)$$

$$\varphi_n^w = [\ w_n^T \quad \cdots \quad w_{n-p+1}^T \quad (H_3(z)w_{n+1})^T \quad \cdots \quad (H_3(z)w_{n-q+2})^T$$

$$w_n^T \quad \cdots \quad w_{n-r+1}^T]^T \tag{6.44}$$

$$\varphi_n(y^*) = [\ y_n^{*T}\ \dots\ y_{n-p+1}^{*T}\quad (H_1(z)y_{n+1}^*)^T\ \dots\ (H_1(z)y_{n-q+2}^*)^T$$
$$0\ \dots\ 0]^T \tag{6.45}$$

$$\varphi_n(\zeta) = [\ \zeta_n^T\ \dots\ \zeta_{n-p+1}^T\quad (H_1(z)\zeta_{n+1})^T\ \dots\ (H_1(z)\zeta_{n-q+2})^T$$
$$0\ \dots\ 0]^T \tag{6.46}$$

where ζ_n is given by (4.68).

Noticing (4.68) and (5.4) we find that

$$y_{n+1} = \zeta_{n+1} + w_{n+1} + y_{n+1}^*. \tag{6.47}$$

Using (6.42)-(6.47) we then can express φ_n^0 defined by (4.17) as follows

$$\varphi_n^0 = \varphi_n^w + \varphi_n(y^*) + \varphi_n(\zeta) \triangleq \varphi_n^1 + \varphi_n(\zeta) \tag{6.48}$$

where φ_n^1 obviously denotes $\varphi_n^w + \varphi_n(y^*)$.

Define

$$\Phi^1(n+1,i) = \left(I - \frac{a\varphi_n^1\varphi_n^{1T}}{r_n^1}\right)\Phi^1(n,i), \quad \Phi^1(i,i) = I, \tag{6.49}$$

$$r_n^1 = 1 + \sum_{i=0}^{n} \|\varphi_i^1\|^2. \tag{6.50}$$

By Lemma 4.4 it follows that

$$\sum_{i=0}^{\infty} \frac{\|\varphi(\zeta)\|^2}{r_i} < \infty \quad a.s. \tag{6.51}$$

which corresponds to (4.106) in Theorem 4.4. Then completely the same argument as that used in Theorem 4.4 leads to that $\Phi^0(n,0) \xrightarrow[n\to\infty]{} 0$ if and only if $\Phi^1(n,0) \xrightarrow[n\to\infty]{} 0$, where $\Phi^0(n,0)$ is defined by (4.104). Then to complete the proof of the theorem by Theorem 4.4 it suffices to show that under the conditions of the theorem $\Phi^1(n,0) \xrightarrow[n\to\infty]{} 0$. For this we first prove

$$\liminf_{n\to\infty} \frac{log^{1/4} n}{n} \lambda_{min}\left(\sum_{i=1}^{n} \varphi_i^1 \varphi_i^{1T}\right) \neq 0. \tag{6.52}$$

Multiplying φ_n^1, φ_n^w and $\varphi_n(y^*)$ from the left by $z^{-1}det(B_1^{-1}B(z))$ and denoting the products by $\overline{\varphi}_n^1$, $\overline{\varphi}_n^w$ and $\overline{\varphi}_n(y^*)$ respectively.

Notice that B_q is of full rank. Then $b_{l(q-1)} \neq 0$ in the expression:

$$z^{-1}det(B_1^{-1}B(z)) = b_0 + b_1 z + \quad \dots \quad + b_{l(q-1)} z^{l(q-1)}.$$

Paying attention to that

$$\begin{aligned}\lambda_{min}\left(\sum_{i=1}^{n}\overline{\varphi}_i^1\overline{\varphi}_i^{1\tau}\right) &= \inf_{\|x\|=1}\sum_{i=1}^{n}(x^\tau\overline{\varphi}_i^1)^2 \\ &= \inf_{\|x\|=1}\sum_{i=1}^{n}\left(\sum_{j=1}^{l(q-1)} b_j^2 x^\tau \varphi_{i-j}^1\right)^2 \\ &\leq (lq+1)\sum_{j=1}^{l(q-1)} b_j^2 \inf_{\|x\|=1}\sum_{i=1}^{n}(x^\tau\varphi_i^1)^2\end{aligned}$$

we find that for (6.52) it suffices to show

$$\liminf_{n\to\infty}\frac{log^{1/4}n}{n}\lambda_{min}\left(\sum_{i=1}^{n}\overline{\varphi}_i^1\overline{\varphi}_i^{1\tau}\right) \neq 0. \tag{6.53}$$

Suppose that the converse were true. Then there would exist $\alpha_{n_k} \in \mathbb{R}^{mp+lq}$, $\beta_{n_k} \in \mathbb{R}^{mr}$ with $\|\alpha_{n_k}\|^2 + \|\beta_{n_k}\|^2 = 1$ such that

$$\frac{log^{1/4}n_k}{n_k}[\alpha_{n_k}^\tau \quad \beta_{n_k}^\tau]\left(\sum_{i=1}^{n_k}\overline{\varphi}_i^1\overline{\varphi}_i^{1\tau}\right)\begin{bmatrix}\alpha_{n_k}\\ \beta_{n_k}\end{bmatrix}\xrightarrow[k\to\infty]{} 0. \tag{6.54}$$

By Theorem 2.8 from (6.54) it follows that

$$\frac{log^{1/4}n_k}{n_k}[\alpha_{n_k}^\tau \quad \beta_{n_k}^\tau]\left(\sum_{i=1}^{n_k}\overline{\varphi}_i^\omega\overline{\varphi}_i^{\omega\tau}\right)\begin{bmatrix}\alpha_{n_k}\\ \beta_{n_k}\end{bmatrix}\xrightarrow[k\to\infty]{} 0 \tag{6.55}$$

$$\frac{log^{1/4}n_k}{n_k}[\alpha_{n_k}^\tau \quad \beta_{n_k}^\tau]\left(\sum_{i=1}^{n_k}\overline{\varphi}_i(y^*)\overline{\varphi}_i^\tau(y^*)\right)\begin{bmatrix}\alpha_{n_k}\\ \beta_{n_k}\end{bmatrix}\xrightarrow[k\to\infty]{} 0. \tag{6.56}$$

Noticing $H_1(z) = z\left[B_1^{-1}B(z)\right]^{-1}B_1^{-1}A(z)$ and identifying $G_1^+F(z)$ and $G_1^+G(z)$ in Lemma 6.2 to $B_1^{-1}A(z)$ and $B_1^{-1}z^{-1}B(z)$ respectively we find that applying Lemma 6.2 to (6.56) leads to that

$$\alpha_{n_k}\xrightarrow[k\to\infty]{} 0. \tag{6.57}$$

Hence, we have

$$\|\beta_{n_k}\|\xrightarrow[k\to\infty]{} 1. \tag{6.58}$$

Denote the first $mp+lq$ components of $\overline{\varphi}_n^{\omega}$ by $\overline{\varphi}_{n1}^{\omega}$ and the rest by $\overline{\varphi}_{n2}^{\omega}$, i.e.

$$\overline{\varphi}_n^{\omega}=[\overline{\varphi}_{n1}^{\omega\tau}\quad \overline{\varphi}_{n2}^{\omega\tau}]^{\tau}.$$

Then (6.55) yields

$$\frac{1}{n_k}\sum_{i=1}^{n_k}(\alpha_{n_k}^{\tau}\overline{\varphi}_{i1}^{\omega}+\beta_{n_k}^{\tau}\overline{\varphi}_{i2}^{\omega})^2\xrightarrow[k\to\infty]{}0. \tag{6.59}$$

From (5.15) it is easy to see

$$\limsup_{n\to\infty}\frac{1}{n}\sum_{i=1}^{n}\|\overline{\varphi}_{i1}^{\omega}\|^2<\infty \quad a.s. \tag{6.60}$$

Then (6.57), (6.59) and (6.60) imply that

$$\lim_{k\to\infty}\frac{1}{n_k}\sum_{i=1}^{n_k}\|\beta_{n_k}^{\tau}\overline{\varphi}_{i2}^{\omega}\|^2=0. \tag{6.61}$$

Write β_{n_k} in the component form

$$\beta_{n_k}\triangleq[\beta_{n_k}^{(1)\tau}\quad\ldots\quad\beta_{n_k}^{(r)\tau}]^{\tau},\quad \beta_{n_k}^{(i)}\in I\!R^m,\quad i=1,\ \ldots,\ r$$

and set

$$\begin{aligned}G_{n_k}(z)&\triangleq \beta_{n_k}^{(1)\tau}z^{-1}det(B_1^{-1}B(z))+\quad\ldots\quad+\beta_{n_k}^{(r)\tau}z^{r-1}det(B_1^{-1}B(z))\\&\triangleq\sum_{i=0}^{l(q-1)+r-1}g_i^{\tau}(n_k)z^i.\end{aligned} \tag{6.62}$$

Then we can rewrite (6.61) as

$$\begin{aligned}0&=\lim_{k\to\infty}\frac{1}{n_k}\sum_{i=1}^{n_k}\left(\sum_{j=0}^{l(q-1)+r-1}g_j^{\tau}(n_k)w_{i-j}\right)^2\\&=\lim_{k\to\infty}\frac{1}{n_k}\sum_{i=1}^{n_k}g_{n_k}^{\tau}[w_i^{\tau}\quad\ldots\quad w_{i-l(q-1)-r+1}^{\tau}]^{\tau}\\&\quad[w_i^{\tau}\quad\ldots\quad w_{i-l(q-1)-r+1}^{\tau}]g_{n_k},\end{aligned} \tag{6.63}$$

where

$$g_{n_k}\triangleq[g_0^{\tau}(n_k)\quad\ldots\quad g_{l(q-1)+r-1}^{\tau}(n_k)]^{\tau}.$$

By (5.15) and Theorem 2.8 from (6.63) we derive

$$g_{n_k}\xrightarrow[k\to\infty]{}0\quad\text{and}\quad G_{n_k}(z)\xrightarrow[k\to\infty]{}0$$

which via (6.62) imply $\beta_{n_k} \xrightarrow[k \to \infty]{} 0$. This contradicts (6.58). Thus (6.53) is verified. Hence for any fixed ω there are $\alpha_0 > 0$, $N_0 > 0$ such that

$$\lambda_{min}\left(\sum_{i=1}^{n} \varphi_i^1 \varphi_i^{1\tau}\right) \geq \frac{\alpha_0 n}{log^{1/4} n}, \qquad \forall n \geq N_0. \tag{6.64}$$

Using Condition 2) of Theorem 5.1, the boundedness of $\{y_n^*\}$ and invoking Theorem 2.8 we find that

$$\liminf_{n\to\infty} \frac{1}{n}\sum_{i=0}^{n} \|\varphi_i^1\|^2 \geq \lim_{n\to\infty} \frac{1}{n}\sum_{i=0}^{n} \|w_i\|^2 tr\, R > 0$$

and

$$\limsup_{n\to\infty} \frac{1}{n}\sum_{i=0}^{n} \|\varphi_i^1\|^2 < \infty \qquad a.s.$$

From this we conclude that there are $c_1 > 0$ and $c_2 > 0$ such that

$$c_1 n \leq r_n^1 \leq c_2 n. \tag{6.65}$$

As we have noted in Remark 4.3 all manipulations in Theorem 4.4 are purely algebraic. Then in Theorem 4.4 setting $\delta = 0$ and identifying φ_n to φ_n^1 for the present case we see that all conditions of that theorem are fulfilled. Therefore, we assert

$$\Phi^1(n, 0) \xrightarrow[n \to \infty]{} 0.$$

This completes the proof of the theorem. Q.E.D.

6.3 Diminishingly Excited Control

If we add some dither with constant variance to the system input, then we can get consistent parameter estimates, but the loss function will differ from its minimal value [CL], [Ca], [Che2], [CCa2]. In other words, by a constant variance dither method the consistency of parameter estimates and the minimality of the loss function cannot be achieved simultaneously. The performance indices of long run average type discussed in Theorem 5.1 and 6.1 as well as in Chapter 3 have a common feature, which consists in that they remain unchanged if an additional quantity $\{f_i\}$ with $\sum_{i=0}^{n} \|f_i\| = o(n)$ is added to the summand. This motivates us to introduce a diminishing dither to the system input or to the reference signal when the adaptive tracking system is concerned in the hope that the diminishing dither gives sufficient excitation to make estimates strongly consistent but does not upset the performance index. A different approach is adopted in [LW3] where a so-called "occasional excitation" technique is applied.

We now define "dither" as it will be applied to the system. Let $\{\varepsilon_i\}$ be a sequence of l-dimensional independent and identically distributed random vectors with continuous distribution and let $\{\varepsilon_i\}$ be independent of $\{w_i\}$ with $E\varepsilon_i = 0$, $E\varepsilon_i\varepsilon_i^\tau = I$ and $\|\varepsilon_i\| \leq \sigma$, $\sigma > 0$. Define

$$v_n = \frac{\varepsilon_n}{n^{\varepsilon/2}}, \qquad \varepsilon \in [0, \frac{1}{2(t+1)}), \tag{6.66}$$

$$t = max(p,q,r) + mp - 1. \tag{6.67}$$

$\{v_i\}$ will serve as the diminishing excitation source for the case where the ELS algorithm is used.

When the SG algorithm is applied, we define the excitation source different from (6.66) and (6.67), namely,

$$v_n = \frac{\varepsilon_n}{log^{\varepsilon/2} n}, \qquad \varepsilon \in (0, \frac{1}{4s(m+2)}), \tag{6.68}$$

$$s = max(p,q,r+1). \tag{6.69}$$

For $\{v_i\}$ defined by (6.66) we note that

$$E\sum_{i=1}^{\infty} \frac{\left\| v_i v_i^\tau - \frac{1}{i^\varepsilon} I \right\|^\gamma}{i^{(1-\varepsilon)\gamma}} < \infty, \qquad \forall \gamma \in (\frac{1}{1-\varepsilon}, 2]$$

Then by Theorem 2.7 we find that

$$\sum_{i=1}^{\infty} \frac{v_i v_i^\tau - \frac{1}{i^\varepsilon} I}{i^{1-\varepsilon}} < \infty \qquad a.s. \tag{6.70}$$

since $\left\{ \frac{v_i v_i^\tau - \frac{1}{i^\varepsilon} I}{i^{1-\varepsilon}}, \mathcal{F}_i^v \right\}$ is a matrix martingale difference sequence, and v_i is independent of $\mathcal{F}_{i-1}^v$, which is the σ-algebra generated by $\{v_j, j \leq i-1\}$.

Applying Lemma 2.4 to (6.70) we see

$$\frac{1}{n^{1-\varepsilon}} \sum_{i=1}^{n} \left(v_i v_i^\tau - \frac{1}{i^\varepsilon} I \right) \xrightarrow[n\to\infty]{} 0. \tag{6.71}$$

From the following chain of equalities and inequalities

$$\frac{(n+1)^{1-\varepsilon}-1}{1-\varepsilon} = \int_1^{n+1} \frac{dx}{x^\varepsilon} = \sum_{i=1}^n \int_i^{i+1} \frac{dx}{x^\varepsilon} \le \sum_{i=1}^n \frac{1}{i^\varepsilon}$$
$$\le \left(1+\sum_{i=2}^n \int_{i-1}^i \frac{dx}{x^\varepsilon}\right) = \left(1+\frac{n^{1-\varepsilon}-1}{1-\varepsilon}\right)$$

we find that

$$\frac{1}{n^{1-\varepsilon}} \sum_{i=1}^n \frac{1}{i^\varepsilon} \xrightarrow[n\to\infty]{} \frac{1}{1-\varepsilon}.$$

Then from (6.71) it follows that

$$\frac{1-\varepsilon}{n^{1-\varepsilon}} \sum_{i=1}^n v_i v_i^T \xrightarrow[n\to\infty]{} I. \tag{6.72}$$

Similarly, for $\{v_n\}$ defined by (6.68) we have

$$\sum_{i=1}^\infty \frac{v_i v_i^T - \frac{1}{\log^\varepsilon i} I}{i^{\frac{3\gamma}{2}}} < \infty \quad \forall \gamma \in \left(\frac{2}{3}, 1\right), \tag{6.73}$$

and

$$\frac{\log^\varepsilon n}{n} \sum_{i=1}^n v_i v_i^T \xrightarrow[n\to\infty]{} I. \tag{6.74}$$

because

$$\frac{\log^\varepsilon n}{n} \sum_{i=2}^n \frac{1}{\log^\varepsilon i} \xrightarrow[n\to\infty]{} 1. \tag{6.75}$$

Hence $\{v_n\}$ has no influence on the performance index of long run average type if it is added to the reference signal for an adaptive tracking system or to the system input for an adaptive control system with quadratic cost.

Let us return to the problem discussed in Section 6.1. When $y_i^* \equiv 0$ the performance that is minimized is

$$\lim_{n\to\infty} \frac{1}{n} \sum_{i=1}^n y_i y_i^T$$

and the system is called the minimum variance adaptive control system.

Let us take

$$y_n^* = v_n,$$

where $\{v_n\}$ is given by (6.68).

By Theorem 2.8 and by (6.74) it is clear that Condition (6.40) is satisfied. Then under the rest conditions of Theorem 6.1 by (6.74) and (6.39) we know that the control defined by (5.4) is the minimum variance adaptive control:

$$\lim_{n\to\infty}\frac{1}{n}\sum_{i=1}^{n} y_i y_i^{\tau} = \lim_{n\to\infty}\frac{1}{n}\sum_{i=1}^{n}(y_i - v_i)(y_i - v_i)^{\tau} = R.$$

Meanwhile, by Theorem 6.1 the parameter estimate is strongly consistent. Thus, we have got both minimality of the system output and consistency of parameter estimate.

We now proceed to deal with the general adaptive control systems, for which we add the excitation source to the system input.

Without loss of generality, we may assume that the family of σ-algebras $\{\mathcal{F}_n\}$ is rich enough such that both w_n and v_n are $\mathcal{F}_n$-measurable, otherwise we need only to extend $\mathcal{F}_n$ appropriately.

Set

$$\mathcal{F}'_n = \sigma\{w_i,\ 0 \le i \le n,\ v_j,\ 0 \le j \le n-1\}.$$

Let $\mathcal{F}'_n$-measurable u^s_n be the desired control at time n. The diminishing excitation technique suggests to take adaptive control as

$$u_n = u^s_n + v_n \tag{6.76}$$

Theorem 6.2 *[CG3]. Suppose that for system (3.91)-(3.94) the following conditions are satisfied.*

1). $\{w_n, \mathcal{F}_n\}$ is a martingale difference sequence with

$$\sup_{n\ge 1} E\{\|w_n\|^{\beta}|\mathcal{F}_{n-1}\} < \infty \quad a.s., \quad \beta \ge 2$$

$$\liminf_{n\to\infty}\lambda_{\min}\left(\frac{1}{n}\sum_{i=1}^{n} w_i w_i^{\tau}\right) > 0 \quad and \quad \limsup_{n\to\infty}\frac{1}{n}\sum_{i=1}^{n}\|w_i\|^2 < \infty \ a.s. \tag{6.77}$$

2). $A(z)$, $B(z)$ and $C(z)$ have no common left factor and $[A_p \quad B_q \quad C_r]$ is of row-full-rank.

3). The diminishingly excited control (6.76) is applied, where $\mathcal{F}'_n$-measurable u^s_n satisfies

$$\frac{1}{n}\sum_{i=1}^{n}\|u^s_i\|^2 = O(n^{\delta}), \quad \delta \in [0, \frac{1-2\varepsilon(t+1)}{2t+3}), \tag{6.78}$$

$$\varepsilon \in [0, \frac{1}{2(t+1)}), \quad t = max(p,q,r) + mp - 1$$

and where $\{v_n\}$ is defined by (6.66) and (6.67).

Then the following assertions are true:

(i) There are constant $c_0 > 0$ and integer n_0 such that

$$\lambda_{\min}\left(\sum_{i=1}^{n} \varphi_i^0 \varphi_i^{0\tau}\right) \geq c_0 n^{\alpha}, \quad \forall n \geq n_0 \tag{6.79}$$

for any $\alpha \in \left(\frac{1}{2}(1+\delta), 1-(1+t)(\varepsilon+\delta)\right]$.

(ii) If $C^{-1}(z) - \frac{1}{2}I$ is SPR and

$$\|y_n\|^2 = O(n^b), \text{for some } \; b > 0, \tag{6.80}$$

then the ELS estimate is strongly consistent with convergence rate

$$\|\theta_n - \theta\|^2 = O\left(\frac{\log n(\log\log n)^{\delta(\beta-2)}}{n^{\alpha}}\right), \tag{6.81}$$

where θ is defined by (4.1) and

$$\delta(x) = \begin{cases} 0, & x \neq 0 \\ c, & x = 0 \end{cases}$$

with c arbitrary but greater than 1.

Proof. We first note that the interval $(\frac{1}{2}(1+\delta), 1-(1+t)(\varepsilon+\delta)]$ is not empty, because

$$(1+t)(\varepsilon+\delta)+\frac{1}{2}\delta \leq (1+t)\left[\varepsilon + \frac{1-2\varepsilon(t+1)}{2t+3}\right] + \frac{1-2\varepsilon(t+1)}{2(2t+3)} = \frac{1}{2}$$

By Theorem 2.8 from (6.78) it follows that

$$\begin{aligned}\sum_{i=1}^{n} u_i^{s\tau} v_i &= O\left(\left(\sum_{i=1}^{n}\|u_i^s\|^2\right)^{\frac{1}{2}}\left(log\left(\sum_{i=1}^{n}\|u_i^s\|^2+e\right)\right)^{\frac{1}{2}+\eta}\right) \\ &= O\left(n^{\frac{1+\delta}{2}}(logn)^{\frac{1}{2}+\eta}\right), \quad \forall \eta > 0\end{aligned} \tag{6.82}$$

Recall that φ_n^0 and r_n^0 are defined by (4.17) and (4.105) respectively, from (6.72), (6.77), (6.78), (6.80) and (6.82) we find that

$$r_n^0 = O\left(n^{1+b+\delta}\right). \tag{6.83}$$

Hence the result (6.81) will immediately follow from Theorem 4.2 if we can show that (6.79) holds.

Clearly, (6.79) is equivalent to

$$\liminf_{n\to\infty} n^{-\alpha}\lambda_{min}\left(\sum_{i=0}^{n}\varphi_i^0\varphi_i^{0\tau}\right) \neq 0. \tag{6.84}$$

We now prove (6.84). Set

$$f_n \triangleq (det A(z))\varphi_n^0 \tag{6.85}$$

$$det A(z) = a_0 + a_1 z + \ \dots \ + a_s z^s, \quad s \leq mp. \tag{6.86}$$

By the Schwarz inequality and the fact that $\varphi_i^0 = 0$ for $i < 0$, it is easy to see

$$\lambda_{min}\left(\sum_{i=1}^{n} f_i f_i^\tau\right) = \inf_{\|x\|=1}\sum_{i=1}^{n}(x^\tau f_i)^2 \leq (s+1)\sum_{j=0}^{s} a_j^2\lambda_{min}^0(n).$$

So for (6.84) it suffices to show that

$$\liminf_{n\to\infty} n^{-\alpha}\lambda_{min}\left(\sum_{i=1}^{n} f_i f_i^\tau\right) \neq 0. \tag{6.87}$$

If (6.87) were not true, then there would exist a vector sequence $\{\eta_{n_k}\}$,

$$\eta_{n_k} = [\ \alpha_{n_k}^{(0)\tau} \ \dots \ \alpha_{n_k}^{(p-1)\tau} \ \ \beta_{n_k}^{(0)\tau} \ \dots \ \beta_{n_k}^{(q-1)\tau}$$

$$\gamma_{n_k}^{(0)\tau} \ \dots \ \gamma_{n_k}^{(r-1)\tau}]^\tau \in I\!R^h, \quad h = mp + lp + mr$$

such that $\|\eta_{n_k}\| = 1$ and

$$\lim_{k\to\infty} n_k^{-\alpha}\left(\sum_{i=1}^{n_k}(\eta_{n_k}^\tau f_i)^2\right) = 0. \tag{6.88}$$

Let

$$\begin{aligned} H_{n_k}(z) &= \sum_{i=1}^{p-1}\alpha_{n_k}^{(i)\tau} z^i (Adj A(z))[B(z) \quad C(z)] \\ &\quad + \sum_{i=1}^{q-1}\beta_{n_k}^{(i)\tau} z^i [(det A(z))I_l \quad 0] \\ &\quad + \sum_{i=1}^{r-1}\gamma_{n_k}^{(i)\tau} z^i [0 \quad (det A(z))I_m] \\ &= \sum_{j=0}^{\mu}[h_{n_k}^{(j)\tau} \quad g_{n_k}^{(j)\tau}]z^j, \quad \mu \leq t = max(p,q,r) + mp - 1, \end{aligned} \tag{6.89}$$

where $h_{n_k}^{(i)\tau}$ and $g_{n_k}^{(j)\tau}$ are l- and m-dimensional vectors respectively and I_x denotes the $x \times x$ identity matrix. Obviously, $h_{n_k}^{(i)\tau}$ and $g_{n_k}^{(j)\tau}$ are bounded in k and ω.

Notice

$$y_n = A^{-1}(z)[B(z)u_n + C(z)w_n],$$

$$(det\, A(z))y_n = (Adj\, A(z))[B(z) \quad C(z)] \begin{bmatrix} u_n \\ w_n \end{bmatrix},$$

$$(det\, A(z))u_n = [(det\, A(z))I_l \quad 0] \begin{bmatrix} u_n \\ w_n \end{bmatrix},$$

and

$$(det\, A(z))w_n = [0 \quad (det\, A(z))I_m] \begin{bmatrix} u_n \\ w_n \end{bmatrix}.$$

Then we have

$$\begin{aligned} \eta_{n_k}^\tau f_i = & \Big\{\alpha_{n_k}^{(0)\tau} z^i (Adj A(z))[B(z) \quad C(z)] + \dots \\ & +\alpha_{n_k}^{(p-1)\tau} z^{i-p+1} (Adj A(z))[B(z) \quad C(z)] \\ & +\beta_{n_k}^{(0)\tau}[(det A(z))I_l \quad 0] + \dots + \beta_{n_k}^{(q-1)\tau} z^{i-q+1}[(det A(z))I_l \quad 0] \\ & +\gamma_{n_k}^{(0)\tau}[0 \quad (det A(z))I_m] + \dots \\ & + \gamma_{n_k}^{(r-1)\tau} z^{i-r+1}[0 \quad (det A(z))I_m]\Big\} \begin{bmatrix} u_i \\ w_i \end{bmatrix} \\ = & \; H_{n_k}(z) \begin{bmatrix} u_i \\ w_i \end{bmatrix} = \sum_{j=0}^{\mu} (h_{n_k}^{(j)\tau} u_{i-j} + g_{n_k}^{(j)\tau} w_{i-j}) \end{aligned}$$

and by (6.88)

$$\begin{aligned} & \lim_{n\to\infty} n_k^{-\alpha} \sum_{i=1}^{n_k} (\eta_{n_k}^\tau f_i)^2 \\ = & \; n_k^{-\alpha} \sum_{i=1}^{n_k} \Big(h_{n_k}^{(0)\tau} u_i + \quad \dots \quad + h_{n_k}^{(\mu)\tau} u_{i-\mu} \\ & + g_{n_k}^{(0)\tau} w_i + \quad \dots \quad + g_{n_k}^{(\mu)\tau} w_{i-\mu} \Big)^2 = 0 \end{aligned} \tag{6.90}$$

By Theorem 2.8 from (6.78) and (6.77) it follows that

$$\left\| \sum_{i=1}^{n} u_{i-j}^s v_i^\tau \right\| = O\left(n^{\frac{1+\delta}{2}} (log n)^{\frac{1}{2}+\eta} \right), \quad j \ge 0, \tag{6.91}$$

$$\left\|\sum_{i=1}^{n} u_{i-j} v_i^{\tau}\right\| = O\left(n^{\frac{1+\delta}{2}} (logn)^{\frac{1}{2}+\eta}\right), \quad j > 0, \tag{6.92}$$

and

$$\left\|\sum_{i=1}^{n} w_{i-j} v_i^{\tau}\right\| = O\left(n^{\frac{1}{2}} (logn)^{\frac{1}{2}+\eta}\right), \quad j \geq 0. \tag{6.93}$$

Because $g_{n_k}^{(j)}$ and $h_{n_k}^{(j)}$ are bounded and $\alpha > \frac{1}{2}(1+\delta)$ from (6.91)-(6.93) we know that

$$n_k^{-\alpha} \left\{ h_{n_k}^{(0)\tau} \sum_{i=1}^{n_k} u_i^s v_i^{\tau} h_{n_k}^{(0)\tau} + \sum_{j=1}^{\mu} h_{n_k}^{(j)\tau} \sum_{i=1}^{n_k} u_{i-j} v_i^{\tau} h_{n_k}^{(0)\tau} \right.$$
$$\left. + \sum_{j=1}^{\mu} g_{n_k}^{(j)\tau} \sum_{i=1}^{n_k} w_{i-j} v_i^{\tau} h_{n_k}^{(0)\tau} \right\} \xrightarrow[k \to \infty]{} 0. \tag{6.94}$$

This together with (6.90) implies

$$n_k^{-\alpha} \sum_{i=1}^{n_k} (h_{n_k}^{(0)\tau} v_i)^2 \xrightarrow[k \to \infty]{} 0 \tag{6.95}$$

$$n_k^{-\alpha} \sum_{i=1}^{n_k} \left(h_{n_k}^{(0)\tau} u_i^s + h_{n_k}^{(1)\tau} u_{i-1} + \quad \dots \quad + h_{n_k}^{(\mu)\tau} u_{i-\mu} \right.$$
$$\left. + g_{n_k}^{(0)\tau} w_i + \quad \dots \quad + g_{n_k}^{(\mu)\tau} w_{i-\mu} \right)^2 \xrightarrow[k \to \infty]{} 0. \tag{6.96}$$

From (6.72) and (6.95) we know

$$\|h_{n_k}^{(0)\tau}\|^2 = o\left(n_k^{-(1-\varepsilon-\alpha)}\right), \tag{6.97}$$

which incorporating (6.78) yields

$$n^{-(1+\delta)+1-\varepsilon-\alpha} \sum_{i=1}^{n_k} (h_{n_k}^{(0)\tau} u_i^s)^2 \xrightarrow[k \to \infty]{} 0. \tag{6.98}$$

From (6.96) and (6.98) we then have

$$n_k^{-\alpha-(\varepsilon+\delta)} \sum_{i=1}^{n_k} \left(h_{n_k}^{(1)\tau} u_{i-1} + h_{n_k}^{(1)\tau} u_{i-1} + \quad \dots \quad + h_{n_k}^{(\mu)\tau} u_{i-\mu} \right.$$
$$\left. + g_{n_k}^{(0)\tau} w_i + \quad \dots \quad + g_{n_k}^{(\mu)\tau} w_{i-\mu} \right)^2 \xrightarrow[k \to \infty]{} 0. \tag{6.99}$$

Comparing (6.99) with (6.90) we find that $n_k^{-\alpha}$ in (6.90) is replaced by $n_k^{-\alpha-(\varepsilon+\delta)}$ and the term $h_{n_k}^{(0)\tau} u_i$ disappears. Continuing the same argument, we obtain

$$\|h_{n_k}^{(i)\tau}\|^2 = o\left(n_k^{-(1-\varepsilon-\alpha-i(\varepsilon+\delta))}\right), \quad 0 \le i \le \mu \tag{6.100}$$

$$n_k^{-\alpha-s(\varepsilon+\delta)} \sum_{i=1}^{n_k} \Big(h_{n_k}^{(s)\tau} u_{i-s} + h_{n_k}^{(s+1)\tau} u_{i-s-1} + \quad \dots \quad + h_{n_k}^{(\mu)\tau} u_{i-\mu}$$
$$+ g_{n_k}^{(0)\tau} w_i + \quad \dots \quad + g_{n_k}^{(\mu)\tau} w_{i-\mu} \Big)^2 \xrightarrow[k\to\infty]{} 0 \tag{6.101}$$

for any integer $s \in [1, \mu+1]$.

Since $\alpha \le 1 - (\mu+1)(\varepsilon+\delta)$, it immediately follows from (6.100) that

$$\|h_{n_k}^{(i)\tau}\|^2 \xrightarrow[k\to\infty]{} 0, \quad 0 \le i \le \mu. \tag{6.102}$$

For $s = t+1$ we have from (6.101)

$$n_k^{-\alpha-(\mu+1)(\varepsilon+\delta)} \sum_{i=1}^{n_k} \left(g_{n_k}^{(0)\tau} w_i + \quad \dots \quad + g_{n_k}^{(\mu)\tau} w_{i-\mu} \right)^2 \xrightarrow[k\to\infty]{} 0. \tag{6.103}$$

Using Theorem 2.8 and (6.77) we see that for $j \ne s$,

$$\left\| g_{n_k}^{(j)\tau} \sum_{i=1}^{n_k} w_{i-j} w_{i-s}^\tau g_{n_k}^{(s)} \right\|$$
$$\le \|g_{n_k}^{(j)}\| \|g_{n_k}^{(s)}\| O\left\{ \left(\sum_{i=1}^{n_k} \|w_i\|^2 \right)^{\frac{1}{2}} \left[\log\left(\sum_{i=1}^{n_k} \|w_i\|^2 + e \right) \right]^{\frac{1}{2}+\eta} \right\}$$
$$\le O\left\{ \left[\|g_{n_k}^{(j)}\|^2 \left(\sum_{i=1}^{n_k} \|w_i\|^2 \right)^\gamma + \|g_{n_k}^{(s)}\|^2 \left(\sum_{i=1}^{n_k} \|w_i\|^2 \right)^\gamma \right] \right), \quad \frac{1}{2} < \gamma < 1,$$

and hence

$$\sum_{i=1}^{n_k} \left(g_{n_k}^{(0)\tau} w_i + \quad \dots \quad + g_{n_k}^{(\mu)\tau} w_{i-\mu} \right)^2$$
$$\ge \sum_{i=1}^{n_k} \left[(g_{n_k}^{(0)\tau} w_i)^2 + \quad \dots \quad + (g_{n_k}^{(\mu)\tau} w_{i-\mu})^2 \right]$$
$$- \sum_{j=0}^{\mu} \sum_{\substack{s=0 \\ s \ne j}}^{\mu} \left\| g_{n_k}^{(j)\tau} \sum_{i=1}^{n_k} w_{i-j} w_{i-s}^\tau g_{n_k}^{(s)} \right\|$$

$$\geq \sum_{i=1}^{n_k}\left[(g_{n_k}^{(0)\tau}w_i)^2+\ \ldots\ +(g_{n_k}^{(\mu)\tau}w_{i-\mu})^2\right]$$

$$-O\left(\sum_{j=0}^{\mu}\|g_{n_k}^{(j)}\|^2\left(\sum_{i=1}^{n_k}\|w_i\|^2\right)^{\gamma}\right)$$

$$\geq [\lambda_{min}(R)n_k+o(n_k)]\sum_{j=0}^{\mu}\|g_{n_k}^{(j)}\|^2-O(n_k^{\gamma})\sum_{j=0}^{\mu}\|g_{n_k}^{(j)}\|^2$$

$$\geq \frac{1}{2}[\lambda_{min}(R)n_k+o(n_k)]\sum_{j=0}^{\mu}\|g_{n_k}^{(j)}\|^2. \tag{6.104}$$

This incorporating (6.103) leads to

$$n_k^{1-\alpha-(\mu+1)(\varepsilon+\delta)}\sum_{j=0}^{\mu}\|g_{n_k}^{(j)}\|^2\xrightarrow[k\to\infty]{}0.$$

However, $1-\alpha-(\mu+1)(\varepsilon+\delta)\geq 0$. Hence we have

$$\|g_{n_k}^{(j)}\|^2\xrightarrow[k\to\infty]{}0\qquad j=0,\ 1,\ \ldots,\ \mu$$

which together with (6.102) imply

$$H_{n_k}(z)\xrightarrow[k\to\infty]{}0. \tag{6.105}$$

Since $\{\eta_{n_k}\}$ is bounded, there exists a convergent subsequence tending to a limit with unit norm

$$[\alpha^{(0)\tau}\ \ldots\ \alpha^{(p-1)\tau}\quad \beta^{(0)\tau}\ \ldots\ \beta^{(q-1)\tau}\quad \gamma^{(0)\tau}\ \ldots\ \gamma^{(r-1)\tau}]^{\tau}.$$

Then by (6.105) from (6.89) we have

$$\sum_{i=0}^{p-1}\alpha^{(i)\tau}z^i(Adj A(z))[B(z)\quad C(z)]+\sum_{i=1}^{q-1}\beta^{(i)\tau}z^i[(det A(z))I_l\quad 0]$$
$$+\sum_{i=1}^{r-1}\gamma^{(i)\tau}z^i[0\quad (det A(z))I_m]=0$$

or

$$\sum_{i=0}^{p-1}\alpha^{(i)\tau}z^i(Adj A(z))B(z)=-\sum_{i=1}^{q-1}\beta^{(i)\tau}z^i(det A(z))I_l \tag{6.106}$$

$$\sum_{i=0}^{p-1}\alpha^{(i)\tau}z^i(Adj A(z))C(z)=-\sum_{i=1}^{r-1}\gamma^{(i)\tau}z^i(det A(z))I_m. \tag{6.107}$$

Since $A(Z)$, $B(z)$ and $C(z)$ have no common left factor, there are polynomial matrices $M(z)$, $N(z)$ and $L(z)$ such that

$$A(z)M(z) + B(z)N(z) + C(z)L(z) = I.$$

Therefore, by (6.106) and (6.107) we derive

$$\begin{aligned}
&\sum_{i=0}^{p-1} \alpha^{(i)\tau} z^i (Adj A(z)) \\
= &\sum_{i=0}^{p-1} \alpha^{(i)\tau} z^i (Adj A(z))[A(z)M(z) + B(z)N(z) + C(z)L(z)] \\
= &(det A(z)) \left[\sum_{i=0}^{p-1} \alpha^{(i)\tau} z^i M(z) + \sum_{i=0}^{q-1} \beta^{(i)\tau} z^i N(z) + \sum_{i=0}^{r-1} \gamma^{(i)\tau} z^i L(z) \right] \\
\triangleq &(det A(z)) \sum_{i=0}^{\lambda} \mu^{(i)\tau} z^i,
\end{aligned} \tag{6.108}$$

where $\mu^{(i)}$ is m-dimensional, $i = 0, 1, \ldots, \lambda$.

Multiplying the equality (6.108) respectively by $A(z)$, $B(z)$ and $C(z)$ from the right and using (6.106) and (6.107) we obtain

$$\sum_{i=0}^{p-1} \alpha^{(i)\tau} z^i = \sum_{i=0}^{\lambda} \mu^{(i)\tau} z^i A(z) \tag{6.109}$$

$$\sum_{i=1}^{q-1} \beta^{(i)\tau} z^i = \sum_{i=0}^{\lambda} \mu^{(i)\tau} z^i B(z) \tag{6.110}$$

and

$$\sum_{i=1}^{r-1} \gamma^{(i)\tau} z^i = \sum_{i=0}^{\lambda} \mu^{(i)\tau} z^i C(z). \tag{6.111}$$

Since $[A_p \quad B_q \quad C_r]$ is of row-full-rank, $\mu^{(i)\tau}[A_p \quad B_q \quad C_r] \neq 0$ for any non-zero $\mu^{(i)\tau}$. If $\mu^{(i)\tau} A_p \neq 0$, then the right-hand side of (6.109) is a polynomial with order greater than or equal to p, while its left-hand side is of order strictly less than p. The contradiction shows $\mu^{(i)\tau} A_p = 0$, and hence $\mu^{(i)\tau}[B_q \quad C_r] \neq 0$. If $\mu^{(i)\tau} B_q \neq 0$, then from (6.110) a similar argument leads to a contradiction. Hence $\mu^{(i)\tau} B_q = 0$; consequently $\mu^{(i)\tau} C_r \neq 0$. But this contradicts (6.111). Therefore, we conclude that

$$\mu^{(i)} = 0, \quad i = 0, \ldots, \lambda \tag{6.112}$$

and from (6.109)-(6.111), (6.112) implies

$$\alpha^{(i)} = 0, \quad \beta^{(j)} = 0, \; i = 0, \ldots, p-1, \; j = 0, \ldots, q-1,$$

$$\gamma^{(k)} = 0, \ k = 0, ..., r-1.$$

However, this is impossible, because

$$\sum_{i=0}^{p-1} \|\alpha^{(i)}\|^2 + \sum_{j=0}^{q-1} \|\beta^{(j)}\|^2 + \sum_{k=0}^{r-1} \|\gamma^{(k)}\|^2 = 1$$

The contradiction shows that (6.87) is valid. Thus the proof is complete.
Q.E.D.

Remark 6.1. Theorem 6.2 remains valid if $\{v_n\}$ in (6.76) is any sequence satisfying the following relaxed conditions:
1) v_n is $\mathcal{F}_n$-measurable and such that (6.91)-(6.93) hold;
2) $\frac{1}{n^{1-\varepsilon}} \sum_{i=1}^{n} v_i v_i^T \geq c_1 I$, $c_1 > 0$ a.s. for sufficiently large n,

$$\varepsilon \in \left[0, \frac{1}{2(t+1)}\right), \quad t = \max(p, q, r) + mp - 1;$$

3) $\|v_n\|^2 = O(n^{-\varepsilon})$ a.s.

For example, if $\{t_n\}$ is a sequence of nonnegative random variables, t_n is $\mathcal{F}_n$-measurable and $t_n = O(n)$, $n = O(t_n)$, then $\frac{v_n}{t_{n-1}^{\varepsilon/2}}$ with $\varepsilon \in \left[0, \frac{1}{2(t+1)}\right)$, $t = \max(p, q, r) + mp - 1$ satisfies Conditions 1)-3) mentioned above. In fact, $\{v_n, \mathcal{F}_n\}$ is a martingale difference sequence, then (6.91) and (6.92) are consequences of Theorem 2.8. By independence between $\{\varepsilon_n\}$ and $\{w_n\}$, $\{v_n, \mathcal{G}_n\}$ is also a martingale difference sequence, where $\mathcal{G}_n = \sigma\{w_{i+1}, \varepsilon_i, i \leq n\}$. Hence by Theorem 2.8 (6.93) is satisfied as well. Conditions 2), 3) are obviously fulfilled.

Theorem 6.2 establishes strong consistency of ELS estimate when the diminishing excited control is applied to the system. We now prove a similar result for SG estimate.

Theorem 6.3 *[CG5]. Suppose that for system (3.91)-(3.94) the following conditions hold.*

1). $\{w_n, \mathcal{F}_n\}$ is a martingale difference sequence with

$$\sup_{n \geq 1} E\{\|w_n\|^2 | \mathcal{F}_{n-1}\} < \infty \quad a.s.$$

$$\lim_{n \to \infty} \frac{1}{n} \sum_{i=1}^{n} w_i w_i^T = R > 0 \quad a.s.; \tag{6.113}$$

2). $C(z) - \frac{a}{2}I$ is SPR for some $a > 0$;

3). $A(z)$, $B(z)$ and $C(z)$ have no common left factor and $[A_p \quad B_q \quad C_r]$ is of row-full-rank;

4). The diminishingly excited control (6.76) is applied to the system, where the $\mathcal{F}'_n$-measurable desired control u^s_n satisfies

$$\frac{1}{n}\sum_{i=1}^{n}\|u^s_i\|^2 = O(log^\delta n), \quad \delta \in \left(0, \frac{\frac{1}{4}-(m+2)s\varepsilon}{2+(m+1)s}\right), \tag{6.114}$$

and where the excitation signal is defined by (6.68) and (6.69). Under such a control the system output satisfies

$$\frac{1}{n}\sum_{i=1}^{n}\|y_i\|^2 = O(log^\delta n) \quad a.s., \quad \delta \in \left(0, \frac{\frac{1}{4}-(m+2)s\varepsilon}{2+(m+1)s}\right). \tag{6.115}$$

Then θ_n given by SG algorithm (4.22) and (4.23) is strongly consistent: $\theta_n \xrightarrow[n\to\infty]{} \theta$ a.s., where θ is defined by (4.1).

Proof. We note that

$$\frac{\frac{1}{4}-(m+2)s\varepsilon}{2+(m+1)s} > 0, \quad for \quad \varepsilon \in \left(0, \frac{1}{4s(m+2)}\right).$$

Hence the interval for δ is nonempty.

By Theorem 2.8 from (6.114) it follows that

$$\begin{aligned}\sum_{i=1}^{n} u^s_i v^\tau_i &= O\left(\left(\sum_{i=1}^{n}\|u^s_i\|^2\right)^{\frac{1}{2}}\left(log\left(\sum_{i=1}^{n}\|u^s_i\|^2+e\right)\right)^{\frac{1}{2}+\eta}\right)\\ &= O\left(n^{\frac{1}{2}}(logn)^{\frac{1}{2}+\delta}\right).\end{aligned} \tag{6.116}$$

Then from (6.113)-(6.116) we see that

$$r^0_n = O(nlog^\delta n). \tag{6.117}$$

On the other hand, using Theorem 2.8 and (6.74) we find that as $n \to \infty$

$$r^0_n \geq \sum_{i=1}^{n}\|u_i\|^2 \geq \frac{1}{2}\sum_{i=1}^{n}\|v_i\|^2 \geq \frac{l}{4}\frac{n}{log^\varepsilon n} \longrightarrow \infty \tag{6.118}$$

where l is the dimension of u_i. From (6.117) and (6.118) we derive

$$\begin{aligned} \frac{r_{n+1}^0}{r_n^0} &= O\left(\frac{\frac{(n+1)log^{\delta}(n+1)}{n}}{log^{\varepsilon} n}\right) = O\left(log^{\delta+\varepsilon} n\right) \\ &= O\left((logr_n^0)^{\delta+\varepsilon}\right). \end{aligned} \tag{6.119}$$

To prove $\theta_n \xrightarrow[n\to\infty]{} \theta$ by Theorem 4.4 we need only to verify $\Phi^0(n,0) \xrightarrow[n\to\infty]{} 0$ w here $\Phi^0(n,0)$ is defined (4.104) and (4.105). For this we apply Remark 4.3 which asserts that if $i)$ $r_n^0 \xrightarrow[n\to\infty]{} \infty$; $ii)$ $r_n^0/r_{n-1}^0 \le M(logr_{n-1}^0)^{\lambda}$, $\forall n \ge N_0$ for some $\lambda \in [0, \frac{1}{4}]$, $M > 0$ and $N_0 > 0$; $iii)$ $\mu_{max}^0(n)/\mu_{min}^0(n) \le M(logr_{n-1}^0)^{\frac{1}{4}-\lambda}$, $\forall n \ge N_0$, then $\Phi^0(n,0) \xrightarrow[n\to\infty]{} 0$, where $\mu_{max}^0(n)$ and $\mu_{min}^0(n)$ respectively denote the maximum and the minimum eigenvalue of $\sum_{i=0}^{n} \varphi_i^0 \varphi_i^{0\tau} + \frac{1}{h}I$, $h = mp + lq + mr$.

Noticing

$$\delta < \frac{\frac{1}{4} - (m+2)s\varepsilon}{2 + (m+1)s} \quad or \quad 2\delta + (m+1)s\delta + (m+2)s\varepsilon < \frac{1}{4} \tag{6.120}$$

we find that $\delta + \varepsilon < \frac{1}{4}$ and by (6.119) Condition ii) holds with $\lambda = \delta + \varepsilon$. Therefore to complete the proof it suffices to verify Condition iii).

It is clear that (6.117) implies

$$\mu_{max}^0(n) = O(nlog^{\delta} n). \tag{6.121}$$

Then for iii) it is sufficient to verify

$$\liminf_{n\to\infty} \frac{(logn)^{\nu}}{n} \mu_{min}^0(n) \triangleq c > 0, \quad \nu = \frac{1}{4} - 2\delta - \varepsilon, \tag{6.122}$$

because (6.121) and (6.122) yield

$$\begin{aligned} \mu_{max}^0(n)/\mu_{min}^0(n) &\le O\left(nlog^{\delta} n \left(\frac{n}{(logn)^{\nu}}\right)^{-1}\right) \\ &= O((logn)^{\nu+\delta}) = O((logr_n^0)^{\nu+\delta}) \end{aligned} \tag{6.123}$$

where the last equality follows from (6.118). The relationship (6.123) means that iii) holds with $\lambda = \varepsilon + \delta$.

We now proceed to prove (6.122). By the argument developed in (6.84)-(6.87) it suffices to show

$$\frac{(logn)^{\nu}}{n}\lambda_{min}\left(\sum_{i=1}^{n} f_i f_i^{\tau}\right) \neq 0. \tag{6.124}$$

Suppose that (6.124) were not true. Comparing (6.124) with (6.87) we see that $n^{-1}(logn)^{\nu}$ in (6.124) replaces $n^{-\alpha}$ in (6.87). With this in mind arguing in the same way as for (6.87) we arrive at (6.90) without change, while by Theorem 2.8 and (6.74), (6.91) and (6.92) turn to

$$\left\|\sum_{i=1}^{n} u_{i-j}^{s} v_i^{\tau}\right\| = O\left(n^{\frac{1}{2}}(logn)^{\frac{1}{2}+\delta}\right) \quad j \geq 0 \tag{6.125}$$

$$\left\|\sum_{i=1}^{n} u_{i-j} v_i^{\tau}\right\| = O\left(n^{\frac{1}{2}}(logn)^{\frac{1}{2}+\delta}\right) \quad j \geq 0 \tag{6.126}$$

respectively, and (6.93) remains unchanged. Hence (6.94)-(6.96) hold true with $n_k^{-\alpha}$ replaced by $n_k^{-1}(logn_k)^{\nu}$.

Therefore, corresponding to (6.94)-(6.99) we have

$$\left\|h_{n_k}^{(0)}\right\|^2 = o\left((logn_k)^{-\nu+\varepsilon}\right) \tag{6.127}$$

which together with (6.114) implies

$$n_k^{-1}(logn_k)^{\nu-(\varepsilon+\delta)}\sum_{i=1}^{n_k}\left(h_{n_k}^{(0)\tau}u_i^{s}\right)^2 \xrightarrow[k\to\infty]{} 0 \tag{6.128}$$

which is the analogue of (6.98) with $n_k^{-\alpha}$ replaced by $n_k^{-1}(logn_k)^{\nu}$. Finally, corresponding to (6.99) we have

$$\frac{(logn_k)^{\nu-(\varepsilon+\delta)}}{n_k}\sum_{i=1}^{n_k}\left(h_{n_k}^{(1)\tau}u_{i-1} + \quad \dots \quad + h_{n_k}^{(\mu)\tau}u_{i-\mu}\right.$$

$$\left. + g_{n_k}^{(0)\tau}w_i + \quad \dots \quad + g_{n_k}^{(\mu)\tau}w_{i-\mu}\right)^2 \xrightarrow[k\to\infty]{} 0. \tag{6.129}$$

Comparing (6.129) with (6.96) with $n_k^{-\alpha}$ replaced by $n_k^{-1}(logn_k)^{\nu}$ we find that in (6.129) $h_{n_k}^{(0)\tau}$ disappears and the power of $logn_k$ is reduced by $(\varepsilon+\delta)$. Repeating the same argument one can be convinced of that

$$\|h_{n_k}^{(i)}\|^2 = o\left((logn_k)^{-\nu+i(\varepsilon+\delta)+\varepsilon}\right), \quad 0 \leq i \leq \mu \tag{6.130}$$

and

$$\frac{(\log n_k)^{\nu-(\mu+1)(\varepsilon+\delta)}}{n_k}\sum_{i=1}^{n_k}\left(g_{n_k}^{(0)\tau}w_i + \quad \dots \quad + g_{n_k}^{(\mu)\tau}w_{i-\mu}\right)^2$$

$$\xrightarrow[k\to\infty]{} 0. \tag{6.131}$$

Noticing from (6.67), (6.69) and (6.89) that $t \leq s+mp-1$, and hence $\mu \leq s+mp-1$ we find that

$$\begin{aligned}\nu-(\mu+1)(\varepsilon+\delta) &= \frac{1}{4}-2\delta-\varepsilon-(\mu+1)(\varepsilon+\delta)\\ \geq\ & \frac{1}{4}-2\delta-\varepsilon-(s+mp)(\varepsilon+\delta)\\ >\ & \frac{1}{4}-2\delta-s\varepsilon-(m+1)(\varepsilon+\delta)s>0\end{aligned} \tag{6.132}$$

where the last inequality follows from (6.120).

Therefore, (6.130) implies

$$h_{n_k}^{(i)}\xrightarrow[k\to\infty]{} 0, \quad 0\leq i\leq\mu.$$

By (6.104) and (6.131) we have

$$(\log n_k)^{\nu-(\mu+1)(\varepsilon+\delta)}\sum_{j=0}^{\mu}\left\|g_{n_k}^{(j)\tau}\right\|^2\xrightarrow[k\to\infty]{} 0$$

and

$$g_{n_k}^{(j)\tau}\xrightarrow[k\to\infty]{} 0, \quad 0\leq j\leq\mu$$

by (6.132). Therefore (6.105) holds.

Starting from (6.105) by completely the same argument as that used in Theorem 6.1 we obtain a contradiction which proves (6.124) and at the same time completes the proof of the theorem. Q.E.D.

Similar to Remark 6.1 we have the following

Remark 6.2. Theorem 6.3 is still true if $\{v_n\}$ in (6.76) satisfies the following relaxed conditions:

1) v_n satisfies (6.93), (6.125) and (6.126) and v_n is $\mathcal{F}_n$-measurable;

2) $\frac{\log^\varepsilon n}{n}\sum_{i=1}^{n}v_iv_i^\tau > c_2 I$, $c_2>0$ a.s. for sufficiently large n,

$$\varepsilon\in\left(0,\frac{1}{4s(m+2)}\right), \qquad s=\max(p,q,r+1);$$

3) $\|v_n\|^2 = O(\log^{-\varepsilon} n)$ a.s.

CHAPTER 7

Order Estimation

The order estimation of time series is a problem of long-standing interest. Various criteria such as multiple decision rule [An3], *AIC* [A2], *BIC* [Ri1], [Sw] and ΦIC [HQ] have been proposed to solve this problem. Among them, the information based criteria usually choose the order to minimize the following quantity

$$\log \widehat{\sigma}_n^2 + a_n/n \tag{7.1}$$

where n is the data size, $\widehat{\sigma}_n^2$ is the residual variance and a_n is a non-negative random variable that reflects the complexity of the nominal model. For example, in the order estimation of a scalar autoregressive model, we have

$$\begin{array}{ll} AIC, & \text{if } a_n = 2p \\ BIC, & \text{if } a_n = p\log n \\ \Phi IC, & \text{if } a_n = pc\log\log n, \quad c > 1. \end{array}$$

For analyzing these criteria, in particular, in the above-mentioned works, some sort of stationarity and ergodicity of the observed data (or stochastic processes involved) are usually assumed. Therefore, results obtained in the time-series analysis cannot be directly applied to feedback control systems, since these assumptions normally are not satisfied for such systems [CG6]. In this Chapter, we will concentrate on the order estimation problem of non-stationary ARMAX models. A different criterion *CIC* will be introduced and analyzed. We shall find that the excitation condition

$$\log \lambda_{\max}(n)/\lambda_{\min}(n) \xrightarrow[n \to \infty]{} 0 \quad \text{a.s.}$$

required for order estimation is essentially the same as that used for LS parameter estimation (see Chapter 4). Finally, we will discuss the connections of *CIC* and *BIC*.

7.1 Order Estimation by Use of a Priori Information

Throughout this chapter, we consider linear feedback control systems described by the following ARMAX model:

$$A(z)y_n = B(z)u_n + C(z)w_n, \quad n \geq 0, \tag{7.2}$$

where y_n, u_n and w_n are respectively the m-, l- and m-dimensional system output, input and noise sequences with initial values $y_i = w_i = 0$, $u_j = 0$, $i < 0$, $j < 0$. $A(z)$, $B(z)$ and $C(z)$ are unknown matrix polynomials in backwards-shift operator z:

$$A(z) = I + A_1 z + \ldots + A_{p_0} z^{p_0}, \quad p_0 \geq 0 \tag{7.3}$$

$$B(z) = B_1 z + B_2 z^2 + \ldots + B_{q_0} z^{q_0}, \quad q_0 \geq 0 \tag{7.4}$$

$$C(z) = I + C_1 z + \ldots + C_{r_0} z^{r_0}, \quad r_0 \geq 0 \tag{7.5}$$

where p_0, q_0 and r_0 are the unknown true orders ($A_{p_0} \neq 0$, $B_{q_0} \neq 0$, $C_{r_0} \neq 0$).

Such an ARMAX model though may not be uniquely defined in the multivariate case [HDe, Section 2.7], it is not critical for a part of results we shall investigate. However, for estimating the true orders (p_0, q_0, r_0) and the parameters $\{A_i, B_j, C_k, i = 1, \ldots, p_0, j = 1, \ldots, q_0, k = 1, \ldots, r_0\}$, we need the following identifiability condition:

H1). $A(z)$, $B(z)$ and $C(z)$ have no common left factor, and A_{p_0} is of row full rank.

It should be noted that in the scalar variable case the rank condition is automatically satisfied. Also, in what follows the rank condition on A_{p_0} can be replaced by that of B_{q_0} or C_{r_0}.

As usual we also require the following standard assumptions:

H2). The driven noise $\{w_n, \mathcal{F}_n\}$ is a martingale difference sequence with respect to a family of non-decreasing σ-algebras $\{\mathcal{F}_n\}$ and such that for some $\beta \geq 2$

$$\sup_n E\left[\|w_{n+1}\|^\beta | \mathcal{F}_n\right] < \infty \quad \text{a.s.}$$

H3). For any $n \geq 1$, u_n is $\mathcal{F}_n$-measurable.

Similar to the parameter estimation problem discussed in Chapter 4, a key step in constructing the order estimate is to formulate the estimates for the noise processes $\{w_i\}$. Since the accuracy of the noise estimates depends on the *a priori* information available to us, we shall separately consider the cases where *a priori* information is or is not available. In this section, we consider the case where the *a priori* information is available.

H4). The true orders (p_0, q_0, r_0) belong to a known finite set M:

$$M \triangleq \{(p,q,r): \quad 0 \le p \le p^*,\ 0 \le q \le q^*,\ 0 \le r \le r^*\}.$$

Thus we may use (4.182)-(4.185) to get the noise estimate $\widehat{w}_i$.

Lemma 7.1. *Let Assumptions H2)-H4) be satisfied, and $\{\widehat{w}_i\}$ be given by (4.182)-(4.185). Suppose also that either of the following two conditions is satisfied:*
i). $C^{-1}(z) - \frac{1}{2}I$ is SPR, and in (4.185) $(\overline{p},\overline{q},\overline{r})$ is taken as (p^*, q^*, r^*);
ii). $\det C(z) \neq 0$, $|z| \le 1$, and in (4.185) $(\overline{p},\overline{q},\overline{r})$ is taken as $k(p^*, q^*, r^*)$, where k satisfies conditions in Theorem 4.7.
Then as $n \to \infty$

$$\sum_{i=0}^{n} \|\widehat{w}_i - w_i\|^2 = O\left(\log s_n \{\log\log s_n\}^{\delta(\beta-2)}\right) \quad a.s.$$

where

$$\delta(\beta-2) = \begin{cases} 0, & \beta > 2 \\ c, & \beta = 2 \end{cases} \quad c > 1,$$

and s_n is defined as

$$s_n = \sum_{i=1}^{n} (\|y_i\|^2 + \|u_i\|^2 + \|w_i\|^2). \tag{7.6}$$

This result can be found in Chapter 4 (see the proofs of Theorems 4.1 and 4.2).

For any $(p,q,r) \in M$, set

$$\theta(p,q,r) = [-A_1 \ \dots \ -A_p \quad B_1 \ \dots \ B_q \quad C_1 \ \dots \ C_r]^\tau, \tag{7.7}$$

where by definition

$$A_i = 0,\ B_j = 0,\ C_k = 0, \quad i > p_0,\ j > q_0,\ k > r_0. \tag{7.8}$$

The ELS estimate

$$\theta_n(p,q,r) = [-A_{1n} \ \dots \ -A_{pn} \quad B_{1n} \ \dots \ B_{qn} \quad C_{1n} \ \dots \ C_{rn}]^\tau \tag{7.9}$$

for $\theta(p,q,r)$ at time n is given by

$$\theta_n(p,q,r) = \left(\sum_{i=0}^{n-1} \varphi_i(p,q,r)\varphi_i^\tau(p,q,r) + \varepsilon I\right)^{-1} \sum_{i=0}^{n-1} \varphi_i(p,q,r) y_{i+1}^\tau \tag{7.10}$$

where $\varepsilon > 0$ and $\varphi_i(p,q,r)$ is defined by

$$\varphi_i(p,q,r) \triangleq [y_i^T \ \dots \ y_{i-p+1}^T \quad u_i^T \ \dots \ u_{i-q+1}^T \quad \widehat{w}_i^T \ \dots \ \widehat{w}_{i-r+1}^T]^T \tag{7.11}$$

and where $\widehat{w}_n$ is given by (4.182)-(4.185).

We introduce an information criterion CIC, where the first "C" emphasizes that the criterion is designed for feedback control systems:

$$CIC(p,q,r)_n = \sigma_n(p,q,r) + (p+q+r)a_n, \tag{7.12}$$

where the subscript n denotes the data size, a_n is a real sequence independent of (p,q,r) and is specified later in theorems, and $\sigma_n(p,q,r)$ is a residual given by

$$\sigma_n(p,q,r) = \sum_{i=0}^{n-1} \|y_{i+1} - \theta_n^T(p,q,r)\varphi_i(p,q,r)\|^2. \tag{7.13}$$

The estimate $(\widehat{p}_n, \widehat{q}_n, \widehat{r}_n)$ for (p_0, q_0, r_0) is given by minimizing $CIC(p,q,r)_n$:

$$\widehat{p}_n = \operatorname*{argmin}_{0 \le p \le p^*} CIC(p,q^*,r^*)_n, \tag{7.14}$$

$$\widehat{q}_n = \operatorname*{argmin}_{0 \le q \le q^*} CIC(\widehat{p}_n,q,r^*)_n, \tag{7.15}$$

$$\widehat{r}_n = \operatorname*{argmin}_{0 \le r \le r^*} CIC(\widehat{p}_n,\widehat{q}_n,r)_n. \tag{7.16}$$

Theorem 7.1 *[CG6], [GCZ]. For the order estimate defined by (7.10)-(7.16) suppose that Assumptions in Lemma 7.1 hold and that in (7.12) the sequence $\{a_n\}$ is positive, nondecreasing and satisfies:*

$$\frac{(\log s_n)(\log\log s_n)^{\delta(\beta-2)}}{a_n} \xrightarrow[n\to\infty]{} 0 \quad a.s. \tag{7.17}$$

and

$$\frac{a_n}{\lambda_{\min}^{(p_0,q^*,r^*)}(n)} \xrightarrow[n\to\infty]{} 0 \quad a.s. \tag{7.18}$$

where s_n and $\delta(\beta-2)$ are defined in Lemma 7.1, and $\lambda_{\min}^{(p,q,r)}(n)$ denotes the minimum eigenvalue of

$$\sum_{i=0}^{n-1} \varphi_i^0(p,q,r)\varphi_i^{0T}(p,q,r) \tag{7.19}$$

with

$$\varphi_i^0(p,q,r) \triangleq [y_i^T \ \dots \ y_{i-p+1}^T \quad u_i^T \ \dots \ u_{i-q+1}^T \quad w_i^T \ \dots \ w_{i-r+1}^T]^T \tag{7.20}$$

Then the order estimate $(\widehat{p}_n, \widehat{q}_n, \widehat{r}_n)$ is strongly consistent:

$$(\widehat{p}_n, \widehat{q}_n, \widehat{r}_n) \xrightarrow[n \to \infty]{} (p_0, q_0, r_0) \quad a.s. \tag{7.21}$$

For the proof of this theorem, we need the following lemma.

Lemma 7.2. *Let Assumption H2) be satisfied, and let the random vector φ_n be $\mathcal{F}_n$-measurable, $\forall n$. Then as $n \to \infty$,*

$$\left\| \left(\sum_{i=0}^{n-1} \varphi_i \varphi_i^T + \varepsilon I \right)^{-\frac{1}{2}} \sum_{i=0}^{n-1} \varphi_i w_{i+1}^T \right\|^2 = O\left([\log r_n][\log\log r_n]^{\delta(\beta-2)} \right) \quad a.s.$$

where $\delta(\beta - 2)$ is defined as in Lemma 7.1, $\varepsilon > 0$, and

$$r_n \triangleq 1 + \sum_{i=0}^{n-1} \|\varphi_i\|^2.$$

Proof. Let us consider the following linear regression model

$$z_{i+1} = \theta^T \varphi_i + w_{i+1}.$$

Applying the least squares recursion (4.13) and (4.14) with $\theta_0 = 0$, $P_0 = \frac{1}{\varepsilon} I$ we get an estimate for θ:

$$\theta_n = \left(\sum_{i=0}^{n-1} \varphi_i \varphi_i^T + \varepsilon I \right)^{-1} \sum_{i=0}^{n-1} \varphi_i z_{i+1}^T$$

or

$$\widetilde{\theta}_n \triangleq \theta - \theta_n = P_n \theta - P_n \sum_{i=0}^{n-1} \varphi_i w_{i+1}^T.$$

Then using the relationship (4.38) derived in the proof of Theorem 4.1, we have

$$\begin{aligned}
&\left\| \left(\sum_{i=0}^{n-1} \varphi_i \varphi_i^T + \varepsilon I \right)^{-\frac{1}{2}} \sum_{i=0}^{n-1} \varphi_i w_{i+1}^T \right\|^2 \\
= &\left\| \left(\sum_{i=0}^{n-1} \varphi_i w_{i+1}^T \right)^T P_n \sum_{i=0}^{n-1} \varphi_i w_{i+1}^T \right\| \\
= &\left\| \left(\theta - P_n^{-1} \widetilde{\theta}_n \right)^T P_n (\theta - P_n^{-1} \widetilde{\theta}_n) \right\| \\
\le &\ 2 tr \theta^T P_n \theta + 2 tr \widetilde{\theta}_n^T P_n^{-1} \widetilde{\theta}_n \\
= &\ O\left([\log r_n][\log\log r_n]^{\delta(\beta-2)} \right) \quad \text{a.s.}
\end{aligned}$$

This completes the proof of the lemma. Q.E.D.

Proof of Theorem 7.1. We start with the proof of strong consistency of $\widehat{p}_n$. For this we first show that

$$\limsup_{n\to\infty} \widehat{p}_n \leq p_0 \quad \text{a.s.} \tag{7.22}$$

By Lemma 7.1 we have

$$\sum_{i=0}^{n-1} \|\varphi_i(p,q,r) - \varphi_i^0(p,q,r)\|^2 = O((\log s_n)(\log\log s_n)^{\delta(\beta-2)}) \tag{7.23}$$

for any $(p,q,r) \in M$, where $\varphi_i(p,q,r)$ is defined by (7.11). By (7.13) we have for $p \geq p_0$,

$$\begin{aligned}
&\sigma_n(p,q^*,r^*)\\
&= tr\widetilde{\theta}_n^\tau(p,q^*,r^*)\sum_{i=0}^{n-1}\varphi_i(p,q^*,r^*)\varphi_i^\tau(p,q^*,r^*)\widetilde{\theta}_n(p,q^*,r^*)\\
&\quad +2tr\widetilde{\theta}_n^\tau(p,q^*,r^*)\sum_{i=0}^{n-1}\varphi_i(p,q^*,r^*)[-\theta^\tau(p,q^*,r^*)\varphi_i^\xi(p,q^*,r^*)+w_{i+1}]^\tau\\
&\quad +\sum_{i=0}^{n-1}\|\theta^\tau(p,q^*,r^*)\varphi_i^\xi(p,q^*,r^*)-w_{i+1}\|^2,
\end{aligned} \tag{7.24}$$

where

$$\begin{aligned}
\widetilde{\theta}_n(p,q,r) &= \theta(p,q,r) - \theta_n(p,q,r)\\
\varphi_n^\xi(p,q,r) &= \varphi_n(p,q,r) - \varphi_n^0(p,q,r)
\end{aligned}$$

By (7.23) and the Schwarz inequality it follows that for any $(p,q,r) \in M$,

$$\begin{aligned}
&\left\|\left(\sum_{i=0}^{n-1}\varphi_i(p,q,r)\varphi_i^\tau(p,q,r)+\varepsilon I\right)^{-\frac{1}{2}}\sum_{i=0}^{n-1}\varphi_i(p,q,r)\varphi_i^{\xi\tau}(p,q,r)\right\|^2\\
&= O\left(\sum_{i=0}^{n}\|\varphi_i^\xi(p,q,r)\|^2\right)\\
&= O\left(\{\log s_n\}\{\log\log s_n\}^{\delta(\beta-2)}\right) \quad \text{a.s.}
\end{aligned} \tag{7.25}$$

Hence, by Lemma 7.2, (7.24) and (7.25) it is easy to see that for any $p \geq p_0$

$$\begin{aligned}
\sigma_n(p,q^*,r^*) &= O\left(\{\log s_n\}\{\log\log s_n\}^{\delta(\beta-2)}\right)\\
&\quad +\sum_{i=0}^{n}\|\theta^\tau(p,q^*,r^*)\varphi_i^\xi(p,q^*,r^*)-w_{i+1}\|^2.
\end{aligned} \tag{7.26}$$

Noting that the last term does not depend on p when $p \geq p_0$, we have from (7.12) and (7.26) that

$$\begin{aligned} & CIC(p,q^*,r^*)_n - CIC(p_0,q^*,r^*)_n \\ = \; & a_n \left\{ p - p_0 + O\left(\frac{(\log s_n)(\log\log s_n)^{\delta(\beta-2)}}{a_n} \right) \right\}. \end{aligned} \tag{7.27}$$

From this and (7.17) we see that for sufficiently large n

$$\min_{p_0 < p \leq p^*} [CIC(p,q^*,r^*)_n - CIC(p_0,q^*,r^*)_n] > 0.$$

Consequently, by the definition of $\widehat{p}_n$, we know that (7.22) is true.

Next we prove that

$$\liminf_{n\to\infty} \widehat{p}_n \geq p_0 \quad \text{a.s.} \tag{7.28}$$

For $p \leq p_0$ let us set

$$\begin{aligned} \widehat{\theta}_n(p) \; = \; & [-A_{1n}(p) \; \ldots \; -A_{pn}(p) \; 0 \ldots 0 \;\; B_{1n}(p) \; \ldots \; B_{q^*n}(p) \\ & C_{1n}(p) \; \ldots \; C_{r^*n}(p)]^{\tau}, \end{aligned} \tag{7.29}$$

where each 0 represents an $m \times m$ zero matrix and there are $p_0 - p$ such matrices in total. The rest entries in $\widehat{\theta}_n(p)$ are given by $\theta_n(p,q^*,r^*)$.

Then it follows that for any $p \leq p_0$,

$$\begin{aligned} & y_{i+1} - \theta_n^{\tau}(p,q^*,r^*)\varphi_i(p,q^*,r^*) \\ = \; & y_{i+1} - \widehat{\theta}_n^{\tau}(p)\varphi_i(p_0,q^*,r^*) \\ = \; & w_{i+1} + \theta^{\tau}(p_0,q^*,r^*)\varphi_i^0(p_0,q^*,r^*) \\ & - \widehat{\theta}_n^{\tau}(p)\varphi_i(p_0,q^*,r^*) \\ = \; & w_{i+1} + \widetilde{\theta}_n^{\tau}(p)\varphi_i(p_0,q^*,r^*) - \theta^{\tau}(p_0,q^*,r^*)\varphi_i^{\xi}(p_0,q^*,r^*) \end{aligned}$$

where $\widetilde{\theta}_n^{\tau}(p) \triangleq \theta^{\tau}(p_0,q^*,r^*) - \widehat{\theta}_n(p)$.

Hence for $p \leq p_0$,

$$\begin{aligned} & \sigma_n(p,q^*,r^*) \\ = \; & tr\widetilde{\theta}_n^{\tau}(p)\sum_{i=0}^{n-1}\varphi_i(p_0,q^*,r^*)\varphi_i^{\tau}(p_0,q^*,r^*)\widetilde{\theta}_n(p) \\ & + 2tr\widetilde{\theta}_n^{\tau}(p)\sum_{i=0}^{n-1}\varphi_i(p_0,q^*,r^*)[w_{i+1} - \theta^{\tau}(p_0,q^*,r^*)\varphi_i^{\xi}(p_0,q^*,r^*)]^{\tau} \\ & + \sum_{i=0}^{n-1}\|w_{i+1} - \theta^{\tau}(p_0,q^*,r^*)\varphi_i^{\xi}(p_0,q^*,r^*)\|^2. \end{aligned} \tag{7.30}$$

By an argument completely similar to that used in the proof of Theorem 4.2, it can be shown that for sufficiently large n

$$\begin{aligned}
&\lambda_{\min}\left(\sum_{i=0}^{n-1}\varphi_i(p,q^*,r^*)\varphi_i^\tau(p,q^*,r^*)\right)\\
\geq\ & \frac{1}{3}\lambda_{\min}\left(\sum_{i=0}^{n-1}\varphi_i^0(p,q^*,r^*)\varphi_i^{0\tau}(p,q^*,r^*)\right)\\
=\ & \frac{1}{3}\lambda_{\min}^{(p,q^*,r^*)}(n). \qquad (7.31)
\end{aligned}$$

Also, in the case of $p < p_0$, it can be seen from the definition of $\widetilde{\theta}_n(p)$ that

$$\|\widetilde{\theta}_n(p)\|^2 \geq \|A_{p_0}\|^2 > 0. \qquad (7.32)$$

Hence, by (7.31) the first term on the right-hand side of (7.30) is bounded from below by

$$\begin{aligned}
& tr\widetilde{\theta}_n^\tau(p)\sum_{i=0}^{n-1}\varphi_i(p_0,q^*,r^*)\varphi_i^\tau(p_0,q^*,r^*)\widetilde{\theta}_n(p)\\
\geq\ & \frac{1}{3}\|A_{p_0}\|^2\lambda_{\min}^{(p_0,q^*,r^*)}(n). \qquad (7.33)
\end{aligned}$$

which diverges to infinity by assumption (7.18).

By Lemma 7.2 and (7.25) we estimate the second term on the right-hand side of (7.30) as follows

$$\begin{aligned}
& 2\left|tr\widetilde{\theta}_n^\tau(p)\sum_{i=0}^{n-1}\varphi_i(p_0,q^*,r^*)[w_{i+1}-\theta^\tau(p_0,q^*,r^*)\varphi_i^\xi(p_0,q^*,r^*)]^\tau\right|\\
\leq\ & O\left(\left\|\widetilde{\theta}_n^\tau(p)\left[\sum_{i=0}^{n-1}\varphi_i(p_0,q^*,r^*)\varphi_i^\tau(p_0,q^*,r^*)+\varepsilon I\right]^{\frac{1}{2}}\right\|\right.\\
& \times\left\|\left[\sum_{i=0}^{n-1}\varphi_i(p_0,q^*,r^*)\varphi_i^\tau(p_0,q^*,r^*)+\varepsilon I\right]^{-\frac{1}{2}}\right.\\
& \left.\left.\times\sum_{i=0}^{n-1}\varphi_i(p_0,q^*,r^*)[w_{i+1}-\theta^\tau(p_0,q^*,r^*)\varphi_i^\xi(p_0,q^*,r^*)]^\tau\right\|\right)\\
=\ & O\left(\left\|\widetilde{\theta}_n^\tau(p)\left[\sum_{i=0}^{n-1}\varphi_i(p_0,q^*,r^*)\varphi_i^\tau(p_0,q^*,r^*)+\varepsilon I\right]^{\frac{1}{2}}\right\|\right.\\
& \left.\times\sqrt{(\log s_n)(\log\log s_n)^{\delta(\beta-2)}}\right). \qquad (7.34)
\end{aligned}$$

Combining (7.33) and (7.34) and noting condition (7.17) we see from (7.30) that for $p < p_0$

$$\begin{aligned}
&\sigma_n(p, q^*, r^*) \\
\geq\ & \frac{1}{3}\|A_{p_0}\|^2 \lambda_{\min}^{(p_0,q^*,r^*)}(n)\left\{1 + O\left(\left[\frac{(\log s_n)(\log\log s_n)^{\delta(\beta-2)}}{\lambda_{\min}^{(p_0,q^*,r^*)}(n)}\right]^{\frac{1}{2}}\right)\right\} \\
& + \sum_{i=0}^{n-1} \|w_{i+1} - \theta^\tau(p_0, q^*, r^*)\varphi_i^\xi(p_0, q^*, r^*)\|^2, \qquad (7.35)
\end{aligned}$$

When $p = p_0$, $\sigma_n(p_0, q^*, r^*)$ may be estimated by (7.26). Hence by (7.12) we have

$$\begin{aligned}
&\min_{0 \leq p < p_0} [CIC(p, q^*, r^*)_n - CIC(p_0, q^*, r^*)_n] \\
\geq\ & \lambda_{\min}^{(p_0,q^*,r^*)}(n)\left\{\frac{\|A_{p_0}\|^2}{3} + O\left(\left[\frac{(\log s_n)(\log\log s_n)^{\delta(\beta-2)}}{\lambda_{\min}^{(p_0,q^*,r^*)}(n)}\right]^{\frac{1}{2}}\right)\right. \\
& \left. + O\left(\frac{(\log s_n)(\log\log s_n)^{\delta(\beta-2)}}{\lambda_{\min}^{(p_0,q^*,r^*)}(n)}\right) - \frac{p_0 a_n}{\lambda_{\min}^{(p_0,q^*,r^*)}(n)}\right\} \\
\geq\ & \lambda_{\min}^{(p_0,q^*,r^*)}(n)\left\{\frac{\|A_{p_0}\|^2}{3} + o(1)\right\} \xrightarrow[n\to\infty]{} \infty \qquad (7.36)
\end{aligned}$$

where for the last inequality we have used (7.17) and (7.18).

From (7.36) and the definition of $\widehat{p}_n$ we known that (7.28) holds. Hence $\widehat{p}_n$ is strongly consistent. Finally, by using $\widehat{p}_n \to p_0$, the strong consistency of $\widehat{q}_n$ and $\widehat{r}_n$ can be established in a completely similar way, details will not be repeated here. Q.E.D.

Remark 7.1. The coprimeness condition and the rank condition in H1) are implicitly used in guaranteeing $\lambda_{\min}^{(p_0,q^*,r^*)}(n) \to \infty$. If the rank condition on A_{p_0} is replaced by that B_{q_0} or C_{r_0}, then similar results hold. For example, if in H1) C_{r_0} is assumed to be of full rank, and (7.14)-(7.16) are replaced by

$$\begin{aligned}
\widehat{r}_n &= \underset{0\leq r\leq r^*}{\text{argmin}}\ CIC(p^*, q^*, r)_n, \\
\widehat{q}_n &= \underset{0\leq q\leq q^*}{\text{argmin}}\ CIC(p^*, q, \widehat{r}_n)_n, \\
\widehat{p}_n &= \underset{0\leq p\leq p^*}{\text{argmin}}\ CIC(p, \widehat{q}_n, \widehat{r}_n)_n,
\end{aligned}$$

then Theorem 7.1 still holds provided that (7.18) is replaced by

$$a_n / \lambda_{\min}^{(p^*,q^*,r_0)}(n) \to \infty \quad \text{a.s.}$$

From Theorem 7.1 we know that if in Assumption H2) the constant β is greater than 2, then condition (7.17) is reduced to

$$\frac{\log s_n}{a_n} \xrightarrow[n\to\infty]{} 0 \quad \text{a.s.} \tag{7.37}$$

This condition can be relaxed to

$$\frac{\log s_n}{a_n} = O(1) \quad \text{a.s.} \tag{7.38}$$

if further conditions are imposed. This is the content of Theorem 7.2. In the following we simply write $\varphi(p,q)$, $\theta_n(p,q)$, $\lambda_{\min}^{(p,q)}(n)$ for $\varphi(p,q,0)$, $\theta_n(p,q,0)$, $\lambda_{\min}^{(p,q,0)}(n)$, etc.

Theorem 7.2. *For system (7.2)-(7.5) with $r_0 = r^* = 0$ and the order estimation procedure defined by (7.10)-(7.16) assume H2)-H4) hold and that for any $0 \le p \le p^*$, $0 \le q \le q^*$*

$$\left\| \left(\sum_{i=0}^{n-1} \varphi_i(p,q)\varphi_i^T(p,q) + \varepsilon I \right)^{-\frac{1}{2}} \sum_{i=0}^{n-1} \varphi_i(p,q) w_{i+1}^T \right\|^2 = o(\log s_n) \quad a.s. \tag{7.39}$$

If in (7.12) the sequence $\{a_n\}$ is positive and satisfies

$$\log s_n = O(a_n), \quad a_n = o\left(\lambda_{\min}^{(p^*,q^*)}(n)\right) \quad a.s. \tag{7.40}$$

then $(\widehat{p}_n, \widehat{q}_n)$ is strongly consistent, i.e.,

$$(\widehat{p}_n, \widehat{q}_n) \xrightarrow[n\to\infty]{} (p_0, q_0) \quad a.s.$$

Proof. The proof is completely similar to that for Theorem 7.1. We need only to note that under condition (7.39), (7.27) becomes

$$CIC(p,q^*)_n - CIC(p_0,q^*)_n = a_n \left\{ p - p_0 + o\left(\frac{\log s_n}{a_n}\right) \right\}, \tag{7.41}$$

while (7.35) should be

$$\sigma_n(p,q^*) \ge \frac{1}{3} \|A_{p_0}\|^2 \lambda_{\min}^{(p_0,q^*)}(n) \left\{ 1 + o\left(\left[\frac{(\log s_n)}{\lambda_{\min}^{(p_0,q^*)}(n)} \right]^{\frac{1}{2}} \right) \right\}$$

$$+\sum_{i=0}^{n-1}\|w_{i+1}\|^2.$$

Q.E.D.

Condition (7.39) plays a crucial role in Theorem 7.2. More detailed discussions on this condition will be given in Section 7.4. We remark that this condition is also crucial (See [HDa]) in establishing the strong consistency of the PLS criterion [Ri2].

7.2 Order Estimation by not Using Upper Bounds for Orders

In this section, we treat the case where the upper bounds for orders p^*, q^* and r^* are not available *a priori*. We need the following conditions:

H2′). $\{w_i, \mathcal{F}_i\}$ is an m-dimensional martingale difference sequence satisfying

$$\sup_i E[(\|w_{i+1}\|^2|\mathcal{F}_i)] < \infty, \quad 0 < \limsup_{n\to\infty}\frac{1}{n}\sum_{i=1}^{n}\|w_i\|^2 < \infty$$

and

$$\|w_i\| = o(d(i)) \quad \text{a.s.}$$

where $\{d(i)\}$ is a positive, deterministic and nondecreasing sequence and satisfies

$$\sup_i d(e^{i+1})/d(e^i) < \infty.$$

H3′). u_i is $\mathcal{F}_i$-measurable, and there is a constant $b > 0$ such that

$$\sum_{i=0}^{n-1}(\|y_i\|^2 + \|u_i\|^2) = O(n^b) \quad \text{a.s.}$$

H4′). $\det C(z) \neq 0, \forall z\colon |z| \leq 1.$

In the present case, since (p^*, q^*, r^*) is not available, Lemma 7.1 can not be used, instead we will use Theorem 4.9 and algorithm (4.196)-(4.199) to construct the noise estimate.

We still keep the notations (7.7)-(7.13), but with $\{\widehat{w}_i\}$ in (7.11) replaced by $\{\widehat{w}_i(n)\}$, i.e.,

$$\begin{aligned}\varphi_i(p,q,r) \triangleq\ & [y_i^T \ \dots \ y_{i-p+1}^T \ \ u_i^T \ \dots \ u_{i-q+1}^T \\ & \widehat{w}_i^T(n) \ \dots \ \widehat{w}_{i-r+1}^T(n)]^T,\end{aligned} \tag{7.42}$$

where $\{\widehat{w}_i(n)\}$ is produced by (4.195)-(4.199). We assume that in (4.195) the regression lag h_n is of order $h_n = O(\log^\alpha n)$, $\alpha > 1$, and $\log n = o(h_n)$.

The order estimation procedure is described as follows:

For any $n \geq 1$, define

$$\widehat{m}_n = \operatorname*{argmin}_{0\leq k\leq[\log n]} CIC(k,k,k)_n, \tag{7.43}$$

$$\widehat{p}_n = \operatorname*{argmin}_{0\leq p\leq\widehat{m}_n} CIC(p,\widehat{m}_n,\widehat{m}_n)_n, \tag{7.44}$$

$$\widehat{q}_n = \operatorname*{argmin}_{0\leq q\leq\widehat{m}_n} CIC(\widehat{p}_n,q,\widehat{m}_n)_n, \tag{7.45}$$

$$\widehat{r}_n = \operatorname*{argmin}_{0\leq r\leq\widehat{m}_n} CIC(\widehat{p}_n,\widehat{q}_n,r)_n. \tag{7.46}$$

It is worth noting that in this procedure the first step (7.43) corresponds to estimating the value of

$$m_0 \triangleq \max\{p_0, q_0, r_0\}, \tag{7.47}$$

while in (7.44)-(7.46), the orders p_0, q_0 and r_0 are searched between at most $3\widehat{m}_n$ points at each time n. This is simpler than (c.f. [Hu2]) the procedure

$$(\widehat{p}_n,\widehat{q}_n,\widehat{r}_n) \triangleq \operatorname*{argmin}_{0\leq p,q,r\leq\widehat{m}_n} CIC(p,q,r)_n,$$

for which (p_0, q_0, r_0) is searched among $(\widehat{m}_n)^3$ points at each time n.

Let us define

$$\lambda^0_{\min}(n) = \lambda_{\min}\left\{\sum_{i=0}^{n-1}\varphi^0_i(m_0,m_0,m_0)\varphi^{0\tau}_i(m_0,m_0,m_0)\right\} \tag{7.48}$$

where m_0 is defined by (7.47) and $\varphi^0_i(p,q,r)$ is given by (7.20).

Theorem 7.3 *[HG]. Under Conditions H2')-H4'), if $\lambda^0_{\min}(n)$ satisfies*

$$h_n \log n + [d(n)\log\log n]^2 = o\left(\lambda^0_{\min}(n)\right)$$

and if in the criterion (7.12), the sequence $\{a_n\}$ is chosen to be positive and to satisfy

$$\{h_n \log n + [d(n)\log\log n]^2\}/a_n \xrightarrow[n\to\infty]{} 0 \quad a.s. \tag{7.49}$$

and

$$a_n/\lambda^0_{\min}(n) \xrightarrow[n\to\infty]{} 0 \quad a.s. \tag{7.50}$$

then for the estimation algorithm defined by (7.43)-(7.46),

$$\widehat{m}_n \xrightarrow[n\to\infty]{} m_0 \quad a.s. \tag{7.51}$$

and

$$(\widehat{p}_n, \widehat{q}_n, \widehat{r}_n) \xrightarrow[n \to \infty]{} (p_0, q_0, r_0) \quad a.s. \tag{7.52}$$

Proof. The first step is to show the strong consistency of $\widehat{m}_n$. For this we first prove that

$$\limsup_{n \to \infty} \widehat{m}_n \le m_0 \quad \text{a.s.} \tag{7.53}$$

In the following, when $p = q = r = k$, we shall simply write (k) for (k, k, k) in $\theta(\cdot)$, $\varphi(\cdot)$, $\varphi^0(\cdot)$, $\varphi^\xi(\cdot)$, etc. Then we have for any $k \ge m_0$,

$$\begin{aligned} y_{i+1}^\tau &= \varphi_i^{0\tau}(k)\theta(k) + w_{i+1}^\tau \\ &= \varphi_i^\tau(k)\theta(k) - \varphi_i^{\xi\tau}(k)\theta(k) + w_{i+1}^\tau, \end{aligned} \tag{7.54}$$

where $\varphi_i^\xi(k) = \varphi(k) - \varphi_i^0(k)$.

Substituting (7.54) into (7.10) we have

$$\begin{aligned} \widetilde{\theta}_n(k) &\triangleq \theta(k) - \theta_n(k) \\ &= \left(\sum_{i=0}^{n-1} \varphi_i(k)\varphi_i^\tau(k) + \varepsilon I\right)^{-1} \\ &\left\{\theta(k)\varepsilon + \sum_{i=0}^{n-1} \varphi_i(k)[\varphi_i^{\xi\tau}(k)\theta(k) + w_{i+1}^\tau]\right\}. \end{aligned} \tag{7.55}$$

Similar to (7.24), by (7.13) we know that for $k \ge m_0$,

$$\begin{aligned} \sigma_n(k) &= tr\widetilde{\theta}_n^\tau(k) \sum_{i=0}^{n-1} \varphi_i(k)\varphi_i^\tau(k)\widetilde{\theta}_n(k) \\ &+ 2tr\widetilde{\theta}_n^\tau(k) \sum_{i=0}^{n-1} \varphi_i(k)[-\theta^\tau(k)\varphi_i^\xi(k) + w_{i+1}]^\tau \\ &+ \sum_{i=0}^{n-1} \|\theta^\tau(k)\varphi_i^\xi(k) - w_{i+1}\|^2. \end{aligned} \tag{7.56}$$

Note that by (7.7) and (7.8),

$$\theta^\tau(k)\varphi_i^\xi(k) \equiv \theta^\tau(m_0)\varphi_i^\xi(m_0). \tag{7.57}$$

Hence applying Theorem 4.9 we get

$$\begin{aligned} &\max_{k \ge m_0} \sum_{i=0}^{n-1} \|\varphi_i^{\xi\tau}(k)\theta(k)\|^2 = O\left(\sum_{i=0}^{n-1} \|\varphi_i^\xi(m_0)\|^2\right) \\ &= O(h_n \log n) + o\left(\{d(n) \log\log n\}^2\right) \quad \text{a.s.} \end{aligned}$$

Consequently, by this and Lemma 4.7(ii),

$$\begin{aligned}
&\max_{m_0\le k\le[\log n]}\left\|\left(\sum_{i=0}^{n-1}\varphi_i(k)\varphi_i^{\tau}(k)+\varepsilon I\right)^{-\frac{1}{2}}\sum_{i=0}^{n-1}\varphi_i(k)\varphi_i^{\xi\tau}(k)\theta(k)\right\|^2\\
\le&\max_{m_0\le k\le[\log n]}\sum_{i=0}^{n-1}\|\varphi_i^{\xi\tau}(k)\theta(k)\|^2\\
=&\ O(h_n\log n)+o(\{d(n)\log\log n\}^2)\quad\text{a.s.}
\end{aligned}\tag{7.58}$$

By Lemma 4.6 we have

$$\begin{aligned}
&\max_{m_0\le k\le[\log n]}\left\|\left(\sum_{i=0}^{n-1}\varphi_i(k)\varphi_i^{\tau}(k)+\varepsilon I\right)^{-\frac{1}{2}}\sum_{i=0}^{n-1}\varphi_i(k)w_{i+1}\right\|^2\\
=&\ O(h_n\log n)+o(\{d(n)\log\log n\}^2)\quad\text{a.s.}
\end{aligned}\tag{7.59}$$

Thus, from (7.55)-(7.58) it is easy to see that

$$\begin{aligned}
&\max_{m_0\le k\le[\log n]}\left|\sigma_n(k)-\sum_{i=0}^{n-1}\|\theta^{\tau}(m_0)\varphi_i^{\xi}(m_0)-w_{i+1}\|^2\right|\\
=&\ O(h_n\log n)+o(\{d(n)\log\log n\}^2)\quad\text{a.s.}
\end{aligned}\tag{7.60}$$

Consequently, by (7.12),

$$\begin{aligned}
&\max_{m_0< k\le[\log n]}\{CIC(m_0)_n-CIC(k)_n\}\\
=&\max_{m_0< k\le[\log n]}\{\sigma_n(m_0)-\sigma_n(k)+3(m_0-k)a_n\}\\
\le&\ O(h_n\log n)+o(\{d(n)\log\log n\}^2)-3a_n\\
<&\ 0,
\end{aligned}$$

for sufficiently large n, where the last inequality follows from (7.49). Hence by the definition of $\widehat{m}_n$, we see that (7.53) holds.

We now show that

$$\liminf_{n\to\infty}\widehat{m}_n\ge m_0\quad\text{a.s.}\tag{7.61}$$

The proof is similar to that for (7.28). Similar to (7.29), we set

$$\widehat{\theta}_n(k)=[-A_{1n}\ \dots\ -A_{m_0n}\quad B_{1n}\ \dots\ B_{m_0n}\quad C_{1n}\ \dots\ C_{m_0n}]^{\tau},$$

where $A_{in}=0$ for $i>k$, and A_{in}, B_{in}, C_{in}, $i\le k$, are entries of $\theta_n(k)$ defined by (7.9).

In the present case, (7.30) holds with "(p,q^*,r^*)", "(p)", "(p_0,q^*,r^*)" replaced by "(k)", "(k)" and "(m_0)" respectively. Also, (7.31) remains valid with "(p,q,r)" replaced by "(m_0)". Further, (7.32) should be replaced by

$$\|\widetilde{\theta}_n(k)\|^2 \geq \min\{\|A_{p_0}\|^2, \|B_{q_0}\|^2, \|C_{r_0}\|^2\} = \delta_0 > 0, \quad \forall k < m_0,$$

and (7.33) is changed to

$$tr\widetilde{\theta}_n^\tau(k)\sum_{i=0}^{n-1}\varphi_i(k)\varphi_i^\tau(k)\widetilde{\theta}_n(k) \geq \frac{\delta_0}{3}\lambda_{\min}^0(n).$$

Finally, (7.35) becomes (for $k < m_0$),

$$\begin{aligned}\sigma_n(k) \quad &\geq \quad \frac{1}{3}\delta_0\lambda_{\min}^0(n)\left\{1+O\left(\left[\frac{h_n\log n + o([d(n)\log\log n]^2)}{\lambda_{\min}^0(n)}\right]^{\frac{1}{2}}\right)\right\}\\ &\quad + \sum_{i=0}^{n-1}\|w_{i+1}-\theta^\tau(m_0)\varphi_i^\xi(m_0)\|^2.\end{aligned}$$

This in conjunction with (7.60) ($k = m_0$) yields

$$\begin{aligned}&\max_{0\leq k<m_0}\{CIC(k)_n - CIC(m_0)_n\}\\ &\geq \lambda_{\min}^0(n)\left\{\frac{1}{3}\delta_0+o(1)\right\}+O(a_n)+O(h_n\log n)\\ &\quad +o(\{d(n)\log\log n\}^2)\\ &= \lambda_{\min}^0(n)\left\{\frac{1}{3}\delta_0+o(1)\right\}\xrightarrow[n\to\infty]{}\infty.\end{aligned}$$

Hence (7.61) holds. Thus, we have proved the strong consistency of $\widehat{m}_n$. With the fact that $\widehat{m}_n \xrightarrow[n\to\infty]{} m_0$ in mind, the proof of (7.52) is completely similar to that for Theorem 7.1, the details will not be repeated here.

Q.E.D.

Remark 7.2. From Theorems 7.1-7.3 we see that the growth rate of $\lambda_{\min}^{(p,q,r)}(n)$ or $\lambda_{\min}^0(n)$ plays a crucial role in establishing consistency of the order selection procedure. The criterion CIC (7.12) depends on the choice of a_n, which in turn depends on the input sequence $\{u_i\}$. This is the key difference between AIC and BIC-like criterion (7.1), since in (7.1) the extra term a_n/n is free of observation data.

Remark 7.3. In adaptive control systems, the requirement on $\lambda_{\min}^{(p_0,q^*,r^*)}(n)$ can be guaranteed by using the diminishing excitation technique discussed in Section 6.3. Under Condition H1) and conditions of Theorem 6.2, it is easy to see that for sufficiently large n,

$$\lambda_{\min}^{(p_0,q^*,r^*)}(n) \geq c_0 n^\alpha \quad \text{a.s.} \tag{7.62}$$

where $c_0 > 0$ is free of n and $\alpha > 0$ is defined in Theorem 6.2. Hence, if the output process $\{y_n\}$ satisfies

$$\|y_n\|^2 = o(n^b), \quad \text{for some } b > 0, \tag{7.63}$$

the sequence $\{a_n\}$ in Theorem 7.1 can be taken as any sequence satisfying

$$\frac{\log n(\log\log)^{\delta(\beta-2)}}{a_n} \xrightarrow[n\to\infty]{} 0, \qquad \frac{a_n}{n^\alpha} \xrightarrow[n\to\infty]{} 0.$$

Remark 7.4. Theorems 7.1 and 7.3 can also be used in the off-line identification of (possibly unstable) ARMAX systems. For example, we have the following corollary:

Corollary 7.1. *For ARMAX model (7.2) assume that all unstable zeros of $A(z)$ are on the unit circle, condition H1) holds, and that the input sequence is a linear process of the form*

$$u_n = \sum_{i=0}^{\infty} F_i v_{n-i}, \quad F_0 = I$$

where $\{F_i\}$ is a sequence of real matrices and

$$\sum_{i=0}^{\infty} \|F_i\|^2 < \infty$$

$\{v_i\}$ is an i.i.d. sequence independent of $\{w_i\}$ with zero mean and $Ev_1v_1^T = \Sigma > 0$. Then in Theorem 7.3 $\{a_n\}$ can be taken as any sequence satisfying

$$\{h_n \log n + [d(n)\log\log n]^2\}/a_n \xrightarrow[n\to\infty]{} 0,$$

and

$$a_n/n \xrightarrow[n\to\infty]{} 0 \quad a.s.$$

Proof. We need only to show that

$$\liminf_{n\to\infty} \lambda_{\min}^0(n)/n > 0 \quad \text{a.s.} \tag{7.64}$$

Set

$$u_n^s = \sum_{i=1}^{\infty} F_i v_{n-i}.$$

Then it is easy to see that $\{u_n^s\}$ satisfies conditions required in Theorem 6.2. Furthermore, the constants δ, ε and α appearing in Theorem 6.2 can respectively be taken as 0, 0 and 1. Hence, (7.64) follows from conclusion (i) of Theorem 6.2.

7.3 Time-Delay Estimation

In practical applications, it is of interest to estimate the time-delay of a system. The time-delay is defined as the smallest d (say d_0) such that $B_d \neq 0$, i.e.

$$d_0 = \min\{d: \ B_d \neq 0\} \tag{7.65}$$

where B_d is defined in (7.4). Obviously, d_0 satisfies $1 \leq d_0 \leq q_0$.

The idea for estimating d_0 is similar to that for estimating the orders [CZ3], [CZ4]. To be precise, let us redefine $\varphi_i(p,q,r)$ appearing in (7.11) as

$$\begin{aligned}\varphi_i(p,q,r,d) \ \triangleq \ & [y_i^T \ \ldots \ y_{i-p+1}^T \quad u_{i-d+1}^T \ \cdots \ u_{i-q+1}^T \\ & \widehat{w}_i^T \ \ldots \ \widehat{w}_{i-r+1}^T]^T.\end{aligned} \tag{7.66}$$

Similar to (7.10) and (7.13) we take

$$\begin{aligned}&\theta_n(p,q,r,d)\\ &= \left(\sum_{i=0}^{n-1} \varphi_i(p,q,r,d)\varphi_i^T(p,q,r,d) + \varepsilon I\right)^{-1} \sum_{i=0}^{n-1} \varphi_i(p,q,r,d) y_{i+1}^T\end{aligned} \tag{7.67}$$

$$\sigma_n(p,q,r,d) = \sum_{i=0}^{n-1} \|y_{i+1} - \theta_n^T(p,q,r,d)\varphi_i(p,q,r,d)\|^2. \tag{7.68}$$

It is easy to see that $\sigma_n(p,q,r,1) \equiv \sigma_n(p,q,r)$ where $\sigma_n(p,q,r)$ is defined by (7.12). Let $(\widehat{p}_n, \widehat{q}_n, \widehat{r}_n)$ be given by (7.14)-(7.16). Then the criterion for time-delay estimation (we call it DIC) is defined as

$$DIC(d)_n = \sigma_n(\widehat{p}_n, \widehat{q}_n, \widehat{r}_n, d) - d a_n \tag{7.69}$$

where $\{a_n\}$ is the same as in (7.12). Finally, the estimate $\widehat{d}_n$ for d is defined by

$$\widehat{d}_n = \operatorname*{argmin}_{1 \leq d \leq \widehat{q}_n} DIC(d)_n. \tag{7.70}$$

Theorem 7.4. *Under conditions of Theorem 7.1, as $n \to \infty$,*

$$\widehat{d}_n \xrightarrow[n \to \infty]{} d_0 \quad a.s.$$

where $\widehat{d}_n$ is defined by (7.70).

Proof. By Theorem 7.1, we know that

$$\sigma_n(\widehat{p}_n, \widehat{q}_n, \widehat{r}_n, d) \equiv \sigma_n(p_0, q_0, r_0, d)$$

for all sufficiently large n. The rest of the proof is completely similar to that of Theorem 7.1, details will not be repeated [CZ4]. Q.E.D.

When no *a priori* upper bounds for the orders (p_0, q_0, r_0) are available, we need only to change the definition of (7.66) to

$$\begin{aligned}\varphi_i(p,q,r,d) \triangleq\ & [y_i^\tau\ \cdots\ y_{i-p+1}^\tau\quad u_{i-d+1}^\tau\ \cdots\ u_{i-q+1}^\tau \\ & \widehat{w}_i^\tau(n)\ \ldots\ \widehat{w}_{i-r+1}^\tau(n)]^\tau, \qquad (7.71)\end{aligned}$$

where $\{\widehat{w}_i(n)\}$ is produced by (4.195)-(4.199). We have the following result.

Theorem 7.5. *Let $DIC(d)_n$ be defined as in (7.67)-(7.69) with $\varphi_i(p,q,r,d)$ defined by (7.71), $(\widehat{p}_n, \widehat{q}_n, \widehat{r}_n)$ given by (7.44)-(7.46) and with $\{a_n\}$ satisfying conditions of Theorem 7.3. Then under conditions of Theorem 7.3, as $n \to \infty$*

$$\widehat{d}_n \xrightarrow[n\to\infty]{} d_0 \quad a.s.$$

where $\widehat{d}_n = \underset{1\le d\le \widehat{q}_n}{\operatorname{argmin}} DIC(d)_n$.

The proof is completely similar to that of Theorem 7.4.

7.4 Connections of CIC and BIC

For simplicity of discussion, we take the ARX(p_0, q_0) model only. The BIC criterion for selecting p_0 and q_0 is defined as

$$BIC(p,q)_n = \log\left(\frac{\sigma_n(p,q)}{n}\right) + (p+q)\frac{\log n}{n}, \qquad (7.72)$$

where $\sigma_n(p,q)$ is defined by (7.13) with $r \equiv 0$.

By the Taylor expansion, one has

$$\begin{aligned}\exp\{BIC(p,q)_n\} &= \frac{\sigma_n(p,q)}{n}\exp\left\{(p+q)\frac{\log n}{n}\right\} \\ &= \frac{\sigma_n(p,q)}{n}\left\{1+(p+q)\frac{\log n}{n}(1+o(1))\right\}\end{aligned}$$

or

$$\begin{aligned}& n\exp\{BIC(p,q)_n\} \\ =\ & \sigma_n(p,q) + (p+q)\frac{\sigma_n(p,q)}{n}(\log n)(1+o(1)) \qquad (7.73)\end{aligned}$$

This expression can be further simplified in the case where $p \ge p_0$ and $q \ge q_0$. Assume that

$$\lim_{n\to\infty}\frac{1}{n}\sum_{i=0}^{n}\|w_i\|^2 = \sigma^2 > 0 \quad \text{a.s.} \qquad (7.74)$$

and

$$\frac{1}{n}\sum_{i=0}^{n}(\|y_i\|^2 + \|u_i\|^2) = O(n^b) \quad \text{a.s.} \tag{7.75}$$

for some $b > 0$.

Then similar to (7.26) we have for all $p \geq p_0$, $q \geq q_0$,

$$\begin{aligned}\sigma_n(p,q) &= O\left(\{\log n\}\{\log\log n\}^{\delta(\beta-2)}\right) + \sum_{i=0}^{n-1}\|w_{i+1}\|^2 \\ &= \sigma^2(1+o(1))n \end{aligned}\tag{7.76}$$

Substituting this into (7.73) we get ($p \geq p_0$, $q \geq q_0$),

$$\begin{aligned}&n\exp\{BIC(p,q)_n\} \\ &= \sigma_n(p,q) + (p+q)(\log n)\sigma^2(1+o(1)).\end{aligned}\tag{7.77}$$

Hence, if in (7.12) we take $r = 0$ and

$$a_n = (\log n)\sigma^2(1+o(1)) \tag{7.78}$$

then (7.77) is nothing but a special form of CIC. Such a criterion CIC has been analyzed in Theorem 7.2, where for strong consistency of $(\widehat{p}_n, \widehat{q}_n)$ we have assumed that (under (7.75)),

$$\begin{aligned}&\left\|\left(\sum_{i=0}^{n}\varphi_i(p,q)\varphi_i^\tau(p,q) + \varepsilon I\right)^{-\frac{1}{2}}\sum_{i=0}^{n}\varphi_i(p,q)w_{i+1}^\tau\right\|^2 \\ &= o(\log n) \quad \text{a.s.}\end{aligned}\tag{7.79}$$

for $0 \leq p \leq p^*$, $0 \leq q \leq q^*$ (see (7.39)).

This condition may also be used to establish the strong consistency of the order estimates given by BIC, this is the content of

Theorem 7.6. *Let Assumptions H2)-H4) be satisfied for $ARX(p_0, q_0)$ model defined by (7.2)-(7.5) with $r_0 = 0$. If (7.79) holds, $\sum_{i=0}^{n}\|w_i\|^2 = O(n)$, and*

$$\log n/\lambda_{\min}\left(\sum_{i=0}^{n-1}\varphi_i(p^*,q^*)\varphi_i^\tau(p^*,q^*)\right)\xrightarrow[n\to\infty]{}0 \quad a.s.,$$

then the estimate given by BIC (7.72) is strongly consistent, i.e.,

$$(\widehat{p}_n, \widehat{q}_n)\xrightarrow[n\to\infty]{}(p_0, q_0) \quad a.s.$$

where

$$\widehat{p}_n = \underset{0\le p\le p^*}{argmin}\, BIC(p,q^*)_n$$
$$\widehat{q}_n = \underset{0\le q\le q^*}{argmin}\, BIC(\widehat{p}_n,q)_n$$

Proof. Let us set

$$L_n(p,q) = \sigma_n(p,q) + (p+q)\frac{\sigma_n(p,q)}{n}(\log n)(1+o(1)). \tag{7.80}$$

Then by (7.73) it is obvious that

$$\widehat{p}_n = \underset{0\le p\le p^*}{\operatorname{argmin}}\, L_n(p,q^*)_n \tag{7.81}$$
$$\widehat{q}_n = \underset{0\le q\le q^*}{\operatorname{argmin}}\, L_n(\widehat{p}_n,q)_n \tag{7.82}$$

The proof technique is similar to that for Theorem 7.1. We first consider the strong consistency of $\widehat{p}_n$. By use of (7.79), and following the proof of (7.26) we know that for $p \ge p_0$

$$\sigma_n(p,q) = o(\log n) + \sum_{i=0}^{n-1}\|w_{i+1}\|^2.$$

Consequently, by (7.74) and (7.80) it is easy to see that for $p_0 \le p \le p^*$,

$$L_n(p,q^*) = \sum_{i=0}^{n-1}\|w_{i+1}\|^2 + (p+q^*)\sigma^2(\log n)(1+o(1)). \tag{7.83}$$

Hence, we obtain for large n,

$$\min_{p_0<p\le p^*}[L_n(p,q^*) - L_n(p_0,q^*)] = (\log n)\{\sigma^2 + o(1)\} > 0$$

which implies that $\limsup_{n\to\infty} \widehat{p}_n \le p_0$. Thus for consistency of $\widehat{p}_n$ we need only to show that

$$\liminf_{n\to\infty} \widehat{p}_n \ge p_0, \quad \text{a.s.} \tag{7.84}$$

Similar to the proof of (7.35), it is easy to obtain that for $p < p_0$

$$\begin{aligned}\sigma_n(p,q^*) &\ge \frac{1}{3}\|A_{p_0}\|^2\lambda_{\min}^{(p_0,q^*)}(n)\left\{1+o\left(\left[\frac{(\log n)}{\lambda_{\min}^{(p_0,q^*)}(n)}\right]^{\frac{1}{2}}\right)\right\}\\ &\quad +\sum_{i=0}^{n}\|w_{i+1}\|^2\\ &\ge \frac{1}{3}\|A_{p_0}\|^2\lambda_{\min}^{(p^*,q^*)}(n)\{1+o(1)\} + \sum_{i=0}^{n}\|w_{i+1}\|^2,\end{aligned} \tag{7.85}$$

where

$$\lambda_{\min}^{(p,q)}(n) = \lambda_{\min}\left(\sum_{i=0}^{n-1} \varphi_i(p,q)\varphi_i^\tau(p,q)\right).$$

Hence by (7.80) we have for $p < p_0$

$$\begin{aligned} L_n(p,q^*) = {} & \sigma_n(p,q^*)\left(1+O\left(\frac{\log n}{n}\right)\right) \\ \geq {} & \frac{1}{3}\|A_{p_0}\|^2 \lambda_{\min}^{(p^*,q^*)}(n)\{1+o(1)\} + \sum_{i=0}^{n} \|w_{i+1}\|^2 + O(\log n) \end{aligned}$$

From this and (7.83) it follows that

$$\begin{aligned} & \min_{0\leq p<p_0} [L_n(p,q^*) - L_n(p_0,q^*)] \\ \geq {} & \frac{1}{3}\|A_{p_0}\|^2 \lambda_{\min}^{(p^*,q^*)}(n)\{1+o(1)\} + O(\log n) \\ = {} & \lambda_{\min}^{(p^*,q^*)}(n)\left\{\frac{1}{3}\|A_{p_0}\|^2 + o(1)\right\} > 0, \end{aligned}$$

for sufficiently large n. Hence, (7.84) is true, and $\widehat{p}_n$ is strongly consistent. In a completely similar way, the strong consistency of $\widehat{q}_n$ can also be proved.

Q.E.D.

From Theorem 7.6, we know that a key condition for strong consistency of the estimate given by BIC is (7.79). This condition is not assumed when establishing strong consistency of the estimate given by CIC discussed in Theorem 7.1. Hence, in this sense CIC is more general than BIC. Let us further discuss condition (7.79).

First, we note that (7.79) is equivalent to

$$\left\|\left(\sum_{i=0}^{n} \varphi_i(p^*,q^*)\varphi_i^\tau(p^*,q^*) + \varepsilon I\right)^{-\frac{1}{2}} \sum_{i=0}^{n} \varphi_i(p^*,q^*) w_{i+1}^\tau\right\|^2 = o(\log n) \quad \text{a.s.} \tag{7.86}$$

To see this, set

$$X_n(p,q) = [\varphi_0(p,q) \; \dots \; \varphi_n(p,q)]^\tau$$

$$W_n = [w_1 \; \dots \; w_{n+1}]^\tau.$$

Then the left-hand side of (7.86) can be written as $W_n^\tau X_n (X_n^\tau X_n)^{-1} X_n^\tau W_n$. Note that $X_n(X_n^\tau X_n)^{-1}X_n^\tau$ is the projection operator on the subspace spanned by the column vectors of X_n. Hence we have, for $0 \leq p \leq p^*$,

$0 \leq q \leq q^*,$

$$W_n^\tau X_n(p,q)[X_n^\tau(p,q)X_n(p,q)]^{-1}X_n^\tau(p,q)W_n$$

$$\leq \quad W_n^\tau X_n(p^*,q^*)[X_n^\tau(p^*,q^*)X_n(p^*,q^*)]^{-1}X_n^\tau(p^*,q^*)W_n$$

From this it is easy to see the equivalence of (7.79) and (7.86).

Second, we remark that (7.86) may not hold if no further assumptions are made on the input sequence $\{u_n\}$. For this, consider the following example of Lai and Wei [LW1]:

$$y_i = \beta_1 + \beta_2 u_i + w_i$$

where w_i are i.i.d. random variables with $Ew_i = 0$, $Ew_i^2 = 1$, and u_i is recursively defined by

$$u_{i+1} = \overline{u}_i + c\overline{w}_i, \quad u_1 = 0, \quad c \neq 0.$$

with $\overline{u}_i = \frac{1}{i}\sum_{j=1}^{i} u_j$, $\overline{w}_i = \frac{1}{i}\sum_{j=1}^{i} w_j$.

In this case, the regression vector is

$$\varphi_i = [1, \ u_i]^\tau$$

and the least squares estimate for $[\beta_1 \ \beta_2]^\tau$ is

$$\begin{pmatrix} b_{1n} \\ b_{2n} \end{pmatrix} = \left(\sum_{i=1}^{n} \varphi_i \varphi_i^\tau \right)^{-1} \sum_{i=1}^{n} \varphi_i y_i$$

By the identity

$$\begin{pmatrix} \sum\limits_{i=1}^{n} 1 & \sum\limits_{i=1}^{n} u_i \\ \sum\limits_{i=1}^{n} u_i & \sum\limits_{i=1}^{n} u_i^2 \end{pmatrix}^{-1} = \frac{1}{n\sum\limits_{i=1}^{n}(u_i - \overline{u}_n)^2} \begin{pmatrix} \sum\limits_{i=1}^{n} u_i^2 & -\sum\limits_{i=1}^{n} u_i \\ -\sum\limits_{i=1}^{n} u_i & n \end{pmatrix}$$

it is easy to see that

$$b_{2n} - \beta_2 = \frac{\left\{ \sum\limits_{i=1}^{n}(u_i - \overline{u}_n)w_i \right\}}{\sum\limits_{i=1}^{n}(u_i - \overline{u}_n)^2}$$

and

$$\left\| \left(\sum_{i=0}^{n} \varphi_i \varphi_i^\tau + \varepsilon I \right)^{-\frac{1}{2}} \sum_{i=0}^{n} \varphi_i w_i \right\|^2$$

$$= \frac{\left\{ \sum_{i=1}^{n} (u_i - \overline{u}_n) w_i \right\}^2}{\sum_{i=1}^{n} (u_i - \overline{u}_n)^2} + \frac{\left(\sum_{i=1}^{n} w_i \right)^2}{n}$$

$$= (b_{2n} - \beta_2)^2 \left[\sum_{i=1}^{n} (u_i - \overline{u}_n)^2 \right] + \frac{\left(\sum_{i=1}^{n} w_i \right)^2}{n}. \tag{7.87}$$

By [LW1] (p.160), we know that

$$b_{2n} - \beta_2 \xrightarrow[n \to \infty]{} -c^{-1} \quad \text{a.s.}$$

$$\sum_{i=1}^{n} (u_i - \overline{u}_n)^2 \sim c^2 \log n \quad \text{a.s..}$$

By this and the law of the iterated logarithm for $\{w_i\}$ we see from (7.87) that

$$\left\| \left(\sum_{i=0}^{n} \varphi_i \varphi_i^\tau + \varepsilon I \right)^{-\frac{1}{2}} \sum_{i=0}^{n} \varphi_i w_i \right\|^2 \sim \log n + O(\log \log n) \sim \log n$$

This shows that (7.86) does not hold.

Finally, we consider some cases where (7.86) does hold. For this we need the following auxiliary result [GHH], [Hu3].

Lemma 7.3. *Suppose that $\{w_n, \mathcal{F}_n\}$ is a vector martingale difference sequence satisfying*

$$\sup_n E[\|w_{n+1}\|^2 | \mathcal{F}_n] < \infty,$$
$$\sup_n E\{\|w_n\|^4 (\log^+ \|w_n\|)^{2+\delta}\} < \infty$$

and that $\{x_n, \mathcal{F}_n\}$ is any adapted random vector sequence satisfying

$$\sum_{i=1}^{n} \|x_i\|^2 = O(n), \quad E\{\|x_n\|^4 (\log^+ \|x_n\|)^{2+\delta}\} = O(1)$$

where $\delta > 0$ is some constant. Then as $n \to \infty$

(i) $$\max_{1\le t\le n}\max_{1\le i\le n}\left\|\sum_{j=1}^{i} x_{j-t}w_j^T\right\| = O\left(\sqrt{n\log n}\right) \quad a.s.;$$

(ii) $$\max_{1\le t\le (\log n)^a}\max_{1\le i\le n}\left\|\sum_{j=1}^{i} x_{j-t}w_j^T\right\| = O\left(\sqrt{n\log\log n}\right) \quad a.s., \forall a > 0$$

Theorem 7.7. *For ARX(p_0, q_0) model defined by (7.2)-(7.5) with $r_0 = 0$ suppose that the driven noise $\{w_n\}$ satisfies (6.77) and conditions in Lemma 7.3 and that the input sequence can be decomposed as*

$$u_n = u_n^s + v_n$$

where $\{v_n\}$ is i.i.d. and independent of $\{w_n\}$ with $E[v_n|\mathcal{F}_{n-1}] = 0$, $E[v_n v_n^T|\mathcal{F}_{n-1}] = \Sigma > 0$ and u_n^s is $\mathcal{F}_{n-1}$-measurable with

$$\sum_{i=1}^{n}\|u_i^s\|^2 = O(n), \quad a.s..$$

If the closed-loop signal $\{y_n, u_n\}$ satisfies

$$\sum_{i=1}^{n}(\|y_i\|^2 + \|u_i\|^2) = O(n), \quad a.s.$$
$$E(\|y_i\|^2 + \|u_i\|^2)^4\{\log^+(\|y_i\|^2 + \|u_i\|^2)\}^{2+\delta} = O(1), \quad \delta > 0,$$

then as $n \to \infty$

$$\left\|\left(\sum_{i=0}^{n}\varphi_i(p^*,q^*)\varphi_i^T(p^*,q^*)\right)^{-\frac{1}{2}}\sum_{i=0}^{n}\varphi_i(p^*,q^*)w_{i+1}^T\right\|^2 = O(\log\log n) \quad a.s.$$

where $p^* \ge p_0$, $q^* \ge q_0$ and

$$\varphi_i(p^*,q^*) = [y_i^T \;\dots\; y_{i-p^*+1}^T \quad u_i^T \;\dots\; u_{i-q^*+1}^T]^T.$$

Proof. By Theorem 6.2(i) it is known that

$$\begin{aligned}\lambda_{\min}^{(p^*,q^*)}(n) &\triangleq \lambda_{\min}\left(\sum_{i=0}^{n}\varphi_i(p^*,q^*)\varphi_i^T(p^*,q^*)\right)\\ &\ge c_0 n \quad \text{for some } c_0 > 0,\end{aligned} \tag{7.88}$$

Applying Lemma 7.3(ii) yields

$$\left\|\sum_{i=0}^{n}\varphi_i(p^*,q^*)w_{i+1}\right\|^2 = O(\sqrt{\log\log n}), \quad \text{a.s.}$$

Hence, we obtain

$$
\begin{aligned}
&\left\| \left(\sum_{i=0}^{n} \varphi_i(p^*, q^*) \varphi_i^{\tau}(p^*, q^*) \right)^{-\frac{1}{2}} \sum_{i=0}^{n} \varphi_i(p^*, q^*) w_{i+1} \right\|^2 \\
\leq\ & \frac{1}{\lambda_{\min}^{(p^*,q^*)}(n)} \left\| \sum_{i=0}^{n} \varphi_i(p^*, q^*) w_{i+1} \right\|^2 \\
=\ & O(\log \log n) \quad \text{a.s.}
\end{aligned}
$$

This proves the desired result of the theorem. Q.E.D.

CHAPTER 8

Optimal Adaptive Control with Consistent Parameter Estimate

In Chapter 6 we have shown that in an adaptive control system the parameter estimate may be inconsistent even if the control performance is optimized. We have also shown that the estimate will be consistent in a feedback control system if a diminishingly excited signal is added to the system and the system input and output grow up not "too fast".

In this chapter we apply the diminishing excitation technique introduced in Chapter 6 to adaptive control systems in order to get the minimality of a loss function and consistency of the parameter estimate simultaneously.

8.1 Simultaneously Gaining Optimality and Consistency in Tracking Systems

We start with the SG-based adaptive tracker discussed in Chapter 5. The basic system under consideration is still (3.90) with $d = 1$.

Let $\{\varepsilon_i\}$ be a sequence of l-dimensional independent and identically distributed random vectors with continuous distribution and let $\{\varepsilon_i\}$ be independent of $\{w_i\}$ and $\{y_i^*\}$ with $E\varepsilon_i = 0$, $E\varepsilon_i\varepsilon_i^T = I$ and $\|\varepsilon_i\| \leq \sigma$, $\sigma > 0$, where $\{y_i^*\}$ is a sequence of a.s. bounded reference signals.

As in the preceding chapters $\{w_i\}$ is a martingale difference sequence with respect to a nondecreasing sequence of σ-algebras $\{\mathcal{F}_n\}$, which, without loss of generality, is now assumed to be

$$\mathcal{F}_n = \sigma\{w_i,\ y_{i+1}^*,\ \varepsilon_i,\ i \leq n\}. \tag{8.1}$$

Set

$$\mathcal{F}'_n = \sigma\{w_i,\ y^*_{i+1},\ \varepsilon_{i-1},\ i \leq n\}. \tag{8.2}$$

Let v_n be defined by (6.68), (6.69) and let det $B_{10} \neq 0$, where for definition of B_{1n} see (6.8).

In what follows we consider the $m = l$ case only.

By Lemma 6.1 the following equation

$$B_{1n}u^s_n = y^*_{n+1} - \theta^\tau_n \psi_n + B_{1n}u_n \tag{8.3}$$

is solvable with respect to u^s_n, where

$$u_i = u^s_i + v_i, \quad \forall i \leq n-1. \tag{8.4}$$

It is worth noting that the right-hand side of (8.3) is free of u_n.

Theorem 8.1 (CG4) . *If Conditions 1)-3) of Theorem 6.3 are satisfied, then the SG-based adaptive tracker (3.90), (4.19), (4.22) and (4.23) with u_n defined by (8.3) and (8.4) provides both the optimal tracking and consistent estimation:*

$$\limsup_{n\to\infty} \frac{1}{n}\sum_{i=0}^{n}(\|u_i\|^2 + \|y_i\|^2) < \infty \tag{8.5}$$

$$\lim_{n\to\infty} \frac{1}{n}\sum_{i=0}^{n}(y_i - y^*_i)(y_i - y^*_i)^\tau = R \tag{8.6}$$

$$\theta_n \xrightarrow[n\to\infty]{} \theta \quad a.s. \tag{8.7}$$

where θ_n is given by the SG algorithm (4.19), (4.22) and (4.23).

Proof. We first note that in the present case u_n is $\mathcal{F}_n$-measurable and Lemma 4.3 remains true. By (4.68) and (8.3), (8.4) we find that

$$y_{n+1} - y^*_{n+1} = \zeta_{n+1} + w_{n+1} + B_{1n}v_n \tag{8.8}$$

By the boundedness of θ_n (Lemma 4.3) and (6.74) it is clear that

$$\lim_{n\to\infty} \frac{1}{n}\sum_{i=0}^{n}\|B_{1i}v_i\|^2 = 0. \tag{8.9}$$

Hence (5.20) (5.22)-(5.28) hold true without any change and (8.5) and (8.6) follow immediately.

Further, (8.5) means that (6.114) (6.115) are satisfied with $\delta = 0$. Then (8.7) follows from Theorem 6.3. Q.E.D.

We now turn to the ELS-based tracker. Keeping notations used in Section 5.4 and replacing (5.116) we define

$$u_n^s \triangleq \widehat{B}_{1n}^{-1}\{y_{n+1}^* + (B_{1n}u_n - \theta_n^\tau \varphi_n)\} \tag{8.10}$$

and the diminishingly excited control u_n as:

$$u_n = u_n^s + v_n, \tag{8.11}$$

where

$$v_n = \frac{\varepsilon_n}{s_{n-1}^{\epsilon/2}}, \quad \epsilon \in \left(0, \frac{1}{2(t+1)}\right), \quad t = \max(p, q, r) + mp - 1 \tag{8.12}$$

with s_n defined by (5.60).

Theorem 8.2 (GC2) . *Suppose that for System (3.91)-(3.94) $d = 1$, $m = l$, Conditions A1-A3 stated in Section 5.3 are satisfied, $A(z)$, $B(z)$ and $C(z)$ have no common left factor and $[A_p \ B_q \ C_k]$ is of row-full-rank and that $\{y_{n+1}^*\}$ is a.s. bounded and y_{n+1}^* is $\mathcal{F}_n$-measurable. Then the ELS-based adaptive tracker consisting of (3.91)-(3.94), (4.13), (4.14), (4.18), (4.20), (8.10) and (8.11) with $\widehat{B}_{1n}$ defined by (5.115) $\forall n \geq 1$ simultaneously gains optimality in tracking and consistency in estimating. To be precise, (8.5) and (8.6) hold, and*

$$\|\theta_n - \theta\|^2 = O\left(\frac{\log n}{n^{1-(t+1)\epsilon}}\right) \quad a.s. \tag{8.13}$$

and

$$\sum_{i=1}^{n} \|y_i - y_i^* - w_i\|^2 = O(n^{1-\epsilon}) + O(d_n) a.s. \tag{8.14}$$

where ϵ is given in (8.12) and $\{d_n\}$ is defined in Theorem 5.3.

Proof. Set

$$\overline{y}_{n+1}^* = y_{n+1}^* + \frac{1}{s_{n-1}^{\epsilon/2}} \widehat{B}_{1n}\varepsilon_n. \tag{8.15}$$

Then by (8.10), (8.11), and (8.15) we have

$$u_n \triangleq \widehat{B}_{1n}^{-1}\{\overline{y}_{n+1}^* + (B_{1n}u_n - \theta_n^\tau \varphi_n)\} \tag{8.16}$$

Clearly, $\overline{y}_{n+1}^*$ is $\mathcal{F}_n$-measurable and bounded a.s. Hence $\overline{y}_n^*$ may serve as a new reference signal. Then by Theorem 5.4, (8.5) and (8.6) follow immediately. Consequently,

$$\frac{1}{n}\sum_{i=1}^{n} \|u_i^s\|^2 = O(1).$$

By (5.122) (5.150) we see that all Conditions 1)-3) required in Remark 6.1 are satisfied by $v_n = \frac{\varepsilon_n}{s_{n-1}^{\varepsilon/2}}$. Hence (8.13) is derived from Theorem 6.2 for the case where $\beta > 2$, and $\delta = 0$.

Paying attention to (6.83) and (5.150) we see that Condition (4.57) holds. Hence by (4.67) (4.68) we assert

$$P_n^{-1} \geq c_0 n^{1-\varepsilon(t+1)} I, \quad \varepsilon \in \left(0, \frac{1}{2(t+1)}\right) \tag{8.17}$$

for some $c_0 > 0$ possibly depending on ω.

By (8.17) and (5.140) it follows that

$$\varphi_n^\tau P_n \varphi_n = O(s_n^{-\delta} d_n) \quad \text{a.s.} \quad \forall \delta \in (0, 1 - \varepsilon(t+1)). \tag{8.18}$$

Using notations applied in Lemma 5.3 and setting

$$S_i = \sum_{j=1}^{i} \alpha_j, \quad S_0 = 0$$

by (5.150) we find that

$$\begin{aligned}
\sum_{i=1}^{n} \alpha_i \varphi_i^\tau P_i \varphi_i &= O\left(\sum_{i=1}^{n} \alpha_i s_i^{-\delta} d_i\right) \\
&= O\left(d_n \sum_{i=1}^{n} [S_i - S_{i-1}] i^{-\delta}\right) \\
&= O\left(d_n \left\{\sum_{i=1}^{n-1} S_i [i^{-\delta} - (i+1)^{-\delta}] + S_n n^{-\delta}\right\}\right) \\
&= O\left(d_n \left\{\sum_{i=1}^{n-1} \log(1+i)[i^{-\delta} - (i+1)^{-\delta}]\right\}\right) \\
&= O(d_n), \quad \forall \delta \in (0, 1 - \varepsilon(t+1)).
\end{aligned}$$

Consequently, by Lemma 5.3 and (5.150)

$$\sum_{i=0}^{n} \|\tilde{\theta}_i^\tau \varphi_i\|^2 = \sum_{i=1}^{n} \alpha_i (1 + \varphi_i^\tau P_i \varphi_i) = O(\log n) + O(d_n) \quad \text{a.s.} \tag{8.19}$$

Note that (8.13) implies $B_{1n} \xrightarrow[n \to \infty]{} B$, which is nondegenerate by A3. Hence from (5.115) it follows that $\widehat{B}_{1n} = B_{1n}$ for sufficiently large n.

Using (8.15), (8.16), and the notation φ_n^0 introduced by (4.17) we see that

$$\begin{aligned}
y_{k+1} &= \theta^\tau \varphi_k + \theta^\tau (\varphi_k^0 - \varphi_k) + w_{k+1} \\
&= \tilde{\theta}_k^\tau \varphi_k - \Delta \widehat{B}_{1k} u_k + \overline{y}_{k+1}^* + \theta^\tau (\varphi_k^0 - \varphi_k) + w_{k+1}
\end{aligned}$$

and

$$\sum_{k=1}^{n}\|y_{k+1}-y_{k+1}^{*}-w_{k+1}\|^{2}$$

$$=\sum_{k=1}^{n}\|\tilde{\theta}_{k}^{\tau}\varphi_{k}+\theta^{\tau}(\varphi_{k}^{0}-\varphi_{k})+\Delta\widehat{B}_{1k}u_{k}+\frac{1}{s_{k-1}^{\varepsilon/2}}\widehat{B}_{1k}\varepsilon_{k}\|^{2}. \tag{8.20}$$

By (8.19) and (5.64) the right-hand side of (8.20) equals

$$O(d_{n})+O(\log n)+O(n^{1-\varepsilon})=O(d_{n})+O(n^{1-\varepsilon}) \quad a.s.$$

which proves (8.14). Q.E.D.

8.2 Adaptive Control for Quadratic Cost

The dynamic system discussed in this section is (3.90) with $d=1$ and with known orders (p,q,r). The matrix coefficient

$$\theta^{\tau}=[-A_{1}\ \dots\ -A_{p}\quad B_{1}\ \dots\ B_{q}\quad C_{1}\ \dots\ C_{r}] \tag{8.21}$$

of the system (3.90) is unknown.

The objective is to design adaptive control $\{u_n\}$ minimizing the quadratic loss function

$$J(u)=\limsup_{n\to\infty}\frac{1}{n}\sum_{i=0}^{n-1}[(y_{i}-y_{i}^{*})^{\tau}Q_{1}(y_{i}-y_{i}^{*})+u_{i}^{\tau}Q_{2}u_{i}] \tag{8.22}$$

where $Q_1 \geq 0$, $Q_2 > 0$ and $\{y_i^*\}$ is a bounded reference signal.

Recalling notations used in Section 3.4, by Theorem 3.5 we know that the optimal control in the set U defined by (3.100) is

$$u_{n}^{0}=Lx_{n}+d_{n},\quad \forall n\geq 0 \tag{8.23}$$

where L and d_n are given by (3.107)-(3.109) while $\{x_n\}$ is recursively defined by (3.98) (3.99). The minimum cost is given by (3.110).

In the present situation θ is unknown, then both the optimal control u_n^0 and the state x_n cannot exactly be defined. Suggested by the certainty equivalency principle we first estimate θ defined by (8.22) using the SG or ELS algorithms, then recursively estimate the solution S of the Riccati equation (3.104), and finally estimate x_n by an adaptive filter. Then we form an estimate $\widehat{u}_n^0$ for u_n^0:

$$\widehat{u}_{n}^{0}=L_{n}\widehat{x}_{n}+\widehat{d}_{n}$$

where L_n, $\widehat{d}_n$ and $\widehat{x}_n$ are the estimates for L, d_n and x_n respectively.

Let us denote by A_{in}, B_{jn}, C_{kn} the estimates given by θ_n for A_i, B_j and C_k respectively $i = 1, ..., p$, $j = 1, ..., q$, $k = 1, ..., r$, where θ_n is calculated either by the SG algorithm (4.19) (4.22) (4.23) or by the ELS algorithm (4.13) (4.14) (4.18) (4.20). Then replacing A_i, B_j and C_k in (3.96) (3.97) by their estimates A_{in}, B_{jn} and C_{kn} we obtain estimates $A(n)$, $B(n)$ and $C(n)$ for A, B and C respectively.

The estimate S_n for S is recursively defined by

$$\begin{aligned} S_n &= A^\tau(n)S_{n-1}A(n) - A^\tau(n)S_{n-1}B(n)(Q_2 + B^\tau(n)S_{n-1}B(n))^{-1} \\ &\quad \cdot B^\tau(n)S_{n-1}A(n) + H^\tau Q_1 H \qquad (8.24) \end{aligned}$$

with arbitrary $S_0 \geq 0$, where H is given in (3.97).

The state x_n in (3.98) is estimated by the following adaptive filter

$$\begin{aligned} \widehat{x}_{n+1} &= A(n)\widehat{x}_n + B(n)u_n + C(n)(y_{n+1} - HA(n)\widehat{x}_n - HB(n)u_n) \\ \widehat{x}_0 &= [y_0^\tau \; 0 \; ... \; 0]^\tau. \qquad (8.25) \end{aligned}$$

Finally, we estimate u_n^0 by

$$\widehat{u}_n^0 = L_n\widehat{x}_n + \widehat{d}_n, \qquad (8.26)$$

where

$$\begin{aligned} L_n &= -(Q_2 + B^\tau(n)S_nB(n))^{-1}B^\tau(n)S_nA(n), \qquad (8.27) \\ \widehat{d}_n &= -(Q_2 + B^\tau(n)S_nB(n))^{-1}B^\tau(n)\widehat{b}_{n+1}, \qquad (8.28) \\ \widehat{b}_n &= -\sum_{j=0}^{n} F^{j\tau}(n-1)H^\tau Q_1 y^*_{n+j}, \qquad (8.29) \\ F(n) &= A(n) - B^\tau(n)(Q_2 + B^\tau(n)S_nB(n))^{-1}B^\tau(n)S_nA(n). \\ & \qquad (8.30) \end{aligned}$$

As is known, $\widehat{u}_n^0$ may not sufficiently excite the system in order to get consistent parameter estimates for θ and this may badly influence the control performance. To avoid this trouble we apply the diminishing excitation technique developed in Chapter 6. To apply theorems derived there the growth rate of input-output data should not be exceed a certain limit. For this we need a sequence of stopping times

$$1 = \tau_1 < \sigma_1 < \tau_2 < \sigma_2 < \; ... \qquad (8.31)$$

which are defined as follows in the case where $\{\theta_n\}$ is given by the SG algorithm (4.19) (4.22) (4.23):

$$\sigma_k = \sup\left\{t > \tau_k : \sum_{i=\tau_k}^{j-1} \|y_i\|^2 \leq (j-1)\log^\delta(j-1) + \|y_{\tau_k}\|^2,\right.$$

$$\forall j \in (\tau_k, t] \Big\}, \tag{8.32}$$

$$\tau_{k+1} = \inf\Big\{ t > \sigma_k : \sum_{i=\sigma_k}^{t} \|y_i\|^2 \le \frac{t \log^\delta t}{2^k}; \quad \sum_{i=\tau_k}^{\sigma_k - 1} \|y_i\|^2 \le \frac{t \log^\delta t}{2^k};$$

$$\sum_{i=0}^{t} \|\widehat{x}_i\|^2 \le t \log^{\delta/2} t \Big\} \tag{8.33}$$

and need a set Λ of integers:

$$\Lambda = \{i : \; \|\widehat{u}_i^0\|^2 \le i \log^\delta i, \quad i \ge 1\} \tag{8.34}$$

where

$$\delta \in \left(0, \; \frac{\frac{1}{4} - (m+2)s\varepsilon}{2 + (m+1)s}\right), \quad \varepsilon = \left(0, \; \frac{1}{4s(m+2)}\right), \quad s = \max(p, q, r+1).$$

Let $\{v_n\}$ be defined by (6.68), (6.69) and let u_n' satisfy

$$B_{1n} u_n' + (\theta_n^\tau \varphi_n - B_{1n} u_n) = 0 \tag{8.35}$$

which is solvable with respect to u_n' by Lemma 6.1, if, $u_n = u_n' + v_n$ and u_i' is $\mathcal{F}_i' \triangleq \sigma\{w_k, \; 0 \le k \le i, \; v_j, \; 0 \le j \le i-1\}$-measurable, $\forall i \le n-1$.

Finally, define adaptive control

$$u_n = u_n^s + v_n \tag{8.36}$$

where

$$u_n^s = \begin{cases} 0, & \text{if } n \in [\tau_k, \; \sigma_k) \bigcap \Lambda^c \text{ for some } k, \\ \widehat{u}_n^0, & \text{if } n \in [\tau_k, \; \sigma_k) \bigcap \Lambda \text{ for some } k, \\ u_n', & \text{if } n \in [\sigma_k, \; \tau_{k+1}) \text{ for some } k. \end{cases} \tag{8.37}$$

We explain the meaning of the adaptive control defined by (8.36) and (8.37). The desired control u_n^s is set to equal $\widehat{u}_n^0$ if the growth rates of $\{y_i\}$ and $\{\widehat{u}_i^0\}$ satisfy the requirements indicated in (8.32)-(8.34), or to equal zero if $\|\widehat{u}_n^0\|^2 > n \log^\delta n$ even though the growth rate of $\{y_i\}$ meets the requirement of (8.32), or to equal u_n' forcing y_{n+1} to follow zero in the rest case.

Theorem 8.3 (CG5) , *[CG7]. For System (3.91)-(3.94) assume that the following Conditions 1)-5) are satisfied:*

1) $\{w_n, \mathcal{F}_n\}$ is a martingale difference sequence with

$$\sup_n E(\|w_n\|^2|\mathcal{F}_{n-1}) < \infty \tag{8.38}$$

$$\lim_{n\to\infty} \frac{1}{n}\sum_{i=1}^{n} w_i w_i^T = R > 0. \tag{8.39}$$

2) $C(z) - \frac{a}{2}I$ is SPR for some $a > 0$.

3) $d = 1$, $m = l$, $p \geq q$ and $\det z^{-1}B(z) \neq 0$, $\forall |z| \leq 1$.

4) (A, B, D) is controllable and observable for some D and H satisfying $D^\tau D = H^\tau Q_1 H$, where A, B and H are given by (3.96) (3.97).

5) $[A_p \;\; B_q \;\; C_r]$ is of row-full-rank.

Then the adaptive control (8.36) is optimal in the set U defined by (3.100):

$$\begin{aligned} &\limsup_{n\to\infty} \frac{1}{n}\sum_{i=0}^{n-1}[(y_i - y_i^*)^\tau Q_1(y_i - y_i^*) + u_i^T Q_2 u_i] \\ = &\limsup_{n\to\infty} \frac{1}{n}\sum_{i=0}^{n-1}[y_i^{*\tau} Q_1 y_i^* - b_{i+1}^T B(Q_2 + B^\tau S B)^{-1} B^\tau b_{i+1}] \\ &+ trSCRC^\tau (= \min) \end{aligned} \tag{8.40}$$

and

$$\theta_n \xrightarrow[n\to\infty]{} \theta \quad a.s.$$

where θ_n is given by the SG algorithm (4.19), (4.22) and (4.23), and C, S, $\{b_i\}$ are defined in (3.104)-(3.109).

Proof.

Step 1. We first show the strong consistency of θ_n. In order Theorem 6.3 to be applicable we proceed to verify that a) Condition 4) implies that $A(z)$, $z^{-1}B(z)$ and $C(z)$ have no common left factor and b)

$$\frac{1}{n}\sum_{i=1}^{n}(\|u_i^s\|^2 + \|y_i\|^2) = O(\log^\delta n), \tag{8.41}$$

$$\delta \in \left(0, \frac{\frac{1}{4} - (m+2)s\varepsilon}{2 + (m+1)s}\right), \quad \varepsilon = \left(0, \frac{1}{4s(m+2)}\right), \quad s = \max(p, q, r+1).$$

a) As a matter of fact, we prove a result stronger than that as required. To be specific, we show that $A(z)$ and $z^{-1}B(z)$ are left-coprime if (A, B) is controllable, where A, B are given by (3.96).

Let (A, B) be controllable. We show that $[I - zA \vdots B]$ is of row-full-rank for any complex number z. It suffices to prove for $z \neq 0$.

Let α be a vector of compatible dimension such that

$$\alpha^\tau [I - zA \vdots B] = 0.$$

Then we have $\alpha^\tau B = 0$, $\alpha^\tau = z\alpha^\tau A$, and hence

$$\alpha^\tau [B\ AB\ A^2B\ \dots\ A^{s-1}B] = 0.$$

This is possible $\alpha = 0$, since $[B\ AB\ A^2B\ \dots\ A^{s-1}B]$ is of row-full-rank by controllability of (A, B).

Notice that A and B are $ms \times ms$ and $ms \times m$ matrices respectively.

Let $U(z)$ be the unimodular matrix corresponding to elementary column operations such that at least m of the last columns on the right-hand side are identically zero:

$$[I - zA \quad B] \begin{bmatrix} \overbrace{U_{11}(z)}^{ms} & \overbrace{U_{12}(z)}^{m} \\ U_{21}(z) & U_{22}(z) \end{bmatrix} \begin{matrix} \}ms \\ \}m \end{matrix} = [\overbrace{R(z)}^{ms} \quad \overbrace{0}^{m}]$$

Since $[I - zA \quad B]$ is of row-full-rank for any z, $R(z)$ must be unimodular. Then $I - zA$ and B are left-coprime:

$$(I - zA)V(z) + BW(z) = I, \tag{8.42}$$

where $V(z) = U_{11}(z)R^{-1}(z)$, $W(z) = U_{21}(z)R^{-1}(z)$ are polynomial matrices.

Noticing (3.96) we write (8.42) in the block form

$$\begin{bmatrix} I + zA_1 & -zI & 0 & \dots & 0 \\ zA_2 & I & -zI & \ddots & \vdots \\ \vdots & \vdots & \ddots & \ddots & 0 \\ \vdots & \vdots & \ddots & \ddots & -zI \\ zA_s & 0 & \dots & \dots & I \end{bmatrix} \begin{bmatrix} V_{11}(z) & \dots & V_{1s}(z) \\ \vdots & \dots & \vdots \\ V_{s1}(z) & \dots & V_{ss}(z) \end{bmatrix}$$

$$+ \begin{bmatrix} B_1 \\ \vdots \\ B_s \end{bmatrix} [W_1(z) \ \dots \ W_s(z)] = \begin{bmatrix} I & \dots & 0 \\ & \ddots & \\ 0 & \dots & I \end{bmatrix} \tag{8.43}$$

where each block is an $m \times m$ matrix.

From the first block column we obtain the following equations

$$\begin{aligned}
&(I+zA_1)V_{11}(z) - zV_{21}(z) + B_1W_1(z) = I\\
&zA_2V_{11}(z) + V_{21}(z) - zV_{31}(z) + B_2W_1(z) = 0\\
&zA_3V_{11}(z) + V_{31}(z) - zV_{41}(z) + B_3W_1(z) = 0\\
&\dots\dots\dots\dots\dots\dots\dots\\
&zA_{s-1}V_{11}(z) + V_{s-11}(z) - zV_{s1}(z) + B_{s-1}W_1(z) = 0\\
&zA_sV_{11}(z) + V_{s1}(z) + B_sW_1(z) = 0.
\end{aligned}$$

Multiplying the ith equation by z^{i-1}, $i = 2, \dots, s$ and summing up all equations we derive that

$$A(z)V_{11}(z) + z^{-1}B(z)W_1(z) = I,$$

which means that $A(z)$ and $z^{-1}B(z)$ are left-coprime.

b) Using the minimum phase condition $\det z^{-1}B(z) \neq 0$, $|z| \leq 1$ for (8.41) we need only to prove

$$\frac{1}{n}\sum_{i=1}^{n}\|y_i\|^2 = O(\log^\delta n). \tag{8.44}$$

If $\tau_k < \infty$, $\sigma_k = \infty$ for some k, then (8.44) follows from (8.32). If $\sigma_k < \infty$, $\tau_{k+1} = \infty$ for some k, then by (8.36) and (8.37)

$$u_n = u_n' + v_n, \quad n \geq \sigma_k$$

and by Theorem 5.1

$$\frac{1}{n}\sum_{i=\sigma_k}^{n}\|y_i\|^2 = O(n).$$

If $\tau_k < \infty$, $\sigma_k < \infty$ for some k, then by (8.32) and (8.33) for any $k \geq 2$

$$\begin{aligned}
&\sup_{\tau_k \leq n < \sigma_k} \frac{1}{n\log^\delta n}\sum_{i=1}^{n}\|y_i\|^2\\
\leq\ &\sup_{\tau_k \leq n < \sigma_k} \frac{1}{n\log^\delta n}\left[\sum_{i=\tau_1}^{\sigma_1-1}\|y_i\|^2 + \sum_{i=\sigma_1}^{\tau_2}\|y_i\|^2 + \dots + \sum_{i=\sigma_{k-1}}^{\tau_k}\|y_i\|^2 + \sum_{i=\tau_k}^{n}\|y_i\|^2\right]\\
\leq\ &\sup_{\tau_k \leq n < \sigma_k} \frac{1}{n\log^\delta n}\left[2\sum_{i=1}^{k-1}\frac{\tau_{i+1}\log^\delta \tau_{i+1}}{2^i} + \sum_{i=\tau_k}^{n}\|y_i\|^2\right]\\
\leq\ &2 + \sup_{\tau_k \leq n < \sigma_k} \frac{1}{n\log^\delta n}\sum_{i=\tau_k}^{n}\|y_i\|^2
\end{aligned}$$

$$\leq \ 2 + \sup_{\tau_k \leq n < \sigma_k} \frac{1}{n \log^\delta n} \left(n \log^\delta n + \|y_{\tau_k}\|^2 \right)$$

$$\leq \ 3 + \frac{\|y_{\tau_k}\|^2}{\tau_k \log^\delta \tau_k} \leq 3 + \frac{1}{2^{k-1}} < 4. \tag{8.45}$$

Hence for (8.44) we need only to show the boundedness of

$$\sup_{\sigma_k \leq n < \tau_{k+1}} \frac{1}{n \log^\delta n} \sum_{i=1}^{n} \|y_i\|^2.$$

Noticing $r_n \xrightarrow[n \to \infty]{} \infty$ by Theorem 2.8, then by Lemmas 4.3 and 2.4 we know that

$$\sum_{i=0}^{n} \|\zeta_{i+1}\|^2 = o(r_n) \tag{8.46}$$

where ζ_{i+1} and r_n are defined by (4.68) and (4.23) respectively. The stability of $z^{-1}B(z)$ and (8.39) imply that

$$\sum_{i=0}^{n} \|u_i\|^2 = O\left(\sum_{i=0}^{n+1} \|y_i\|^2 \right) + O(n). \tag{8.47}$$

Combining (8.46) and (8.47) leads to

$$\sum_{i=0}^{n+1} \|\zeta_i\|^2 = o\left(\sum_{i=0}^{n+1} \|y_i\|^2 \right) + o(n). \tag{8.48}$$

From (8.35)-(8.37) it is easy to see

$$y_{i+1} = \zeta_{i+1} + w_{i+1} + B_{1i} v_i \quad \text{for} \quad i \in [\sigma_k, \ \tau_{k+1}). \tag{8.49}$$

Then, if k is large enough, for any $n \in [\sigma_k, \ \tau_{k+1})$ by (8.48), (8.49), (8.39) and $v_i \xrightarrow[i \to \infty]{} 0$,

$$\sum_{i=1}^{n} \|y_i\|^2 = \sum_{i=1}^{\sigma_k - 1} \|y_i\|^2 + \|y_{\sigma_k}\|^2 + \sum_{i=\sigma_k+1}^{n} \|y_i\|^2$$

$$\leq \ 4\sigma_k \log^\delta \sigma_k + \|y_{\sigma_k}\|^2 + 2 \sum_{i=\sigma_k+1}^{n} \|\zeta_i\|^2 + 2 \sum_{i=\sigma_k+1}^{n} \|w_i + B_{i-1} v_{i-1}\|^2$$

$$\leq \ \alpha \left(\sum_{i=1}^{n} \|y_i\|^2 + n \right) + 4\sigma_k \log^\delta \sigma_k + \|y_{\sigma_k}\|^2 + cn \tag{8.50}$$

where $\alpha \in (0,1)$ may be arbitrarily small but possibly depends on sample.

In the following for notational simplicity by the same c we denote various magnitudes that are independent of n but possibly depend on sample.

For sufficiently large k from (8.50) we have

$$\sup_{\sigma_k \le n < \tau_{k+1}} \frac{1}{n \log^\delta n} \sum_{i=1}^{n} \|y_i\|^2 \le \frac{\|y_{\sigma_k}\|^2}{(1-\alpha)\sigma_k \log^\delta \sigma_k} + \frac{4+\alpha+c}{1-\alpha} \tag{8.51}$$

By (3.90) it follows that

$$\|y_{\sigma_k}\|^2 \le c \left(\sum_{i=1}^{\sigma_k - 1} \|y_i\|^2 + \|u_{\sigma_k - 1}\|^2 + \sum_{i=0}^{\sigma_k - 2} \|u_i\|^2 + \sum_{i=0}^{\sigma_k} \|w_i\|^2 \right). \tag{8.52}$$

We now estimate each term on the right-hand side of (8.52).

By (8.47) and then (8.45) we have

$$\begin{aligned} \sum_{i=0}^{\sigma_k - 2} \|u_i\|^2 &\le c \left(\sum_{i=0}^{\sigma_k - 1} \|y_i\|^2 + \sigma_k - 2 \right) \\ &< c \left(4\sigma_k \log^\delta \sigma_k + \sigma_k \right) \end{aligned} \tag{8.53}$$

and by (8.39)

$$\sum_{i=0}^{\sigma_k} \|w_i\|^2 \le c\sigma_k. \tag{8.54}$$

Notice that by (8.37) $u^s_{\sigma_k - 1}$ equals either 0 or $\widehat{u}^0_{\sigma_k - 1}$ with $\|\widehat{u}^0_{\sigma_k - 1}\|^2 \le (\sigma_k - 1)\log^\delta(\sigma_k - 1)$. Hence

$$\|u^s_{\sigma_k - 1}\|^2 < \sigma_k \log^\delta \sigma_k \quad \text{and} \quad \|u_{\sigma_k - 1}\|^2 \le \sigma_k \log^\delta \sigma_k \tag{8.55}$$

for suitably large k.

By (8.53)-(8.55) from (8.52) we conclude that

$$\|y_{\sigma_k}\|^2 \le c\sigma_k \log^\delta \sigma_k$$

which incorporating with (8.51) proves the boundedness of

$$\sup_{\sigma_k \le n < \tau_{k+1}} \frac{1}{n \log^\delta n} \sum_{i=1}^{n} \|y_i\|^2.$$

This together with (8.45) shows that

$$\sup_n \frac{1}{n \log^\delta n} \sum_{i=1}^{n} \|y_i\|^2 < \infty \quad a.s.$$

Then Theorem 6.3 yields the strong consistency of θ_n.

Step 2. We now show that for any fixed sample there is k_0 such that

$$\tau_{k_0} < \infty, \quad \sigma_{k_0} = \infty. \tag{8.56}$$

For this we need only to prove the impossibility of the following two cases:

1) $\sigma_k < \infty$, $\tau_{k+1} = \infty$ for some k,
2) $\tau_k < \infty$ and $\sigma_k < \infty$ for any k.

In Case 1) by (8.37) we have

$$u_n = u_n^1 + v_n, \quad \forall n \geq \sigma_k.$$

Then Theorem 8.1 implies that

$$\sum_{i=\sigma_k}^{n} (\|y_i\|^2 + \|u_i\|^2) = O(n). \tag{8.57}$$

If we can show

$$\sum_{i=1}^{n} \|\widehat{x}_i\|^2 = O(n), \quad \text{as} \quad n \to \infty, \tag{8.58}$$

then this together with (8.57) implies $\tau_{k+1} < \infty$. The contradiction will show the impossibility of Case 1).

Let us write x_k appearing in (3.98) in the block form:

$$x_k^\tau = [x_k^{1\tau} \ldots x_k^{s\tau}] \tag{8.59}$$

where x_k^i, $i = 1, \ldots, s$ is an m-dimensional vector.

From (3.96)-(3.99) it follows that

$$x_{k+1}^s = -A_s x_k^1 + B_s u_k + C_{s-1} w_{k+1} \tag{8.60}$$

$$x_{k+1}^i = -A_i x_k^1 + x_k^{i+1} + B_i u_k + C_{i-1} w_{k+1}, \quad i = 1, \ldots, s-1 \tag{8.61}$$

$$x_k^1 = y_k. \tag{8.62}$$

By (8.39), (8.57), (8.60) and (8.62) it is clear that

$$\limsup_{n\to\infty} \frac{1}{n} \sum_{i=1}^{n} \|x_i^s\|^2 < \infty. \tag{8.63}$$

Starting from $i = s$, by induction from (8.61) we see that

$$\limsup_{n\to\infty} \frac{1}{n} \sum_{k=1}^{n} \|x_k^i\|^2 < \infty \quad i = 1, \ldots, s.$$

Consequently,

$$\limsup_{n\to\infty} \frac{1}{n}\sum_{k=1}^{n} \|x_k\|^2 < \infty. \tag{8.64}$$

Let us write $\widehat{x}_n$ in the block form

$$\widehat{x}_n^\tau = [\widehat{x}_n^{1\tau} \ldots \widehat{x}_n^{s\tau}] \tag{8.65}$$

where $\widehat{x}_n^i$, $i = 1, \ldots, s$ is m-dimensional.

Noticing that the first m rows of the matrix product is $[I\ 0\ \ldots\ 0]$, from (8.25) we find that

$$\widehat{x}_n^1 = x_n^1, \quad \forall n \geq 0. \tag{8.66}$$

Set

$$z_n^\tau = [x_n^{2\tau} \ldots \widehat{x}_n^{s\tau}], \quad \widehat{z}_n^\tau = [\widehat{x}_n^{2\tau} \ldots \widehat{x}_n^{s\tau}]. \tag{8.67}$$

Then

$$x_n^\tau = [x_n^{1\tau}\ z_n^\tau], \quad \widehat{x}_n^\tau = [\widehat{x}_n^{1\tau}\ \widehat{z}_n^\tau]. \tag{8.68}$$

Let us write the recursive equation for $\begin{bmatrix} \widehat{x}_n \\ z_n - \widehat{z}_n \end{bmatrix}$.

In what follows X' denotes the matrix obtained from X by deleting its first m rows.

Rewrite u_n defined by (8.36) as

$$u_n = L_n^s \widehat{x}_n + d_n^s + v_n \tag{8.69}$$

where

$$L_n^s = \begin{cases} 0, & \text{if } n \in [\tau_k, \sigma_k) \bigcap \Lambda^c \text{ for some } k, \\ L_n, & \text{if } n \in [\tau_k, \sigma_k) \bigcap \Lambda \text{ for some } k, \\ 0, & \text{if } n \in [\sigma_k, \tau_{k+1}) \text{ for some } k. \end{cases}$$

$$d_n^s = \begin{cases} 0, & \text{if } n \in [\tau_k, \sigma_k) \bigcap \Lambda^c \text{ for some } k, \\ \widehat{d}_n, & \text{if } n \in [\tau_k, \sigma_k) \bigcap \Lambda \text{ for some } k, \\ u_n^1, & \text{if } n \in [\sigma_k, \tau_{k+1}) \text{ for some } k. \end{cases}$$

From (3.98), (8.25) it is easy to deduce that

$$\begin{aligned} & z_{n+1} - \widehat{z}_{n+1} \\ = \ & G_n(z_n - \widehat{z}_n) + [A' - A'(n) + (B' - B'(n))L_n^s \\ & -C'(n)H(A - A(n) - C'(n)H(B - B(n))L_n^s]\widehat{x}_n \\ & +[B' - B'(n) - C'(n)H(B - B(n))](d_n^s + v_n) \\ & +(C' - C'(n))w_{n+1} \end{aligned} \tag{8.70}$$

where

$$G_n = \left[-C'(n) \vdots \begin{array}{c} I \\ 0 \end{array} \right] \begin{array}{l} \}(s-2)m \\ \}m \end{array}.$$

Consequently, we have

$$\begin{bmatrix} \widehat{x}_{n+1} \\ z_{n+1} - \widehat{z}_{n+1} \end{bmatrix}$$

$$= \Phi_n \begin{bmatrix} \widehat{x}_n \\ z_n - \widehat{z}_n \end{bmatrix} + \begin{bmatrix} B(n) + C(n)H(B - B(n)) \\ B' - B'(n) - C'(n)H(B - B(n)) \end{bmatrix} (d_n^s + v_n)$$

$$+ \begin{bmatrix} C(n) \\ C - C(n) \end{bmatrix} w_{n+1}, \tag{8.71}$$

where

$$\Phi_n = \begin{bmatrix} \begin{matrix} A(n) + B(n)L_n^s \\ +C(n)H(A - A(n)) \\ +C(n)H(B - B(n))L_n^s \end{matrix} & C^0(n) \\ \begin{matrix} A' - A'(n) + (B' - B'(n))L_n^s \\ -C'(n)H(A - A(n)) \\ -C'(n)H(B - B(n))L_n^s \end{matrix} & G_n \end{bmatrix} \tag{8.72}$$

where $C^0(n) = [C(n) \vdots 0]$.

Since $\theta_n \xrightarrow[n \to \infty]{} \theta$, we see

$$G_n \xrightarrow[n \to \infty]{} \begin{bmatrix} -C_1 & I & 0 & \dots & 0 \\ -C_2 & 0 & I & \ddots & \vdots \\ \vdots & \vdots & \ddots & \ddots & 0 \\ \vdots & \vdots & \ddots & \ddots & I \\ -C_{s-1} & 0 & \dots & \dots & 0 \end{bmatrix} \triangleq G. \tag{8.73}$$

It is immediate to verify that

$$\det(\lambda I - G) = \lambda^{(s-1)m} \det\left(I + \lambda^{-1}C_1 + \ \dots\ + \lambda^{-(s-1)}C_{s-1}\right)$$

$$= \lambda^{(s-1)m} \det\left(I + \lambda^{-1}C_1 + \ \dots\ + \lambda^{-(s-1)}C_r\right) \tag{8.74}$$

where the last equality follows from the fact that $s - 1 \geq r$ and by definition $C_i = 0$ for $i > r$.

By Corollary 4.1 Condition 2) implies stability of $C(z)$, i.e. $\det C(z) \neq 0$, $|z| \leq 1$. Therefore, from (8.74) we find that all eigenvalues of G are inside the open unit disk, i.e. G is stable.

We now prove a fact which is needed in the sequel.

Let $\{N_k\}$ be a sequence of square matrices and N_k converges to a stable matrix N. Then

$$\|N_k\ N_{k-1}\ \dots\ N_1\| < c\rho^k, \quad \rho \in (0,1), \quad \forall k \geq 0 \tag{8.75}$$

where c is a positive constant.

Since N is stable, there exists a norm $\|\cdot\|_1$ such that $\|N\|_1 < 1$. Hence by the convergence of N_k to N, we know that for all suitably large k

$$\|N_k\|_1 \leq \|N\|_1 + \|N_k - N\|_1 \leq \rho < 1. \tag{8.76}$$

From this it is easy to see that (8.75) holds with the norm $\|\cdot\|$ replaced by $\|\cdot\|_1$. Hence (8.75) is true.

Consequently, by stability of G there are $c > 0$ and $\rho \in (0, 1)$ such that

$$\|G_k\, G_{k-1} \dots G_1\| < c\rho^k, \quad \forall k \geq 0. \tag{8.77}$$

Noticing that $[L_n^s \;\; d_n^s] = [0 \;\; u_n^1]$ for $n \geq \sigma_k$ in Case 1) and that $\theta_n \xrightarrow[n\to\infty]{} \theta$ from (8.70) we find

$$\begin{aligned}
&\|x_{n+1} - \widehat{x}_{n+1}\| \\
\leq\;& c\rho^{n+1}\|x_0 - \widehat{x}_0\| + c_1 \sum_{j=1}^{n+1} \rho^{n-j+1}[\|A - A(j-1)\|\|\widehat{x}_{j-1}\| \\
&+\|B - B(j-1)\|\|u_{j-1}\| + \|C - C(j-1)\|\|w_j\|]
\end{aligned} \tag{8.78}$$

and by the Schwarz inequality

$$\begin{aligned}
&\|x_{n+1} - \widehat{x}_{n+1}\|^2 \\
\leq\;& 2c_1^2\rho^{2(n+1)}\|x_0 - \widehat{x}_0\|^2 + c_2 \sum_{j=1}^{n+1} \rho^{n-j+1}[\|A - A(j-1)\|^2\|\widehat{x}_{j-1}\|^2 \\
&+\|B - B(j-1)\|^2\|u_{j-1}\|^2 + \|C - C(j-1)\|^2\|w_j\|^2]
\end{aligned} \tag{8.79}$$

where c_1 and c_2 are constants.

Notice that $A - A(j-1) \xrightarrow[j\to\infty]{} 0$, $B - B(j-1) \xrightarrow[j\to\infty]{} 0$ and $C - C(j-1) \xrightarrow[j\to\infty]{} 0$. Then by (8.57) and (8.39) from (8.79) it follows that

$$\sum_{i=1}^{n} \|x_i - \widehat{x}_i\|^2 = o(n) + o\left(\sum_{i=1}^{n} \|\widehat{x}_i\|^2\right). \tag{8.80}$$

This together with (8.64) yields

$$\sum_{i=1}^{n} \|\widehat{x}_i\|^2 = O(n).$$

Thus we have proved that Case 1) cannot occur.

In Case 2) set

$$t_k = \sup\{n : \; j \in [\tau_k, \; \sigma_k) \cap \Lambda, \quad \forall j \in [\tau_k, \; n]\}. \tag{8.81}$$

Since $\theta_n \xrightarrow[n \to \infty]{} \theta$, Condition 4) implies that Theorem 3.4 is applicable to the recursive equation (8.24). Hence

$$S_n \xrightarrow[n \to \infty]{} S$$

where S satisfies equation (3.104). Then L_n given by (8.27) tends to

$$L \triangleq -(Q_2 + B^\tau SB)^{-1} B^\tau SA$$

and $F(n)$ given by (8.30) converges to

$$F \triangleq A - B(Q_2 + B^\tau SB)^{-1} B^\tau SA$$

which by Theorem 3.5 is stable.

Consequently, $\{\widehat{b}_n\}$ is bounded, and hence $\{\widehat{d}_n\}$ is also bounded. Then from (8.26) we conclude that

$$\|\widehat{u}_n^0\|^2 = O(\|\widehat{x}_n\|^2) + O(1). \tag{8.82}$$

By (8.33) it is obvious that

$$\|\widehat{x}_{\tau_k}\|^2 \leq \tau_k \log^{\delta/2} \tau_k \tag{8.83}$$

which by (8.82) implies

$$\|\widehat{u}_{\tau_k}^0\|^2 = O(\tau_k \log^{\delta/2} \tau_k) \quad \text{as} \quad k \to \infty. \tag{8.84}$$

Hence as $k \to \infty$ there exists a t_k defined by (8.80) so that $t_k > \tau_k$.

Noticing $L_n^s = L_n$ for any $n \in [\tau_k, t_k]$ we then find

$$\Phi_n \xrightarrow[\substack{n \in [\tau_k, t_k] \\ k \to \infty}]{} \begin{bmatrix} F & C^0 \\ 0 & G \end{bmatrix} \triangleq \Phi, \tag{8.85}$$

where $C^0 \triangleq [C \vdots 0]$.

It is clear that Φ is stable, because both F and G are stable. Then using (8.37), the boundedness of $\{d_n^s\}$ and $\theta_n \xrightarrow[n \to \infty]{} \theta$, $v_n \xrightarrow[n \to \infty]{} 0$ by an argument similar to that used for deriving (8.79) from (8.71) we obtain

$$\sum_{i=\tau_k}^{n} \|\widehat{x}_{i+1}\|^2 \leq c(\|\widehat{x}_{\tau_k}\|^2 + \|z_{\tau_k} - \widehat{z}_{\tau_k}\|^2 + n) \tag{8.86}$$

for any $n \in [\tau_k, t_k]$.

By the stability of $z^{-1}B(z)$ from (8.44) it follows that

$$\sum_{i=1}^{n} \|u_i\|^2 = O(n \log^\delta n). \tag{8.87}$$

Similar to (8.78) from (8.70) we obtain

$$\begin{aligned} \|z_{\tau_k} - \widehat{z}_{\tau_k}\| \leq\ & c_3 \sum_{j=0}^{\tau_k} \rho^{\tau_k - j} f_{j-1}(\|\widehat{x}_{j-1}\| + \|d_{j-1}^s + v_{j-1}\| + \|w_j\|) \\ & + c\rho^{\tau_k} \|z_0 - \widehat{z}_0\| \end{aligned} \tag{8.88}$$

where

$$\begin{aligned} f_j =\ & \|A' - A'(j) + (B' - B'(j))L_j^s \\ & + C'(j)H(A - A(j)) - C'(j)H(B - B(j))L_j^s\| \\ & + \|B' - B'(j) - C'(j)H(B - B(j))\| + \|C' - C'(j)\| \\ & \xrightarrow[j \to \infty]{} 0 \end{aligned}$$

where c_3 is a constant.

By (8.69) it is clear that

$$\|d_j^s + v_j\| \leq \|u_j\| + \|L_j^s \widehat{x}_j\| \leq \|u_j\| + c_4 \|\widehat{x}_j\| \tag{8.89}$$

where c_4 is a constant.

Putting (8.89) into (8.88) and using the Schwarz inequality we get

$$\begin{aligned} \|z_{\tau_k} - \widehat{z}_{\tau_k}\|^2 & \leq o(\tau_k) + o\left(\sum_{i=0}^{\tau_k} \|\widehat{x}_i\|^2\right) + o\left(\sum_{i=0}^{\tau_k} \|u_i\|^2\right) \\ & = o(\tau_k) + o\left(\sum_{i=0}^{\tau_k} \|\widehat{x}_i\|^2\right) + o\left(\tau_k \log^\delta \tau_k\right) \end{aligned} \tag{8.90}$$

where the last equality is derived by using (8.87).

Combining (8.90) and (8.86) and noticing

$$\sum_{i=0}^{\tau_k} \|\widehat{x}_i\|^2 \leq \tau_k \log^{\delta/2} \tau_k$$

from (8.33) we find that

$$\sum_{i=\tau_k}^{n} \|\widehat{x}_{i+1}\|^2 = o(n \log^\delta n). \tag{8.91}$$

Since $\{\widehat{d}_n\}$ is bounded, $\{L_n\}$ is convergent, from (8.26) and (8.91) we see that

$$||\widehat{u}_n^0||^2 = O(1) + O(||\widehat{x}_n||^2) = o(n \log^\delta n).$$

Hence by definition of Λ it follows that

$$\Lambda = \{i : i \geq n\}$$

for some suitably large n.

This together with (8.81) implies that $t_k = \sigma_k - 1$ for sufficiently large k. Then by the definition (8.69) we have

$$L_n^s = L_n, \quad d_n^s = \widehat{d}_n, \quad \forall n \in [\tau_k, \sigma_k - 1].$$

Similar to (8.79) from (8.70) it is easy to derive

$$\sum_{i=\tau_k}^{n} ||x_{i+1} - \widehat{x}_i||^2 = o(n) + o\left(\sum_{i=\tau_k}^{n} ||\widehat{x}_i||^2\right), \quad \forall n \in [\tau_k, \sigma_k - 1]$$

and by (8.91)

$$\sum_{i=\tau_k}^{n} ||x_{i+1}||^2 = o(n \log^\delta n), \quad \forall n \in [\tau_k, \sigma_k - 1].$$

In particular,

$$\sum_{i=\tau_k}^{\sigma_k - 1} ||y_{i+1}||^2 = o(\sigma_k \log^\delta \sigma_k)$$

or

$$\sum_{i=\tau_k}^{\sigma_k} ||y_i||^2 = o(\sigma_k \log^\delta \sigma_k) + ||y_{\tau_k}||^2$$

which contradicts the definition of σ_k.

Hence Case 2) is also impossible, and (8.56) has been verified.

Step 3. We now show that Λ contains all integers starting from some sufficiently large n_k.

Since $\sigma_{k_0} = \infty$ from (8.56), by (8.32) and the minimum phase condition (i.e. $\det z^{-1}B(z) \neq 0$, $|z| \leq 1$) we have

$$\sum_{i=1}^{n} ||u_i||^2 = O(n \log^\delta n). \tag{8.92}$$

Using (8.39), (8.92) from (8.60)-(8.62) we conclude that

$$\sum_{i=1}^{n} ||x_i^j||^2 = O(n \log^\delta n), \quad \forall j = 1, \dots s$$

and hence

$$\sum_{i=1}^{n} \|x_i\|^2 = O(n \log^\delta n). \tag{8.93}$$

Therefore, there is a subsequence

$$\|x_{n_k}\|^2 = O(n_k \log^\delta n_k) \quad \text{as} \quad k \to \infty. \tag{8.94}$$

Noticing that (8.90) is also true for $\tau_k = n$ we then have

$$\|x_n - \widehat{x}_n\|^2 = o(n) + o\left(\sum_{i=0}^{n} \|\widehat{x}_i\|^2\right) + o(n \log^\delta n). \tag{8.95}$$

From this and (8.93) it is immediate that

$$\sum_{i=0}^{n} \|\widehat{x}_i\|^2 = O(n \log^\delta n) \tag{8.96}$$

which together with (8.95) implies

$$\|x_n - \widehat{x}_n\|^2 = o(n \log^\delta n). \tag{8.97}$$

Paying attention to (8.94) from (8.97) we then have

$$\|\widehat{x}_{n_k}\|^2 = o(n_k \log^\delta n_k) \quad \text{and} \quad \|\widehat{u}^0_{n_k}\|^2 = o(n_k \log^\delta n_k). \tag{8.98}$$

Hence for all sufficiently large k, $n_k \in \Lambda$ and $\sup\{i : \ i \in \Lambda\} = \infty$. Let $n_k \in \Lambda$ and set

$$e_k = \sup\{n : \ j \in \Lambda, \quad \forall j \in [n_k, \ n]\}.$$

We need to show that $e_k = \infty$ for some n_k. Suppose the converse were true, i.e. $e_k < \infty$ for all k.

Similar to (8.86) as $k \to \infty$ we have

$$\sum_{i=n_k}^{e_k} \|\widehat{x}_{i+1}\|^2 \leq c\left(\|\widehat{x}_{n_k}\|^2 + \|x_{n_k} - \widehat{x}_{n_k}\|^2 + e_k\right).$$

This combining with (8.94) and (8.98) leads to

$$\|\widehat{x}_{e_k+1}\|^2 = o(e_k \log^\delta e_k) \quad \text{as} \quad k \to \infty$$

which means that $e_k+1 \in \Lambda$ for all sufficiently large k and thus it contradicts the definition of e_k. Hence there is a k for which $e_k = \infty$ and

$$\Lambda = \{i : \ i \geq n\} \tag{8.99}$$

for some suitably large n.

Step 4. From the assertions proved in Steps 2 and 3 we know that

$$u_n = \widehat{u}_n^0 + v_n \tag{8.100}$$

for all sufficiently large n.

We now show that u_n defined by (8.100) belongs to U given by (3.100). Since Φ_n converges to the stable Φ, from (8.71) and (8.75) we find that

$$\begin{aligned} &\left\| \begin{bmatrix} \widehat{x}_{n+1} \\ z_{n+1} - \widehat{z}_{n+1} \end{bmatrix} \right\| \\ \leq\ & O\left(\sum_{j=1}^{n+1} \rho^{n-j+1} \left(\|d_{j-1}^s\| + \|v_{j-1}\| + \|w_j\| \right) \right) \\ &+o(1) \end{aligned} \tag{8.101}$$

and

$$\begin{aligned} &\|\widehat{x}_{n+1}\|^2 + \|z_{n+1} - \widehat{z}_{n+1}\|^2 \\ \leq\ & O\left(\sum_{j=1}^{n+1} \rho^{n-j+1} \left(\|d_{j-1}^s\|^2 + \|v_{j-1}\|^2 + \|w_j\|^2 \right) \right) \\ &+o(1). \end{aligned} \tag{8.102}$$

$$\begin{aligned} &\frac{1}{n} \sum_{k=1}^{n} \left(\|\widehat{x}_{k+1}\|^2 + \|z_{k+1} - \widehat{z}_{n+1}\|^2 \right) \\ =\ & O(1) + O\left(\frac{1}{n} \sum_{j=0}^{n} \left[\|d_j^s\|^2 + \|v_j\|^2 + \|w_{j+1}\|^2 \right] \right) \\ =\ & O(1), \end{aligned} \tag{8.103}$$

where for the last equality we have invoked (8.39) and the boundedness of d_j^s and v_j.

Noticing by (8.39)

$$\frac{1}{n}\|w_n\|^2 = \frac{1}{n}\sum_{i=1}^{n}\|w_i\|^2 - \frac{n-1}{n(n-1)}\sum_{i=1}^{n-1}\|w_i\|^2 \xrightarrow[n\to\infty]{} 0 \tag{8.104}$$

and recalling $\widehat{x}_k^1 = x_k^1$ we then conclude from (8.102) and (8.103) that

$$\|x_n\|^2 = o(n) \quad \text{and} \quad \frac{1}{n}\sum_{i=1}^{n}\|x_i\|^2 = O(1) \tag{8.105}$$

$$\|\widehat{x}_n\|^2 = o(n) \quad \text{and} \quad \frac{1}{n}\sum_{i=1}^{n}\|\widehat{x}_i\|^2 = O(1). \tag{8.106}$$

By (8.105), (8.106) and from (8.100) and (8.26) we finally find that for u_n defined by (8.36)

$$\sum_{i=1}^{n}\|u_i\|^2 = O(n) \tag{8.107}$$

and $\{u_i\} \in U$. From (3.115) (3.117) and (3.118) it is easy to see that for any $u \in U$

$$\begin{aligned} J(u) &= trSCRC^\tau \\ &+ \limsup_{n\to\infty}\left\{\frac{1}{n}\sum_{i=0}^{n-1}[y_i^{*\tau}Q_1y_i^* - b_{i+1}^\tau B(Q_2 + B^\tau SB)^{-1}B^\tau b_{i+1}]\right. \\ &+ \frac{1}{n}\sum_{i=0}^{n-1}[u_i + (Q_2 + B^\tau SB)^{-1}B^\tau(SAx_i + b_{i+1})]^\tau(Q_2 + B^\tau SB) \\ &\left.[u_i + (Q_2 + B^\tau SB)^{-1}B^\tau(SAx_i + b_{i+1})]\right\}. \end{aligned} \tag{8.108}$$

Hence to complete the proof it suffices to show that for u_n defined by (8.36)

$$\lim_{n\to\infty}\frac{1}{n}\sum_{i=0}^{n-1}\|u_i + (Q_2 + B^\tau SB)^{-1}B^\tau(SAx_i + b_{i+1})\|^2 = 0. \tag{8.109}$$

Since $v_n \xrightarrow[n\to\infty]{} 0$, by (8.100), (8.26) and (8.105) for (8.109) we need only to prove

$$\lim_{n\to\infty}\frac{1}{n}\sum_{i=0}^{n-1}\|L_i\widehat{x}_i + (Q_2 + B^\tau SB)^{-1}B^\tau SAx_i\|^2 = 0 \tag{8.110}$$

and

$$\lim_{n\to\infty}\frac{1}{n}\sum_{i=0}^{n-1}\|\widehat{d}_i + (Q_2 + B^\tau SB)^{-1}B^\tau b_{i+1}\|^2 = 0. \tag{8.111}$$

Notice that by (8.106), (8.107) and consistency of θ_n from (8.79) it is clear that

$$\frac{1}{n}\sum_{i=1}^{n}\|x_i - \widehat{x}_i\|^2 = 0. \tag{8.112}$$

Paying attention to (8.29) and using consistency of θ_n and stability of F we see

$$\|\widehat{b}_n - b_n\| \xrightarrow[n\to\infty]{} 0. \tag{8.113}$$

Then (8.110) and (8.111) immediately follow from (8.112) and (8.113) and consistency of θ_n. Thus we have established (8.109) and hence (8.40). The proof of the theorem is completed. Q.E.D.

Theorem 8.3 establishes the optimality of the adaptive control based on the SG algorithm. We now give the optimal adaptive control using the ELS algorithm for estimating θ.

Let θ_n be given by (4.13), (4.14), (4.18) and (4.20).

Corresponding to (8.32), (8.33) and (8.34) we define

$$\sigma_k = \sup\left\{ t > \tau_k : \sum_{i=\tau_k}^{j-1} \|y_i\|^2 \le (j-1)^{1+\delta} + \|y_{\tau_k}\|^2, \quad \forall j \in (\tau_k, t] \right\} \tag{8.114}$$

$$\tau_{k+1} = \inf\left\{ t > \sigma_k : \sum_{i=\tau_k}^{\sigma_k - 1} \|y_i\|^2 \le \frac{t^{1+\delta}}{2^k}; \quad \sum_{i=\sigma_k}^{t} \|y_i\|^2 \le \frac{t^{1+\delta}}{2^k}; \quad \sum_{i=0}^{t} \|\widehat{x}_i\|^2 \le t^{1+\delta/2} \right\} \tag{8.115}$$

and

$$\Lambda = \{ i : \ \|\widehat{u}_i^0\|^2 \le i^{1+\delta}, \quad i \ge 1 \} \tag{8.116}$$

where

$$\delta \in \left[0, \ \frac{1-2\varepsilon(t+1)}{2t+3}\right), \quad \varepsilon = \left[0, \ \frac{1}{2(t+1)}\right),$$

$$t = \max(p, q, r) + mp - 1.$$

Let $\{v_n\}$ be defined by (8.12) and let u_n^1 be defined by

$$u_n^1 \triangleq \widehat{B}_{1n}^{-1}(B_{1n} u_n - \theta_n^{\tau} \varphi_n) \tag{8.117}$$

where $\widehat{B}_{1n}$ satisfies (5.110) and (5.111).

We retain the definitions (8.36) and (8.37) for u_n and u_n^s respectively but with u_n' replaced by u_n^1 in (8.37).

Theorem 8.4. *For System (3.91)-(3.94) assume that*

1) $\{w_n, \mathcal{F}_n\}$ is a martingale difference sequence with

$$\sup_n E(\|w_n\|^\beta | \mathcal{F}_{n-1}) < \infty, \quad \beta > 2 \tag{8.118}$$

$$\lim_{n\to\infty} \frac{1}{n} \sum_{i=1}^{n} w_i w_i^{\tau} = R > 0; \tag{8.119}$$

2) $C^{-1}(z) - \frac{1}{2}I$ is strictly positive real and that Conditions 3)-5) of Theorem 8.3 are satisfied.

Then the adaptive control (8.36) with u_n^1 defined by (8.117) is optimal in the set U defined by (3.100):

$$\lim_{n\to\infty} \frac{1}{n}\sum_{i=0}^{n-1}[(y_i - y_i^*)^\tau Q_1(y_i - y_i^*) + u_i^\tau Q_2 u_i] = \min \tag{8.120}$$

and

$$\|\theta_n - \theta\|^2 = O\left(\frac{\log n}{n^{1-(t+1)\varepsilon}}\right) \quad a.s. \tag{8.121}$$

where θ_n is given by (4.13), (4.14), (4.18) and (4.20).

Proof. The proof is almost the same as that for Theorem 8.3. We now revisit the proof of Theorem 8.3 and point out modifications we should do for the present case.

In Step 1 we need apply Theorem 6.2 and Remark 6.1 rather than Theorem 6.3. For this instead of (8.44) we need to verify

$$\frac{1}{n}\sum_{i=0}^{n}\|y_i\|^2 = O(n^\delta), \quad \delta \in \left[0, \frac{1-2\varepsilon(t+1)}{2t+3}\right) \tag{8.122}$$

$$\varepsilon = \left[0, \frac{1}{2(t+1)}\right), \quad t = \max(p, q, r) + mp - 1.$$

In the case where $\sigma_k < \infty$, $\tau_{k+1} = \infty$ Theorem 5.4 (instead of 5.1) implies

$$\sum_{i=\sigma_k}^{n}\|y_i\|^2 = O(n).$$

If $\tau_k < \infty$, $\sigma_k < \infty$, then by (8.114) and (8.115) we derive

$$\sup_{\tau_k \le n < \sigma_k} \frac{1}{n^{1+\delta}}\sum_{i=1}^{n}\|y_i\|^2 < 4 \tag{8.123}$$

which corresponds to (8.45) with $\log^\delta n$ replaced by n^δ. Hence for (8.122) it suffices to show the boundedness of

$$\sup_{\sigma_k \le n < \tau_{k+1}} \frac{1}{n^{1+\delta}}\sum_{i=1}^{n}\|y_i\|^2. \tag{8.124}$$

For large k by (8.14) and Remark 5.3 it follows that

$$\sum_{i=1}^{n}\|y_i\|^2 = \sum_{i=1}^{\sigma_k - 1}\|y_i\|^2 + \|y_{\sigma_k}\|^2 + \sum_{i=\sigma_k+1}^{n}\|y_i\|^2$$

$$\le 4\sigma_k^{1+\delta} + \|y_{\sigma_k}\|^2 + 2\sum_{i=\sigma_k+1}^{n}\|w_i\|^2 + o(n)$$

and

$$\sup_{\sigma_k \le n < \tau_{k+1}} \frac{1}{n^{1+\delta}} \sum_{i=1}^{n} \|y_i\|^2 \le \frac{\|y_{\sigma_k}\|^2}{n^{1+\delta}} + c \tag{8.125}$$

where and hereafter c denotes a positive number that is independent of n but may vary from place to place.

We now estimate $\|y_{\sigma_k}\|^2$. It is clear that (8.52) remains unchanged, while (8.53) and (8.55) are valid with $\log^\delta \sigma_k$ replaced by σ_k^δ. Hence we have

$$\|y_{\sigma_k}\|^2 \le c\sigma_k^{1+\delta}$$

which incorporating (8.125) proves (8.124). Then Theorem 6.2 and Remark 6.1 lead to strong consistency of the ELS estimate θ_n.

In Steps 2 and 3 of the proof for Theorem 8.3 "$\log^\delta n$" should be replaced by "n^δ" and (8.57) is derived by Theorem 8.2 rather than 8.1. All conclusions of Steps 2, 3 and 4 remain valid. Thus (8.120) is verified.

Finally, (8.107) and the minimum-phase condition imply that (6.78) and (6.79) hold with $\delta = 0$. Hence (8.121) follows from Theorem 6.2 and Remark 6.1. Q.E.D.

Remark 8.1. In Theorems 8.3 and 8.4 stability of $B(z)$ has been used. If this condition is replaced by the stability of $A(z)$, then similar results are also obtainable (c.f. [CG5]). However, neither stability of $B(z)$ nor stability of $A(z)$ seems to be natural in the LQ control problem. More detailed analysis giving the convergence rate of (8.120) is available in [Gu1].

8.3 Connection Between Adaptive Controls for Tracking and Quadratic Cost

In the last section we have considered the quadratic loss function (8.22), which turns to a criterion for pure tracking, if $Q_1 = I$, $Q_2 = 0$. However, comparing (8.3) with (8.37) and (8.6) with (8.40) we find that the control laws as well as the minimum values of the loss functions for these two cases greatly differ from each other.

In order to connect the tracking with the quadratic cost control we put $Q_1 = I$ and $Q_2 = \varepsilon I$ with $\varepsilon > 0$ in (8.22) and rewrite (8.22) as

$$J^\varepsilon(u) = \limsup_{n\to\infty} \frac{1}{n} \sum_{i=0}^{n-1} [\|y_i - y_i^*\|^2 + \varepsilon\|u_i\|^2]. \tag{8.126}$$

By Theorem 8.3 or 8.4, for $u \in U$ with U given by (3.100) the minimum of (8.126) is

$$\inf_{u\in U} J^\varepsilon(u) \;=\; \limsup_{n\to\infty} \frac{1}{n} \sum_{i=0}^{n-1} [\|y_i^*\|^2 - b_{i+1}^{\varepsilon\tau}(\varepsilon I + B^\tau S^\varepsilon B)^{-1} B^\tau b_{i+1}^\varepsilon]$$

$$+trS^\varepsilon CRC^\tau \tag{8.127}$$

where S^ε is defined by (3.104) with Q_1, Q_2 replaced by I and εI, respectively, and b_i^ε is defined by

$$b_i^\varepsilon = -\sum_{j=0}^{\infty}(F^\varepsilon)^{j\tau} H^\tau y_{i+j}^* \tag{8.128}$$

with

$$F^\varepsilon = A - B(\varepsilon I + B^\tau SB)^{-1}B^\tau S^\varepsilon A \tag{8.129}$$

and with A, B and H given by (3.96) and (3.97).

By Theorem 8.1 or 8.2 the minimum tracking error is trR.

In this section we show that as $\varepsilon \to 0$ the minimum of the quadratic cost (8.127) tends to the minimum tracking error trR, i.e. the minimum value of $J^\varepsilon(u)$ is continuous at $\varepsilon = 0$.

Theorem 8.5 (Gul) , *[CG9]. For System (3.91)-(3.94) assume that Conditions 1) 3) and 4) of Theorem 8.3 are satisfied. Then as* $\varepsilon \to 0$

$$trS^\varepsilon CRC^\tau + \limsup_{n\to\infty}\frac{1}{n}\sum_{i=0}^{n-1}[\|y_i^*\|^2 - b_{i+1}^{\varepsilon\tau}(\varepsilon I + B^\tau S^\varepsilon B)^{-1}B^\tau b_{i+1}^\varepsilon] \to trR.$$

Proof. We prove that the first term on the right-hand side of (8.127) tends to trR, while the last term goes to zero.

By the last assertion of Theorem 3.3 we know that as $\varepsilon \to 0$ S^ε nonincreasingly converges to a finite limit S^0

$$S^\varepsilon \xrightarrow[\varepsilon \to 0]{} S^0. \tag{8.130}$$

By (3.62) we can express S^ε as

$$S^\varepsilon = A^\tau\left[(S^\varepsilon)^{-1} + \frac{1}{\varepsilon}BB^\tau\right]^{-1} A + H^\tau H \tag{8.131}$$

which implies that

$$B^\tau S^\varepsilon B \geq B^\tau H^\tau HB = B_1^\tau B_1 > 0$$

and

$$B^\tau S^0 B > 0. \tag{8.132}$$

Then we have

$$\|S^\varepsilon B(\varepsilon I + B^\tau S^\varepsilon B)^{-1}B^\tau S^\varepsilon - S^0B(B^\tau S^0B)^{-1}B^\tau S^0\| \xrightarrow[\varepsilon \to 0]{} 0. \tag{8.133}$$

Letting $\varepsilon \to 0$ in the equation satisfied by S^ε, from (8.133) we find that S^0 fulfills the following equation

$$S^0 = A^\tau S^0 A - A^\tau S^0 B(B^\tau S^0 B)^{-1} B^\tau S^0 A + H^\tau H. \tag{8.134}$$

Set

$$\begin{aligned} L^0 &= -(B^\tau S^0 B)^{-1} B^\tau S^0 A \\ L^* &= -(B^\tau H^\tau H B)^{-1} B^\tau H^\tau H A. \end{aligned}$$

Noticing that $HB = B_1$,

$$H(A + BL^*) = HA - HB(B^\tau H^\tau HB)^{-1} B^\tau H^\tau HA = 0 \tag{8.135}$$

and

$$(A + BL^0)^\tau S^0 BL^* = A^\tau S^0 BL^* - L^{0\tau} B^\tau S^0 BL^* = 0 \tag{8.136}$$

from (8.134) we find that

$$S^0 - H^\tau H = (A + BL^0)^\tau S^0 A = (A + BL^0)^\tau (S^0 - H^\tau H)(A + BL^*)$$

which implies

$$S^0 - H^\tau H = (A + BL^0)^{n\tau} (S^0 - H^\tau H)(A + BL^*)^n, \ \forall n \geq 1. \tag{8.137}$$

Note that

$$A + BL^* = A - BB_1^{-1} HA$$

$$= \begin{bmatrix} 0 & 0 & 0 & \dots & 0 \\ B_2 B_1^{-1} A_1 - A_2 & -B_2 B_1^{-1} & I & \ddots & \vdots \\ \vdots & \vdots & \ddots & \ddots & 0 \\ \vdots & \vdots & \ddots & \ddots & I \\ B_s B_1^{-1} A_1 - A_s & -B_s B_1^{-1} & \dots & \dots & 0 \end{bmatrix}$$

and

$$\begin{aligned} &\det(\lambda I - (A + BL^*)) \\ = \ & \lambda^m \det \left(\lambda I - \begin{bmatrix} B_2 B_1^{-1} & I & 0 & \dots & 0 \\ \vdots & \vdots & \ddots & \ddots & 0 \\ \vdots & \vdots & \ddots & \ddots & I \\ -B_s B^{-1} & 0 & \dots & \dots & 0 \end{bmatrix} \right) \\ = \ & \lambda^m \det \left(\lambda^{s-1} I + \lambda^{s-2} B_2 B_1^{-1} + \ \dots \ + B_s B_1^{-1} \right) \\ = \ & \det B_1^{-1} \lambda^m \det \left(\lambda^{s-1} B_1 + \lambda^{s-2} B_2 + \ \dots \ + B_s \right). \end{aligned} \tag{8.138}$$

By Condition 3 of Theorem 8.3 all zeros of the right-hand side of (8.138) are inside the open unit disk, i.e. $A + BL^*$ is stable. Therefore

$$\|(A+BL^*)^n\| = O(\gamma^n) \quad \text{for some} \quad \gamma \in (0,1). \tag{8.139}$$

From (8.130) and (8.132) we see that

$$A + BL^0 = \lim_{\varepsilon \to 0} F^\varepsilon \tag{8.140}$$

where F^ε is given by (8.129) and by Theorem 3.5 it is a stable matrix for any $\varepsilon > 0$. Hence all eigenvalues of $A + BL^0$ must be less than or equal to 1. Then we have

$$\|(A+BL^0)^n\| = O(n^{m_s}). \tag{8.141}$$

Consequently, from (8.137), (8.139), and (8.141) it follows that

$$\|S^0 - H^\tau H\| = O(n^{m_s})O(\gamma^n) \xrightarrow[n \to \infty]{} 0$$

and

$$S^0 = H^\tau H, \quad L^* = L^0.$$

This yields

$$trS^\varepsilon CRC^\tau \xrightarrow[\varepsilon \to 0]{} trR.$$

To complete the proof of the theorem it remains to show that the last term on the right-hand side of (8.127) tends to zero as $\varepsilon \to 0$.

From (8.128) we have

$$b^\varepsilon_{i+1} = -\sum_{j=0}^{\infty} (F^\varepsilon)^{j\tau} H^\tau y^*_{i+j+1} = H^\tau y^*_{i+1} - \sum_{j=1}^{\infty} (F^\varepsilon)^{j\tau} H^\tau y^*_{i+j+1}$$

and by (8.135)

$$\begin{aligned} &\sup_{i \geq 0} \|b^\varepsilon_{i+1} - H^\tau y^*_{i+1}\| \\ &= \sup_{i \geq 0} \left\| \sum_{j=1}^{\infty} \left[(F^\varepsilon)^{j\tau} - (A+BL^0)^{j\tau} \right] H^\tau y^*_{i+j+1} \right\| \end{aligned} \tag{8.142}$$

which tends to zero as $\varepsilon \to 0$ because $\{y^*_i\}$ is bounded, and by (8.140)

$$\sum_{j=1}^{n} \|(F^\varepsilon)^{j\tau} - (A+BL^0)^{j\tau}\| \xrightarrow[\varepsilon \to 0]{} 0$$

for any fixed n and

$$\sum_{j=n+1}^{\infty} \left[\|(F^\varepsilon)^j\| + \|(A+BL^0)^j\| \right]$$

can be made arbitrarily small for any $\varepsilon \in [0, a]$ with some small $a > 0$, if n is sufficiently large. The last assertion is true since we can take the same c and $\rho \in (0, 1)$ for any $\varepsilon \in [0, a]$ in the estimate

$$(F^\varepsilon)^j \leq c\rho^j,$$

using the fact that F^ε is stable and F^ε converges to the stable matrix $A + BL^0$.

Applying (8.142) and noticing that

$$HB(\varepsilon I + B^\tau S^\varepsilon B)^{-1} B^\tau H^\tau \xrightarrow[\varepsilon \to 0]{} I$$

we immediately conclude that

$$\begin{aligned}
&\sup_{i \geq 0} \|y_i^{*\tau} y_i^* - b_{i+1}^{\varepsilon\tau} B(\varepsilon I + B^\tau S^\varepsilon B)^{-1} B^\tau b_{i+1}^\varepsilon\| \\
= &\sup_{i \geq 0} \|y_i^{*\tau}(I - HB(\varepsilon I + B^\tau S^\varepsilon B)^{-1} B^\tau H^\tau) y_i^* \\
&+ y_i^{*\tau} HB(\varepsilon I + B^\tau S^\varepsilon B)^{-1} B^\tau H^\tau y_i^* \\
&- b_{i+1}^{\varepsilon\tau} B(\varepsilon I + B^\tau S^\varepsilon B)^{-1} B^\tau b_{i+1}^\varepsilon\| \xrightarrow[\varepsilon \to 0]{} 0.
\end{aligned}$$

This completes the proof. Q.E.D.

We have shown the continuity of the minimum value of the quadratic loss function as the weighting matrix εI for control goes to zero. However, it is still not known if the control law itself converges in some sense.

8.4 Model Reference Adaptive Control With Consistent Estimate

We now consider the system described by (5.152) and (5.153), for which the coefficient

$$\theta^\tau = [-A_1 \ \dots \ -A_p \quad B_1 \ \dots \ B_q \quad C_1 \ \dots \ C_r] \tag{8.143}$$

is unknown while the orders (p, q, r) and the polynomial $D(z)$ are assumed to be known.

Let

$$A^0(z) y_n = B^0(z) y_n^* \tag{8.144}$$

be the reference model where $A^0(z)$, $B^0(z)$ are given matrix polynomials and $\{y_n^*\}$ is a bounded reference input with y_n^* being $\mathcal{F}_{n-d}$ measurable.

The problem discussed here is to design adaptive control that reduces (5.152) as close to (8.144) as possible, meanwhile to estimate θ consistently. To be precise, the loss function to be minimized is

$$\limsup_{n \to \infty} \frac{1}{n} \sum_{i=0}^{n} \left(A^0(z) y_i - B^0(z) y_i^*\right) \left(A^0(z) y_i - B^0(z) y_i^*\right)^\tau. \tag{8.145}$$

The unknown coefficient θ is estimated by the ELS algorithm:

$$\theta_{n+1} = \theta_n + a_n P_n \varphi_n (D(z) y_{n+1}^T - \varphi_n^T \theta_n) \quad (8.146)$$
$$P_{n+1} = P_n - a_n P_n \varphi_n \varphi_n^T P_n, \quad a_n = (1 + \varphi_n^T P_n \varphi_n)^{-1}, \quad (8.147)$$
$$P_0 = \alpha_0 I, \quad \frac{1}{e} > \alpha_0 > 0$$

$$\begin{aligned}\varphi_n^T = {} & [D(z)y_n^T \dots D(z)y_{n-p+1}^T \quad D(z)u_{n-d+1}^T \dots D(z)u_{n-d-q+2}^T \\ & D(z)y_n^T - \varphi_{n-1}^T\theta_n \dots D(z)y_{n-r+1}^T - \varphi_{n-r}^T\theta_{n-r+1}]. \end{aligned} \quad (8.148)$$

Comparing the problem with the one considered in Section 5.5 we find that (8.145) is the special case of (5.154) with $Q(z) \equiv 0$. However, in Section 5.5 we did not concern consistency of the estimate for θ, while here we want not only to minimize (8.145) but also to consistently estimate θ.

Set

$$A_n(z) = I + A_{1n} z + \dots + A_{pn} z^p, \quad (8.149)$$
$$B_n(z) = B_{1n} + B_{2n} z + \dots + B_{qn} z^{q-1}, \quad (8.150)$$
$$C_n(z) = I + C_{1n} z + \dots + C_{rn} z^r \quad (8.151)$$

where the coefficients of $A_n(z)$, $B_n(z)$ and $C_n(z)$ are given by the estimate θ_n

$$\theta_n^T = [-A_{1n} \dots - A_{pn} \quad B_{1n} \dots B_{qn} \quad C_{1n} \dots C_{rn}].$$

Further, corresponding to (3.120) we assume that $F_n(z) (= I + F_{1n} z + \dots + F_{d-1n} z^{d-1})$ and $G_n(z)$ are the solution of the Diophantine equation

$$(\det C_n(z)) I = F_n(z)(\mathrm{Adj} C_n(z)) A_n(z) D(z) + G_n(z) z^d \quad (8.152)$$

and $N_n(z)$ is the matrix polynomial satisfying

$$A^0(z) F_n(z) = M_{0n} + M_{1n} z + \dots + M_{d-1n} z^{d-1} + N_n(z) z^d. \quad (8.153)$$

Using (5.152) and (5.153) we can rewrite the equation (3.143) satisfied by the optimal control as follows

$$\begin{aligned}&(\det C(z))(N(z)w_n - B^0(z)y_{n+d}^*) \\ &+ A^0(z)[G(z)y_n + F(z)(\mathrm{Adj}C(z))B(z)D(z)u_n] = 0.\end{aligned} \quad (8.154)$$

Then according to the certainty equivalency principle, the adaptive control may be defined as the solution of the following equation:

$$\begin{aligned}&(\det C_n(z))(N_n(z)w_n - B^0(z)y_{n+d}^*) \\ + \; & A^0(z)[G_n(z)y_n + F_n(z)(\mathrm{Adj}C_n(z))B_n(z)D(z)u_n] = 0.\end{aligned} \quad (8.155)$$

Replacing w_n in (8.155) by its estimate $D(z)y_n - \theta_n^{\tau}\varphi_{n-1}$ we get

$$\begin{aligned} & (\det C_n(z))[N_n(z)(D(z)y_n - \theta_n^{\tau}\varphi_{n-1}) - B^0(z)y_{n+d}^*] \\ + \quad & A^0(z)[G_n(z)y_n + F_n(z)(\mathrm{Adj}C_n(z))B_n(z)D(z)u_n] = 0 \end{aligned}$$

which is equivalent to

$$\begin{aligned} & A^0(z)B_{1n}D(z)u_n \\ = \quad & -(\det C_n(z))[N_n(z)(D(z)y_n - \theta_n^{\tau}\varphi_{n-1}) - B^0(z)y_{n+d}^*] \\ & -A^0(z)[G_n(z)y_n + F_n(z)(\mathrm{Adj}C_n(z))B_n(z)D(z)u_n \\ & \quad - B_{1n}D(z)u_n]. \end{aligned} \tag{8.156}$$

Note that the last term does not contain u_n.

By the diminishing excitation technique developed in Chapter 6 we can use (8.156) to define the desired control. Let us denote it by $u_n^{(2)}$. To be specific, with arbitrary initial values $u_i^{(2)}$, $0 \le i \le s-1$, $u_n^{(2)}$ is defined by

$$\begin{aligned} & B_{1n}D(z)u_n^{(2)} + [A^0(z) - I]B_{1n}D(z)u_n \\ = \quad & -(\det C_n(z))[N_n(z)(D(z)y_n - \theta_n^{\tau}\varphi_{n-1}) - B^0(z)y_{n+d}^*] \\ & -A^0(z)[G_n(z)y_n + F_n(z)(\mathrm{Adj}C_n(z))B_n(z)D(z)u_n \\ & -B_{1n}D(z)u_n], \end{aligned} \tag{8.157}$$

if $\det B_{1n} \ne 0$ a.s., otherwise, we simply set $u_n^{(2)} = 0$.

As required in Theorem 6.2, to guarantee strong consistency of the parameter estimates the input and output of the system should not grow too fast. For this we will invoke the control equation (5.158). To be specific, assume $Q(z) \equiv 0$ in (5.157) but retain all notations like $G_0(z)$, $G_1(z)$, $G_2(z)$, θ_n^* and φ_n^* unchanged. Let $\{v_n\}$ be defined by (6.66) and (6.67) and be independent of $\{y_n^*\}$.

By Lemma 6.1 we can define u_n such that

$$\theta_n^{*\tau}\varphi_n^* - G_{20n}v_n = 0, \quad \forall n \ge 0 \tag{8.158}$$

if $\det G_{210} \ne 0$ where θ_n^* and φ_n^* are given by (5.160) (5.161) and G_{20n} is the estimate given by θ_n^* for the first coefficient G_{20} of $G_2(z)$.

Set

$$u_n^{(1)} = u_n - v_n$$

with u_n given by (8.158).

Define the desired control

$$u_n^o = \begin{cases} u_n^{(2)}, & \text{if } n \in [\tau_k, \ \sigma_k)\bigcap\Lambda \text{ for some } k, \\ -\sum_{i=1}^{s}\alpha_i u_{n-i}, & \text{if } n \in [\tau_k, \ \sigma_k)\bigcap\Lambda^c \text{ for some } k, \\ u_n^{(1)}, & \text{if } n \in [\sigma_k, \ \tau_{k+1}) \text{ for some } k \end{cases} \tag{8.159}$$

and the adaptive control u_n as the disturbed version of u_n^o:

$$u_n = u_n^o + v_n, \tag{8.160}$$

where

$$\Lambda = \{j : \ \|u_j^{(2)} + \sum_{i=1}^{s} \alpha_i u_{j-i}\|^2 \le j^{1+\delta}\}, \tag{8.161}$$

$$\delta \in \left[0, \ \frac{1-2\varepsilon(t+1)}{2t+3}\right), \quad \varepsilon \in \left(0, \ \frac{1}{2(t+1)}\right),$$

$$t = \max(p, q, r) + mp - 1,$$

and $\{\tau_k\}$ and $\{\sigma_k\}$ are stopping times defined as follows

$$\begin{aligned} 1 &= \tau_1 < \sigma_1 < \tau_2 < \sigma_2 < \dots \\ \sigma_k &= \sup\left\{t > \tau_k : \sum_{i=\tau_k}^{j-1} \|y_i\|^2 \le (j-1)^{1+\delta/2} + \|y_{\tau_k}\|^2, \right. \\ &\qquad \left. \forall j \in (\tau_k, t]\right\} \end{aligned} \tag{8.162}$$

$$\begin{aligned} \tau_{k+1} &= \inf\left\{t > \sigma_k : \sum_{i=\sigma_k}^{t} \|y_i\|^2 \le \frac{t\log t}{2^k}; \ \sum_{i=\tau_k}^{\sigma_k - 1} \|y_i\|^2 \le \frac{t\log t}{2^k}; \right. \\ &\qquad \sum_{i=\sigma_k}^{t} \|D(z)u_i\|^2 \le \frac{t\log t}{2^k}; \\ &\qquad \left. \sum_{i=\tau_k}^{\sigma_k - 1} \|D(z)u_i\|^2 \le \frac{t\log t}{2^k}; \right\}. \end{aligned} \tag{8.163}$$

Similar to Lemma 6.1 we can show that $\det B_{1n} \neq 0$ a.s . $\forall n > 0$ if $\det B_{10} \neq 0$. Hence (8.157) is solvable with respect to $u_n^{(2)}$.

Theorem 8.6 (CZ2) . *Assume that*

1) (5.14) and (5.15) hold;

2) both $\det C(z) - \frac{a}{2}$ *and* $C^{-1}(z) - \frac{1}{2}I$ *are SPR;*

3) $B(z)$ *and* $A^0(z)$ *are stable with* $A^0(0) = I$*;*

4) $A(z)$, $B(z)$ *and* $C(z)$ *have no common left-factor and* $[A_p \ B_q \ C_r]$ *is of row-full-rank.*

Then adaptive control (8.160) leads to that

a) the system is stabilized in the sense that

$$\limsup_{n\to\infty} \frac{1}{n}\sum_{i=0}^{n} (\|y_i\|^2 + \|D(z)u_i\|^2) < \infty \quad a.s. \tag{8.164}$$

b) the cost is minimized, i.e.

$$\begin{aligned}&\lim_{n\to\infty} \frac{1}{n}\sum_{i=0}^{n}(A^0(z)y_i - B^0(z)y_i^*)(A^0(z)y_i - B^0(z)y_i^*)^\tau \\ &= \sum_{j=0}^{d-1} M_j R M_j^\tau \end{aligned}\tag{8.165}$$

with the rate of convergence

$$\begin{aligned}&\Bigg\| \frac{1}{n}\sum_{i=0}^{n}(A^0(z)y_i - B^0(z)y_i^*)(A^0(z)y_i - B^0(z)y_i^*)^\tau \\ &-\frac{1}{n}\sum_{i=0}^{n}(M(z)w_i)(M(z)w_i)^\tau \Bigg\| = O\left(n^{-\frac{\varepsilon}{2}}\right)\end{aligned} \tag{8.166}$$

c) the parameter estimates are strongly consistent with the rate of convergence

$$\|\theta - \theta_n\|^2 = O\left(\frac{(\log n)(\log\log n)^{\delta(\beta-2)}}{n^{1-(t+1)\varepsilon}}\right), \tag{8.167}$$

where

$$\delta(x) = \begin{cases} 0, & x > 0 \\ c, & x = 0 \end{cases}, \quad c > 1 \text{ is arbitrary but greater than } 1.$$

Proof. We first show

$$\sum_{i=0}^{n} \|y_i\|^2 = O\left(n^{\delta}\right). \tag{8.168}$$

If $\tau_k < \infty$ and $\sigma_k = \infty$ for some k, then (8.168) is automatically satisfied because of (8.162). Also, if $\sigma_k < \infty$ and $\tau_{k+1} = \infty$ for some k, then (8.168) holds by Theorem 5.5. Thus, we need only to consider the case where $\tau_k < \infty$ and $\sigma_k < \infty$ for all k.

Similar to (8.123) or (8.45) we have

$$\sup_{\tau_k \le n < \sigma_k} \frac{1}{n^{1+\delta}} \sum_{i=1}^{n} \|y_i\|^2 < 4 \tag{8.169}$$

and

$$\begin{aligned}
&\sum_{i=0}^{\tau_k} \|D(z)u_i\|^2 \\
= &\left(\sum_{i=\tau_1}^{\sigma_1 - 1} \|D(z)u_i\|^2 + \sum_{i=\sigma_1}^{\tau_2 - 1} \|D(z)u_i\|^2 + \ \dots \ + \sum_{i=\sigma_{k-1}}^{\tau_k} \|D(z)u_i\|^2 \right) \\
\leq &\ 2\sum_{i=1}^{k-1} \frac{\tau_{i+1} \log \tau_{i+1}}{2^i} \leq 2\tau_k \log \tau_k, \qquad (8.170)
\end{aligned}$$

$$\sum_{i=0}^{\tau_k} \|y_i\|^2 \leq 2\tau_k \log \tau_k. \qquad (8.171)$$

We now show

$$\sup_{\sigma_k \leq n < \tau_{k+1}} \frac{1}{n^{1+\delta}} \sum_{i=1}^{n} \|y_i\|^2 = O(1) \qquad (8.172)$$

which incorporating (8.169) implies

$$\frac{1}{n} \sum_{i=1}^{n} \|y_i\|^2 = O(n^\delta). \qquad (8.173)$$

We prove (8.172) for $n \in [\sigma_k, \ \sigma_k + d - 1] \cap [\sigma_k, \ \tau_{k+1})$ and $n \in [\sigma_k + d, \ \tau_{k+1})$ separately.

If $n \in [\sigma_k, \ \sigma_k + d - 1] \cap [\sigma_k, \ \tau_{k+1})$, then by (8.169) we have

$$\sum_{i=1}^{n} \|y_i\|^2 = O(\sigma_k^{1+\delta}) + \sum_{i=\sigma_k}^{n} \|y_i\|^2 = O(n^{1+\delta}) + \sum_{i=\sigma_k}^{n} \|y_i\|^2. \qquad (8.174)$$

By Condition 2) det $C(z)$ is SPR and by Definition in Section 4.2 det $C(z)$ is stable. Then by (5.15) and (3.146) it follows that

$$\sum_{i=\sigma_k}^{n} \|y_i\|^2 = O(n) + O\left(\sum_{i=0}^{n-d} \|y_i\|^2 + \sum_{i=0}^{n-d} \|D(z)u_i\|^2 \right). \qquad (8.175)$$

Note that for $n \in [\sigma_k, \ \sigma_k + d - 1] \cap [\sigma_k, \ \tau_{k+1})$ by (8.169) we have

$$\sum_{i=0}^{n-d} \|y_i\|^2 = O(n^{1+\delta})$$

hence from (8.175)

$$\sum_{i=\sigma_k}^{n} \|y_i\|^2 = O(n^{1+\delta}) + O\left(\sum_{i=0}^{n-d} \|D(z)u_i\|^2\right). \tag{8.176}$$

If $\tau_k \geq \sigma_k - d$, then noticing $n - d \leq \sigma_k - 1$ we have $n - d - \tau_k \leq \sigma_k - 1 - \tau_k \leq d - 1$.

For any i: $\tau_k + 1 \leq i \leq n - d$ $(\leq \sigma_k - 1)$ by (8.159) we see that

$$\|D(z)u_i\|^2 \leq \begin{cases} 2\left\|u_i^{(2)} + \sum_{j=1}^{s} \alpha_j u_{i-j}\right\|^2 + 2\|v_i\|^2 \\ \quad \leq 2i^{1+\delta} + 2\|v_i\|^2 = O(i^{1+\delta}), \text{ if } i \in \Lambda \\ \left\|(u_i^s + v_i) + \sum_{j=1}^{s} \alpha_j u_{i-j}\right\|^2 = \|v_i\|^2 = o(1), \quad \text{if } i \in \Lambda^c \end{cases} \tag{8.177}$$

and hence by (8.170)

$$\begin{aligned} \sum_{i=0}^{n-d} \|D(z)u_i\|^2 &= \sum_{i=0}^{\tau_k} \|D(z)u_i\|^2 + \sum_{i=\tau_k+1}^{n-d} \|D(z)u_i\|^2 \\ &= O(n^{1+\delta}). \end{aligned} \tag{8.178}$$

If $\tau_k < \sigma_k - d$, then by stability of $B(z)$ we find that

$$\begin{aligned} &\sum_{i=0}^{n-d} \|D(z)u_i\|^2 \\ &= \sum_{i=0}^{\sigma_k-d-1} \|D(z)u_i\|^2 + \sum_{i=\sigma_k-d}^{n-d} \|D(z)u_i\|^2 \\ &\leq O\left(\sum_{i=0}^{\sigma_k-1} \|y_i\|^2 + \sigma_k\right) + \sum_{i=\sigma_k-d}^{n-d} \|D(z)u_i\|^2, \end{aligned}$$

which by (8.169) and (8.177) yields

$$\sum_{i=0}^{n-d} \|D(z)u_i\|^2 = O(n^{1+\delta}). \tag{8.179}$$

Combining (8.178), (8.179) and (8.176) we are convinced of that

$$\sum_{i=\sigma_k}^{n} \|y_i\|^2 = O(n^{1+\delta}), \quad \forall n \in [\sigma_k,\ \sigma_k + d - 1] \cap [\sigma_k,\ \tau_{k+1})$$

which together with (8.174) implies that (8.172) is valid for any $n \in [\sigma_k, \sigma_k + d-1] \cap [\sigma_k, \tau_{k+1})$.

We now show (8.172) for $n \in [\sigma_k + d, \tau_{k+1})$.

Write

$$\sum_{i=0}^{n} \|y_i\|^2 = \sum_{i=0}^{\sigma_k+d-1} \|y_i\|^2 + \sum_{i=\sigma_k+d}^{n} \|y_i\|^2, \tag{8.180}$$

where the first term on the right-hand side has just been shown to be of order $O(n^{1+\delta})$. So for (8.172) we need only to prove that the last term of (8.180) is of order $O(n^{1+\delta})$.

Noticing that for $n \in [\sigma_k, \tau_{k+1})$ from (8.159) and (8.158) it follows that $\theta_n^{*\tau} \varphi_n^* = G_{20n} v_n$, from (5.156) with $Q(z) \equiv 0$ and (5.164) we find

$$\begin{aligned} &\sum_{i=\sigma_k+d}^{n} \|A^0(z) y_i\|^2 \\ = &\sum_{i=\sigma_k+d}^{n} \|B^0(z) y_i^* + M(z) w_i + G_{20i-d} v_{i-d} + z_{i-d}\|^2 \end{aligned} \tag{8.181}$$

where z_i is defined by (5.164).

Note that G_{20i} is bounded by (5.163). Then by (5.174), the boundedness of $\{y_i^*\}$ and (5.15) it is easy to see that the right-hand side of (8.181) is of order $O(n)$. Using stability of $A^0(z)$ we then conclude that

$$\sum_{i=\sigma_k+d}^{n} \|y_i\|^2 = O(n). \tag{8.182}$$

This combining (8.180) proves (8.173). Then from Theorem 6.2 it follows that

$$\|\theta - \theta_n\|^2 = O\left(\frac{(\log n)(\log\log n)^{\delta(\beta-2)}}{n^{1-(t+1)(\varepsilon+\delta)}} \right). \tag{8.183}$$

Since B_1 is nonsingular by Condition 3) and $B_{1n} \xrightarrow[n\to\infty]{} B_1$ a.s., we then have $B_{1n} \neq 0$ starting from some n a.s.. Then for any sample after a finite number of steps $u_n^{(2)}$ is purely defined by (8.157).

We now show that there is an integer k (possibly depending upon sample) so that

$$\tau_k < \infty, \quad \sigma_k = \infty \quad \text{a.s.} \tag{8.184}$$

If for some k, $\sigma_k < \infty$ and $\tau_{k+1} = \infty$, then Theorem 5.5 applies and (5.162) contradicts $\tau_{k+1} = \infty$. So for (8.184) it suffices to show the impossibility of that $\tau_k < \infty$, $\sigma_k < \infty$ for all $k \geq 0$.

Define

$$t_k = \sup\{n : \ j \in [\tau_k, \sigma_k) \cap \Lambda, \ \forall j \in [\tau_k, n]\}. \tag{8.185}$$

The existence of t_k for sufficiently large k is justified because of the definition (8.161) for Λ and

$$\|D(z)u_{\tau_k}\|^2 \leq \frac{\tau_k \log \tau_k}{2^{k-1}} \tag{8.186}$$

from (8.163).

By stability of $\det C(z)$ and (8.170), (8.171) from (3.146) we have

$$\sum_{i=0}^{\tau_k+d} \|y_i\|^2 = O(\tau_k \log \tau_k) \tag{8.187}$$

and hence

$$\sum_{i=0}^{\tau_k+d} \|A^0(z)y_i\|^2 = O(\tau_k \log \tau_k). \tag{8.188}$$

Set

$$\begin{aligned} \varphi_n^0 &= [D(z)y_n^T \; \dots \; D(z)y_{n-p+1}^T \quad D(z)u_{n-d+1}^T \; \dots \; D(z)u_{n-d-q+2}^T \\ &\quad w_n^T \; \dots \; w_{n-r+1}^T]^T, & (8.189) \\ \xi_n &= D(z)y_n - \theta_n^T \varphi_{n-1} - w_n. & (8.190) \end{aligned}$$

It is clear that Theorems 4.1, 4.2 remain true if we identify φ_n φ_n^0 and ξ_n defined by (8.148), (8.189) and (8.190) to those used in Chapter 4. Then by (4.62)

$$\sum_{i=0}^{n} \|\xi_i\|^2 = O\left((\log r_n)(\log\log r_n)^{\delta(\beta-2)}\right). \tag{8.191}$$

This together with (8.168) by stability of $B(z)$ implies

$$\sum_{i=0}^{n} \|\xi_i\|^2 = O(\log^2 n). \tag{8.192}$$

In the sequel for any matrix polynomials $F_n(z)$ and $B_n(z)$ with coefficients depending on n by $(F_n B_n)(z)$ we mean the product $F_n(z)$ and $B_n(z)$, while

$$(F_n B)_n(z) = \sum_{i,j} B_{in} F_{jn-i} z^{i+j} \tag{8.193}$$

where B_{in} (or F_{in}) denotes the matrix coefficient for z^i in $B_n(z)$ (or in $F_n(z)$).

From (5.152) (5.153) we have

$$\begin{aligned} D(z)y_{n+d} &= \theta^T \varphi_{n+d-1}^0 + w_{n+d} \\ &= \tilde{\theta}_n^T \varphi_{n+d-1}^0 + w_{n+d} + \theta_n^T \varphi_{n+d-1}^0 & (8.194) \end{aligned}$$

or

$$\widetilde{\theta}_n^{\tau}\varphi_{n+d-1}^0 + C_n(z)w_{n+d} = A_n(z)D(z)y_{n+d} - B_n(z)D(z)u_n. \tag{8.195}$$

Note that (8.157) and (8.190) imply

$$\begin{aligned} & A^0(z)[G_n(z)y_n + (F_n(\mathrm{Adj}C_n)B_n)(z)D(z)u_n] \\ = \ & B_{1n}D(z)v_n \\ & +(\det C_n(z))[B^0(z)y_{n+d}^* - N_n(z)\xi_n - N_n(z)w_n]. \end{aligned} \tag{8.196}$$

Multiplying both sides of (8.195) by $A^0(z)(F_n(\mathrm{Adj}C_n))(z)$ and using (8.152) we obtain

$$\begin{aligned} & A^0(z)(F_n(\mathrm{Adj}C_n))(z)(\widetilde{\theta}_n^{\tau}\varphi_{n+d-1}^0 + C_n(z)w_{n+d}) \\ = \ & A^0(z)[(\det C_n(z))I - G_n(z)z^d]y_{n+d} \\ & +A^0(z)[((F_n(\mathrm{Adj}C_n))A)_n(z) - (F_n(\mathrm{Adj}C_n)A_n)(z)]D(z)y_{n+d} \\ & -A^0(z)((F_n(\mathrm{Adj}C_n))B)_n(z)D(z)u_n \\ = \ & A^0(z)(\det C_n(z))y_{n+d} - A^0(z)G_n(z)y_n \\ & +A^0(z)[((F_n(\mathrm{Adj}C_n))A)_n(z) - (F_n(\mathrm{Adj}C_n)A_n)(z)]D(z)y_{n+d} \\ & -A^0(z)(F_n(\mathrm{Adj}C_n)B_n)(z)D(z)u_n \\ & +A^0(z)[(F_n(\mathrm{Adj}C_n)B_n)(z) - ((F_n(\mathrm{Adj}C_n))B)_n(z)]D(z)u_n. \end{aligned}$$

Substituting (8.196) into this leads to

$$\begin{aligned} & A^0(z)(F_n(\mathrm{Adj}C_n))(z)(\widetilde{\theta}_n^{\tau}\varphi_{n+d-1}^0 + C_n(z)w_{n+d}) \\ = \ & A^0(z)(\det C_n(z))y_{n+d} - B_{1n}D(z)v_n \\ & -(\det C_n(z))(B^0(z)y_{n+d}^* - N_n(z)\xi_n - N_n(z)w_n) \\ & +A^0(z)[((F_n(\mathrm{Adj}C_n))A)_n(z) - (F_n(\mathrm{Adj}C_n)A_n)(z)]D(z)y_{n+d} \\ & +A^0(z)[(F_n(\mathrm{Adj}C_n)B_n)(z) - ((F_n(\mathrm{Adj}C_n))B)_n(z)]D(z)u_n. \\ = \ & (\det C(z))[A^0(z)y_{n+d} - B^0(z)y_{n+d}^*] \\ & +A^0(z)[\det C_n(z) - \det C(z)]y_{n+d} - B_{1n}D(z)v_n \\ & +[\det C(z) - \det C_n(z)]B^0(z)y_{n+d}^* + (\det C_n(z))(N_n(z)\xi_n) \\ & +(\det C_n(z))(N_n(z)w_n) \\ & +A^0(z)[((F_n(\mathrm{Adj}C_n))A)_n(z) - (F_n(\mathrm{Adj}C_n)A_n)(z)]D(z)y_{n+d} \\ & +A^0(z)[(F_n(\mathrm{Adj}C_n)B_n)(z) \\ & \qquad - ((F_n(\mathrm{Adj}C_n))B)_n(z)]D(z)u_n. \end{aligned} \tag{8.197}$$

By (3.135) we express the second term on the left-hand side of (8.197) as

$$\begin{aligned} & A^0(z)((F_n(\mathrm{Adj}C_n))C)_n(z)w_{n+d} \\ = \ & (\det C(z))M(z)w_{n+d} + (\det C(z))N(z)w_n \\ & +A^0(z)[((F_n(\mathrm{Adj}C_n))C)_n(z) - F(z)(\mathrm{Adj}C(z))C(z)]w_{n+d}. \end{aligned}$$

Using this from (8.197) we derive

$$\begin{aligned}&(\det C(z))[A^0(z)y_{n+d}-B^0(z)y^*_{n+d}]\\ =\ &(\det C(z))M(z)w_{n+d}+e_n,\end{aligned} \tag{8.198}$$

where

$$\begin{aligned} e_n \ =\ & A^0(z)(F_n(\mathrm{Adj}C_n))(z)(\widetilde{\theta}^\tau_n\varphi^0_{n+d-1})+B_{1n}D(z)v_n\\ &+[(\det C(z))N(z)-((\det C_n)N)_n(z)]w_n\\ &-(\det C_n(z))(N_n(z)\xi_n)\\ &-A^0(z)[\det C_n(z)-\det C(z)]y_{n+d}\\ &-[\det C(z)-\det C_n(z))B^0(z)y^*_{n+d}\\ &-A^0(z)[((F_n(\mathrm{Adj}C_n))A)_n(z)-(F_n(\mathrm{Adj}C_n)A_n)(z)]D(z)y_{n+d}\\ &-A^0(z)[(F_n(\mathrm{Adj}C_n)B_n)(z)-((F_n(\mathrm{Adj}C_n))B)_n(z)]D(z)u_n\\ &+A^0(z)[((F_n(\mathrm{Adj}C_n))C)_n(z)-F(z)(\mathrm{Adj}C(z))C(z)]w_{n+d}.\end{aligned}$$

Paying attention to (8.183), (8.192) and (6.66) we find that

$$\begin{aligned}&\sum_{i=0}^{n}\|e_i\|^2=O(n^{1-\varepsilon})\\ &+O\left(\sum_{i=1}^{n+d}\frac{\log^2 i}{i^\beta}(\|\varphi^0_i\|^2+\|w_i\|^2+\|y_i\|^2+1)\right)\end{aligned} \tag{8.199}$$

where $\beta=1-(t+1)(\varepsilon+\delta)$.

We now show that the second term of (8.199) is also of order $O(n^{1-\varepsilon})$. Set

$$s_i=\sum_{j=1}^{i}(\|\varphi^0_j\|^2+\|w_j\|^2+\|y_j\|^2+1),\quad s_i=0,\ i<1.$$

By (5.15), (8.168) and stability of $B(z)$ we know that $s_i=O(i^{1+\delta})$. Then, noticing

$$\begin{aligned}&(\log^2 i)(1+i)^\beta-i^\beta\log^2(1+i)\\ =\ & i^\beta(\log^2 i)\left[\left(1+\frac{1}{i}\right)^\beta-\left(1+\frac{\log\left(1+\frac{1}{i}\right)}{\log i}\right)^2\right]\\ =\ & i^\beta(\log^2 i)\left(\frac{\beta}{i}+o\left(\frac{1}{i}\right)\right)\end{aligned}$$

we find that

$$\begin{aligned}
&\sum_{i=1}^{n}\frac{\log^2 i}{i^\beta}(\|\varphi_i^0\|^2+\|w_i\|^2+\|y_i\|^2+1)\\
&= \frac{\log^2 n}{n^\beta}s_n+\sum_{i=1}^{n-1}\left(\frac{\log^2 i}{i^\beta}-\frac{\log^2(i+1)}{(i+1)^\beta}\right)s_i\\
&= O\left(\frac{\log^2 n}{n^\beta}n^{1+\delta}\right)+O\left(\sum_{i=1}^{n-1}\frac{\log^2 i}{i(i+1)^\beta}i^{1+\delta}\right)\\
&= O(n^{1-\varepsilon})
\end{aligned}$$

and

$$\sum_{i=1}^{n}\|e_i\|^2=O(n^{1-\varepsilon}). \tag{8.200}$$

Since $\det C(z)$ is stable, from (8.198) and (8.200) it follows that

$$\sum_{i=\tau_k}^{t_k}\|A^0(z)y_{i+d}\|^2=O(t_k) \tag{8.201}$$

which incorporating with (8.188), by stability of $A^0(z)$, implies

$$\sum_{i=0}^{t_k}\|y_{i+d}\|^2=O(t_k\log t_k). \tag{8.202}$$

From (8.157) it is easy to see that

$$\begin{aligned}
\|D(z)u_{t_k+1}^{(2)}\|^2 &= O\left(t_k+\sum_{i=0}^{t_k+1}\|y_i\|^2+\sum_{i=0}^{t_k}\|D(z)u_i\|^2\right)\\
&= O\left(t_k+\sum_{i=0}^{t_k+d}\|y_i\|^2\right)=O(t_k\log t_k).
\end{aligned} \tag{8.203}$$

By the definition of t_k from (8.203) we know that

$$t_k=\sigma_k-1. \tag{8.204}$$

Hence from (8.202) we have

$$\sum_{i=0}^{\sigma_k}\|y_i\|^2=O(\sigma_k\log\sigma_k).$$

This means that for all sufficiently large k

$$\sum_{i=\tau_k}^{\sigma_k} \|y_i\|^2 \leq \sigma_k^{1+\delta/2}, \tag{8.205}$$

which contradicts the definition (8.162) for σ_k.

The obtained contradiction shows the impossibility of $\tau_k < \infty, \sigma_k < \infty$ for all k.

From (8.185) and (8.204) it is clear that $\Lambda \supset [\tau_k, \sigma_k)$ for all sufficiently large k. Therefore, (8.184) implies that Λ^c is a finite set.

Now let k be large enough so that $\tau_k < \infty$, $\sigma_k = \infty$. Then (8.198) is valid for all $n \geq \tau_k$, and from (8.198), similar to (3.96)-(3.99), we have

$$\begin{aligned} & A^0(z)y_{n+d} - B^0(z)y^*_{n+d} \\ = \; & M(z)w_{n+d} + HC^{n-\tau_k}x_{\tau_k+d} \\ & +H \sum_{j=\tau_k+d}^{n} C^{n-j}H^\tau e_j \end{aligned} \tag{8.206}$$

where

$$C = \underbrace{\begin{bmatrix} -c_1 I & I & 0 & \dots & 0 \\ \vdots & \vdots & \ddots & \ddots & 0 \\ \vdots & \vdots & \ddots & \ddots & I \\ -c_{mr} I & 0 & \dots & \dots & 0 \end{bmatrix}}_{m^2 r}, \quad H = \underbrace{[I \; 0 \; \dots \; 0]}_{m^2 r}, \quad I = I_{m\times m},$$

c_i, $i = 1, ..., mr$ are coefficients of $\det C(z)$, i.e.

$$\det C(z) = 1 + c_1 z + ... + c_{mr} z^{mr}$$

and x_{τ_k} is defined by $\{y_i, \; y_i^*, \; w_i, \; i \leq \tau_k\}$.

By Condition 2) of the theorem $\det C(z)$ is SPR; hence it is stable by Lemma 4.1. Then, as shown by (5.19) C^n exponentially tends to zero as $n \to \infty$. Therefore, applying Theorem 2.8 to (8.206) and paying attention to (8.200) we readily derive (8.166). It is clear that (8.165) follows from (8.166) by Theorem 2.8 and (5.15), while (8.164) follows from (8.165) by using the stability of $A^0(z)$ and $B^0(z)$. Hence, (8.168) and consequently, (8.183) are valid for $\delta = 0$. This proves (8.167). Q.E.D.

Remark 8.2. The adaptive control scheme applied in Theorem 8.6 is rather complicated: it involves two parallel algorithms and a logic switching (8.159). It is desirable to simplify the scheme. Besides, (8.164) does not imply the boundedness of the long run average of $\|u_i\|^2$ unless $D(z)$ is stable.

Remark 8.3. It is easy to generalize the control cost (8.165) to the one considered in Section 5.5, i.e. to (5.154).

8.5 Adaptive Control With Unknown Orders, Time-Delay and Coefficients

In this section we continue considering the problem discussed in the last section, but the orders and the time-delay together with the system coefficients are assumed to be unknown. As is noted in [GCZ], the solution of this problem is a straightforward combination of the order estimation procedure (Chapter 7) with the adaptive control law discussed in Sections 8.1-8.4, and the resulting control law is fairly complicated [HDa], [CZ4].

To be precise, let the l-input m-output system be described by

$$\begin{aligned} A(z)y_n &= B(z)u_n + e_n, && (8.207)\\ D(z)e_n &= C(z)w_n, \quad y_i = w_i = 0, \;\; u_i = 0, \quad i < 0 && (8.208)\end{aligned}$$

where

$$\begin{aligned} A(z) &= I + A_1 z + \ldots + A_{p_0} z^{p_0}, \quad p_0 \ge 0 && (8.209)\\ B(z) &= B_{d_0} z^{d_0} + \ldots + B_{q_0} z^{q_0}, \; q_0 \ge 0, \; d_0 \ge 1 && (8.210)\\ C(z) &= I + C_1 z + \ldots + C_{r_0} z^{r_0}, \quad r_0 \ge 0 && (8.211)\end{aligned}$$

with orders (p_0, q_0, r_0), time-delay d_0 and coefficients A_i, $i = 1, \ldots, p_0$, B_j, $j = d_0, \ldots, q_0$ and C_k, $k = 1, \ldots, r_0$ all unknown, while $D(z)$ is a given scalar polynomial expressed by (5.153).

Throughout this section we assume that the upper bounds (p^*, q^*, r^*) for system orders and the lower bound d^* for time-delay are available, i.e. p^*, $q^* \ge d^* \ge 1$ and r^* are given such that

$$(p_0, q_0, r_0) \in M_0 \triangleq \{(p, q, r) : \quad 0 \le p \le p^*, \; 0 \le q \le q^*, \; 0 \le r \le r^*\} \qquad (8.212)$$

and

$$d_0 \in M_1 \triangleq \{d : \; d^* \le d \le q^*\}. \qquad (8.213)$$

The loss function is the same as (5.154) with d replaced by its lower bound d^*, i.e.

$$\limsup_{n\to\infty} \frac{1}{n} \sum_{i=0}^{n} \left\| A^0(z)y_i - B^0(z)y_i^* + Q(z)D(z)u_{i-d^*} \right\|^2 \qquad (8.214)$$

where $A^0(z)$, $B^0(z)$ and $Q(z)$ are the given matrix polynomials, and $\{y_n^*\}$ is a given bounded reference signal with y_n^* being $\mathcal{F}_{n-d^*}$-measurable.

The task we face here is to minimize the loss function and simultaneously to consistently estimate not only the system coefficients but also the orders (p_0, q_0, r_0) and the time-delay d_0.

Let us estimate w_n by $\widehat{w}_n$

$$\widehat{w}_n = D(z)y_n - \overline{\theta}_n^{\tau}\overline{\varphi}_{n-1}, \ n \geq 0, \ \widehat{w}_n = 0, \ n < 0 \tag{8.215}$$

where

$$\overline{\theta}_{n+1} = \overline{\theta}_n + \overline{a}_n \overline{P}_n \overline{\varphi}_n (D(z)y_{n+1}^{\tau} - \overline{\varphi}_n^{\tau}\overline{\theta}_n) \tag{8.216}$$

$$\overline{P}_{n+1} = \overline{P}_n - \overline{a}_n \overline{P}_n \overline{\varphi}_n \overline{\varphi}_n^{\tau} \overline{P}_n, \quad \overline{a}_n = (1 + \overline{\varphi}_n^{\tau} \overline{P}_n \overline{\varphi}_n)^{-1} \tag{8.217}$$

$$\begin{aligned}\overline{\varphi}_n^{\tau} = {} & [D(z)y_n^{\tau} \ \dots \ D(z)y_{n-p^*+1}^{\tau} \quad D(z)u_{n-d^*+1}^{\tau} \ \dots \ D(z)u_{n-q^*+1}^{\tau} \\ & \widehat{w}_n^{\tau} \ \dots \ \widehat{w}_{n-r^*+1}^{\tau}]\end{aligned} \tag{8.218}$$

with $\overline{P}_0 = I$ and arbitrary $\overline{\theta}_0$.

In other words, $\overline{\theta}_n$ is the ELS-estimate for the system corresponding to the largest possible orders (p^*, q^*, r^*) and the smallest possible time-delay d^*.

For any $(p, q, r) \in M_0$ and $d \in M_1$ let us set

$$\theta^{\tau}(p, q, r, d) = [-A_1 \ \dots \ -A_p \quad B_d \ \dots \ B_q \quad C_1 \ \dots \ C_r] \tag{8.219}$$

$$\begin{aligned}\varphi_i^{\tau}(p, q, r, d) = {} & [D(z)y_i^{\tau} \ \dots \ D(z)y_{i-p+1}^{\tau} \quad D(z)u_{i-d+1}^{\tau} \ \dots \ D(z)u_{i-q+1}^{\tau} \\ & \widehat{w}_i^{\tau} \ \dots \ \widehat{w}_{i-r+1}^{\tau}]\end{aligned} \tag{8.220}$$

where it is assumed that

$$A_i = 0, \ B_j = 0, \ C_k = 0, \ \text{for } i > p_0, \ j > q_0 \text{ or } j < d_0 \text{ and } k > r_0.$$

For brevity of notation set $(g) = (p, q, r, d)$. Then the estimate

$$\begin{aligned}\theta_n^{\tau}(g) \triangleq {} & [-A_{1n} \ \dots \ -A_{pn} \quad B_{dn} \ \dots \ B_{qn} \\ & C_{1n} \ \dots \ C_{rn}]\end{aligned} \tag{8.221}$$

for $\theta(g)$ is given by

$$\begin{aligned}\theta_{n+1}(g) = {} & \theta_n(g) + a_n(g)P_n(g)\varphi_n(g) \\ & (D(z)y_{n+1}^{\tau} - \varphi_n^{\tau}(g)\theta_n(g))\end{aligned} \tag{8.222}$$

$$P_{n+1}(g) = P_n(g) - a_n(g)P_n(g)\varphi_n(g)\varphi_n^{\tau}(g)P_n(g) \tag{8.223}$$

$$a_n(g) = (1 + \varphi_n^{\tau}(g)P_n(g)\varphi_n(g))^{-1} \tag{8.224}$$

with $P_0(g) = I$ and $\theta_0(g)$ arbitrary.

According to (7.14)-(7.16) and (7.69) we define estimates $\widehat{r}_n$, $\widehat{q}_n$, $\widehat{p}_n$ and $\widehat{d}_n$ for r_0, q_0, p_0 and d_0 respectively as follows

$$\widehat{p}_n = \underset{0 \leq p \leq p^*}{\operatorname{argmin}} \, CIC(p, q^*, r^*)_n, \tag{8.225}$$

$$\widehat{q}_n = \underset{0 \leq q \leq q^*}{\operatorname{argmin}} \, CIC(\widehat{p}_n, q, r^*)_n, \tag{8.226}$$

$$\widehat{r}_n = \underset{0 \leq r \leq r^*}{\operatorname{argmin}} \, CIC(\widehat{p}_n, \widehat{q}_n, r)_n \tag{8.227}$$

and

$$\widehat{d}_n = \operatorname*{argmin}_{d^* \le d \le \widehat{q}_n} DIC(d)_n \tag{8.228}$$

where

$$CIC(p,q,r)_n = \sigma_n(p,q,r,d^*) + (p+q+r)a_n, \tag{8.229}$$

$$DIC(d)_n = \sigma_n(\widehat{p}_n, \widehat{q}_n, \widehat{r}_n, d) - da_n \tag{8.230}$$

$$\sigma_n(g) = \sum_{i=0}^{n-1} \|D(z)y_{i+1} - \theta_n^T(g)\varphi_i(g)\|^2 \tag{8.231}$$

and $\{a_n\}$ is such that $a_n > 0$,

$$\frac{(\log n)(\log\log n)^c}{a_n} \xrightarrow[n\to\infty]{} 0, \quad \frac{a_n}{n^{1-(t+1)(\varepsilon+\delta)}} \xrightarrow[n\to\infty]{} 0 \quad a.s. \tag{8.232}$$

for some $c > 1$ with

$$\varepsilon \in \left(0, \frac{1}{2(t+1)}\right), \quad \delta \in \left[0, \frac{1-2\varepsilon(t+1)}{2t+3}\right),$$

$$t = mp^* + \max(p^*, q^*, r^*) \ (\ge 1).$$

Let $F(z)$ and $G(z)$ be the solution of the Diophantine equation (3.120) with d replaced by d^* and let $M(z)$ and $N(z)$ satisfy (3.135) with d replaced by d^*.

We keep the notations used in Section 5.5, but in the present situation we should replace d by d^* and $B(z)$ by $B(z)z^{-d^*}$ because of the notational difference for $B(z)$ between (3.93) and (8.210). To be precise, we now have

$$\overline{y}_n = A^0(z)y_n - B^0(z)y_n^* + Q(z)D(z)u_{n-d^*}$$

$$\begin{aligned} &(\det C(z))(\overline{y}_n - M(z)w_n) \\ &= G_0(z)y_{n-d} + G_1(z)y_n^* + G_2(z)D(z)u_{n-d^*}, \end{aligned} \tag{8.233}$$

$$G_0(z) = N(z)D(z)(\mathrm{Adj}C(z))A(z) + A^0(z)G(z), \tag{8.234}$$

$$G_1(z) = -(\det C(z))B^0(z), \tag{8.235}$$

$$\begin{aligned} G_2(z) = {} & A^0(z)F(z)(\mathrm{Adj}C(z))B(z)z^{-d^*} \\ & -N(z)(\mathrm{Adj}C(z))B(z) + (\det C(z))Q(z), \end{aligned} \tag{8.236}$$

$$\begin{aligned} \varphi_n^* = {} & [y_n^T \ \dots \ y_{n-g_0}^T \ \ y_{n+d^*}^{*T} \ \dots \ y_{n+d^*-g_1}^{*T} \\ & D(z)u_n^T \ \dots \ D(z)u_{n-g_2}^T \\ & -\varphi_{n-1}^{*T}\theta_{n-1}^* \ \dots \ -\varphi_{n-mr^*}^{*T}\theta_{n-mr^*}^*]^T, \end{aligned}$$

where g_i denotes the upper bound for degree of $G_i(z)$, $i = 0, 1, 2$.

It is easy to calculate that the highest possible degree is

$$\begin{aligned} g^* &\triangleq mr^* \vee (mr^* + p^* - r^* + d^* - 1) - d^* \quad \text{for} \quad G(z), \\ n^* &\triangleq (\deg A^0(z)) - 1 \quad \text{for} \quad N(z), \\ g_0 &\triangleq (p^* + (m-1)r^* + n^*) \vee (g^* + \deg A^0(z)) \quad \text{for} \quad G_0(z), \\ g_1 &\triangleq (m-1)r^* + \deg B^0(z) \quad \text{for} \quad G_1(z) \quad \text{and} \\ g_2 &\triangleq [(\deg A^0(z)) + (m-1)r^* + q^* - 1] \vee (mr^* + \deg Q(z)) \end{aligned}$$

for $G_2(z)$.

Let $\{v_n\}$ be defined by (6.66) and (6.67) with p, q, r replaced by p^*, q^* and r^* respectively and let it be independent of $\{y_n^*\}$.

Similar to (8.158) by taking nonsingular initial estimate G_{210} we define u_n from

$$\theta_n^{*\tau} \varphi_n^* - G_{20n} v_n = 0 \quad \forall n \geq 0 \tag{8.237}$$

$$u_n^{(1)} = u_n - v_n \tag{8.238}$$

where θ_n^* is the estimate given by (5.161) for

$$\begin{aligned} \theta^* \;=\; & [G_{00} \; \dots \; G_{0g_0} \quad G_{10} \; \dots \; G_{1g_1} \\ & G_{20} \; \dots \; G_{2g_2} \quad c_1 I \; \dots \; c_{mr} I]^\tau \end{aligned} \tag{8.239}$$

and G_{20n} is the estimate given by θ_n^* for G_{20}.

We now formulate the following theorem.

Theorem 8.7. *Assume that*

1) (5.14) (5.15) hold;

2) $\det C(z) - \frac{a}{2}$ *is SPR;*

3) $m = l$, $(\det A^0(z))B(z)z^{-d^*} + A(z)(AdjA^0(z))Q(z)$ *and* $A^0(z)$ *are stable.*

Then u_n *defined by (8.238) leads to stability in the sense that*

$$\limsup_{n\to\infty} \frac{1}{n} \sum_{i=0}^{n} (\|y_i\|^2 + \|D(z)u_i\|^2) < \infty \quad a.s. \tag{8.240}$$

and to the minimality of the loss function:

$$\limsup_{n\to\infty} \left\| \frac{1}{n} \sum_{i=0}^{n} (A^0(z)y_i - B^0(z)y_i^* + Q(z)D(z)u_{i-d^*}) \right.$$

$$\cdot (A^0(z)y_i - B^0(z)y_i^* + Q(z)D(z)u_{i-d^*})^\tau$$
$$- \frac{1}{n}\sum_{i=0}^{n}(M(z)w_i)(M(z)w_i)^\tau \Bigg\| = 0 \quad a.s. \qquad (8.241)$$

If, in addition, assume that

4) $\liminf_{n\to\infty} \frac{1}{n^{1-\varepsilon^*}}\lambda_{\min}\left(\sum_{i=0}^{n} w_i w_i^\tau\right) > 0$ *a.s., where* $\varepsilon^* = \frac{1}{2t+3}$;

5) $A(z)$, $B(z)$ *and* $C(z)$ *have no common left-factor and* A_{p_0} *is of row-full-rank;*

6) $C^{-1}(z) - \frac{1}{2}I$ *is SPR,*

then $\widehat{r}_n$, $\widehat{q}_n$, $\widehat{p}_n$, $\widehat{d}_n$ *given by (8.225)-(8.228) are strongly consistent*

$$(\widehat{p}_n, \widehat{q}_n, \widehat{r}_n, \widehat{d}_n) \xrightarrow[n\to\infty]{} (p_0, q_0, r_0, d_0) \quad a.s., \qquad (8.242)$$

moreover, the coefficient estimate given by (8.221) has the following convergence rate:

$$\|\theta_n(\widehat{p}_n, \widehat{q}_n, \widehat{r}_n, \widehat{d}_n) - \theta(p_0, q_0, r_0, r_0)\|^2$$
$$= O\left(\frac{(\log n)(\log\log n)^{\delta(\beta-2)}}{n^{1-(t+1)\varepsilon}}\right) \quad a.s. \qquad (8.243)$$

where $\delta(x)$ *is given by (4.35).*

Proof. (8.240) and (8.241) follow from Theorem 5.5, while (8.242) follows from Theorems 7.1, 7.4 and Remark 7.2.

It remains to show (8.243). By (8.207) and (8.208), it is easy to see that

$$A(z)\overline{y}_n = B(z)\overline{u}_n + C(z)w_n \qquad (8.244)$$

where $\overline{y}_n = D(z)y_n$ and $\overline{u}_n = D(z)u_n$.

Hence (8.222)-(8.224) can be regarded as the ELS algorithm for (8.244), save that the noise estimate is generated by (8.215). Therefore, similar to the proof of Theorem 4.8, we have

$$\|\theta_n(g_0) - \theta(g_0)\|^2 = O\left(\frac{(\log n)(\log\log n)^{\delta(\beta-2)}}{\lambda^0_{\min}(n)}\right) \quad a.s. \qquad (8.245)$$

where we have used (8.240) and where $(g_0) \triangleq (p_0, q_0, r_0, d_0)$,

$$\lambda^0_{\min}(n) = \lambda_{\min}\left\{\sum_{i=1}^{n}\varphi_i^0\varphi_i^{0\tau}\right\}$$

$$\varphi_i^0 = [\overline{y}_i \ \dots \ \overline{y}_{i-p_0+1} \quad \overline{u}_i \ \dots \ \overline{u}_{i-q_0+1} \quad w_i \ \dots \ w_{i-r_0+1}]^\tau.$$

By Theorem 6.2(i) it is known that

$$\liminf_{n\to\infty} \lambda_{\min}^0(n)/n^{1-(t+1)\varepsilon} > 0. \tag{8.246}$$

Finally, (8.243) follows by combining (8.245) and (8.246) with (8.242).

Q.E.D.

In Theorem 8.7 adaptive control is realized by an implicit approach and the consistency of the parameter estimates is achieved by an algorithm different from that used in forming adaptive control.

We now use adaptive control similar to that developed in Section 8.4 and apply the estimation algorithm that provides consistent estimates for both order and coefficient to form adaptive control.

We consider the special case where $Q(z) \equiv 0$, $A^0(z) = I$ and d_0 is known for the adaptive control problem (8.208)-(8.214).

We retain all notations in the last section, but the time-delay d, its estimated value $\widehat{d}_n$ and the lower bound d^* should be set to equal the known value d_0 and $Q(z)$ should be replaced by 0.

Set

$$A_n(z) = I + \sum_{i=1}^{\widehat{p}_n} A_{in} z^i, \quad B_n(z) = \sum_{i=d_0}^{\widehat{q}_n} B_{in} z^i, \quad C_n(z) = I + \sum_{i=1}^{\widehat{r}_n} C_{in} z^i,$$

where coefficients A_{in}, B_{in}, C_{in} are given by (8.222)-(8.221).

Let $F_n(z) = \sum_{i=0}^{d_0-1} F_{in} z^i$ and $G_n(z)$ be the solution of the following Diophantine equation

$$(\det C_n(z))I = F_n(z)(\mathrm{Adj}C_n(z))A_n(z) + G_n(z)z^{d_0} \tag{8.247}$$

and let

$$M_n(z) = \sum_{i=0}^{d_0-1} M_{in} z^i \quad \text{and} \quad N_n(z)$$

be given by

$$A^0(z)F_n(z) = M_n(z) + N_n(z)z^{d_0}. \tag{8.248}$$

Similar to (8.157) let $u_n^{(2)}$ be defined by

$$\begin{aligned} &B_{d_0 n} D(z)u_n^{(2)} + (A^0(z) - I)B_{d_0 n} D(z)u_n \\ = \ &(\det C_n(z))[B^0(z)y^*_{n+d_0} - N_n(z)(D(z)y_n \\ &\qquad - \theta_n^\tau(\widehat{p}_n, \widehat{q}_n, \widehat{r}_n, d_0)\varphi_{n-1}(\widehat{p}_n, \widehat{q}_n, \widehat{r}_n, d_0))] \\ &-A^0(z)[G_n(z)y_n + (F_n(z)(\mathrm{Adj}C_n(z))B_n(z))z^{-d_0} D(z)u_n \\ &\qquad - B_{d_0 n} D(z)u_n] \end{aligned} \tag{8.249}$$

if $\det B_{d_0 n} \neq 0$, otherwise let $u_n^{(2)} = 0$.

We still use $u_n^{(1)}$ to denote the control defined by (8.238) and (8.237) with d and d^* replaced by d_0 and with $Q(z)$ set to equal 0 in (8.233)-(8.236).

With new definitions of $u_n^{(1)}$ and $u_n^{(2)}$ given here in mind we keep all notations (8.159)-(8.163) with $t = mp^* + \max(p^*, q^*, r^*)$ (≥ 1).

Theorem 8.8. *Assume that Conditions 1), 2), 4), 5), 6) of Theorem 8.7 hold and that $m = l$, $B(z)z^{-d_0}$ is stable. Then adaptive control (8.160) with modification mentioned above leads to that*

$$(\widehat{p}_n, \widehat{q}_n, \widehat{r}_n) \xrightarrow[n\to\infty]{} (p_0, q_0, r_0) \quad a.s., \tag{8.250}$$

$$\limsup_{n\to\infty} \frac{1}{n}\sum_{i=0}^{n}(\|y_i\|^2 + \|D(z)u_i\|^2) < \infty \quad a.s. \tag{8.251}$$

$$\left\| \frac{1}{n}\sum_{i=0}^{n}(y_i - B^0(z)y_i^*)(y_i - B^0(z)y_i^*)^{\tau} - \frac{1}{n}\sum_{i=0}^{n}(M(z)w_i)(M(z)w_i)^{\tau} \right\| = O(n^{-\frac{\varepsilon}{2}}) \quad a.s. \tag{8.252}$$

$$\|\theta_n(\widehat{p}_n, \widehat{q}_n, \widehat{r}_n, \widehat{d}_n) - \theta(p_0, q_0, r_0, r_0)\|^2 = O\left(\frac{(\log n)(\log\log n)^{\delta(\beta-2)}}{n^{1-(t+1)\varepsilon}}\right) \quad a.s. \tag{8.253}$$

Proof. If we review the proof of Theorem 8.6, then we find that (8.169)-(8.182) still hold true in the present case. Hence (8.168) is verified, and by stability of $B(z)z^{-d_0}$

$$\frac{1}{n}\sum_{i=0}^{n}(\|D(z)y_i\|^2 + \|D(z)u_i\|^2) = O(n^{\delta}) \quad a.s. \tag{8.254}$$

$$\delta \in \left[0, \frac{1-2\varepsilon(t+1)}{2t+3}\right), \quad \varepsilon \in \left(0, \frac{1}{2(t+1)}\right),$$

$$t = mp^* + \max(p^*, q^*, r^*) - 1.$$

Then by Theorem 7.1 and Remark 7.3 the order estimates (8.225)-(8.227) are strongly consistent. Similar to the proof of (8.243) we have

$$\|\theta_n(p_0, q_0, r_0, d_0) - \theta(p_0, q_0, r_0, d_0)\|^2 = O\left(\frac{(\log n)(\log\log n)^{\delta(\beta-2)}}{n^{1-(t+1)(\varepsilon+\delta)}}\right) \quad a.s. \tag{8.255}$$

The next step is to show that (8.184) is also true for the present case. As a matter of fact, the argument used for (8.184)-(8.205) is still true, if the following modifications are taken into account: φ_n^0 defined by (8.189) should change to that defined by (8.245); ξ_n defined by (8.190) changes to

$$\bar{\xi}_n \triangleq D(z)y_n - w_n - \bar{\theta}_n^T \bar{\varphi}_{n-1},$$

and (8.192) corresponds to

$$\sum_{i=0}^{n} \|\bar{\xi}_i\|^2 = O(\log^2 n)$$

because $\bar{s}_n = O(n^{1+\delta})$. θ, θ_n and $\tilde{\theta}_n$ in (8.194) and (8.195) should be understood as $\theta(p_0, q_0, r_0, d_0)$, $\theta_n(p_0, q_0, r_0, d_0)$ and $\varphi_n(p_0, q_0, r_0, d_0)$ respectively. B_1 changes to B_{d_0}, B_{1n} to $B_{d_0 n}$, $B(z)$ to $B(z)z^{-d_0}$, $B_n(z)$ to $B_n(z)z^{-d_0}$.

Thus, for any sample there is a k such that

$$\tau_k < \infty \quad \text{and} \quad \sigma_k = \infty.$$

Then, along the lines of the proof of Theorem 8.6 we derive (8.252). By stability of $A^0(z)$ and $B(z)z^{-d_0}$, (8.251) follows from (8.252). Hence (8.255) is valid with $\delta = 0$. This implies (8.253). Q.E.D.

Remark 8.4. Up to now, we have exclusively considered ARMAX models without unmodeled dynamics. However, in practice it is important to know whether or not the adaptive algorithm still works when unmodeled dynamics are present. In [CG8], it is shown that the ELS algorithm has a certain degree of robustness if the persistence of excitation condition holds. A class of robust stochastic adaptive control algorithms can be designed. Related discussions will be given in Section 11.3 and Chapter 12.

Remark 8.5. The adaptive pole-assignment problem is not of concern in this chapter. For systems with known parameters, the pole-assignment controller not only has a simple structure, but also can stabilize the system without requiring stability of $A(z)$ or $B(z)$. However, the corresponding stochastic adaptive result is yet to be established.

CHAPTER 9

ARX(∞) Model Approximation

9.1 Statement of Problem

In preceding chapters, we have studied finite dimensional linear systems of the form (3.91), or

$$y_n = \left[I - C^{-1}(z)A(z)\right] y_n + C^{-1}(z)B(z)u_n + w_n, \tag{9.1}$$

where $C^{-1}(z)A(z)$ and $C^{-1}(z)B(z)$ are rational transfer matrices.

However, physical systems, in general, are bound to be more complex than (9.1), i.e. it is hardly realistic to assume that (9.1) is the true process generating the observation $\{y_i,\ u_i,\ i = 0,\ 1,\ \ldots\}$. It seems preferable, as in [H] to regard (9.1) as no more than a model on which an approximation procedure for the true structure is based. Then the orders of the polynomial matrices $A(z)$, $B(z)$ and $C(z)$ will depend on the data size n. Thus it is necessary to investigate the nature of this approximation procedure. We assume that the m-dimensional process $\{y_n\}$ is generated by

$$\begin{aligned} y_n &= \sum_{i=1}^{\infty}(A_i y_{n-i} + B_i u_{n-i}) + w_n,\ n \geq 0; \\ y_n &= w_n = 0,\ u_n = 0,\ \forall n < 0, \end{aligned} \tag{9.2}$$

where

$$\sum_{i=1}^{\infty}(\|A_i\| + \|B_i\|) < \infty. \tag{9.3}$$

However, the following matrix functions of z:

$$a)\ A(z) \triangleq -\sum_{i=0}^{\infty} A_i z^i,\ (A_0 \triangleq -I), \quad A_i \in \mathbb{R}^{m\times m}, \tag{9.4}$$

and

$$b) \quad B(z) \triangleq \sum_{i=1}^{\infty} B_i z^i, \quad B_i \in \mathbb{R}^{m\times l},$$

are no longer assumed to be rational.

Let us denote the "transfer function" matrix associated with (9.2) as

$$G(z) = [A(z), \quad B(z)]. \tag{9.5}$$

In this chapter, we first estimate and/or approximate $G(z)$ by the increasing lag (denoted by h_n) least squares method based on the observed process $\{y_i,\ u_i\}$. But this procedure may result a model (say $\widehat{G}(z)$) of very high order and may contain $(m^2+lm)h_n$ parameters at each step n. One way to overcome this problem is through further approximation to get a simple model $\widehat{G}_s(z)$ which may be obtained by either (i) minimizing $\|\widehat{G}_s - \widehat{G}\|$ with given complexity (number of parameters in $\widehat{G}_s$), or (ii) minimizing the complexity of $\widehat{G}_s$ with given misfit (for example, an error bound). When the complexity is given, the procedure of the balanced truncation or the optimal Hankel norm approximation, discussed, for example, in [Gl], may be applied. An alternative approach is to fit an ARMAX model directly to the data $\{y_i,\ u_i\}$. In this case estimating the noise w_n is an important step, and it is carried out also by the increasing lag least squares method. The asymptotic behavior of the estimation error is presented in this chapter.

9.2 Transfer Function Approximation

We need the following two norms for measuring the accuracy of the transfer function approximation:

$$\|F(z)\|_2 = \left\{\lambda_{\max}\left[\frac{1}{2\pi}\int_0^{2\pi} F(e^{i\theta})F^*(e^{i\theta})d\theta\right]\right\}^{\frac{1}{2}}, \tag{9.6}$$

$$\|F(z)\|_\infty = \operatorname*{ess\,sup}_{\theta\in[0,2\pi]} \left\{\lambda_{\max}\left[F(e^{i\theta})F^*(e^{i\theta})\right]\right\}^{\frac{1}{2}}, \tag{9.7}$$

where the first is the H_2-norm of a measurable complex matrix $F(z)$ defined in $|z| \le 1$, analytic in $|z| < 1$ and such that (9.6) is finite. The second is the H^∞-norm of a complex matrix $F(z)$ analytic in $|z| < 1$ and bounded almost everywhere on the unit circle.

As usual, we assume that the system noise $\{w_n, \mathcal{F}_n\}$ is a martingale difference sequence with respect to a family $\{\mathcal{F}_n\}$ of non-decreasing σ-algebras, and that the input u_n is an $\mathcal{F}_n$-measurable input (vector) for any $n \geq 0$, i.e.

$$E[w_{n+1}|\mathcal{F}_n] = 0, \quad u_n \in \mathcal{F}_n, \quad n \geq 0. \tag{9.8}$$

Clearly, system (9.2) under (9.3) and (9.8) is nonstationary in general because: (i) there are no restrictions on the location of zeros of $\det A(z)$; (ii) the system input $\{u_n\}$ may be nonstationary.

Let us now describe the approximation algorithm.

Let $\{h_n\}$ be any non-decreasing sequence of positive integers, $h_n \leq n$, $\forall n$. Set

$$\theta(n) = [A_1 \; ... \; A_{h_n} \quad B_1 \; ... \; B_{h_n}]^\tau, \tag{9.9}$$

and

$$\varphi_i(n) = [y_i^\tau \; ... \; y_{i-h_n+1}^\tau \quad u_i^\tau \; ... \; u_{i-h_n+1}^\tau]^\tau, \quad 1 \leq i \leq n. \tag{9.10}$$

The least-squares estimate $\widehat{\theta}(n)$ for $\theta(n)$ at time n is given by

$$\widehat{\theta}(n) = \left[\sum_{i=0}^{n-1} \varphi_i(n)\varphi_i^\tau(n) + \gamma I\right]^{-1} \sum_{i=0}^{n-1} \varphi_i(n) y_{i+1}^\tau \tag{9.11}$$

with real number $\gamma > 0$ arbitrarily chosen.

Let us then write $\widehat{\theta}(n)$ in its component form

$$\widehat{\theta}(n) = [A_1(n) \; ... \; A_{h_n}(n) \quad B_1(n) \; ... \; B_{h_n}(n)]^\tau, \tag{9.12}$$

and set

$$\widehat{A}_n(z) = I - \sum_{i=1}^{h_n} A_i(n) z^i, \quad \widehat{B}_n(z) = \sum_{i=1}^{h_n} B_i(n) z^i. \tag{9.13}$$

Then the estimate $\widehat{G}_n(z)$ for $G(z)$ at time n can be formed as

$$\widehat{G}_n(z) = [\widehat{A}_n(z) \quad \widehat{B}_n(z)]. \tag{9.14}$$

The convergence (or divergence) rate of $\widehat{G}_n(z)$ is summarized in the following theorems [GHH].

Theorem 9.1. *Consider the system (9.2)-(9.3) and the approximation algorithm (9.9)-(9.14) with regression lag $h_n = O(\log^\alpha n)$, $\alpha > 0$. Suppose that (9.8) holds and that*

$$\sup_i E(\|w_{i+1}\|^2|\mathcal{F}_i) < \infty, \quad \liminf_{n\to\infty} \frac{1}{n}\sum_{i=0}^{n-1} \|w_i\|^2 \neq 0 \quad a.s. \tag{9.15}$$

and

$$\|w_i\|^2 = o(d(i)) \quad a.s. \tag{9.16}$$

where $\{d(i)\}$ is a positive, deterministic and nondecreasing sequence and satisfies

$$\sup_i d(e^{i+1})/d(e^i) < \infty. \tag{9.17}$$

Then as $n \to \infty$, the following asymptotic expansions hold:

(i)

$$\begin{aligned} &\|\widehat{G}_n(z) - G(z)\|_\infty^2 \\ &= O\left(\frac{h_n}{\lambda_{\min}(n)}\{h_n \log r_n + \delta_n r_n\}\right) \\ &\quad + o\left(\frac{h_n[d(n)\log\log n]^2}{\lambda_{\min}(n)}\right) \quad a.s., \end{aligned} \tag{9.18}$$

(ii)

$$\begin{aligned} &\|\widehat{G}_n(z) - G(z)\|_2^2 \\ &= O\left(\frac{1}{\lambda_{\min}(n)}\{h_n \log r_n + \delta_n r_n\}\right) + o\left(\frac{[d(n)\log\log n]^2}{\lambda_{\min}(n)}\right) \quad a.s. \end{aligned} \tag{9.19}$$

where r_n, δ_n and $\lambda_{\min}(n)$ are defined by

$$r_n \triangleq 1 + \sum_{i=0}^{n-1}(\|y_i\|^2 + \|u_i\|^2), \tag{9.20}$$

$$\delta_n \triangleq \left(\sum_{i=h_n+1}^{\infty}\|A_i\|\right)^2 + \left(\sum_{i=h_n+1}^{\infty}\|B_i\|\right)^2, \tag{9.21}$$

$$\lambda_{\min}(n) \triangleq \lambda_{\min}\left(\sum_{i=0}^{n-1}\varphi_i(n)\varphi_i^\tau(n) + \gamma I\right). \tag{9.22}$$

Proof. Let us denote

$$\begin{aligned} G_n(z) &= [A_n(z) \quad B_n(z)] \\ A_n(z) &= I - \sum_{i=1}^{h_n} A_i z^i, \quad B_n(z) = \sum_{i=1}^{h_n} B_i z^i. \end{aligned} \tag{9.23}$$

Then by (9.4), (9.5), (9.9), (9.11)-(9.13) and (9.21) we know that

$$\begin{aligned} &\|\widehat{G}_n(z) - G(z)\|_\infty^2 \\ &\le 2\|\widehat{G}_n(z) - G_n(z)\|_\infty^2 + 2\|G_n(z) - G(z)\|_\infty^2 \end{aligned}$$

$$
\begin{aligned}
&\leq 2\left\{\sum_{i=1}^{h_n} \|[A_i(n) - A_i, \quad B_i(n) - B_i]\|\right\}^2 \\
&\quad +2\left\{\sum_{i=h_n+1}^{\infty} (\|A_i\| + \|B_i\|)\right\}^2 \\
&\leq 2h_n \sum_{i=1}^{h_n} \|[A_i(n) - A_i, \quad B_i(n) - B_i]\|^2 \\
&\quad +4\left(\sum_{i=h_n+1}^{\infty} \|A_i\|\right)^2 + 4\left(\sum_{i=h_n+1}^{\infty} \|B_i\|\right)^2 \\
&\leq 2h_n tr\left\{\sum_{i=1}^{h_n} [A_i(n) - A_i, \quad B_i(n) - B_i]\right. \\
&\quad \left. [A_i(n) - A_i, \quad B_i(n) - B_i]^\tau\right\} + 4\delta_n \\
&= 2h_n tr\left\{[\widehat{\theta}(n) - \theta(n)]^\tau [\widehat{\theta}(n) - \theta(n)]\right\} + 4\delta_n \\
&\leq 2mh_n \|\widehat{\theta}(n) - \theta(n)\|^2 + 4\delta_n.
\end{aligned} \tag{9.24}
$$

Similarly, for the H_2-norm we have the following estimation:

$$
\begin{aligned}
&\|\widehat{G}_n(z) - G(z)\|_2^2 \\
&\leq 2\|\widehat{G}_n(z) - G_n(z)\|_2^2 + 2\|G_n(z) - G(z)\|_2^2 \\
&\leq 2\left\{\lambda_{\max}\left[\frac{1}{2\pi}\int_0^{2\pi} [\widehat{G}_n(e^{it}) - G_n(e^{it})]\right.\right. \\
&\quad \left.\left. [\widehat{G}_n(e^{-it}) - G_n(e^{-it})]^\tau dt\right]\right\} \\
&\quad +2\left\{\lambda_{\max}\left[\frac{1}{2\pi}\int_0^{2\pi} [G_n(e^{it}) - G(e^{it})]\right.\right. \\
&\quad \left.\left. [G_n(e^{-it}) - G(e^{-it})]^\tau dt\right]\right\} \\
&= 2\lambda_{\max}\left\{\sum_{i=1}^{h_n} [A_i(n) - A_i, \quad B_i(n) - B_i]\right. \\
&\quad \left. [A_i(n) - A_i, \quad B_i(n) - B_i]^\tau\right\}
\end{aligned}
$$

$$+2\lambda_{\max}\left\{\sum_{i=h_n+1}^{\infty}[A_i,\quad B_i][A_i,\quad B_i]^{\tau}\right\}$$

$$\leq\ 2\|\widehat{\theta}(n)-\theta(n)\|^2+2\delta_n. \tag{9.25}$$

We now estimate the error $\|\widehat{\theta}(n)-\theta(n)\|$.
Set

$$\varepsilon_i(n)=\sum_{j=h_n+1}^{\infty}[A_j y_{i-j+1}+B_j u_{i-j+1}]. \tag{9.26}$$

Then by (9.2), (9.9)-(9.11) and (9.22) we have

$$\|\widehat{\theta}(n)-\theta(n)\|^2=\left\|\left[\sum_{i=0}^{n-1}\varphi_i(n)\varphi_i^{\tau}(n)+\gamma I\right]^{-1}\right.$$

$$\left.\times\left\{\sum_{i=0}^{n-1}\varphi_i(n)[y_{i+1}^{\tau}-\varphi_i^{\tau}(n)\theta(n)]-\gamma\theta(n)\right\}\right\|^2$$

$$=\left\|\left[\sum_{i=0}^{n-1}\varphi_i(n)\varphi_i^{\tau}(n)+\gamma I\right]^{-1}\right.$$

$$\left.\times\left\{\sum_{i=0}^{n-1}\varphi_i(n)[w_{i+1}^{\tau}+\varepsilon_i^{\tau}(n)]-\gamma\theta(n)\right\}\right\|^2$$

$$\leq\frac{3}{\lambda_{\min}(n)}\left\{\left\|\left[\sum_{i=0}^{n-1}\varphi_i(n)\varphi_i^{\tau}(n)+\gamma I\right]^{-\frac{1}{2}}\sum_{i=0}^{n-1}\varphi_i(n)w_{i+1}^{\tau}\right\|^2\right.$$

$$\left.+\left\|\left[\sum_{i=0}^{n-1}\varphi_i(n)\varphi_i^{\tau}(n)+\gamma I\right]^{-\frac{1}{2}}\sum_{i=0}^{n-1}\varphi_i(n)\varepsilon_i^{\tau}(n)\right\|^2+O(1)\right\}. \tag{9.27}$$

Now for the second term on the right-hand side of (9.27) we apply first Lemma 4.7(ii) and then the Schwarz inequality; this yields under (9.20), (9.21) and (9.26),

$$\left\|\left[\sum_{i=0}^{n-1}\varphi_i(n)\varphi_i^{\tau}(n)+\gamma I\right]^{-\frac{1}{2}}\sum_{i=0}^{n-1}\varphi_i(n)\varepsilon_i^{\tau}(n)\right\|^2$$

$$\leq\sum_{i=0}^{n-1}\|\varepsilon_i(n)\|^2$$

$$
\begin{aligned}
&\le 2\sum_{i=0}^{n-1}\left\{\sum_{j=h_n+1}^{\infty}\|A_j\|\sum_{j=h_n+1}^{\infty}\|A_j\|\|y_{i-j+1}\|^2\right.\\
&\qquad\left.+\sum_{j=h_n+1}^{\infty}\|B_j\|\sum_{j=h_n+1}^{\infty}\|B_j\|\|u_{i-j+1}\|^2\right\}\\
&\le 2\left(\sum_{j=h_n+1}^{\infty}\|A_j\|\right)^2\sum_{i=0}^{n-1}\|y_i\|^2+2\left(\sum_{j=h_n+1}^{\infty}\|B_j\|\right)^2\sum_{i=0}^{n-1}\|u_i\|^2\\
&\le 2\delta_n r_n. \qquad (9.28)
\end{aligned}
$$

As for the first term on the right-hand side of (9.27) we use Lemma 4.6 to estimate it:

$$
\begin{aligned}
&\left\|\left[\sum_{i=0}^{n-1}\varphi_i(n)\varphi_i^\tau(n)+\gamma I\right]^{-\frac{1}{2}}\sum_{i=0}^{n-1}\varphi_i(n)w_{i+1}^\tau\right\|^2\\
&= O\left(h_n\log^+\lambda_{\max}\left[\sum_{i=0}^{n-1}\varphi_i(n)\varphi_i^\tau(n)+\gamma I\right]\right)+o([d(n)\log\log n]^2)\\
&= O\left(h_n\log^+\{h_n r_n+h_n\}\right)\\
&\quad +o([d(n)\log\log n]^2)\\
&= O\left(h_n\log r_n\right)+o([d(n)\log\log n]^2) \qquad (9.29)
\end{aligned}
$$

where for the last relation we have used the fact that

$$\liminf_{n\to\infty}\frac{r_n}{n}>0 \text{ a.s.} \qquad (9.30)$$

which follows from (9.2), (9.15) and the Schwarz inequality ($A_0 \triangleq I$, $B_0 \triangleq 0$):

$$
\begin{aligned}
0 &\ne \liminf_{n\to\infty}\frac{1}{n}\sum_{i=0}^{n-1}\|w_i\|^2\\
&\le \liminf_{n\to\infty}\frac{1}{n}\sum_{i=0}^{n-1}\left\{\sum_{j=0}^{\infty}(\|A_j y_{i-j}\|+\|B_j u_{i-j}\|)\right\}^2\\
&\le 2\left\{\sum_{j=0}^{\infty}(\|A_j\|+\|B_j\|)\right\}^2\liminf_{n\to\infty}\frac{r_n}{n}.
\end{aligned}
$$

Substituting (9.28) and (9.29) into (9.27) we get

$$\|\widehat{\theta}(n)-\theta(n)\|^2$$

$$= O\left(\frac{1}{\lambda_{\min}(n)}\{h_n \log r_n + o([d(n)\log\log n]^2) + \delta_n r_n\}\right). \tag{9.31}$$

Putting (9.31) into (9.24) and (9.25) we obtain the desired results (9.18) and (9.19). Q.E.D.

If in Theorem 9.1 the regression lag h_n is of $O(n^\alpha)$, $0 < \alpha \leq 1$, then similar results can be derived. This is the content of

Theorem 9.1′. *Under Conditions of Theorem 9.1, if the regression lag h_n in (9.10) is taken as $h_n = O(n^\alpha)$, $0 < \alpha \leq 1$, then (9.18) and (9.19) hold with "$\log\log n$" in them replaced by "$\log n$".*

Proof. The proof is essentially the same as that for Theorem 9.1. The only difference is that in the present case (9.29) becomes

$$\left\| \left[\sum_{i=0}^{n-1} \varphi_i(n)\varphi_i^T(n) + \gamma I\right]^{-\frac{1}{2}} \sum_{i=0}^{n-1} \varphi_i(n) w_{i+1}^T \right\|^2$$
$$= O\left(h_n \log^+ \lambda_{\max}\left[\sum_{i=0}^{n-1} \varphi_i(n)\varphi_i^T(n) + \gamma I\right]\right) + o([d(n)\log n]^2).$$

This change occurs because $h_n = O(\log^\alpha n)$ is replaced by $h_n = O(n^\alpha)$ (see the remark following Theorem 2.9). Details are not repeated here.

Q.E.D.

Remark 9.1 In order to keep generality of Theorems 9.1 and 9.1′, we have tried to impose as few restrictions as possible. Of course, with some further conditions, "simple" expansion may be deduced immediately from Theorem 9.1. For example, if in (9.2) the random disturbance $\{w_n\}$ is a Gaussian white noise (i.i.d.) sequence and

$$\|A_i\| + \|B_i\| = O(\lambda^i), \quad 0 < \lambda < 1, \quad i \geq 0 \tag{9.32}$$

$$\sum_{i=0}^{n-1} (\|y_i\|^2 + \|u_i\|^2) = O(n^b), \quad \text{for some} \quad b \geq 1, \tag{9.33}$$

then by taking $h_n = \log^\alpha n$, $\alpha > 1$, and taking account of $\|w_n\| = O(\log^{\frac{1}{2}} n)$, we see from Theorem 9.1 that

$$\|\widehat{G}_n(z) - G(z)\|_\infty^2 = O\left(\frac{\log^{2\alpha+1} n}{\lambda_{\min}(n)}\right)$$

$$\|\widehat{G}_n(z) - G(z)\|_2^2 = O\left(\frac{\log^{\alpha+1} n}{\lambda_{\min}(n)}\right).$$

From Theorems 9.1 and 9.1′, it is seen that the growth rate of $\lambda_{\min}(n)$ plays a crucial role in the convergence of the approximation algorithm. It is clear that $\lambda_{\min}(n)$ depends essentially on the two input signals $\{u_i, w_i\}$, even though it is defined via the observation data $\{y_i, u_i\}$. In the rest of this section, we will study how the growth rate of $\lambda_{\min}(n)$ depends explicitly on the input signals $\{u_i, w_i\}$.

In this chapter we use $\varphi_i^0(n)$ and $\lambda_{\min}^0(n)$ to denote the following vector and eigenvalue respectively:

$$\begin{aligned} \varphi_i^0(n) &= [u_i^\tau \ \dots \ u_{i-2h_n+1}^\tau \quad w_i^\tau \ \dots \ w_{i-2h_n+1}^\tau]^\tau, \\ & \quad 1 \le i \le n, \end{aligned} \tag{9.34}$$

$$\lambda_{\min}^0(n) \triangleq \lambda_{\min}\left\{\sum_{i=0}^{n-1} \varphi_i^0(n)\varphi_i^{0\tau}(n)\right\}. \tag{9.35}$$

They should not be confused with those defined by (7.21) and (7.57).

The next theorem [GHH] establishes the connection between $\lambda_{\min}(n)$ and $\lambda_{\min}^0(n)$.

Theorem 9.2. *Suppose that in System (9.2)-(9.4) the number of zeros of* $\det A(z)$ *on the unit circle* $|z| = 1$ *is finite (with the largest multiplicity of these zeros denoted by d) and that*

$$\sum_{k=1}^{\infty} k^{\frac{1}{2}+d}(\|A_k\| + \|B_k\|) < \infty. \tag{9.36}$$

Then as $n \to \infty$

$$\lambda_{\min}(n) \ge c_0(h_n)^{-2d}\lambda_{\min}^0(n) + O(\delta_n^0 r_n^0) \quad a.s. \tag{9.37}$$

where $c_0 > 0$ *is a constant,* $\lambda_{\min}(n)$ *and* $\lambda_{\min}^0(n)$ *are respectively defined by (9.22) and (9.35), and*

$$r_n^0 = \sum_{i=0}^{n-1}(\|u_i\|^2 + \|w_i\|^2)$$

$$\delta_n^0 = h_n\left\{\sum_{i=[h_n/2^m]-1}^{\infty} (\|A_i\| + \|B_i\|)\right\}^2$$

with m being the dimension of y_n.

Before we prove the theorem, we first show a corollary of it.

Corollary 9.1. *Under conditions of Theorem 9.2, if either*

(i) $r_n^0 = O(n), \quad \sup_n n/\lambda_{\min}^0(n) < \infty$

or

(ii) $r_n^0 = O(n^b), \quad b \geq 1, \; h_n = \log^\alpha n, \; \alpha > 1$ *and (9.32) holds,*

then as $n \to \infty$

$$\lambda_{\min}(n) \geq c_1 (h_n)^{-2d} \lambda_{\min}^0(n) \quad a.s. \tag{9.38}$$

for some constant $c_1 > 0$.

Proof. In the case of (i), we need only to note that by Condition (9.36),

$$\begin{aligned} h_n^{2d} \delta_n^0 &= \left\{ h_n^{\frac{1}{2}+d} \sum_{i=[h_n/2^m]-1}^{\infty} (\|A_i\| + \|B_i\|) \right\}^2 \\ &\leq \left\{ \sum_{i=[h_n/2^m]-1}^{\infty} i^{\frac{1}{2}+d} (\|A_i\| + \|B_i\|) \right\}^2 \xrightarrow[n \to \infty]{} 0, \end{aligned}$$

while in the case of (ii) it suffices to mention that

$$\delta_n^0 r_n^0 \to 0, \quad \text{as} \quad n \to \infty.$$

Q.E.D.

We now proceed to prove Theorem 9.2. We start with lemmas.

Lemma 9.1. *Let* $f_i(z)$, $1 \leq i \leq p$, *be analytic functions expanded as*

$$f_i(z) = \sum_{j=0}^{\infty} f_j^{(i)} z^j, \quad 1 \leq i \leq p,$$

and let the product of $f_i(z)$ *be expanded as*

$$\prod_{i=1}^{p} f_i(z) = \sum_{j=0}^{\infty} c_j z^j.$$

Then

(i) $\sum_{j=0}^{\infty}|f_j^{(i)}| < \infty$, $1 \le i \le p$, *implies that*

$$\sum_{j=n}^{\infty}|c_j| = O\left(\sum_{j=[n/2^{p-1}]-1}^{\infty}\sum_{i=1}^{p}|f_j^{(i)}|\right)$$

(ii) $\sum_{j=0}^{\infty} j^r|f_j^{(i)}| < \infty$, $r \ge 0$, $1 \le i \le p$ *implies that* $\sum_{j=0}^{\infty} j^r|c_j| < \infty$.

Proof. (i) Let us first consider the case of $p = 2$. In this case we have

$$c_i = \sum_{j=0}^{i} f_j^{(1)} f_{i-j}^{(2)} \tag{9.39}$$

and then

$$\begin{aligned}
\sum_{i=n}^{\infty}|c_i| &\le \sum_{i=n}^{\infty}\sum_{j=0}^{i}|f_j^{(1)}||f_{i-j}^{(2)}| \\
&= \sum_{j=0}^{n}\sum_{i=n}^{\infty}|f_j^{(1)}||f_{i-j}^{(2)}| + \sum_{j=n}^{\infty}\sum_{i=j}^{\infty}|f_j^{(1)}||f_{i-j}^{(2)}| \\
&\le \sum_{j=0}^{[\frac{n}{2}]-1}|f_j^{(1)}|\sum_{i=n}^{\infty}|f_{i-j}^{(2)}| + \sum_{j=[\frac{n}{2}]}^{n}|f_j^{(1)}|\sum_{i=n}^{\infty}|f_{i-j}^{(2)}| \\
&\quad + O\left(\sum_{j=n}^{\infty}|f_j^{(1)}|\right) \\
&\le O\left(\sum_{i=[\frac{n}{2}]}^{\infty}|f_i^{(2)}|\right) + O\left(\sum_{j=[\frac{n}{2}]}^{\infty}|f_j^{(1)}|\right) + O\left(\sum_{j=n}^{\infty}|f_j^{(1)}|\right).
\end{aligned}$$

Thus

$$\sum_{i=n}^{\infty}|c_i| = O\left(\sum_{i=[\frac{n}{2}]}^{\infty}(|f_i^{(1)}| + |f_i^{(2)}|)\right).$$

This proves the (i) part for the case of $p = 2$. For the general $p > 2$ case it can be proved by induction.

(ii) By induction we need only to prove the case of $p = 2$. Note that $i^r \le 2^r\{(i-j)^r + j^r\}$, $\forall i \ge j \ge 0$; we have by (9.39)

$$\sum_{i=0}^{\infty} i^r|c_i| \le \sum_{i=0}^{\infty} 2^r\{(i-j)^r + j^r\}\sum_{j=0}^{i}|f_j^{(1)}||f_{i-j}^{(2)}|$$

$$
\begin{aligned}
&\le 2^r \sum_{j=0}^{\infty}\sum_{i=j}^{\infty}\{(i-j)^r + j^r\}|f_j^{(1)}||f_{i-j}^{(2)}| \\
&= 2^r \sum_{j=0}^{\infty}|f_j^{(1)}|\sum_{i=j}^{\infty}(i-j)^r|f_{i-j}^{(2)}| + 2^r\sum_{j=0}^{\infty} j^r|f_j^{(1)}|\sum_{i=j}^{\infty}|f_{i-j}^{(2)}| \\
&= 2^r\left(\sum_{j=0}^{\infty}|f_j^{(1)}|\right)\sum_{i=0}^{\infty} i^r|f_i^{(2)}| \\
&\quad +2^r\left(\sum_{j=0}^{\infty} j^r|f_j^{(1)}|\right)\sum_{i=0}^{\infty}|f_i^{(2)}| < \infty.
\end{aligned}
$$

Q.E.D.

Write $x_n \in \mathbb{R}^{(m+l)h_n}$ with

$$\|x_n\| = 1, \quad \forall n \ge 1 \tag{9.40}$$

in the component form

$$x_n = [\alpha_n^{(0)\tau} \ \dots \ \alpha_n^{(h_n-1)\tau} \quad \beta_n^{(0)\tau} \ \dots \ \beta_n^{(h_n-1)\tau}]^\tau, \tag{9.41}$$

with $\alpha_n^{(i)} \in \mathbb{R}^m$, $\beta_n^{(i)} \in \mathbb{R}^l$ (l is the dimension of u_n), and introduce the vector complex function

$$
\begin{aligned}
H_n(z) &\triangleq \sum_{i=0}^{h_n-1}\{\alpha_n^{(i)\tau}[\mathrm{Adj}A(z)][B(z) \quad I] \\
&\qquad +\beta_n^{(i)\tau}[\det A(z) I_l \quad 0]\}z^i, &(9.42)\\
&= \sum_{i=0}^{\infty}[f_n^{(i)\tau} \quad g_n^{(i)\tau}]z^i. &(9.43)
\end{aligned}
$$

Obviously $f_n^{(i)}$ and $g_n^{(i)}$ are functions of x_n.

Lemma 9.2. *Under the conditions of Theorem 9.2*

$$\liminf_{n\to\infty} \inf_{\|x_n\|=1} h_n^{2d}\sum_{i=0}^{2h_n-1}\left(\|f_n^{(i)}\|^2 + \|g_n^{(i)}\|^2\right) \ne 0 \quad a.s.$$

Proof. Let

$$\det A(z) = \sum_{i=0}^{\infty} a_i z^i, \quad \mathrm{Adj}[A(z)] = \sum_{i=0}^{\infty}\overline{A}_i z^i, \tag{9.44}$$

$$[\mathrm{Adj}A(z)]B(z) = \sum_{i=1}^{\infty}\overline{B}_i z^i. \tag{9.45}$$

By Lemma 9.1 (ii) and (9.36) it is easy to convince oneself that

$$\sum_{i=0}^{\infty} i^{\frac{1}{2}+d}(\|\overline{A}_i\| + \|\overline{B}_i\| + |a_i|) < \infty. \tag{9.46}$$

Note that by (9.42)-(9.45),

$$\begin{aligned} &\sum_{i=0}^{h_n-1} \{\alpha_n^{(i)\tau}[\mathrm{Adj}A(z)]B(z) + \beta_n^{(i)\tau}[\det A(z)]\}z^i \\ &= \sum_{i=0}^{\infty} f_n^{(i)\tau} z^i, \end{aligned} \tag{9.47}$$

$$\sum_{i=0}^{h_n-1} \alpha_n^{(i)\tau}[\mathrm{Adj}A(z)]z^i = \sum_{i=0}^{\infty} g_n^{(i)\tau} z^i, \tag{9.48}$$

which imply that

$$f_n^{(i)\tau} = \sum_{j=0}^{i} \left[\alpha_n^{(j)\tau}\overline{B}_{i-j} + \beta_n^{(j)\tau} a_{i-j}\right], \tag{9.49}$$

$$g_n^{(i)\tau} = \sum_{j=0}^{i} \alpha_n^{(j)\tau}\overline{A}_{i-j}, \quad i \geq 0, \tag{9.50}$$

where by definition $\alpha_n^{(i)} = 0$, $\beta_n^{(i)} = 0$ for any $i \geq h_n$, $(n \geq 1)$. Therefore by (9.40) and (9.41) it follows that

$$\begin{aligned} \sum_{i=2h_n}^{\infty} \|f_n^{(i)}\|^2 &\leq 2\sum_{i=2h_n}^{\infty} \left\{ \left(\sum_{j=0}^{h_n-1} \|\alpha_n^{(j)}\| \|\overline{B}_{i-j}\| \right)^2 + \left(\sum_{j=0}^{h_n-1} \|\beta_n^{(j)}\| |a_{i-j}| \right)^2 \right\} \\ &\leq 2\sum_{i=2h_n}^{\infty} \left\{ \sum_{j=0}^{h_n-1} (\|\alpha_n^{(j)}\|^2 + \|\beta_n^{(j)}\|^2) \right\} \\ &\quad \times \left\{ \sum_{j=0}^{h_n-1} (\|\overline{B}_{i-j}\|^2 + |a_{i-j}|^2) \right\} \\ &= 2\sum_{i=2h_n}^{\infty} \sum_{j=0}^{h_n-1} (\|\overline{B}_{i-j}\|^2 + |a_{i-j}|^2) \end{aligned}$$

$$\leq 2\sum_{j=0}^{h_n-1}\sum_{i=h_n+1}^{\infty}(\|\overline{B}_i\|^2+|a_i|^2)$$

$$\leq 2h_n\left\{\sum_{i=h_n+1}^{\infty}(\|\overline{B}_i\|+|a_i|)\right\}^2$$

Consequently, by (9.46) we know that as $n\to\infty$

$$\sup_{\|x_n\|=1} h_n^{2d}\sum_{i=2h_n}^{\infty}\|f_n^{(i)}\|^2 \leq 2\left\{\sum_{i=h_n+1}^{\infty} i^{\frac{1}{2}+d}(\|\overline{B}_i\|+|a_i|)\right\}^2 \to 0.$$

Similarly,

$$\sup_{\|x_n\|=1} h_n^{2d}\sum_{i=2h_n}^{\infty}\|g_n^{(i)}\|^2 \to 0, \quad \text{as} \quad n\to\infty.$$

Hence, for the lemma it suffices to show that

$$\liminf_{n\to\infty}\inf_{\|x_n\|=1} h_n^{2d}\sum_{i=0}^{\infty}(\|f_n^{(i)}\|^2+\|g_n^{(i)}\|^2)\neq 0 \quad \text{a.s.} \tag{9.51}$$

If (9.51) were not true, we would find a set D with $P(D)>0$, and a subsequence of $\{n\}$, which without loss of generality, is also denoted by $\{n\}$ for notational simplicity, such that

$$h_n^{2d}\sum_{i=0}^{\infty}(\|f_n^{(i)}\|^2+\|g_n^{(i)}\|^2)\xrightarrow[n\to\infty]{} 0, \quad \forall\omega\in D. \tag{9.52}$$

Now substituting (9.48) into (9.47) we see that

$$\sum_{i=0}^{h_n-1}\beta_n^{(i)\tau}z^i[\det A(z)]=\sum_{i=0}^{\infty}[f_n^{(i)\tau}-g_n^{(i)\tau}B(z)]z^i. \tag{9.53}$$

Taking account of $[\text{Adj}A(z)]A(z)=[\det A(z)]I$, we have by (9.48),

$$\sum_{i=0}^{h_n-1}\alpha_n^{(i)\tau}z^i[\det A(z)]=\sum_{i=0}^{\infty}g_n^{(i)\tau}A(z)z^i. \tag{9.54}$$

Let $e^{j\theta_i}$, $i=1,\dots,s$, $j^2\triangleq-1$, $\theta_i\in[0,2\pi]$, be distinct zeros of $\det A(z)$ on the unit circle $|z|=1$, and let their multiplicities be $d_1,\dots,d_s$ respectively. Then we have $d=\max\{d_1,\dots,d_s\}$, and

$$\det A(e^{j\theta})=f(e^{j\theta})(e^{j\theta}-e^{j\theta_1})^{d_1}\dots(e^{j\theta}-e^{j\theta_s})^{d_s}, \tag{9.55}$$

where $f(e^{j\theta}) \neq 0$, $\forall \theta \in [0, 2\pi]$.

Without loss of generality assume that

$$\theta_0 \triangleq 0 < \theta_1 < \theta_2 < \dots < \theta_s < 2\pi \triangleq \theta_{s+1}. \tag{9.56}$$

It is easy to convince oneself from (9.55) that there exists a constant $c_1 > 0$ such that

$$\min_{k\in[0,s]} \min_{\theta\in[\theta_k+\varepsilon,\theta_{k+1}-\varepsilon]} |\det A(e^{j\theta})|^2 \geq c_1\varepsilon^{2d} \tag{9.57}$$

for all appropriately small $\varepsilon > 0$.

Thus, by (9.54) and (9.57) it follows that for any small $\varepsilon > 0$,

$$\begin{aligned}
\sum_{i=0}^{h_n-1} \|\alpha_n^{(i)}\|^2 &= \frac{1}{2\pi}\int_0^{2\pi} \left\| \sum_{i=0}^{h_n-1} \alpha_n^{(i)\tau} e^{ij\theta} \right\|^2 d\theta \\
&= \frac{1}{2\pi}\sum_{k=0}^{s} \int_{\theta_k+\varepsilon/h_n}^{\theta_{k+1}-\varepsilon/h_n} \left\| \sum_{i=0}^{h_n-1} \alpha_n^{(i)\tau} e^{ij\theta} \right\|^2 d\theta \\
&\quad + \frac{1}{2\pi}\left\{ \int_{\theta_0}^{\theta_0+\varepsilon/h_n} \left\| \sum_{i=0}^{h_n-1} \alpha_n^{(i)\tau} e^{ij\theta} \right\|^2 d\theta \right. \\
&\quad + \sum_{k=1}^{s} \int_{\theta_k-\varepsilon/h_n}^{\theta_k+\varepsilon/h_n} \left\| \sum_{i=0}^{h_n-1} \alpha_n^{(i)\tau} e^{ij\theta} \right\|^2 d\theta \\
&\quad \left. + \int_{\theta_{s+1}-\varepsilon/h_n}^{\theta_{s+1}} \left\| \sum_{i=0}^{h_n-1} \alpha_n^{(i)\tau} e^{ij\theta} \right\|^2 d\theta \right\} \\
&\leq \frac{h_n^{2d}}{2\pi c_1\varepsilon^{2d}} \sum_{k=0}^{s} \int_{\theta_k+\varepsilon/h_n}^{\theta_{k+1}-\varepsilon/h_n} \left\| \sum_{i=0}^{h_n-1} \alpha_n^{(i)\tau} e^{ij\theta} [\det A(e^{j\theta})] \right\|^2 d\theta \\
&\quad + \frac{1}{2\pi}\left\{ 2(s+1)\frac{\varepsilon}{h_n} \right\} h_n \sum_{i=0}^{h_n-1} \|\alpha_n^{(i)}\|^2 \\
&\leq \frac{h_n^{2d}}{2\pi c_1\varepsilon^{2d}} \int_0^{2\pi} \left\| \sum_{i=0}^{\infty} g_n^{(i)\tau} e^{ij\theta} A(e^{i\theta}) \right\|^2 d\theta + (s+1)\varepsilon/\pi \\
&\leq \frac{h_n^{2d}\|A(e^{i\theta})\|_\infty^2}{2\pi c_1\varepsilon^{2d}} \int_0^{2\pi} \left\| \sum_{i=0}^{\infty} g_n^{(i)\tau} e^{ij\theta} \right\|^2 d\theta + (s+1)\varepsilon/\pi \\
&= \frac{\|A(e^{i\theta})\|_\infty^2}{c_1\varepsilon^{2d}} \left\{ h_n^{2d} \sum_{i=0}^{\infty} \|g_n^{(i)}\|^2 \right\} + (s+1)\varepsilon/\pi.
\end{aligned}$$

Consequently, it follows from (9.52) that

$$\limsup_{n\to\infty} \sum_{i=0}^{h_n-1} \|\alpha_n^{(i)\tau}\|^2 \le (s+1)\varepsilon/\pi, \ \forall\omega \in D$$

which by the arbitrariness of ε yields

$$\limsup_{n\to\infty} \sum_{i=0}^{h_n-1} \|\alpha_n^{(i)}\|^2 = 0, \quad \forall\omega \in D. \tag{9.58}$$

Similarly, by (9.52), (9.53) and (9.57) it can be shown that

$$\limsup_{n\to\infty} \sum_{i=0}^{h_n-1} \|\beta_n^{(i)}\|^2 = 0, \quad \forall\omega \in D. \tag{9.59}$$

Finally, combining (9.40), (9.41), (9.58) and (9.59) we get a contradiction:

$$1 = \|x_n\|^2 = \sum_{i=0}^{h_n-1} (\|\alpha_n^{(i)}\|^2 + \|\beta_n^{(i)}\|^2) \xrightarrow[n\to\infty]{} 0, \quad \forall\omega \in D. \tag{9.60}$$

Hence, the assertion (9.51) holds. This completes the proof of the lemma.

Lemma 9.3. *With $f_n^{(i)}$ and $g_n^{(i)}$ defined by (9.43)*

$$\sum_{i=0}^{n-1} \left\{ \sum_{j=2h_n}^{\infty} \left[f_n^{(j)\tau} u_{i-j} + g_n^{(j)\tau} w_{i-j} \right] \right\}^2 = O(\delta_n^0 r_n^0) \quad a.s.,$$

where δ_n^0 and r_n^0 are defined in Theorem 9.2.

Proof. By (9.44), (9.49), (9.50) and Lemma 9.1(i) it follows that

$$\begin{aligned} \sum_{i=2h_n}^{\infty} \left\| g_n^{(i)} \right\| &\le \sum_{j=0}^{h_n-1} \sum_{i=2h_n}^{\infty} \left\| \alpha_n^{(j)} \right\| \|\overline{A}_{i-j}\| \\ &\le \sum_{j=0}^{h_n-1} \left\| \alpha_n^{(j)} \right\| \sum_{i=h_n}^{\infty} \|\overline{A}_i\| \\ &= O\left([h_n]^{1/2} \sum_{i=[h_n/2^m]-1}^{\infty} \|A_i\| \right), \end{aligned} \tag{9.61}$$

and similarly,

$$\sum_{i=2h_n}^{\infty} \left\| f_n^{(i)} \right\| = O\left([h_n]^{1/2} \sum_{i=[h_n/2^m]-1}^{\infty} [\|A_i\| + \|B_i\|] \right). \tag{9.62}$$

Hence, by (9.62) and the Schwarz inequality,

$$\begin{aligned}
&\sum_{i=0}^{n-1} \left\{ \sum_{j=2h_n}^{\infty} f_n^{(j)\tau} u_{i-j} \right\}^2 \\
\le\ & \sum_{i=0}^{n-1} \sum_{j=2h_n}^{\infty} \left\| f_n^{(j)} \right\| \sum_{j=2h_n}^{\infty} \left\| f_n^{(j)} \right\| \|u_{i-j}\| \\
\le\ & \left\{ \sum_{j=2h_n}^{\infty} \left\| f_n^{(j)} \right\| \right\}^2 \sum_{i=0}^{n-1} \|u_i\|^2 \\
=\ & O\left(h_n \left\{ \sum_{i=[h_n/2^m]-1}^{\infty} \|A_i\| + \|B_i\| \right\}^2 \sum_{i=0}^{n-1} \|u_i\|^2 \right) \\
=\ & O(\delta_n^0 r_n^0).
\end{aligned} \tag{9.63}$$

Similarly, by (9.61),

$$\sum_{i=0}^{n-1} \left\{ \sum_{j=2h_n}^{\infty} g_n^{(j)\tau} w_{i-j} \right\}^2 = O(\delta_n^0 r_n^0). \tag{9.64}$$

Finally, the desired result follows from (9.63) and (9.64). Q.E.D.

We are now in a position to prove Theorem 9.2.

Proof of Theorem 9.2.
Let us define

$$\psi_i(n) \triangleq [\det A(z)] \varphi_i(n). \tag{9.65}$$

with $\varphi_i(n)$ given by (9.10). Then we have

$$\begin{aligned}
\lambda_{\min} \left\{ \sum_{i=0}^{n-1} \psi_i(n) \psi_i^\tau(n) \right\} &= \inf_{\|x\|=1} \sum_{i=0}^{n-1} [x^\tau \psi_i(n)]^2 \\
&= \inf_{\|x\|=1} \sum_{i=0}^{n-1} \left[\sum_{j=0}^{\infty} a_j x^\tau \varphi_{i-j}(n) \right]^2
\end{aligned}$$

$$
\begin{aligned}
&\le \left(\sum_{j=0}^{\infty}|a_j|\right)\inf_{\|x\|=1}\sum_{i=0}^{n-1}\sum_{j=0}^{\infty}|a_j|[x^\tau\varphi_{i-j}(n)]^2 \\
&\le \left(\sum_{j=0}^{\infty}|a_j|\right)^2\inf_{\|x\|=1}\sum_{i=0}^{n-1}[x^\tau\varphi_i(n)]^2 \\
&\le \left(\sum_{j=0}^{\infty}|a_j|\right)^2\lambda_{\min}(n). \qquad (9.66)
\end{aligned}
$$

Multiplying both sides of (9.2) by $z^i\text{Adj}A(z)$ and noting (9.4), we see that

$$
[\det A(z)]y_{n-i} = z^i[\text{Adj}A(z)]B(z)u_n + z^i[\text{Adj}A(z)]w_n, \ \ \forall i \ge 0, \ n \ge 0. \qquad (9.67)
$$

Now, let $x_n \in I\!R^{(m+l)h_n}$ be the unit eigenvector corresponding to the minimum eigenvalue of the matrix $\sum_{i=0}^{n-1}\psi_i(n)\psi_i^\tau(n)$, and let x_n be written as (9.41), and $f_n^{(i)}$ and $g_n^{(i)}$ be defined via (9.43). Then similar to the finite order ARMAX case (see the proof of Theorem 6.2), it is easy to see from (9.10), (9.41)-(9.43), (9.65) and (9.67) that

$$
\begin{aligned}
\lambda_{\min}\left\{\sum_{i=0}^{n-1}\psi_i(n)\psi_i^\tau(n)\right\} &= \sum_{i=0}^{n-1}[x_n^\tau\psi_i(n)]^2 \\
&= \sum_{i=0}^{n-1}\{H_n(z)[u_i^\tau \quad w_i^\tau]^\tau\}^2 \\
&= \sum_{i=0}^{n-1}\left\{\sum_{j=0}^{\infty}[f_n^{(j)\tau}u_{i-j}^\tau + g_n^{(j)\tau}w_{i-j}]\right\}^2.
\end{aligned}
$$

Consequently, by Lemmas 9.2 and 9.3, and the elementary inequality $(x+y)^2 \ge \frac{1}{2}x^2 - y^2$, we know that

$$
\begin{aligned}
&\lambda_{\min}\left\{\sum_{i=0}^{n-1}\psi_i(n)\psi_i^\tau(n)\right\} \\
\ge\ &\frac{1}{2}\sum_{i=0}^{n-1}\left\{\sum_{j=0}^{2h_n-1}[f_n^{(j)\tau}u_{i-j}^\tau + g_n^{(j)\tau}w_{i-j}]\right\}^2.
\end{aligned}
$$

$$-\sum_{i=0}^{n-1}\left\{\sum_{j=2h_n}^{\infty}[f_n^{(j)\tau}u_{i-j}^{\tau}+g_n^{(j)\tau}w_{i-j}]\right\}^2.$$

$$= \frac{1}{2}\sum_{i=0}^{n-1}\left\{[f_n^{(0)\tau}\dots f_n^{(2h_n-1)\tau} \quad g_n^{(0)\tau}\dots g_n^{(2h_n-1)\tau}]^\tau\varphi_i^0(n)\right\}^2$$
$$+O(\delta_n^0 r_n^0)$$
$$\geq \frac{1}{2}\lambda_{\min}^0(n)\sum_{i=0}^{2h_n-1}(\|f_n^{(i)}\|^2+\|g_n^{(i)}\|^2)+O(\delta_n^0 r_n^0)$$
$$= \frac{1}{2}h_n^{-2d}\lambda_{\min}^0(n)\left\{h_n^{2d}\sum_{i=0}^{2h_n-1}(\|f_n^{(i)}\|^2+\|g_n^{(i)}\|^2)\right\}+O(\delta_n^0 r_n^0)$$
$$\geq c_2 h_n^{-2d}\lambda_{\min}^0(n)+O(\delta_n^0 r_n^0), \quad \text{a.s.} \quad \text{as} \quad n\to\infty$$

where $c_2>0$ is a constant.

Finally, combining this inequality with (9.66) yields the assertion of Theorem 9.2 with $c_0 = c_2/\left(\sum_{i=0}^{\infty}|a_i|\right)^2$. Q.E.D.

Theorem 9.2 shows that to study the growth rate of $\lambda_{\min}(n)$ it suffices to study that of $\lambda_{\min}^0(n)$. We now give two examples illustrating the case where

$$\liminf_{n\to\infty}\lambda_{\min}^0(n)/n \neq 0 \quad \text{a.s.}. \tag{9.68}$$

Example 9.1. Suppose that in addition to (9.8) the input sequence $\{u_i\}$ and the noise sequence $\{w_i\}$ are independent and that

$$\liminf_{n\to\infty}\lambda_{\min}\{E[w_n w_n^\tau|\mathcal{F}_{n-1}]\}>0, \quad \sup_n E[\|w_{n+1}\|^2|\mathcal{F}_n]<\infty, \tag{9.69}$$

$$\sup_n E[\|w_n\|^4(\log^+\|w_n\|)^{2+\delta}]<\infty \quad \text{a.s.} \tag{9.70}$$

$$\sum_{i=0}^{n-1}\|u_i\|^2 = O(n) \quad a.s., \quad E\{\|u_n\|^4(\log^+\|u_n\|)^{2+\delta}\}=O(1) \tag{9.71}$$

where $\delta>0$ is a constant. Assume further that $h_n = O(\sqrt{n}/\log n)$ a.s. Then there exists $\alpha>0$ such that as $n\to\infty$,

$$\lambda_{\min}^0(n)\geq\alpha\min\{n, \ \lambda_{\min}^1(n)\}+O(h_n\sqrt{n\log n}) \tag{9.72}$$

where

$$\lambda_{\min}^1(n)\triangleq\lambda_{\min}\left\{\sum_{i=0}^{n-1}\varphi_i^1(n)\varphi_i^{1\tau}(n)\right\}, \tag{9.73}$$

$$\varphi_i^1(n) = [u_i^\tau \ u_{i-1}^\tau \ \dots \ u_{i-2h_n+1}^\tau]^\tau. \tag{9.74}$$

In particular, if u_n takes the form

$$u_n = u_n^0 + v_n \tag{9.75}$$

where $\{u_n^0\}$ and $\{v_n\}$ are two independent sequences satisfying $E[v_n|\mathcal{F}_{n-1}] = 0$, and

$$\begin{aligned} 0 &< \liminf_{n\to\infty} \lambda_{\min}\{E[v_n v_n^\tau|\mathcal{F}_{n-1}]\} \\ &\le \limsup_{n\to\infty} E[\|v_n\|^2|\mathcal{F}_{n-1}] < \infty \quad \text{a.s.} \end{aligned} \tag{9.76}$$

$$E\left(\|u_n^0\| + \|v_n\|\right)^4 \left\{\log^+\left(\|u_n^0\| + \|v_n\|\right)\right\}^{2+\delta} = O(1), \quad \delta > 0. \tag{9.77}$$

$$\sum_{i=0}^{n-1} \|u_i^0\|^2 = O(n) \quad \text{a.s.} \tag{9.78}$$

then

$$\liminf_{n\to\infty} \lambda_{\min}^0(n)/n \neq 0 \quad \text{a.s.} \tag{9.79}$$

Proof. By Lemma 7.3, we know that

$$\begin{aligned} &\max_{1\le t\le n}\max_{1\le s\le n}\left\|\sum_{i=0}^{n} w_{i-s}u_{i-t}^\tau\right\| \\ = &\max_{1\le t\le n}\max_{1\le s\le n}\left\|\sum_{l=-s}^{n-s} w_l u_{l+s-t}^\tau\right\| \\ \le &\max_{1\le t\le n}\max_{1\le s\le n}\max_{0\le i\le n}\left\|\sum_{l=0}^{i} w_l u_{l-(t-s)}^\tau\right\| \\ \le &\max_{-n\le j\le n}\max_{0\le i\le n}\left\|\sum_{l=0}^{i} w_l u_{l-j}^\tau\right\| \\ = &\ O(\sqrt{n\log n}) \quad \text{a.s.} \end{aligned} \tag{9.80}$$

and similarly,

$$\max_{0\le t\le n}\max_{t<s\le n}\left\|\sum_{i=0}^{n-1} w_{i-s}w_{i-t}^\tau\right\| = O(\sqrt{n\log n}) \quad \text{a.s.} \tag{9.81}$$

By (9.69) and (9.70) it is not difficult to see that for some $c_w > 0$, which possibly depending on the sample,

$$\liminf_{n\to\infty}\min_{0\le j\le h_n}\lambda_{\min}\left\{\frac{1}{n}\sum_{i=0}^{n-1} E[w_{i-j}w_{i-j}^\tau|\mathcal{F}_{i-j-1}]\right\} \ge c_w \tag{9.82}$$

By (9.70) it is easy to see that

$$\|w_n\|^2 + E[\|w_n\|^2|\mathcal{F}_{n-1}] = O(\sqrt{n}).$$

Consequently, we have

$$\begin{aligned}
&\max_{0\le j\le h_n}\left\|\frac{1}{n}\sum_{i=0}^{n-1}[w_{i-j}w_{i-j}^\tau - E(w_{i-j}w_{i-j}^\tau|\mathcal{F}_{i-j-1})]\right\| \\
\le\ & \max_{0\le j\le h_n}\left\|\frac{1}{n}\sum_{l=0}^{n-j}[w_l w_l^\tau - E(w_l w_l^\tau|\mathcal{F}_{l-1})]\right\| \\
\le\ & \frac{1}{n}\left\|\sum_{l=0}^{n}\{w_l w_l^\tau - E[w_l w_l^\tau|\mathcal{F}_{l-1}]\}\right\| \\
& + \max_{0\le j\le h_n}\left\|\frac{1}{n}\sum_{l=n-j+1}^{n}\{w_l w_l^\tau - E[w_l w_l^\tau|\mathcal{F}_{l-1}]\}\right\| \\
=\ & o(1) + O\left(\frac{1}{n}(h_n\sqrt{n})\right) = o(1) \quad \text{a.s.}
\end{aligned} \tag{9.83}$$

Hence, (9.82) and (9.83) yield

$$\begin{aligned}
&\liminf_{n\to\infty}\min_{1\le j\le h_n}\lambda_{\min}\left\{\frac{1}{n}\sum_{i=0}^{n-1}w_{i-j}w_{i-j}^\tau\right\} \\
\ge\ & \liminf_{n\to\infty}\min_{1\le j\le h_n}\lambda_{\min}\left\{\frac{1}{n}\sum_{i=0}^{n-1}E[w_{i-j}w_{i-j}^\tau|\mathcal{F}_{i-j-1}]\right\} \\
& - \limsup_{n\to\infty}\max_{1\le j\le h_n}\left\|\frac{1}{n}\sum_{i=0}^{n-1}\{w_{i-j}w_{i-j}^\tau - E[w_{i-j}w_{i-j}^\tau|\mathcal{F}_{i-j-1}]\}\right\| \\
\ge\ & c_\omega > 0 \quad \text{a.s.}
\end{aligned} \tag{9.84}$$

For any vector $[\alpha^\tau,\ \ \beta^\tau]^\tau$ with $\|\alpha\|^2 + \|\beta\|^2 = 1$ and

$$\alpha \triangleq [\alpha_0^\tau\ \alpha_1^\tau\ ...\alpha_{2h_n-1}^\tau]^\tau \in \mathbb{R}^{2lh_n},$$

$$\beta \triangleq [\beta_0^\tau\ \beta_1^\tau\ ...\beta_{2h_n-1}^\tau]^\tau \in \mathbb{R}^{2mh_n},$$

we have by (9.34), (9.73), (9.74), (9.80), (9.81) and (9.84) that

$$\begin{aligned}
&[\alpha^\tau\ \ \beta^\tau]\sum_{i=0}^{n-1}\varphi_i^0(n)\varphi_i^{0\tau}(n)[\alpha^\tau\ \ \beta^\tau]^\tau \\
=\ & \alpha^\tau\sum_{i=0}^{n-1}\varphi_i^1(n)\varphi_i^{1\tau}(n)\alpha + 2\alpha\sum_{i=0}^{n-1}\varphi_i^1(n)[w_i^\tau\ ...w_{i-2h_n+1}^\tau]\beta
\end{aligned}$$

$$
\begin{aligned}
&+\sum_{j=0}^{2h_n-1}\beta_j^T\sum_{i=0}^{n-1}w_{i-j}w_{i-j}^T\beta_j\\
&+2\sum_{t=0}^{2h_n-2}\sum_{s=t+1}^{2h_n-1}\beta_t^T\sum_{i=0}^{n-1}w_{i-t}w_{i-s}^T\beta_s\\
\geq\ & \lambda_{\min}^1(n)\|\alpha\|^2+O(h_n\{n\log n\}^{1/2})+c_w n\|\beta\|^2\\
&+O\left(\sqrt{n\log n}\left\{\sum_{j=0}^{2h_n-1}|\beta_j|\right\}^2\right)\\
\geq\ & \min\{1,c_\omega\}\cdot\min\{n,\lambda_{\min}^1(n)\}+O\left(h_n\sqrt{n\log n}\right)
\end{aligned}
$$

This proves the assertion (9.72). In a similar way, it can be shown that under (9.74)-(9.78), there is a $c_\gamma > 0$, such that

$$
\lambda_{\min}\left\{\sum_{i=0}^{n-1}\varphi_i^1(n)\varphi_i^{1\tau}(n)\right\}\geq c_\gamma n+o(n) \tag{9.85}
$$

which in conjunction with (9.72) yields (9.79). Q.E.D.

Next, we give another example where the two exogenous sequences $\{u_i\}$ and $\{w_i\}$ may be correlated.

Example 9.2. Suppose that $\eta_i \triangleq [u_i^T \quad w_i^T]^\tau$ is a stationary sequence with spectral density matrix uniformly positive definite and with autocovariances satisfying

$$
\begin{aligned}
&\max_{0\leq t,k\leq 2h_n}\left\|\frac{1}{n}\sum_{i=0}^{n-1}\{\eta_{i-k}\eta_{i-t}^T-E[\eta_{i-k}\eta_{i-t}^T]\}\right\|\\
=\ & o(h_n^{-1})\quad \text{a.s.}
\end{aligned} \tag{9.86}
$$

Then with $\lambda_{\min}^0(n)$ defined by (9.35),

$$
\liminf_{n\to\infty}\lambda_{\min}^0(n)/n\neq 0\quad \text{a.s.} \tag{9.87}
$$

Proof. We need the following simple fact: The norm of any matrix $A=(A_{ij})$ with matrix blocks, A_{ij} $1\leq i,j\leq m$ is bounded by

$$
\|A\|\leq m\max_{1\leq i,j\leq m}\|A_{ij}\|. \tag{9.88}
$$

Let us denote

$$
\varphi_i^\eta(n)=[\eta_i^T \ \dots\ \eta_{i-2h_n+1}^T]^\tau. \tag{9.89}
$$

By (9.86) and (9.88) we know that

$$\left\| \frac{1}{n} \sum_{i=0}^{n-1} \{\varphi_i^\eta(n)\varphi_i^{\eta\tau}(n) - E[\varphi_0^\eta(n)\varphi_0^{\eta\tau}(n)]\} \right\|$$

$$\leq 2h_n \max_{0\leq k,t\leq 2h_n} \left\| \frac{1}{n} \sum_{i=0}^{n-1} \{\eta_{i-k}\eta_{i-t}^\tau - E[\eta_{i-k}\eta_{i-t}^\tau]\} \right\|$$

$$= o(1), \quad \text{a.s.} \quad \text{as} \quad n \to \infty$$

It is known that the autocovariance matrix and the spectral density function are related by (see, Section 1.5),

$$E\eta_i\eta_k^\tau = \int_{-\pi}^{\pi} e^{(i-k)jx} f(x)dx, \quad (j = \sqrt{-1})$$

where $f(x)$ is the spectral density of $\{\eta_i\}$. Then for any $x = [x_0^\tau \ \dots \ x_{2h_n-1}^\tau]^\tau$, with $\|x\| = 1$, we have

$$x^\tau E[\varphi_0^\eta(n)\varphi_0^{\eta\tau}(n)]x = E\left(\sum_{k=0}^{2h_n-1} x_k^\tau \eta_{-k}\right)^2$$

$$= \sum_{0\leq i,k\leq 2h_n-1} x_i^\tau (E\eta_{-i}\eta_{-k}^\tau)x_k$$

$$= \int_{-\pi}^{\pi} \sum_{0\leq i,k\leq 2h_n-1} x_i^\tau e^{-ijx} f(x) e^{kjx} x_k dx$$

$$\geq \min_{x\in[-\pi,\pi]} f(x) \int_{-\pi}^{\pi} \sum_{0\leq i,k\leq 2h_n-1} x_i^\tau e^{(k-i)jx} x_k dx$$

$$= 2\pi \min_{x\in[-\pi,\pi]} f(x) \sum_{i=0}^{2h_n-1} \|x_i\|^2 = 2\pi \min_{x\in[-\pi,\pi]} f(x) > 0,$$

and hence,

$$\inf_n \lambda_{\min} \{E[\varphi_0^\eta(n)\varphi_0^{\eta\tau}(n)]\} > 0. \tag{9.90}$$

Consequently, by noting that there is an orthogonal matrix T_n such that $\varphi_i^\eta(n) = T_n\varphi_i^0(n)$, we get

$$\liminf_{n\to\infty} \lambda_{\min}^0(n)/n = \liminf_{n\to\infty} \lambda_{\min} \left\{ \sum_{i=0}^{n-1} \varphi_i^0(n)\varphi_i^{0\tau}(n) \right\}$$

$$= \liminf_{n\to\infty} \lambda_{\min} \left\{ T_n^\tau \sum_{i=0}^{n-1} \varphi_i^\eta(n)\varphi_i^{\eta\tau}(n) T_n \right\}$$

$$
\begin{aligned}
&= \liminf_{n\to\infty} \lambda_{\min}\left\{\sum_{i=0}^{n-1} \varphi_i^{\eta}(n)\varphi_i^{\eta\tau}(n)\right\} \\
&= \liminf_{n\to\infty} \lambda_{\min}\left\{E[\varphi_0^{\eta}(n)\varphi_0^{\eta\tau}(n)]\right\} > 0.
\end{aligned}
$$

Hence (9.87) is true. Q.E.D.

Corollary 9.2. *Consider the system (9.2)-(9.4). Suppose that* $\det A(z) \neq 0,\ |z| = 1$, *and that*

$$\sum_{k=0}^{\infty} \sqrt{k}(\|A_k\| + \|B_k\|) < \infty.$$

If $\{u_i,\ w_i\}$ *satisfies conditions either in Example 9.1 or in Example 9.2, then*

$$\liminf_{n\to\infty} \lambda_{\min}(n)/n > 0 \quad a.s.$$

where $\lambda_{\min}(n)$ *is defined by (9.22) with* $\varphi_i(n)$ *given by (9.10).*

The result is a simple combination of Corollary 9.1 with Examples 9.1 and 9.2.

9.3 Estimation of Noise Process

The estimation of the noise process $\{w_n\}$ is a key step in solving many problems. In particular, if we try to fit the data $\{y_i,\ u_i\}$ generated by (9.2) to an ARMAX model, then such a step seems to be necessary.

Let us write $\widehat{\theta}(n)$ defined by (9.11) in its recursive form ($i \leq n$):

$$\widehat{\theta}_{i+1}(n) = \widehat{\theta}_i(n) + b_i(n)P_i(n)\varphi_i(n)[y_{i+1}^{\tau} - \varphi_i^{\tau}(n)\widehat{\theta}_i(n)] \tag{9.91}$$

$$P_{i+1}(n) = P_i(n) - b_i(n)P_i(n)\varphi_i(n)\varphi_i^{\tau}(n)P_i(n) \tag{9.92}$$

$$b_i(n) = [1 + \varphi_i^{\tau}(n)P_i(n)\varphi_i(n)]^{-1}, \tag{9.93}$$

where the initial values $\widehat{\theta}_0(n) = 0,\ P_0(n) = \gamma^{-1}I$, and $\varphi_i(n)$ is defined by (9.10).

For any $n \geq 1$, the noise estimate $\{\widehat{w}_i(n), 1 \leq i \leq n\}$ for $\{w_i,\ 1 \leq i \leq n\}$ is defined as

$$\widehat{w}_i(n) = y_i - \widehat{\theta}_i^{\tau}(n)\varphi_{i-1}(n), \quad 1 \leq i \leq n. \tag{9.94}$$

Theorem 9.3 *[HG]. For system (9.2)-(9.4) let (9.8) hold. If* $\{w_n\}$ *satisfies conditions required in Theorem 9.1, and if in (9.10) the regression lag* h_n

is taken as $h_n = O(\log^\alpha n)$, $\alpha > 0$, *then as* $n \to \infty$, *the estimate* $\{\widehat{w}_i(n), 1 \le i \le n\}$ *produced by (9.91)-(9.94) has the following property:*

$$\begin{aligned}&\sum_{i=0}^{n} \|\widehat{w}_i(n) - w_i\|^2 \\ = \; & O(h_n \log r_n) + O(n r_n \delta_n) \\ & + o([d(n) \log\log n]^2), \quad a.s., \end{aligned} \tag{9.95}$$

where r_n, δ_n *and* $d(n)$ *are all defined in Theorem 9.1.*

Proof. The proof is similar to that of Theorem 4.9. We note that G_i, H_i, $\alpha(n)$, $\alpha_i(n)$, β and $\psi_i(n)$ there are now replaced by A_i, B_i, $\theta(n)$, $\widehat{\theta}_i(n)$, γ and $\varphi_i(n)$ respectively. Hence (4.212) now reads as follows

$$\widehat{w}_i(n) - w_i = [\theta^\tau(n) - \widehat{\theta}_i^\tau(n)]\varphi_{i-1}(n) + \varepsilon_{i-1}(n), \tag{9.96}$$

where $\varepsilon_i(n)$ is defined by (4.210) with G_i, H_i replaced by A_i, B_i.

By (4.213) we know that

$$\begin{aligned}&\sum_{i=1}^{n} \|\varphi_{i-1}^\tau(n)[\widehat{\theta}_i(n) - \theta(n)]\|^2 \\ = \; & O(h_n \log r_n) + o([d(n) \log\log n]^2) \\ & + \left(\sum_{i=0}^{n} \sum_{j=0}^{i-1} \|\varepsilon_j(n)\|^2 \right). \end{aligned} \tag{9.97}$$

For the last term, following the derivation of (4.214), we have

$$\sum_{i=0}^{n-1} \|\varepsilon_i(n)\|^2 = O(\delta_n r_n). \tag{9.98}$$

Hence, from (9.96)-(9.98) it follows that

$$\begin{aligned}&\sum_{i=0}^{n} \|\widehat{w}_i(n) - w_i\|^2 \\ = \; & O(h_n \log r_n) + o([d(n) \log\log n]^2) + O(n\delta_n r_n) + O(\delta_n r_n)\end{aligned}$$

which implies the desired result (9.95). Q.E.D.

Certainly, if in Theorem 9.3 we only assume h_n to satisfy $h_n = O(n^\alpha)$, $0 \le \alpha \le 1$, similar results also hold.

Theorem 9.3′. *If in Theorem 9.3, the assumption $h_n = O(\log^\alpha n)$ is replaced by $h_n = O(n^\alpha)$, $0 \le \alpha \le 1$, then (9.95) becomes:*

$$\begin{aligned}&\sum_{i=0}^{n} \|\widehat{w}_i(n) - w_i\|^2 \\ = \; & O(h_n \log r_n) + O(n r_n \delta_n) \\ & +o([d(n)\log n]^2) \quad a.s. \end{aligned} \tag{9.99}$$

We now discuss the tightness of the bound in the result of Theorem 9.3. Let us assume that $\|w_n\| = O\left(\frac{\log n}{\log\log n)}\right)$, and $h_n/\log n$ is sufficiently large. Then under conditions (9.32) and (9.33) the last two terms in (9.95) are negligible. Hence (9.99) means that

$$\sum_{i=0}^{n} \|\widehat{w}_i(n) - w_i\|^2 = O(h_n \log n) \quad \text{a.s.} \tag{9.100}$$

The rate $O(h_n \log n)$ can really be achieved in some cases. To show this, we need to introduce the "persistence of excitation" (PE) concept for the current ARX(∞) model. A natural modification of the PE condition appeared in Section 4.2 is that

$$\liminf_{n\to\infty} \lambda_{\min}\left(\sum_{i=0}^{n} \varphi_i(n)\varphi_i^\tau(n)\right)/n > 0 \quad \text{a.s.} \tag{9.101}$$

where $\varphi_i(n)$ is defined by (9.10). This condition is used in transfer function approximation problems discussed in the last section. Corollary 9.2 shows how the PE condition (9.101) can be guaranteed. It is worth noting that Theorem 9.3 requires no PE condition on the system.

Note that the dimension of $\varphi_i(n)$ is $(m+l)h_n$. Another form of PE condition in the ARX(∞) case is the following uniform persistence of excitation (UPE) condition:

$$\liminf_{n\to\infty} \inf_{h_n^\beta \le t \le n} \lambda_{\min}\left\{\frac{1}{t}\sum_{i=0}^{t} \varphi_i(n)\varphi_i^\tau(n)\right\} > 0 \quad \text{a.s.} \tag{9.102}$$

where $\beta > 2$. Obviously, (9.102) is stronger than (9.101). When (9.102) is satisfied, one says that the system (9.2) is "uniformly persistently excited". Later in this section we shall give an example that satisfies (9.102).

The next theorem which is a slight extension of [HG] shows that for a class of systems with "UPE" condition satisfied, the upper bound $O(h_n \log n)$ in (9.100) is the best possible.

Theorem 9.4 *[HG]. Consider the ARX(∞) model (9.2) with output $\{y_i\}$ and input $\{u_i\}$ being m- and l-dimensional, respectively. Suppose that the following conditions hold:*

a) $\{w_i, \mathcal{F}_i\}$ *is a martingale difference sequence satisfying*

$$E[w_{i+1}w_{i+1}^{\tau}|\mathcal{F}_i] = \Sigma,\ \forall i; \quad E[\|w_{n+1}\|^4|\mathcal{F}_n] = O(1) \ a.s.$$
$$E[\|w_n\|^4(\log^+\|w_n\|)^{2+\delta}] = O(1) \quad \text{for some} \quad \delta > 0,$$

and

$$\|w_{n+1}\| = o(\sqrt{h_n \log n}/\log\log n) \quad a.s.;$$

b) The regression lag h_n monotonically increases to infinity and $h_n = O(\log^{\alpha} n)$, $\alpha > 0$;

c) The residual δ_n defined by (9.21) has the following order: $\delta_n = o(h_n \log n/n^2)$;

d) The UPE condition (9.102) holds and

$$\sum_{i=1}^{n}(\|y_i\|^2 + \|u_i\|^2) = O(n) \quad a.s.$$
$$E(\|y_n\| + \|u_n\|)^{4+\delta} = O(1), \quad \text{for some} \quad \delta > 0.$$

Then the noise estimate $\{\widehat{w}_i(n),\ i \le i \le n\}$ defined by (9.91)-(9.94) satisfies:

$$\lim_{n\to\infty}\sum_{i=0}^{n}(\widehat{w}_i(n) - w_i)(\widehat{w}_i(n) - w_i)^{\tau}/h_n \log n = (m+l)\Sigma \quad a.s.$$

Proof. By (9.96) it is known that

$$\begin{aligned}
&(\widehat{w}_i(n) - w_i)(\widehat{w}_i(n) - w_i)^{\tau}\\
= \ &[\theta(n) - \widehat{\theta}_i(n)]^{\tau}\varphi_{i-1}(n)\varphi_{i-1}^{\tau}(n)[\theta(n) - \widehat{\theta}_i(n)]\\
&+O\left(\|[\theta^{\tau}(n) - \widehat{\theta}_i^{\tau}(n)]\varphi_{i-1}(n)\| \cdot \|\varepsilon_{i-1}(n)\|\right)\\
&+O\left(\|\varepsilon_{i-1}(n)\|^2\right). \qquad (9.103)
\end{aligned}$$

By the assumption $r_n = O(n)$, $\delta_n = o(h_n \log n/n^2)$ we know from (9.98) that

$$\sum_{i=0}^{n}\|\varepsilon_{i-1}(n)\|^2 = o(h_n \log n/n) \quad \text{a.s.} \qquad (9.104)$$

Substituting this into (9.97) and taking account of $d(n) = O(\sqrt{h_n \log n}/\log\log n)$, we get

$$\begin{aligned}&\sum_{i=1}^{n} \|\varphi_{i-1}^{T}(n)[\widehat{\theta}_i(n) - \theta(n)]\|^2 \\ = \;& O(h_n \log n) + o(h_n \log n) = O(h_n \log n). \qquad (9.105)\end{aligned}$$

Substituting (9.104) and (9.105) into (9.103) and applying the Schwarz inequality it is easy to see that

$$\begin{aligned}&\sum_{i=1}^{n} (\widehat{w}_i(n) - w_i)(\widehat{w}_i(n) - w_i)^{\tau} \\ = \;& \sum_{i=1}^{n} [\theta(n) - \widehat{\theta}_i(n)]^{\tau} \varphi_{i-1}(n)\varphi_{i-1}^{T}(n)[\theta(n) - \widehat{\theta}_i(n)] \\ & + o(h_n \log n) \quad \text{a.s.} \qquad (9.106)\end{aligned}$$

We now proceed to analyse the first term on the right-hand side of (9.106).

Set

$$M_i(k) = \sum_{j=0}^{i} \varphi_j(k)\varphi_j^{\tau}(k) + \gamma I, \qquad (9.107)$$

$$S_i(k) = \sum_{j=0}^{i} \varphi_j(k) w_{j+1}^{\tau}, \quad S_{-1}(k) = 0. \qquad (9.108)$$

Similar to the proof of Theorem 4.9, it is not difficult to see that

$$\begin{aligned}&\widehat{\theta}_i(n) - \theta(n) \\ = \;& M_{i-1}^{-1}(n) \left\{ S_{i-1}(n) + \sum_{j=0}^{i-1} \varphi_j(n)\varepsilon_j^{\tau}(n) - \gamma\theta(n) \right\}\end{aligned}$$

and therefore,

$$\begin{aligned}&\varphi_{i-1}^{T}(n)[\widehat{\theta}_i(n) - \theta(n)] \\ = \;& \varphi_{i-1}^{T}(n) M_{i-1}^{-1}(n) S_{i-1}(n) + \varphi_{i-1}^{T}(n) M_{i-1}^{-1}(n) \sum_{j=0}^{i-1} \varphi_j(n)\varepsilon_j^{\tau}(n) \\ & - \varphi_{i-1}^{T}(n) M_{i-1}^{-1}(n)\gamma\theta(n). \qquad (9.109)\end{aligned}$$

Note that $\|\varphi_{i-1}^{T}(n) M_{i-1}^{-1}(n)\varphi_{i-1}(n)\| \le 1$. It follows from Lemma 4.7(ii) and (9.104) that

$$\sum_{i=1}^{n} \left\| \varphi_{i-1}^{T}(n) M_{i-1}^{-1}(n) \sum_{j=0}^{i-1} \varphi_j(n)\varepsilon_j^{\tau}(n) \right\|^2$$

$$\leq \sum_{i=1}^{n}\left\| M_{i-1}^{-\frac{1}{2}}(n)\sum_{j=0}^{i-1}\varphi_j(n)\varepsilon_j^{\tau}(n)\right\|^2$$
$$\leq \sum_{i=1}^{n}\sum_{j=0}^{i-1}\|\varepsilon_j(n)\|^2 = \sum_{i=1}^{n} o(h_n\log n/n)$$
$$= o(h_n\log n) \quad \text{a.s.} \tag{9.110}$$

By Lemma 4.7(i) and the assumption $r_n = O(n)$, and $\inf\limits_{i\geq (h_n)\beta}\lambda_{\min}(M_i(n)) \xrightarrow[n\to\infty]{} \infty$ it is not difficult to see that

$$\sum_{i=1}^{n}\|\varphi_{i-1}^{\tau}(n)M_{i-1}^{-1}(n)\|^2$$
$$\leq O\left(\sum_{i=1}^{h_n^{\beta}}\varphi_{i-1}^{\tau}(n)M_{i-1}^{-1}(n)\varphi_{i-1}(n)\right)$$
$$+\sum_{i=h_n^{\beta}+1}^{n}\varphi_{i-1}^{\tau}(n)M_{i-1}^{-1}(n)\varphi_{i-1}(n)/\lambda_{\min}(M_{i-1}(n))$$
$$= o(h_n\log n) \quad \text{a.s.} \tag{9.111}$$

For the first term on the right-hand side of (9.109) we apply Lemma 4.6 to estimate it. Hence we have

$$\sum_{i=1}^{n}\left\|\varphi_{i-1}^{\tau}(n)M_{i-1}^{-\frac{1}{2}}(n)S_{i-1}(n)\right\|^2$$
$$\leq O\left(h_n\log^{+}\lambda_{\min}(M_n(h_n))\right) + o([d(n)\log\log n]^2)$$
$$= O(h_n\log n) + o(h_n\log n)$$
$$= O(h_n\log n) \quad \text{a.s.} \tag{9.112}$$

By (9.110)-(9.112) and the Schwarz inequality we deduce from (9.109) that

$$\sum_{i=1}^{n}[\theta(n)-\widehat{\theta}_i(n)]^{\tau}\varphi_{i-1}(n)\varphi_{i-1}^{\tau}(n)[\theta(n)-\widehat{\theta}_i(n)]$$
$$= \sum_{i=1}^{n}S_{i-1}^{\tau}(n)M_{i-1}^{-1}(n)\varphi_{i-1}(n)\varphi_{i-1}^{\tau}(n)M_{i-1}^{-1}(n)S_{i-1}(n)$$
$$+o(h_n\log n).$$

Substituting this into (9.106) we obtain

$$\sum_{i=1}^{n}(\widehat{w}_i(n)-w_i)(\widehat{w}_i(n)-w_i)^{\tau}$$

$$= \sum_{i=1}^{n} S_{i-1}^{\tau}(n)M_{i-1}^{-1}(n)\varphi_{i-1}(n)\varphi_{i-1}^{\tau}(n)M_{i-1}^{-1}(n)S_{i-1}(n) + o(h_n \log n). \quad (9.113)$$

The remaining task is to estimate the second term on the right-hand side of (9.113).

Similar to the derivation of (4.201), by an expansion of the Lyapunov function $S_i^{\tau}(n)M_i^{-1}(n)S_i(n)$, we have

$$\begin{aligned} & S_i^{\tau}(n)M_i^{-1}(n)S_i(n) \\ = \; & S_{i-1}^{\tau}(n)M_{i-1}^{-1}(n)S_{i-1}(n) + c_i(n)w_{i+1}\varphi_i^{\tau}(n)M_{i-1}^{-1}(n)S_{i-1}(n) \\ & + c_i(n)[w_{i+1}\varphi_i^{\tau}(n)M_{i-1}^{-1}(n)S_{i-1}(n)]^{\tau} \\ & - c_i(n)S_{i-1}^{\tau}(n)M_{i-1}^{-1}(n)\varphi_i(n)\varphi_i^{\tau}(n)M_{i-1}^{-1}(n)S_{i-1}(n) \\ & + c_i(n)\varphi_i^{\tau}(n)M_{i-1}^{-1}(n)\varphi_i(n)w_{i+1}w_{i+1}^{\tau} \end{aligned} \quad (9.114)$$

where $c_i(n) = [1 + \varphi_i^{\tau}(n)M_{i-1}^{-1}(n)\varphi_i(n)]^{-1}$.

By the formula

$$M_i^{-1}(n) = M_{i-1}^{-1}(n) - c_i(n)M_{i-1}^{-1}(n)\varphi_i(n)\varphi_i^{\tau}(n)M_{i-1}^{-1}(n),$$

it is easy to verify that $c_i(n)M_{i-1}^{-1}(n)\varphi_i(n) = M_i^{-1}(n)\varphi_i(n)$. Hence from (9.114) it follows that

$$\begin{aligned} & S_i^{\tau}(n)M_i^{-1}(n)S_i(n) \\ = \; & S_{i-1}^{\tau}(n)M_{i-1}^{-1}(n)S_{i-1}(n) + w_{i+1}\varphi_i^{\tau}(n)M_i^{-1}(n)S_{i-1}(n) \\ & + S_{i-1}^{\tau}(n)M_i^{-1}(n)\varphi_i(n)w_{i+1}^{\tau} \\ & - S_{i-1}^{\tau}(n)M_i^{-1}(n)\varphi_i(n)\varphi_i^{\tau}(n)M_{i-1}^{-1}(n)S_{i-1}(n) \\ & + \varphi_i^{\tau}(n)M_i^{-1}(n)\varphi_i(n)w_{i+1}w_{i+1}^{\tau}. \end{aligned} \quad (9.115)$$

By the definitions of $c_i(n)$ and $S_i(n)$, we have

$$\begin{aligned} & S_{i-1}^{\tau}(n)M_i^{-1}(n)\varphi_i(n)\varphi_i^{\tau}(n)M_{i-1}^{-1}(n)S_{i-1}(n) \\ = \; & S_{i-1}^{\tau}(n)M_i^{-1}(n)\varphi_i(n)\varphi_i^{\tau}(n)M_i^{-1}(n)S_{i-1}(n)c_i^{-1}(n) \\ = \; & S_{i-1}^{\tau}(n)M_i^{-1}(n)\varphi_i(n)\varphi_i^{\tau}(n)M_i^{-1}(n)S_{i-1}(n) \\ & + \varphi_i^{\tau}(n)M_{i-1}^{-1}(n)\varphi_i(n)S_{i-1}^{\tau}(n) \\ & M_i^{-1}(n)\varphi_i(n)\varphi_i^{\tau}(n)M_i^{-1}(n)S_{i-1}(n) \\ = \; & [S_i(n) - \varphi_i(n)w_{i+1}^{\tau}]^{\tau}M_i^{-1}(n)\varphi_i(n) \\ & \varphi_i^{\tau}(n)M_i^{-1}(n)[S_i(n) - \varphi_i(n)w_{i+1}^{\tau}] \\ & + O\left(\varphi_i^{\tau}(n)M_{i-1}^{-1}(n)\varphi_i(n)\|\varphi_i^{\tau}(n)M_i^{-1}(n)S_{i-1}(n)\|^2\right) \\ = \; & S_i^{\tau}(n)M_i^{-1}(n)\varphi_i(n)\varphi_i^{\tau}(n)M_i^{-1}(n)S_i(n) \end{aligned}$$

$$
\begin{aligned}
&+O\left(\|\varphi_i^T(n)M_i^{-1}(n)S_i(n)\|\cdot\|w_{i+1}\|\cdot\varphi_i^T(n)M_i^{-1}(n)\varphi_i(n)\right)\\
&+O\left([\varphi_i^T(n)M_i^{-1}(n)\varphi_i(n)]^2\|w_{i+1}\|^2\right)\\
&+O\left(\varphi_i^T(n)M_i^{-1}(n)\varphi_i(n)\|\varphi_i^T(n)M_i^{-1}(n)S_{i-1}(n)\|^2\right).
\end{aligned}
\tag{9.116}
$$

We now analyze the last three terms on the right-hand side of (9.116). By the moment assumption on $\|y_n\|+\|u_n\|$, it is seen that

$$
\begin{aligned}
& P\left([\|y_n\|+\|u_n\|]^4 \geq n^{\frac{4+\delta/2}{4+\delta}}\right)\\
\leq\ & P\left([\|y_n\|+\|u_n\|]^{4+\delta} \geq n^{1+\delta/8}\right) \leq O\left(\frac{1}{n^{1+\delta/8}}\right).
\end{aligned}
$$

So by the Borel-Cantelli lemma,

$$
\|y_n\|^4+\|u_n\|^4 = O\left(n^{1-\frac{\delta}{2(4+\delta)}}\right). \tag{9.117}
$$

Without loss of generality we may assume that in the UPE condition (9.102), β satisfies

$$
\beta = 2\delta^{-1}(8+2\delta). \tag{9.118}
$$

Then by (9.117) and (9.102) we have

$$
\begin{aligned}
& \sum_{i=[h_n^\beta]+1}^{n} [\varphi_i^T(n)M_i^{-1}(n)\varphi_i(n)]^2\\
\leq\ & \sum_{i=[h_n^\beta]+1}^{n} \frac{\|\varphi_i(n)\|^4}{\lambda_{\min}^2(M_i(n))}\\
=\ & O\left(\sum_{i=[h_n^\beta]+1}^{n} \frac{h_n^2 i^{1-\frac{\delta}{2(4+\delta)}}}{i^2}\right)\\
=\ & O\left(h_n^2 \sum_{i=[h_n^\beta]+1}^{n} \frac{1}{i^{1+\frac{\delta}{2(4+\delta)}}}\right)\\
=\ & O\left(h_n^{2-\frac{\delta\beta}{8+2\delta}}\right) = O(1) \quad \text{a.s.}
\end{aligned}
\tag{9.119}
$$

Hence by using the upper bounds on $\{w_i\}$ we can estimate the second term on the right-hand side of (9.116) as follows

$$
\sum_{i=[h_n^\beta]+1}^{n} [\varphi_i^T(n)M_i^{-1}(n)\varphi_i(n)]^2\|w_{i+1}\|^2
$$

$$
\begin{aligned}
&\leq \max_{[h_n^\beta]\leq i\leq n} \|w_{i+1}\|^2 \sum_{i=[h_n^\beta]+1}^{n} [\varphi_i^T(n)M_i^{-1}(n)\varphi_i(n)]^2 \\
&= o(h_n \log n)\cdot O(1) = o(h_n \log n) \quad \text{a.s.} \qquad (9.120)
\end{aligned}
$$

By (9.117), (9.118) and the UPE condition we have

$$
\begin{aligned}
&\max_{[h_n^\beta]+1\leq i\leq n} \varphi_i^T(n)M_{i-1}^{-1}(n)\varphi_i(n) \\
&= O\left(\max_{[h_n^\beta]+1\leq i\leq n} \frac{h_n i^{\frac{1}{2}-\frac{\delta}{4(4+\delta)}}}{i}\right) \\
&= O\left(\frac{h_n}{h_n^{\delta\beta/(8+2\delta)}}\right) \xrightarrow[n\to\infty]{} 0. \qquad (9.121)
\end{aligned}
$$

By this and (4.206) we obtain

$$
\begin{aligned}
&\sum_{i=[h_n^\beta]+1}^{n} \varphi_i^T(n)M_{i-1}^{-1}(n)\varphi_i(n)\|\varphi_i^T(n)M_i^{-1}(n)S_{i-1}(n)\|^2 \\
&= o\left(\sum_{i=[h_n^\beta]+1}^{n} \|\varphi_i^T(n)M_i^{-1}(n)S_{i-1}(n)\|^2\right) \\
&= o(h_n \log n) + o([d(n)\log\log n]^2) \\
&= o(h_n \log n) \quad \text{a.s.} \qquad (9.122)
\end{aligned}
$$

By Lemma 4.6, we know that

$$
\sum_{i=1}^{n} \|\varphi_i^T(n)M_i^{-1}(n)S_i(n)\|^2 = O(h_n \log n).
$$

Hence by (9.120) and (9.122) we see from (9.116) that

$$
\begin{aligned}
&\sum_{i=[h_n^\beta]+1}^{n} S_{i-1}^T(n)M_i^{-1}(n)\varphi_i(n)\varphi_i^T(n)M_{i-1}^{-1}(n)S_{i-1}(n) \\
&= \sum_{i=[h_n^\beta]+1}^{n} S_i^T(n)M_i^{-1}(n)\varphi_i(n)\varphi_i^T(n)M_i^{-1}(n)S_i(n) \\
&\quad + o(h_n \log n).
\end{aligned}
$$

By this and (9.115) we have

$$
S_n^T(n)M_n^{-1}(n)S_n(n)
$$

$$
\begin{aligned}
&+\sum_{i=[h_n^\beta]+1}^{n} S_i^\tau(n)M_i^{-1}(n)\varphi_i(n)\varphi_i^\tau(n)M_i^{-1}(n)S_i(n) \\
&+o(h_n\log n) \\
=\;& S^\tau_{[h_n^\beta]}(n)M^{-1}_{[h_n^\beta]}(n)S_{[h_n^\beta]}(n) \\
&+O\left(\left\|\sum_{i=[h_n^\beta]+1}^{n} S_{i-1}^\tau(n)M_i^{-1}(n)\varphi_i(n)w_{i+1}\right\|\right) \\
&+\sum_{i=[h_n^\beta]+1}^{n} \varphi_i^\tau(n)M_i^{-1}(n)\varphi_i(n)w_{i+1}w_{i+1}^\tau.
\end{aligned}
\tag{9.123}
$$

Similar to the proof of Lemma 4.6, it is easy to show that

$$
\begin{aligned}
&\left\|S^\tau_{[h_n^\beta]}(n)M^{-\frac{1}{2}}_{[h_n^\beta]}(n)\right\|^2 \\
=\;& O\left(h_n\log^+\lambda_{\max}\left(M_{[h_n^\beta]}(n)\right)\right)+o([d(n)\log\log n]^2) \\
=\;& O(h_n\log n)+o(h_n\log n)=o(h_n\log n) \quad \text{a.s.}
\end{aligned}
\tag{9.124}
$$

and

$$
\sum_{j=0}^{[h_n^\beta]}\|\varphi_j^\tau(n)M_j^{-1}(n)S_j(n)\|^2=o(h_n\log n) \quad \text{a.s.} \tag{9.125}
$$

By Lemma 7.3(ii), we have

$$
\|S_n(n)\|^2=\left\|\sum_{j=0}^{n}\varphi_j(n)w_{j+1}^\tau\right\|^2=O(h_n n\log\log n) \quad \text{a.s.}
$$

Consequently, the UPE condition ensures

$$
S_n^\tau(n)M_n^{-1}(n)S_n(n)=O(h_n\log\log n) \quad \text{a.s.} \tag{9.126}
$$

Next, we consider the second term on the right-hand side of (9.123). By (4.206) it is known that

$$
\sum_{i=1}^{n}\|S_{i-1}^\tau(n)M_i^{-1}(n)\varphi_i(n)\|^2=O(h_n\log n) \quad \text{a.s.}
$$

Applying Theorem 2.10 we get

$$
\begin{aligned}
&\max_{1\le k\le n}\left\|\sum_{i=1}^{k}S_{i-1}^\tau(n)M_i^{-1}(n)\varphi_i(n)w_{i+1}\right\| \\
=\;& O\left(\sqrt{h_n\log n}\log\log n\right)+o\left(\sqrt{h_n\log n}\,d(n)\log\log n\right) \\
=\;& o(h_n\log n) \quad \text{a.s.}
\end{aligned}
\tag{9.127}
$$

Finally, we estimate the last term on the right-hand side of (9.123). By Lemma 4.7(i), it is obvious that

$$\sum_{i=0}^{[h_n^\beta]} \|\varphi_i^T(n)M_i^{-1}(n)\varphi_i(n)\|^2 = O(h_n \log\log n) \quad \text{a.s.}$$

From this and (9.119) it follows that

$$\sum_{i=0}^{n} [\varphi_i^T(n)M_i^{-1}(n)\varphi_i(n)]^2 = O(h_n \log\log n) \quad \text{a.s.}$$

Hence applying Theorem 2.9(i) we get

$$\begin{aligned}
&\max_{1\le k\le n} \left\| \sum_{i=0}^{k} \varphi_i^T(n)M_i^{-1}(n)\varphi_i(n)[w_{i+1}w_{i+1}^T - \Sigma] \right\| \\
&= O\left(h_n \log\log n\right) + o\left(d^2(n)\log\log n\right) \\
&= o(h_n \log n) \quad \text{a.s.} \qquad (9.128)
\end{aligned}$$

Similar to (4.204), it is easy to show that

$$\begin{aligned}
&\sum_{i=[h_n^\beta]+1}^{n} \varphi_i^T(n)M_i^{-1}(n)\varphi_i(n) \\
&= \sum_{i=[h_n^\beta]+1}^{n} \frac{\det[M_i(n)] - \det[M_{i-1}(n)]}{\det[M_i(n)]} \\
&\le \int_{\det[M_{[h_n^\beta]}(n)]}^{\det[M_n(n)]} \frac{dx}{x} \le \log\det[M_n(n)] \\
&\le (m+l)h_n \log \lambda_{\max}[M_n(n)] \\
&\le (m+l)h_n \log\{O(h_n\cdot n)\} \\
&= (m+l)h_n \log n + o(h_n \log n). \qquad (9.129)
\end{aligned}$$

On the other hand, by the UPE condition we know that there exists $\varepsilon > 0$ such that for sufficiently large n,

$$\lambda_{\min}[M_i(n)] \ge \varepsilon i, \quad \forall i \in \left[[h_n^\beta],\ n\right]$$

and therefore

$$\sum_{i=[h_n^\beta]+1}^{n} \varphi_i^T(n)M_{i-1}^{-1}(n)\varphi_i(n)$$

$$
\begin{aligned}
&= \sum_{i=[h_n^\beta]+1}^{n} \frac{\det[M_i(n)] - \det[M_{i-1}(n)]}{\det[M_{i-1}(n)]} \\
&\geq \int_{\det[M_{[h_n^\beta]}(n)]}^{\det[M_n(n)]} \frac{dx}{x} \\
&\geq \log\det[M_n(n)] - \log\det\left[M_{[h_n^\beta]}(n)\right] \\
&\geq \log(\varepsilon n)^{(m+l)h_n} + o(h_n \log\log n) \\
&= (m+l)h_n \log n + o(h_n \log n). \qquad (9.130)
\end{aligned}
$$

Similar to (9.119), it is easy to show that

$$
\sum_{i=[h_n^\beta]+1}^{n} [\varphi_i^T(n) M_{i-1}^{-1}(n)\varphi_i(n)]^2 = O(1) \quad \text{a.s.} \qquad (9.131)
$$

Hence by (9.130), (9.131) and the elementary inequality $\frac{x}{1+x} \geq (1-x)x$, $(x > 0)$ we have

$$
\begin{aligned}
&\sum_{i=[h_n^\beta]+1}^{n} \varphi_i^T(n) M_i^{-1}(n)\varphi_i(n) \\
&= \sum_{i=[h_n^\beta]+1}^{n} \frac{\varphi_i^T(n) M_{i-1}^{-1}(n)\varphi_i(n)}{1+\varphi_i^T(n) M_{i-1}^{-1}(n)\varphi_i(n)} \\
&\geq \sum_{i=[h_n^\beta]+1}^{n} \left\{1 - \varphi_i^T(n) M_{i-1}^{-1}(n)\varphi_i(n)\right\} \varphi_i^T(n) M_{i-1}^{-1}(n)\varphi_i(n) \\
&\geq h_n(m+l)\log n + o(h_n \log n) \\
&\quad - \sum_{i=[h_n^\beta]+1}^{n} [\varphi_i^T(n) M_{i-1}^{-1}(n)\varphi_i(n)]^2 \\
&= (m+l)h_n \log n + o(h_n \log n).
\end{aligned}
$$

From this and (9.129), we arrive at

$$
\begin{aligned}
&\sum_{i=[h_n^\beta]+1}^{n} \varphi_i^T(n) M_i^{-1}(n)\varphi_i(n) \\
&= (m+l)h_n \log n + o(h_n \log n). \qquad (9.132)
\end{aligned}
$$

Combining (9.128) with (9.132), we get

$$
\begin{aligned}
&\sum_{i=[h_n^\beta]+1}^{n} \varphi_i^T(n) M_i^{-1}(n)\varphi_i(n) w_{i+1} w_{i+1}^T \\
&= (m+l)(h_n \log n)\Sigma + o(h_n \log n). \qquad (9.133)
\end{aligned}
$$

Substituting (9.124), (9.126), (9.127) and (9.133) into (9.123) leads to

$$\begin{aligned}&\sum_{i=[h_n^{\beta}]+1}^{n} S_i^{\tau}(n)M_i^{-1}(n)\varphi_i(n)\varphi_i^{\tau}(n)M_i^{-1}(n)S_i(n)\\ = \;&(m+l)(h_n\log n)\Sigma + o(h_n\log n).\end{aligned}$$

This in conjunction with (9.125) gives

$$\begin{aligned}&\sum_{i=0}^{n} S_i^{\tau}(n)M_i^{-1}(n)\varphi_i(n)\varphi_i^{\tau}(n)M_i^{-1}(n)S_i(n)\\ = \;&(m+l)(h_n\log n)\Sigma + o(h_n\log n).\end{aligned} \tag{9.134}$$

By (9.126) it is seen that

$$\begin{aligned}&\left\|S_n^{\tau}(n)M_n^{-1}(n)\varphi_n(n)\right\|^2\\ \le\;&\left\|S_n^{\tau}(n)M_n^{-\frac{1}{2}}(n)\right\|^2 = O(h_n\log\log n)\end{aligned}$$

Hence (9.134) implies

$$\begin{aligned}&\sum_{i=0}^{n-1} S_i^{\tau}(n)M_i^{-1}(n)\varphi_i(n)\varphi_i^{\tau}(n)M_i^{-1}(n)S_i(n)\\ = \;&(m+l)(h_n\log n)\Sigma + o(h_n\log n).\end{aligned}$$

Substituting this into (9.113) we finally get the desired result.

Q.E.D.

We now discuss the conditions used in Theorem 9.4. Condition a) is easy to understand. It is satisfied if for example $\{w_i\}$ is a zero mean i.i.d. sequence with $E\exp(|w_1|^{1+\delta}) < \infty$ for some $\delta > 0$, and if $\log n/h_n$ is bounded. Condition b) is not a condition on the system itself, since the lag h_n appears in the estimation algorithm. The result of Theorem 9.4 indicates that the smaller h_n the smaller the tracking error $\{\widehat{w}_i(n) - w_i\}$. However, h_n cannot be made arbitrarily small, since Condition c) contains a restriction on h_n. For example, if $\|A_i\| + \|B_i\| = O(\lambda^i)$, $\forall i$, $\lambda \in (0,1)$, then by (9.21) we know that $\delta_n = O(\lambda^{2h_n})$. To satisfy Condition c) we solve the following equations:

$$\lambda^{2h_n} = o(h_n\log n/n^2), \quad h_n = O(\log^{\alpha} n), \quad \alpha > 0,$$

and get

$$(\alpha+1)\log\log n + O(1) - 2\log n + 2h_n\log\left(\frac{1}{\lambda}\right) \to \infty.$$

This shows that the smallest α is 1, and that h_n may be taken as any sequence that satisfies

$$h_n - \frac{\log\left(\frac{n}{\log n}\right)}{\log(\lambda^{-1})} \to \infty. \tag{9.135}$$

For Condition d), we illustrate the UPE condition (9.102) by the following example.

Example 9.3. Assume in model (9.2)-(9.4) that $A(z)$ and $B(z)$ are rational, $\det A(z) \neq 0$, $|z| \leq 1$, the input sequence $\{u_i\}$ is an ARMA process of the form

$$\begin{aligned} &C(z)u_i = D(z)v_i, \quad \det C(z) \neq 0, \quad |z| \leq 1 \\ &C^{-1}(z)D(z)[C^{-1}(z)D(z)]^* > 0, \quad |z| = 1. \end{aligned} \tag{9.136}$$

Assume further that $\{v_i\}$ is independent of $\{w_i\}$, and that $\varepsilon_i \triangleq [w_i^\tau \quad v_i^\tau]^\tau$ is i.i.d. with zero mean and positive definite covariance matrix, and satisfies

$$E\|\varepsilon_1\|^4(\log^+\|\varepsilon_1\|)^{2+\delta} < \infty \quad \text{for some} \quad \delta > 0.$$

Then the UPE condition (9.102) holds for any $h_n = O(\log^\alpha n)$, ($\alpha > 0$, $h_n \to \infty$), and $\beta > 2$.

Proof. Denote $x_i = [y_i^\tau, \quad u_i^\tau]^\tau$. Then by (9.2) and (9.136)

$$x_i = \begin{bmatrix} A^{-1}(z), & A^{-1}(z)B(z)C^{-1}(z)D(z) \\ 0, & C^{-1}(z)D(z) \end{bmatrix} \varepsilon_i.$$

Hence $\{x_i\}$ is a stationary ARMA process with innovations ε_i. Therefore, we have the following uniform convergence rate for autocovariances [HDe, Theorem 5.3.2], [Hu1, Theorem 4]:

$$\begin{aligned} &\max_{0 \leq i \leq \log^a n} \left\| \frac{1}{n} \sum_{j=1}^{n-i} x_j x_{i+j}^\tau - E x_0 x_i^\tau \right\| \\ &= O\left(\sqrt{\frac{\log\log n}{n}}\right), \quad \forall a > 0 \quad \text{a.s.} \end{aligned} \tag{9.137}$$

Put

$$\psi_j(n) = [x_j^\tau \ \dots \ x_{j-h_n+1}^\tau]^\tau \in \mathbb{R}^{(m+l)h_n}$$

Then by (9.89) and (9.137) we have

$$\left\| \frac{1}{i} \sum_{j=0}^{i} \psi_j(n)\psi_j^\tau(n) - E\psi_0(n)\psi_0^\tau(n) \right\|$$

$$
\begin{aligned}
&\le (m+l)h_n \max_{0\le k,s\le h_n} \left\| \frac{1}{i}\sum_{j=0}^{i} x_{j-k}x_{j-s}^{\tau} - Ex_{-k}x_{-s}^{\tau} \right\| \\
&= (m+l)h_n O\left(\sqrt{\frac{\log\log i}{i}}\right)
\end{aligned}
$$

and for $\beta > 2$

$$
\begin{aligned}
&\sup_{i\ge h_n^{\beta}} \left\| \frac{1}{i}\sum_{j=0}^{i} \psi_j(n)\psi_j^{\tau}(n) - E\psi_0(n)\psi_0^{\tau}(n) \right\| \\
&\le (m+l)h_n O\left(\sqrt{\frac{\log\log n}{h_n^{\beta}}}\right) \to 0 \quad \text{a.s.}
\end{aligned}
$$

Since there is an orthogonal matrix T_n such that $\psi_j(n) = T_n\varphi_j(n)$, we conclude that

$$
\begin{aligned}
&\sup_{i\ge h_n^{\beta}} \left\| \frac{1}{i}\sum_{j=0}^{i} \varphi_j(n)\varphi_j^{\tau}(n) - E\varphi_0(n)\varphi_0^{\tau}(n) \right\| \\
&= O(1) \quad \text{a.s.} \qquad (9.138)
\end{aligned}
$$

By Theorem 9.2 and Example 9.1, we know that as $n \to \infty$

$$
\lambda_{\min}(n) \ge \alpha \min\left\{n,\ \lambda_{\min}^{1}(n)\right\} + O\left(\sqrt{n}\log^{\alpha+\frac{1}{2}} n\right), \qquad (9.139)
$$

where $\lambda_{\min}^{1}(n)$ is defined by (9.73). Since $\{u_i\}$ is a stationary process with spectral density matrix being uniformly positive definite and since by (9.137)

$$
\max_{0\le t,k\le 2h_n} \left\| \frac{1}{n}\sum_{i=0}^{n-1} u_{i-k}u_{i-t}^{\tau} - Eu_{i-k}u_{i-t}^{\tau} \right\| = o\left(h_n^{-1}\right),
$$

similar to Example 9.2, it can be shown that

$$
\liminf_{n\to\infty} \lambda_{\min}^{1}(n)/n \ne 0 \quad \text{a.s.}
$$

Substituting this into (9.139) we get

$$
\liminf_{n\to\infty} \lambda_{\min}(n)/n \ne 0 \quad \text{a.s.}
$$

By this and (9.138) we have

$$
\liminf_{n\to\infty} \lambda_{\min}\left(E\varphi_0(n)\varphi_0^{\tau}(n)\right)
$$

$$
\begin{aligned}
&\geq \liminf_{n\to\infty} \lambda_{\min}(n)/n \\
&\quad - \limsup_{n\to\infty} \left\| \frac{1}{n} \sum_{j=0}^{n} \varphi_j(n)\varphi_j^\tau(n) - E\varphi_0(n)\varphi_0^\tau(n) \right\| \\
&= \liminf_{n\to\infty} \lambda_{\min}(n)/n > 0 \quad \text{a.s.}
\end{aligned}
$$

Hence, by (9.138) we get the desired result:

$$
\begin{aligned}
&\liminf_{n\to\infty} \inf_{t\geq h_n^\beta} \lambda_{\min}\left(\frac{1}{t}\sum_{i=0}^{t} \varphi_i(n)\varphi_i^\tau(n)\right) \\
&\geq \liminf_{n\to\infty} \lambda_{\min}\left(E\varphi_0(n)\varphi_0^\tau(n)\right) \\
&\quad - \limsup_{n\to\infty} \sup_{t\geq h_n^\beta} \left\| \frac{1}{t}\sum_{j=0}^{t} \varphi_j(n)\varphi_j^\tau(n) - E\varphi_0(n)\varphi_0^\tau(n) \right\| \\
&> 0 \quad \text{a.s.}
\end{aligned}
$$

Q.E.D.

We remark that in Example 9.3 if $A(z)$ and $B(z)$ are not rational but their coefficients satisfy a certain summable condition, then the UPE condition (9.102) can also be verified. In this case the only nontrivial change in the proof is that the right-hand side of (9.137) is replaced by $O\left(\sqrt{\frac{\log n}{n}}\right)$ (see [Hu1, Theorem 4]).

CHAPTER 10

Estimation for Time-Varying Parameters

Tracking or estimating a system or a signal whose properties vary with time is a fundamental problem in system identification as well as in signal processing. The basic time-varying model is that of a regression:

$$y_k = \varphi_k^\tau \theta_k + v_k, \quad \forall k \geq 0 \tag{10.1}$$

where y_k and v_k are the scalar output and the noise respectively, and φ_k and θ_k are, respectively, the r-dimensional *stochastic* regressor and the unknown time-varying parameter. It is convenient to denote the parameter variation at time k by Δ_k:

$$\Delta_k \triangleq \theta_k - \theta_{k-1}, \quad \forall k \geq 1. \tag{10.2}$$

It is clear that if $\Delta_k \equiv 0$, and φ_k and v_k respectively have the following forms

$$\begin{aligned} \varphi_k &= [y_{k-1} \dots y_{k-t} \quad u_{k-1} \dots u_{k-s}]^\tau \\ v_k &= w_k + c_1 w_{k-1} + \dots + c_l w_{k-l}, \end{aligned}$$

then (10.1) is reduced to the time-invariant ARMAX model studied in the preceding chapters. In the time-invariant case, the adaptation gain in the estimation algorithm is usually diminishing (for example, the gain $\frac{a\psi_n}{r_n}$ in SG algorithm (4.22)). Algorithms for estimating constant parameters will fail in the time varying case, since the parameter variations Δ_k are not expected to be vanishing as $k \to \infty$. Hence algorithms with non-vanishing gains are naturally used. The analysis for such algorithms is no longer similar to that applied in Chapter 4 and hinges on the stability analysis for time-varying linear equations. The latter problem is known by its difficulty.

Section 10.1 present some results on this topic, which are necessary for our study. A key excitation condition in the time-varying case, called the "conditional richness condition" is introduced in Section 10.2. Two typical algorithms for tracking time-varying parameters, the Kalman filtering and the LMS algorithms, are studied in Sections 10.3 and 10.4. The material of this chapter is essentially based on [Gu2].

10.1 Stability of Random Time-Varying Equations

We start with the following simple time-varying linear equation:

$$x_{k+1} = (1 - a_k)x_k + \xi_{k+1}, \quad k \geq 0, \tag{10.3}$$

where $a_k \in [0, 1)$ are random variables, and where the initial value x_0 satisfies $E|x_0| < \infty$.

The solution of (10.3) can be expressed as

$$x_{k+1} = \prod_{i=0}^{k}(1 - a_i)x_0 + \sum_{i=0}^{k}\left[\prod_{j=i+1}^{k}(1 - a_j)\right]\xi_{i+1}, \tag{10.4}$$

where as usual $\prod_{j=i+1}^{k}(\cdot) \triangleq 1$, if $k \leq i$.

For convenience of discussion, we introduce the following definition.

Definition 10.1. A random sequence $\{x_k,\ k \geq 0\}$ defined on the basic probability space $(\Omega, \mathcal{F}, P)$ is called L_p-stable ($p > 0$) if $\sup_k E\|x_k\|^p < \infty$. If such a sequence is generated by (10.3), then the equation (10.3) is called L_p-stable.

In the sequel, we shall use the following notation

$$L_p = \left\{\{x_k\} : \sup_k E\|x_k\|^p < \infty\right\}. \tag{10.5}$$

A natural question is: under which conditions equation (10.3) is L_p-stable.

A commonly used condition is that there exist constants $c > 0$, $\gamma \in (0, 1)$ such that

$$E\prod_{j=i+1}^{k}(1 - a_j) \leq c\gamma^{k-i}, \quad \forall k \geq i, \tag{10.6}$$

which, as easily to be seen, guarantees the L_p-stability of (10.3) provided that $\{x_0,\ \xi_k\}$ satisfies some moment conditions. The next result shows that (10.6) is also a necessary condition in some sense.

Proposition 10.1 *Let $\{a_k\}$ be a sequence of mutually independent random variables. Then for any $\{\xi_k\} \in \mathcal{M}$, the equation (10.3) is L_1-stable if and only if (10.6) holds, where $\mathcal{M}$ is the set defined as*

$$\mathcal{M} \triangleq \{\{\xi_k\} : \{\xi_k\} \in L_1 \text{ and } \{\xi_k\} \text{ is independent of } \{a_k\}\}.$$

Proof. The sufficiency is straightforward from the expression (10.4). We need only to prove the necessity.

For this we take $\xi_k \equiv 1$. Obviously $\{\xi_k\} \in \mathcal{M}$. Hence, by the assumption $\{x_k\} \in L_1$ from (10.4) we have

$$\sup_k \sum_{i=0}^{k} E\left[\prod_{j=i+1}^{k}(1-a_j)\right] \le c < \infty. \tag{10.7}$$

Setting

$$b_k = \prod_{j=0}^{k} E(1-a_j),$$

by the independency of $\{a_j\}$ we have from (10.7) that for any $n \ge k \ge 0$,

$$b_n \sum_{i=k}^{n} b_i^{-1} = \sum_{i=k}^{n} \prod_{j=i+1}^{n} E(1-a_j) \le c, \tag{10.8}$$

and hence

$$\begin{aligned}
\sum_{i=k}^{n} b_i^{-1} &= \sum_{i=k}^{n-1} b_i^{-1} + b_n^{-1} \\
&\ge \sum_{i=k}^{n-1} b_i^{-1} + \frac{1}{c}\sum_{i=k}^{n-1} b_i^{-1} \ge \left(1+\frac{1}{c}\right)\sum_{i=k}^{n-1} b_i^{-1} \\
&\ge \dots \ge \left(1+\frac{1}{c}\right)^{n-k} b_k^{-1}, \quad \forall n \ge k \ge 0.
\end{aligned}$$

Therefore, by (10.8) again, we see that

$$b_n \le c\left(\sum_{i=k}^{n} b_i^{-1}\right)^{-1} \le c\left(\frac{c}{1+c}\right)^{n-k} b_k$$

which is tantamount to (10.6). Q.E.D.

Since (10.6) plays a key role in the stability study of equation (10.3) it is important to investigate for which kind of (possibly strongly correlated) random variables a_k, (10.6) does hold. The results presented here will be used in later sections.

Theorem 10.1 *Let $\{a_m, \mathcal{F}_m\}$ be an adapted random sequence satisfying*

$$a_m \in [0,1], \quad E[a_{m+1}|\mathcal{F}_m] \geq \frac{1}{\alpha_m}, \quad \forall m \geq 0 \tag{10.9}$$

with $\{\alpha_m, \mathcal{F}_m\}$ being an adapted nonnegative sequence, satisfying

$$\alpha_{m+1} \leq a\alpha_m + \eta_{m+1}, \quad \forall m \geq 0, \quad M_0 \triangleq E\alpha_0^{1+\delta} < \infty, \tag{10.10}$$

where $\{\eta_m, \mathcal{F}_m\}$ is an adapted nonnegative sequence with

$$\sup_{m\geq 0} E\left[\eta_{m+1}^{1+\delta}|\mathcal{F}_m\right] \leq M \quad a.s. \tag{10.11}$$

and where $a \in [0,1)$, $0 < \delta < \infty$ and $0 \leq M < \infty$. Then there exist two constants $c > 0$ and $\gamma \in (0,1)$ such that

$$E\prod_{k=m}^{n}(1 - a_{k+1}) \leq c\gamma^{n-m+1}, \quad \forall n \geq m \geq 0, \tag{10.12}$$

where c and γ only depend on a, M, M_0 and δ.

For the proof we need the following auxiliary result.

Lemma 10.1 *Let $\{\alpha_m, \mathcal{F}_m\}$ be an adapted nonnegative sequence satisfying (10.10) and (10.11). Then there is an adapted sequence $\{\beta_m, \mathcal{F}_m\}$ such that*

$$\beta_m \geq \alpha_m, \quad \beta_m \geq 1, \quad \forall m \geq 0 \tag{10.13}$$

and that

$$\beta_{m+1} = b\beta_m + \xi_{m+1}, \quad 0 < b < 1, \quad E\beta_0^{1+\delta} < \infty, \tag{10.14}$$

where $\{\xi_k, \mathcal{F}_k\}$ is an adapted sequence satisfying

$$\xi_k \geq 0, \quad \sup_{k\geq 0} E[\xi_{k+1}^{1+\delta}|\mathcal{F}_k] \leq \overline{M} < \infty \quad a.s. \tag{10.15}$$

and

$$N \triangleq \sup_{k\geq 0} E[\xi_{k+1}|\mathcal{F}_k] \leq b \quad a.s. \tag{10.16}$$

We remark that property (10.16) characterizes the key difference between $\{\alpha_k\}$ and $\{\beta_k\}$

Proof. Define a constant L by

$$L = \max\left\{1, (2M)^{1/\delta}\right\} \tag{10.17}$$

where δ and M are the same as in (10.11) and take b as

$$b = \frac{L + \frac{1-a}{2}}{L + (1-a)} \tag{10.18}$$

where a is the same as in (10.10). Obviously, $b \in (0,1)$ and $b > a$.

Next, let us introduce two processes $\{\alpha_k^{(1)}\}$ and $\{\alpha_k^{(2)}\}$ as follows

$$\begin{aligned} \alpha_{k+1}^{(1)} &= a\alpha_k^{(1)} + \eta_{k+1} I(\eta_{k+1} < L), \quad \alpha_0^{(1)} = 0; \\ \alpha_{k+1}^{(2)} &= b\alpha_k^{(2)} + \eta_{k+1} I(\eta_{k+1} \geq L), \quad \alpha_0^{(2)} = \alpha_0. \end{aligned} \tag{10.19}$$

Obviously, $\alpha_k^{(1)} \leq \frac{L}{1-a}$, $\forall k \geq 0$, and by (10.10) and the fact that $b \geq a$ we have

$$\alpha_k \leq \alpha_k^{(1)} + \alpha_k^{(2)} \leq \frac{L}{1-a} + \alpha_k^{(2)}.$$

Now, define $\beta_k = \frac{L}{1-a} + \alpha_k^{(2)}$. Then (10.13) holds. Further, from (10.19) it follows that

$$\beta_{k+1} = b\beta_k + \xi_{k+1}$$

where $\xi_{k+1} = \frac{(1-b)L}{1-a} + \eta_{k+1} I(\eta_{k+1} \geq L)$. Hence (10.14) is true, and we need only to verify (10.15) and (10.16).

By (10.11) and (10.17) we know that

$$E\left[\eta_{k+1} I(\eta_{k+1} \geq L) | \mathcal{F}_k\right] \leq E\left[\left.\frac{\eta_{k+1}^{1+\delta}}{L^\delta}\right| \mathcal{F}_k\right] \leq \frac{M}{L^\delta} < \frac{1}{2}$$

and by (10.18)

$$b = \frac{1}{2} + \frac{(1-b)L}{1-a}.$$

Hence (10.16) holds, while (10.15) is obvious. Q.E.D.

Proof of Theorem 10.1.

Let us define a sequence $\{x_k\ k \in [m,n]\}$ by

$$x_{k+1} = (1 - a_{k+1})x_k, \quad x_m = 1, \quad \forall k \geq m. \tag{10.20}$$

By (10.9) and Lemma 10.1, it follows that

$$\begin{aligned} & E\left[\beta_{k+1} x_{k+1} | \mathcal{F}_k\right] \leq E\left[(b\beta_k + \xi_{k+1})(1 - a_{k+1}) x_k | \mathcal{F}_k\right] \\ \leq\ & b\beta_k \left[1 - E(a_{k+1}|\mathcal{F}_k)\right] x_k + x_k E[\xi_{k+1}|\mathcal{F}_k] \\ \leq\ & b\beta_k \left(1 - \frac{1}{\beta_k}\right) x_k + N x_k \\ =\ & b\beta_k x_k + (N - b)x_k \leq b\beta_k x_k. \end{aligned}$$

Consequently, we have

$$E\beta_{n+1}x_{n+1} \le bE\beta_n x_n \le \ldots \le b^{n-m+1}E\beta_m x_m$$
$$= b^{n-m+1}E\beta_m \le b^{n-m+1}\left[E\beta_0 + \frac{b}{1-b}\right], \qquad (10.21)$$

where for the last inequality we have used (10.14) and (10.16). This completes the proof of Theorem 10.1. Q.E.D.

Theorem 10.2 *Let $\{x_k, \mathcal{F}_k\}$ be a nonnegative adapted process, $x_k \ge 1$, and satisfy*

$$x_{k+1} \le (1-a_{k+1})x_k + c, \quad \forall k \ge 0, \quad Ex_0^2 < \infty, \qquad (10.22)$$

where $c > 0$ is a constant, and $\{a_k\}$ is defined as in Theorem 10.1. Furthermore, assume that $a_k \in [0, \overline{a}]$, $\overline{a} < 1$. Then there exist constants $N > 0$ and $\lambda \in (0,1)$ such that

$$E\prod_{k=m}^{n}\left(1-\frac{1}{x_k}\right) \le N\lambda^{n-m+1}, \forall n \ge m \ge 0. \qquad (10.23)$$

Proof. Without loss of generality assume that

$$x_{k+1} = (1-a_{k+1})x_k + c, \quad \forall k \ge 0, \quad Ex_0^2 < \infty. \qquad (10.24)$$

We first prove (10.23) for the case where $c \le 1$. Defining a sequence $\{y_k, k \in [m,n]\}$ by

$$y_k = \left(1-\frac{1}{x_k}\right)y_{k-1}, \quad y_{m-1} = 1, \qquad (10.25)$$

we have

$$x_k y_k = (x_k - 1)y_{k-1} \le (1-a_k)x_{k-1}y_{k-1}$$

and

$$y_n \le x_n y_n \le \prod_{k=m}^{n}(1-a_k)\cdot x_{m-1}. \qquad (10.26)$$

Note that by Theorem 10.1 and (10.24),

$$\sup_k Ex_k^2 < \infty.$$

Hence by this, Theorem 10.1 and the Schwarz inequality we obtain

$$E\prod_{k=m}^{n}\left(1-\frac{1}{x_k}\right) = Ey_k \le c'\gamma^{(n-m+1)/2}, \quad n \ge m,$$

where $c' > 0$, $\gamma \in (0,1)$. This shows that (10.23) is valid.

Next, we consider the case where $c > 1$.

Let us take $\varepsilon = c^{-1}$. Then $\varepsilon \in (0,1)$,

$$\varepsilon x_{k+1} = (1 - a_{k+1})(\varepsilon x_k) + 1, \quad \varepsilon x_k \geq 1, \quad \forall k \geq 1 \tag{10.27}$$

and the argument used above leads to

$$E \prod_{k=m}^{n} \left(1 - \frac{1}{\varepsilon x_k}\right) \leq c' \gamma^{(n-m+1)/2}, \quad n \geq m, \tag{10.28}$$

where, again, $c' > 0$ and $\gamma \in (0,1)$.

Since $a_k \in [0, \overline{a}]$, $\overline{a} < 1$, we see from (10.27) that

$$\varepsilon x_{k+1} \geq (1 - \overline{a}) + 1, \quad \forall k \geq 1.$$

Next we prove the following inequality which is needed here:

$$1 - x \leq (1 - dx)^{(1-t)/d}, \quad 0 \leq dx \leq t < 1, \quad d > 1. \tag{10.29}$$

To see this, let $f(x) = x + (1 - dx)^{(1-t)/d} - 1$. Then $f(0) = 0$, and the derivative of $f(x)$ satisfies:

$$\begin{aligned} f'(x) &= 1 - (1-t)(1-dx)^{\frac{(1-t)}{d}-1} \\ &\geq 1 - (1-t)(1-t)^{\frac{(1-t)}{d}-1} = 1 - (1-t)^{(1-t)/d} \geq 0. \end{aligned}$$

Hence (10.29) holds. Substituting $x = \frac{1}{x_k}$, $d = \frac{1}{\varepsilon}$, $t = \frac{1}{2 - \overline{a}}$ into (10.29) yields

$$1 - \frac{1}{x_k} \leq \left(1 - \frac{1}{\varepsilon x_k}\right)^{(1-\overline{a})\varepsilon/(2-\overline{a})}, \quad \forall k \geq 2.$$

Consequently, by the Hölder inequality and (10.28) it follows that

$$\begin{aligned} E \prod_{k=m}^{n} \left(1 - \frac{1}{x_k}\right) &\leq \left\{ E \prod_{k=m}^{n} \left(1 - \frac{1}{\varepsilon x_k}\right) \right\}^{(1-\overline{a})\varepsilon/(2-\overline{a})} \\ &\leq c' \lambda^{n-m+1}, \quad \text{for some } c' > 0 \quad \gamma \in (0,1), \quad \forall n \geq m \geq 2. \end{aligned}$$

By suitably adjusting c' it is easy to see that this inequality holds for all $n \geq m \geq 0$. Q.E.D.

Up to now we have studied some cases where (10.6) holds. As we mentioned before (10.6) is a key condition in guaranteeing the L_p-stability of (10.3). However, from the L_p-stability of $\{x_k\}$, we cannot directly infer the boundedness of sample averages of $\{x_n\}$. This issue is addressed in the following theorem.

Theorem 10.3 *Let $\{f_k, \mathcal{F}_k\}$ be an adapted nonnegative sequence satisfying*

$$f_{k+1} \le (1 - a_{k+1})f_k + \bar{\xi}_{k+1}, \quad \forall k \ge 0, \quad Ef_0^\alpha < \infty \tag{10.30}$$

for some $\alpha > 0$, where $\{a_k\}$ is the same as in Theorem 10.1, and $\{\bar{\xi}_k, \mathcal{F}_k\}$ is nonnegative and satisfies

$$\sup_{k\ge 0} E\bar{\xi}_{k+1}^\alpha < \infty, \quad B \triangleq \limsup_{n\to\infty} \frac{1}{n}\sum_{i=0}^n \bar{\xi}_i^\alpha < \infty \quad a.s. \tag{10.31}$$

Then there exists a constant $L \ge 0$ such that

$$\limsup_{n\to\infty} \frac{1}{n}\sum_{i=0}^n f_i^\beta \le LB^{\frac{\beta}{\alpha}} \quad a.s., \quad \forall \beta \in (0, \alpha) \tag{10.32}$$

whenever δ appearing in (10.11) satisfies $\delta > \dfrac{2\beta}{\alpha - \beta}$.

Proof. First of all, it is easy to see that for any $\varepsilon > 0$ and $d > 0$, there is a constant $g(\varepsilon, d)$ such that

$$(a + b)^d \le (1 + \varepsilon)a^d + g(\varepsilon, d)b^d, \quad \forall a \ge 0, b \ge 0. \tag{10.33}$$

Applying this to (10.30) leads to that for any $\beta \in (0, \alpha)$,

$$f_{k+1}^{\frac{\alpha+\beta}{2}} \le (1 - ta_{k+1})(1 + \varepsilon)f_k^{\frac{\alpha+\beta}{2}} + g(\varepsilon, \frac{\alpha+\beta}{2})\bar{\xi}_{k+1}^{\frac{\alpha+\beta}{2}}, \quad \forall k \ge 0 \tag{10.34}$$

and

$$f_{k+1}^\beta \le (1 - ta_{k+1})(1 + \varepsilon)f_k^\beta + g(\varepsilon, \beta)\bar{\xi}_{k+1}^\beta, \quad \forall k \ge 0, \tag{10.35}$$

where the following inequality has been invoked

$$(1 - a_{k+1})^{\frac{\alpha+\beta}{2}} \le (1 - a_{k+1})^\beta \le (1 - ta_{k+1}) \quad t = \min(1, \beta).$$

It is clear that Theorem 10.1 still holds if $\{a_m, \mathcal{F}_m\}$ is replaced by $\{ta_m, \mathcal{F}_m\}$. Hence applying Theorem 10.1, and taking ε suitably small in (10.34) we see that

$$\sup_{k\ge 0} Ef_k^{\frac{\alpha+\beta}{2}} < \infty. \tag{10.36}$$

By Lemma 10.1, we know that there is an adapted sequence $\{\beta_n, \mathcal{F}_n\}$ satisfying (10.14)-(10.16) and

$$\beta_m \ge \frac{1}{t}\alpha_m, \quad \forall m \ge 0. \tag{10.37}$$

Now, let us set

$$\tilde{a}_k \triangleq a_k - E(a_k|\mathcal{F}_{k-1}), \quad \tilde{\xi}_k \triangleq \xi_k - E(\xi_k|\mathcal{F}_{k-1}), \tag{10.38}$$

where $\{\xi_k\}$ is the process appearing in (10.14).

By (10.37), (10.9) and (10.16) it is easy to see that

$$a_{k+1} \geq \tilde{a}_{k+1} + \frac{1}{t\beta_k}, \quad \xi_k \leq \tilde{\xi}_k + b. \tag{10.39}$$

By this, (10.14) and (10.35) it follows that

$$\begin{aligned}
\beta_{k+1} f_{k+1}^{\beta} &\leq (b\beta_k + \xi_{k+1})(1 - ta_{k+1})(1+\varepsilon) f_k^{\beta} \\
&\quad + g(\varepsilon, \beta)\overline{\xi}_{k+1}^{\beta} \beta_{k+1} \\
&\leq b(1+\varepsilon)(1 - ta_{k+1}) f_k^{\beta} \beta_k + (1+\varepsilon)\xi_{k+1} f_k^{\beta} \\
&\quad + g(\varepsilon, \beta)\overline{\xi}_{k+1}^{\beta} \beta_{k+1} \\
&\leq b(1+\varepsilon)\left[1 - t\tilde{a}_{k+1} - \frac{t}{t\beta_k}\right] f_k^{\beta} \beta_k \\
&\quad + (1+\varepsilon)(\tilde{\xi}_{k+1} + b) f_k^{\beta} \\
&\quad + g(\varepsilon, \beta)\overline{\xi}_{k+1}^{\beta} \beta_{k+1} \\
&= b(1+\varepsilon) f_k^{\beta} \beta_k - b(1+\varepsilon) t\tilde{a}_{k+1} f_k^{\beta} \beta_k \\
&\quad + (1+\varepsilon)\tilde{\xi}_{k+1} f_k^{\beta} + g(\varepsilon, \beta)\overline{\xi}_{k+1}^{\beta} \beta_{k+1}.
\end{aligned} \tag{10.40}$$

We now proceed to estimate the last three terms on the right-hand side of (10.40).

By (10.14)-(10.16) it is easy to see that

$$\sup_{k\geq 0} E\beta_k^{1+\delta} < \infty, \quad \sup_{k\geq 0} E|\xi_k|^{1+\delta} < \infty. \tag{10.41}$$

Taking account of $|\tilde{a}_k| \leq 1$, from (10.41), (10.36) and the Hölder inequality (with $p = \dfrac{\alpha + 3\beta + 2\beta\delta}{2\beta(1+\delta)}$, $q = \dfrac{\alpha + 3\beta + 2\beta\delta}{\alpha + \beta}$) we see that for any $\lambda \in \left(0, \dfrac{(\alpha - \beta)\delta - 2\beta}{\alpha + 3\beta + 2\beta\delta}\right]$,

$$\sup_{k\geq 0} E|\tilde{a}_{k+1} f_k^{\beta} \beta_k|^{1+\lambda} < \infty,$$

and

$$\sup_{k\geq 0} E|\tilde{\xi}_{k+1} f_k^{\beta}|^{1+\lambda} < \infty.$$

Therefore, by Corollary 2.6 we have

$$\frac{1}{n}\sum_{k=1}^{n}\left(-b(1+\varepsilon)+\tilde{a}_{k+1}f_k^{\beta}\beta_k+(1+\varepsilon)\tilde{\xi}_{k+1}f_k^{\beta}\right) \xrightarrow[n\to\infty]{} 0 \quad a.s. \tag{10.42}$$

By (10.40) and $\beta_k \geq 1$ it is clear that if ε is small enough such that $b(1+\varepsilon)<1$ then

$$\begin{aligned}\frac{1}{n}\sum_{k=0}^{n} f_k^{\beta} &\leq \frac{1}{n}\sum_{k=0}^{n}\beta_k f_k^{\beta}\\ &\leq \frac{1}{[1-b(1+\varepsilon)]n}\Bigg[\beta_0 f_0^{\beta}+\sum_{k=0}^{n-1}(-b(1+\varepsilon)t\tilde{a}_{k+1}f_k^{\beta}\beta_k\\ &\quad +(1+\varepsilon)\tilde{\xi}_{k+1}f_k^{\beta})\\ &\quad +\sum_{k=0}^{n-1}g(\varepsilon,\beta)\overline{\xi}_{k+1}^{\beta}\beta_{k+1}\Bigg].\end{aligned} \tag{10.43}$$

By (10.42) and (10.43) we know that to complete the proof it remains to analyse the last term on the right-hand side of (10.43).

By (10.15) and Corollary 2.6, it is easy to see that for any $c\in(0,\delta)$

$$\begin{aligned}\frac{1}{n}\sum_{k=1}^{n}\xi_k^{1+c} &= \frac{1}{n}\sum_{k=1}^{n}E(\xi_k^{1+c}|\mathcal{F}_{k-1})+\frac{1}{n}\sum_{k=1}^{n}[\xi_k^{1+c}-E(\xi_k^{1+c}|\mathcal{F}_{k-1})]\\ &= O(1) \quad a.s.\end{aligned}$$

which in conjunction with (10.14) yields

$$\frac{1}{n}\sum_{k=1}^{n}\beta_k^{1+c}=O(1), \quad \forall c<\delta.$$

By this, (10.31) and the Hölder inequality (with $p=\frac{\alpha}{\beta}$, $q=\frac{\alpha}{\alpha-\beta}$) we have

$$\limsup_{n\to\infty}\frac{1}{n}\sum_{k=0}^{n-1}\overline{\xi}_{k+1}^{\beta}\beta_{k+1}\leq L'B^{\frac{\beta}{\alpha}} \quad a.s.$$

Hence the desired result (10.32) is true. Q.E.D.

In a completely similar way, one can prove the following result.

Theorem 10.3′. *Let $\{x_k, \mathcal{F}_k\}$ be an adapted nonnegative random process satisfying*

$$x_{k+1}\leq(1-a_{k+1})\alpha x_k+c, \quad x_0=1, \quad \forall k\geq 0, \tag{10.44}$$

where $\{a_k\}$ *is defined as in Theorem 10.1, and* α *and* c *are finite, positive constants. Then there exists a constant* $\alpha^* > 1$ *such that whenever* $\alpha \in [1, \alpha^*)$,

$$\limsup_{n\to\infty} \frac{1}{n}\sum_{i=0}^{n} x_i \le c' \quad a.s. \tag{10.45}$$

where c' *is a constant.*

To conclude this section, we remark that up to now there is no universal result on stability of time-varying equations, all the more when the time-varying coefficients are random processes. This section is by no means an exposition of a general stability theory, since we restricted ourselves to present only results needed in later sections.

10.2 Conditional Richness Condition

Like the constant parameter case, some kind of excitation or richness on φ_i is necessary in estimating the unknown time-varying parameters. The stationarity assumption on φ_i is widely used in the adaptive signal processing area for studying LMS algorithms (more discussions on LMS will be given in Section 10.4). Without any doubt it is reasonable in certain circumstances, but restrictive in general. For example, the stationarity assumption on φ_i excludes the feedback control systems from consideration. Thus, finding a weaker richness condition on φ_i, which includes both stationary and nonstationary signals, is important in both theory and application. The following condition will be used in the next two sections.

Conditional Richness (CR) Condition

We say that an adapted sequence $\{\varphi_k, \mathcal{F}_k\}$ (i.e., φ_k is $\mathcal{F}_k$-measurable for any k, where $\{\mathcal{F}_k\}$ is a family of nondecreasing σ-algebras), satisfies the CR condition, if there exists an integer $h > 0$ such that

$$E\left[\sum_{k=m+1}^{m+h} \frac{\varphi_k\varphi_k^\tau}{1+\|\varphi_k\|^2} | \mathcal{F}_m\right] \ge \frac{1}{\alpha_m} I \ a.s., \ \forall m \ge 0, \tag{10.46}$$

where $\{\alpha_m, \mathcal{F}_m\}$ is an adapted nonnegative sequence satisfying

$$\alpha_{m+1} \le a\alpha_m + \eta_{m+1}, \quad \forall m \ge 0, \quad M_0 \triangleq E\alpha_0^{1+\delta} < \infty, \tag{10.47}$$

with $\{\eta_m, \mathcal{F}_m\}$ being an adapted nonnegative sequence satisfying

$$\sup_{m\ge 0} E\left[\eta_{m+1}^{1+\delta} | \mathcal{F}_m\right] \le M \quad a.s. \tag{10.48}$$

and where $a \in [0, 1)$, $0 < \delta < \infty$ and $0 \le M < \infty$ are constants.

At first glance, the CR condition looks rather complicated, however, it does have a clear meaning and is satisfied by a large class of stochastic signals. An important special case of (10.47) is when $a = 0$, $\eta_n \equiv \alpha^{-1}$ for some $\alpha > 0$. In this case (10.46) reduces to

$$E\left[\sum_{k=m+1}^{m+h} \frac{\varphi_k\varphi_k^{\tau}}{1+\|\varphi_k\|^2}|\mathcal{F}_m\right] \ge \alpha I \quad a.s., \quad \forall m \ge 0. \tag{10.49}$$

What (10.46) effectively means is that the matrix on the left-hand side of (10.46) may not be uniformly positive definite, since the sequence $\{\alpha_k\}$ may not be bounded in sample path. We now give some examples to illustrate the CR condition.

Example 10.1 If there are constants $0 < \alpha < \beta < \infty$ and integer $h > 0$ such that

$$\alpha I < \sum_{k=m+1}^{m+h} \varphi_k\varphi_k^{\tau} \le \beta I \quad a.s. \quad \forall m \ge 0, \tag{10.50}$$

then the CR condition (10.46) holds.

The proof is straightforward since in this case (10.49) holds. Note that if we regard (10.1) and (10.2) as a state space equation with state θ_k, then (10.50) is nothing but the uniform observability condition for θ_k. In the area of deterministic adaptive control, (10.50) is sometimes called "sufficient richness condition". Notwithstanding the fairly wide use of (10.50), its verification appears to be very difficult if not impossible. Actually, (10.50) is mainly a deterministic hypothesis and excludes many standard signals including any unbounded signals φ_k. Hence (10.50) is an undesirable condition. A different generalization of (10.50) is given in [GXM].

Example 10.2 Let $\{\varphi_k\}$ be an r-dimensional ϕ-mixing process. This means that there is a deterministic sequence $\{\phi(h),\, h \ge 0\}$ such that

(i). $\phi(h) \to 0$, as $h \to \infty$; and

(ii). $\sup_{A \in \mathcal{F}_{s+h}^{\infty},\, B \in \mathcal{F}_0^{s}} |P(A|B) - P(A)| \le \phi(h), \quad \forall s \ge 0, \quad \forall h \ge 0,$

where, for any nonnegative integers $s \ge 0$ and $h \ge 0$, $\mathcal{F}_0^s \triangleq \sigma\{\varphi_k, 0 \le k \le s\}$ $\mathcal{F}_{s+h}^{\infty} \triangleq \sigma\{\varphi_k, s+h \le k < \infty\}$.

Suppose further that

$$\inf_k \lambda_{min}(E\varphi_k\varphi_k^{\tau}) > 0 \quad \text{and} \quad \sup_k E\|\varphi_k\|^4 < \infty. \tag{10.51}$$

Then the CR condition (10.46) holds with $\mathcal{F}_m = \mathcal{F}_0^m$.

Proof. We need the following two simple inequalities:

a). For any $\mathcal{F}_{m+h}^{\infty}$-measurable scalar function x_{m+h} with $|x_{m+h}| \leq 1$,

$$|E[x_{m+h}|\mathcal{F}_m] - Ex_{m+h}| \leq 2\phi(h). \tag{10.52}$$

b). For any random vector φ

$$\lambda_{min}\left(E\frac{\varphi\varphi^{\tau}}{1+\|\varphi\|^2}\right) \geq \frac{[\lambda_{min}(E[\varphi\varphi^{\tau}])]^2}{E\left[\|\varphi\|^2+\|\varphi\|^4\right]}. \tag{10.53}$$

We first prove (10.52) and start with the simple case where x_{m+h} is an indicator function $x_{m+h} = I_A$, where $A \in \mathcal{F}_{m+h}^{\infty}$. Suppose that there is a set $B \in \mathcal{F}_m$ with $P(B) > 0$ such that $|E[x_{m+h}|\mathcal{F}_m] - Ex_{m+h}| > \phi(h)$, $\omega \in B$. Without loss of generality assume that

$$E[x_{m+h}|\mathcal{F}_m] - Ex_{m+h} > \phi(h), \quad \omega \in B.$$

Using this and the ϕ-mixing property (ii) gives

$$\begin{aligned}\phi(h) &< \left|\frac{1}{P(B)}\int_B [E[x_{m+h}|\mathcal{F}_m] - P(A)]dP\right| \\ &= \left|\frac{1}{P(B)}[P(AB) - P(B)P(A)]\right| \leq \phi(h)\end{aligned}$$

which is a contradiction. This proves that for each indicator function x_{m+h} the strengthened inequality (10.52) with 2 replaced by 1 holds.

In the general case, since x_{m+h} can be approximated by step (random) functions, we need only to consider the case where

$$x_{m+h} = \sum_i a_i I_{A_i}$$

where $\{A_i\}$ is a finite decomposition of the sample space Ω into disjoint element of $\mathcal{F}_{m+h}^{\infty}$. Note that the inequality $|x_{m+h}| \leq 1$ implies $|a_i| \leq 1$. Then we have

$$\begin{aligned}|E[x_{m+h}|\mathcal{F}_m] - Ex_{m+h}| &= \left|\sum_i a_i[P(A_i|\mathcal{F}_m) - P(A_i)]\right| \\ &\leq \sum_i |P(A_i|\mathcal{F}_m) - P(A_i)|.\end{aligned} \tag{10.54}$$

Further, let us denote

$$\begin{aligned}C^+ &= \{A_i : P(A_i|\mathcal{F}_m) - P(A_i) > 0\}, \\ C^- &= \{A_i : P(A_i|\mathcal{F}_m) - P(A_i) \leq 0\}.\end{aligned}$$

Obviously both C^+ and C^- belong to $\mathcal{F}_{m+h}^{\infty}$; hence

$$\begin{aligned}&\sum_i |P(A_i|\mathcal{F}_m) - P(A_i)| \\ = \; & [P(C^+|\mathcal{F}_m) - P(C^+)] + [P(C^-) - P(C^-|\mathcal{F}_m)] \\ \leq \; & 2\phi(h),\end{aligned}$$

which in conjunction with (10.54) yields (10.52).

To prove (10.53), let x be the unit eigenvector corresponding to $\lambda_{\min}\left(\frac{\varphi\varphi^{\tau}}{1+\|\varphi\|^2}\right)$. Then by the Schwarz inequality we see that

$$\begin{aligned}[\lambda_{\min}(E\varphi\varphi^{\tau})]^2 &\leq \left\{E\frac{|x^{\tau}\varphi|}{\sqrt{1+\|\varphi\|^2}}\cdot\|\varphi\|\sqrt{1+\|\varphi\|^2}\right\}^2 \\ &\leq E\frac{x^{\tau}\varphi\varphi^{\tau}x}{1+\|\varphi\|^2}E[\|\varphi\|^2(1+\|\varphi\|^2)].\end{aligned}$$

Hence (10.53) is true.

Now, by inequalities (9.88) and (10.52) we know that for any $m \geq 0$,

$$\begin{aligned}&\left\|E\left[\frac{\varphi_{m+h}\varphi_{m+h}^{\tau}}{1+\|\varphi_{m+h}\|^2}|\mathcal{F}_m\right] - E\left[\frac{\varphi_{m+h}\varphi_{m+h}^{\tau}}{1+\|\varphi_{m+h}\|^2}\right]\right\| \\ \leq \; & 2r\phi(h), \quad \forall m \geq 0.\end{aligned}$$

Therefore by (10.51), (10.53) and $\phi(h) \xrightarrow[h\to\infty]{} 0$, there is a constant $\alpha > 0$ such that

$$E\left[\frac{\varphi_{m+h}\varphi_{m+h}^{\tau}}{1+\|\varphi_{m+h}\|^2}\middle|\mathcal{F}_m\right] \geq \alpha I$$

for all $m \geq 0$ and large h. This shows that (10.49) is valid and hence the CR condition (10.46) holds. Q.E.D.

We remark that the ϕ-mixing process contains a large class of random processes. In particular, any h-dependent random process (including moving average processes of order h) is ϕ-mixing.

Example 10.3. Let $\{\varphi_k\}$ be the output of the following linear stochastic model:

$$x_k = Ax_{k-1} + B\xi_k, \quad \forall k \geq 1, \quad E\|x_0\|^5 < \infty, \tag{10.55}$$

$$\varphi_k = Cx_k + \zeta_k, \quad \forall k \geq 0, \tag{10.56}$$

where $A \in \mathbb{R}^{n\times n}$, $B \in \mathbb{R}^{n\times q}$ and $C \in \mathbb{R}^{r\times n}$ are deterministic matrices, A is stable and (A, B, C) is output controllable in the sense that

$$\sum_{i=0}^{n-1} CA^iB(CA^iB)^{\tau} > 0. \tag{10.57}$$

Suppose that $\{\xi_k\}$ and $\{\zeta_k\}$ are independent processes which are also mutually independent, and satisfy

$$E\xi_k = 0, \qquad E\zeta_k = 0, \tag{10.58}$$
$$E[\xi_k\xi_k^\tau] \geq \varepsilon I, \quad \forall k \geq 0, \tag{10.59}$$
$$E\left[\|\xi_k\|^{4(1+\mu)} + \|\zeta_k\|^4\right] \leq M < \infty, \ \forall k \geq 0, \tag{10.60}$$

for some constants $\varepsilon > 0$, $\mu > 0$ and $M > 0$. Then the CR condition (10.46) is fulfilled.

Proof. Let us denote $\mathcal{F}_m = \sigma\{x_0, \xi_i, \zeta_i, i \leq m\}$. Similar to (10.53), it is not difficult to see that

$$\lambda_{min}\left(E\left[\frac{\varphi_{m+h}\varphi_{m+h}^\tau}{1+\|\varphi_{m+h}\|^2}|\mathcal{F}_m\right]\right)$$
$$\geq \frac{\left[\lambda_{min}(E[\varphi_{m+h}\varphi_{m+h}^\tau|\mathcal{F}_m])\right]^2}{[E(\|\varphi_{m+h}\|^2+\|\varphi_{m+h}\|^4)|\mathcal{F}_m]}. \tag{10.61}$$

We now proceed to estimate the numerator and denominator on the right-hand side of (10.61).

From (10.55) and (10.56) it follows that

$$\varphi_{m+h} = CA^h x_m + \sum_{i=m+1}^{m+h} CA^{m+h-i}B\xi_i + \zeta_{m+h}. \tag{10.62}$$

From this, (10.57), (10.58) and (10.59) we then have

$$\begin{aligned} &E\{\varphi_{m+h}\varphi_{m+h}^\tau|\mathcal{F}_m\} \\ &\geq \sum_{i=m+1}^{m+h} CA^{m+h-i}BE[\xi_i\xi_i^\tau](CA^{m+h-i}B)^\tau \\ &\geq \varepsilon\sum_{i=0}^{h-1} CA^iB(CA^iB)^\tau \geq \alpha I > 0, \quad \forall m \geq 0, \quad \forall h \geq n, \end{aligned} \tag{10.63}$$

where α is a constant.

Since A is a stable matrix, there exists a norm $\|\cdot\|_1$ on $I\!R^n$ such that its induced matrix norm (also denoted by $\|\cdot\|_1$) satisfies $\|A\|_1 \stackrel{\Delta}{=} a < 1$. Consequently, by (10.55) we have

$$\|x_{m+1}\|_1 \leq a\|x_m\|_1 + \|B\xi_{m+1}\|_1$$

and

$$\|x_{m+1}\|_1^i \leq a\|x_m\|_1^i + (1-a)\left\|\frac{B\xi_{m+1}}{1-a}\right\|_1^i$$

since x^i is convex for $x \geq 0$, $i \geq 1$. From this it is easy to find a constant $b > 0$ such that

$$\begin{aligned} & \|x_{m+1}\|_1^2 + \|x_{m+1}\|_1^4 \\ \leq\ & a(\|x_m\|_1^2 + \|x_m\|_1^4) + b(\|\xi_{m+1}\|_1^2 + \|\xi_{m+1}\|_1^4). \end{aligned} \tag{10.64}$$

Again from (10.62) and (10.58)-(10.60) it is easy to see that there exists a constant $d > 0$ such that

$$E\{\|\varphi_{m+1}\|^2 + \|\varphi_{m+1}\|^4 | \mathcal{F}_m\} \leq d(\|x_m\|^2 + \|x_m\|^4) + d.$$

Consequently, by the equivalency of the norms $\|\cdot\|$ and $\|\cdot\|_1$, we know that there is $d_1 > 0$ such that

$$\begin{aligned} & E\{\|\varphi_{m+1}\|^2 + \|\varphi_{m+1}\|^4 | \mathcal{F}_m\} \\ \leq\ & d_1(\|x_m\|_1^2 + \|x_m\|_1^4) + d. \end{aligned} \tag{10.65}$$

Set

$$\alpha_m = \alpha^{-1} d_1(\|x_m\|_1^2 + \|x_m\|_1^4) + \alpha^{-1} d.$$

Then from (10.61), (10.63) and (10.65) we see that (10.46) holds, while (10.47) and (10.48) can be seen from (10.64) and (10.60). Q.E.D.

10.3 Analysis of Kalman Filter Based Algorithms

This and the next section deal with the estimation problem of the time-varying parameter $\{\theta_k\}$.

Note that if we regard (10.1) and (10.2) as a state space model with state θ_k, then it is natural to use the Kalman filter to estimate the time-varying parameter θ_k. (see, e.g. [KG], [BR], [GL]). The Kalman filter takes the following form:

$$\hat{\theta}_{k+1} = \hat{\theta}_k + \frac{P_k \varphi_k}{R + \varphi_k^\tau P_k \varphi_k}(y_k - \varphi_k^\tau \hat{\theta}_k) \tag{10.66}$$

$$P_{k+1} = P_k - \frac{P_k \varphi_k \varphi_k^\tau P_k}{R + \varphi_k^\tau P_k \varphi_k} + Q \tag{10.67}$$

where $P_0 \geq 0$, $R > 0$, $Q > 0$ and $\hat{\theta}_0$ are deterministic, and can arbitrarily be chosen (here R and Q may be regarded as the *a priori* estimates for the variances of v_k and Δ_k, respectively).

From Theorem 3.1 it is known that if φ_k is $\mathcal{F}_{k-1}$-measurable, where $\mathcal{F}_{k-1} = \sigma\{y_i, i \leq k-1\}$ and $\{\Delta_k, v_k\}$ is a Gaussian white noise process, then θ_k generated by (10.66) and (10.67) is the minimum variance estimate for θ_k and P_k is the estimation error covariance, i.e.,

$$\hat{\theta}_k = E[\theta_k|\mathcal{F}_{k-1}], \quad P_k = E\left[\tilde{\theta}_k\tilde{\theta}_k^T|\mathcal{F}_{k-1}\right], \quad (\tilde{\theta}_k = \theta_k - \hat{\theta}_k) \tag{10.68}$$

where $\widetilde{\theta}_k \triangleq \theta_k - \widehat{\theta}_k$, provided that $Q = E\Delta_k\Delta_k^T$, $R = Ev_k^2$, $\hat{\theta}_0 = E\theta_0$ and $P_0 = E\left[\tilde{\theta}_0\tilde{\theta}_0^T\right]$.

In studying asymptotic properties of the algorithm (10.66) and (10.67), the primary issue is to establish boundedness (in some sense) of the tracking error $\widetilde{\theta}_k$. This problem is reminiscent of the stability theory of the Kalman filter, and the standard condition for such a stability (boundedness) is (10.50) (see, e.g. [J]). As we mentioned before, (10.50) is no longer suitable for the stability study of the Kalman filter (10.66) because in the present case $\{\varphi_k\}$ is a random process rather than a deterministic sequence.

The first concrete stability result concerning the algorithm (10.66) and (10.67), allowing $\{\varphi_k\}$ to be a large class of stochastic processes, seems to be that in [Gu2]. The only assumption there on the stochastic regressor φ_k is (10.49). This assumption is later relaxed to (10.46) in [ZGC]. We first analyse properties of $\{P_k\}$ defined by (10.67), then prove stability of $\{\widetilde{\theta}_k\}$.

Lemma 10.2 *Let $\{P_k\}$ be generated by (10.67). Then*

$$T_{m+1} \leq (1 - a_{m+1})T_m + d, \tag{10.69}$$

where

$$T_m = \sum_{k=(m-1)h}^{mh-1} trP_{k+1}, \quad T_0 = 0, \tag{10.70}$$

$$a_{m+1} = \frac{tr\left\{(P_{mh} + hQ)^2 \sum_{k=mh}^{(m+1)h-1} \frac{\varphi_k\varphi_k^T}{1+\|\varphi_k\|^2}\right\}}{h(R+1)[1+\lambda_{max}(P_{mh}+hQ)]tr(P_{mh}+hQ)} \tag{10.71}$$

$$d = \frac{3}{2}h(h+1)trQ \tag{10.72}$$

and where h is the constant appearing in the CR condition (10.46).

Proof. Note that by (10.67)

$$P_k \leq P_{k-1} + Q \leq \ldots \leq P_{mh} + hQ \tag{10.73}$$

for any $k \in [mh, (m+1)h]$. Hence, by the matrix inverse formula (see (3.62)) it follows that for any $k \in [mh, (m+1)h]$

$$\begin{aligned} P_{k+1} &= \left(P_k^{-1} + R^{-1}\varphi_k\varphi_k^T\right)^{-1} + Q \\ &\leq \left\{(P_{mh}+hQ)^{-1} + R^{-1}\varphi_k\varphi_k^T\right\}^{-1} + Q \\ &= P_{mh} - \frac{(P_{mh}+hQ)\varphi_k\varphi_k^T(P_{mh}+hQ)}{R+\varphi_k^T(P_{mh}+hQ)\varphi_k} + (h+1)Q \\ &\leq P_{mh} - \frac{(P_{mh}+hQ)\frac{\varphi_k\varphi_k^T}{1+\|\varphi_k\|^2}(P_{mh}+hQ)}{(R+1)\left[1+\lambda_{max}(P_{mh}+hQ)\right]} + (h+1)Q \\ &\leq P_{mh} + (h+1)Q \\ &\quad - \frac{(P_{mh}+hQ)\frac{\varphi_k\varphi_k^T}{1+\|\varphi_k\|^2}(P_{mh}+hQ)}{h(R+1)\left[1+\lambda_{max}(P_{mh}+hQ)\right] tr(P_{mh}+hQ)} htrP_{mh} \end{aligned}$$

Summing both sides of this inequality and paying attention to (10.70)-(10.72) we obtain

$$T_{m+1} \leq htrP_{mh} - a_{m+1}htrP_{mh} + h(h+1)trQ. \tag{10.74}$$

Again, by (10.70) and (10.73),

$$\begin{aligned} htrP_{mh} &= \sum_{k=(m-1)h}^{mh-1} trP_{mh} \leq \sum_{k=(m-1)h}^{mh-1} tr\{P_{k+1} + (mh-k)Q\} \\ &\leq T_m + \frac{1}{2}h(h+1)trQ. \end{aligned}$$

Substituting this into (10.74) we get the desired result (10.69). Q.E.D.

The following Theorem shows that under the CR condition (10.46) the moment generating functions of $\|P_k\|$, $k = 0, 1, \ldots$, exist in a small neighborhood of the origin and are uniformly bounded in k.

Theorem 10.4 *For $\{P_k\}$ recursively defined by (10.67) if the CR condition (10.46) holds, then there exists a constant $\varepsilon^* > 0$ such that for any $\varepsilon \in [0, \varepsilon^*)$*

$$\sup_{k\geq 0} E\exp\{\varepsilon\|P_k\|\} \leq C \tag{10.75}$$

and

$$\limsup_{k\to\infty} \frac{1}{k}\sum_{i=0}^{k} \exp\{\varepsilon\|P_i\|\} \leq C' \quad a.s. \tag{10.76}$$

where C and C' are constants.

Proof. Denote $\mathcal{Y}_m = \mathcal{F}_{mh-1}$, where $\{\mathcal{F}_m\}$ is the same as in Condition (10.46). It is clear that T_m and a_m defined by (10.70) and (10.71) are $\mathcal{Y}_m$-measurable, moreover, by (10.71) and Condition (10.46)

$$a_{m+1} \in [0, \frac{1}{R+1}] \tag{10.77}$$

and

$$\begin{aligned} & E\left[a_{m+1}|\mathcal{Y}_m\right] \\ \geq \; & \frac{tr(P_{mh}+hQ)^2}{\alpha_{mh-1}h(R+1)[1+\lambda_{max}(P_{mh}+hQ)]tr(P_{mh}+hQ)} \\ \geq \; & \frac{r^{-1}\{tr(P_{mh}+hQ)\}^2}{\alpha_{mh-1}h(R+1)[1+\lambda_{max}(P_{mh}+hQ)]tr(P_{mh}+hQ)} \\ \geq \; & \frac{\|Q\|}{r(R+1)(1+h\|Q\|)} \times \frac{1}{\alpha_{mh-1}}. \end{aligned}$$

Set

$$\beta_m = r\|Q\|^{-1}(R+1)(1+h\|Q\|)\alpha_{mh-1}.$$

Then we have

$$E\left[a_{m+1}|\mathcal{Y}_m\right] \geq \frac{1}{\beta_m}.$$

It is easy to verify that $\{\beta_m, \mathcal{Y}_m\}$ is an adapted sequence and by (10.47) and (10.48) satisfies

$$\beta_{m+1} \leq \overline{a}\beta_m + \overline{\eta}_{m+1}, \quad m \geq 0, \quad E\beta_0 < \infty, \tag{10.78}$$

where $\overline{a} = a^h$ and $\{\overline{\eta}_m, \mathcal{Y}_m\}$ is an adapted sequence satisfying

$$\sup_{m\geq 0} E\left[\overline{\eta}_{m+1}^{1+\delta}|\mathcal{Y}_m\right] \leq \overline{M} \tag{10.79}$$

for some constant $\overline{M} > 0$.

Consequently, by applying Theorem 10.1 we obtain

$$E\prod_{k=m}^{n}(1-a_{k+1}) \leq c\gamma^{n-m+1}, \quad \forall n \geq m \geq 0, \tag{10.80}$$

for some constants $c > 0$, $\gamma \in (0, 1)$.

Next, from Lemma 10.2 it follows that for any $\varepsilon > 0$

$$\exp\{\varepsilon T_{m+1}\} \leq \exp\{(1-a_{m+1})\varepsilon T_m\}e^{d\varepsilon}. \tag{10.81}$$

Consequently, noticing the obvious inequality

$$\exp(\alpha x) - 1 \leq \alpha\exp(x), \quad 0 < \alpha < 1, \quad x > 0,$$

we get

$$\exp\{\varepsilon T_{m+1}\} \leq e^{d\varepsilon}\{(1-a_{m+1})\exp(\varepsilon T_m)+1\}. \tag{10.82}$$

Hence from this and (10.80) it is easy to convince oneself that if $\varepsilon^* > 0$ is taken small enough such that $e^{d\varepsilon^*}\gamma < 1$, then

$$\sup_{m\geq 0} E\exp\{\varepsilon T_m\} < \infty, \quad \forall \varepsilon \in (0, \varepsilon^*).$$

This proves the first assertion of the theorem, while the second assertion follows immediately by applying Theorem 10.3′ with $x_k = exp\{\varepsilon T_k\}$, $\alpha = e^{d\varepsilon}$ to (10.82). Q.E.D.

Corollary 10.1. *For $\{P_k\}$ generated by (10.67) if the CR condition (10.46) is verified, then the following properties hold:*

(i) $\quad \sup_{k\geq 0} E\|P_k\|^m < \infty, \forall m > 0,$

(ii) $\quad \limsup_{k\to\infty} \frac{1}{k}\sum_{i=0}^{k}\|P_i\|^m \leq c < \infty,$ *a.s.*, $\forall m > 0,$

(iii) $\quad \|P_k\| = O(\log k),$ *a.s. as* $k \to \infty$.

We now proceed to analyse the tracking error $\widehat{\theta}_k - \theta_k$. We first present some lemmas.

Denote

$$\tilde{\theta}_k = \theta_k - \hat{\theta}_k, \qquad V_k = \tilde{\theta}_k^\tau P_k^{-1}\tilde{\theta}_k, \quad \forall k \geq 0. \tag{10.83}$$

V_k may be regarded as a stochastic Lyapunov function. Although it has the same form as that used in Section 4.2, here the analysis for it is completely different from there (Theorem 4.1) due to different definitions of P_k: There is an additional $Q > 0$ in (10.67), which prevents P_k from tending to zero.

Lemma 10.3 (Gu2) . *For the algorithm (10.66) and (10.67), the Lyapunov function V_k defined by (10.83) has the following property:*

$$V_{k+1} \leq V_k - \frac{V_k}{4+atr(P_k)} + c(\|P_k\|Z_k^2), \quad \forall k \geq 0, \tag{10.84}$$

where $a = 2\|Q^{-1}\|$, $Z_k \triangleq \|v_k\| + \|\Delta_{k+1}\|$ and c is a constant.

Proof.

Let us denote

$$K_k = \frac{P_k\varphi_k}{R+\varphi_k^\tau P_k\varphi_k}, \quad G_k = I - K_k\varphi_k$$

and rewrite (10.67) as

$$P_{k+1} = G_k P_k G_k^T + RK_k K_k^T + Q. \tag{10.85}$$

Note that by (10.1), (10.2) and (10.66) the error equation is

$$\widetilde{\theta}_{k+1} = G_k \widetilde{\theta}_k + \xi_{k+1}, \quad \xi_{k+1} = -K_k v_k + \Delta_{k+1}. \tag{10.86}$$

So we have by (10.83)

$$\begin{aligned} V_{k+1} &= [G_k\widetilde{\theta}_k + \xi_{k+1}]^T P_{k+1}^{-1} [G_k\widetilde{\theta}_k + \xi_{k+1}] \\ &= \widetilde{\theta}_k^T G_k P_{k+1}^{-1} G_k \widetilde{\theta}_k + 2\xi_{k+1}^T P_{k+1}^{-1} G_k \widetilde{\theta}_k \\ &\quad + \xi_{k+1}^T P_{k+1}^{-1} \xi_{k+1}. \end{aligned} \tag{10.87}$$

By (10.85) and the matrix inverse formula (3.62), we know that

$$\begin{aligned} & G_k^T P_{k+1}^{-1} G_k = G_k^T \left\{ G_k^T P_k^{-1} G_k + K_k R K_k^T + Q \right\}^{-1} G_k \\ = \; & P_k^{-1} - \left[P_k + P_k G_k^T (K_k R K_k^T + Q)^{-1} G_k P_k \right]^{-1} \\ = \; & P_k^{-\frac{1}{2}} \left\{ I - \left[I + (P_k)^{\frac{1}{2}} G_k^T (K_k R K_k^T + Q)^{-1} G_k (P_k)^{\frac{1}{2}} \right]^{-1} \right\} P_k^{-\frac{1}{2}} \\ \leq \; & \left\{ 1 - \left[1 + \|(K_k R K_k^T + Q)^{-1} G_k P_k G_k^T\| \right]^{-1} \right\} P_k^{-1} \\ \leq \; & \left\{ 1 - \left[1 + \|(K_k R K_k^T + Q)^{-1} P_{k+1}\| \right]^{-1} \right\} P_k^{-1} \\ \leq \; & \left\{ 1 - \left[1 + \|Q^{-1}(P_k + Q)\| \right]^{-1} \right\} P_k^{-1} \\ \leq \; & P_k^{-1} - \frac{1}{2 + \|Q^{-1}\| \|P_k\|} P_k^{-1}. \end{aligned} \tag{10.88}$$

Putting this into (10.87) we get

$$\begin{aligned} V_{k+1} \;\leq\; & V_k - \frac{1}{2\|Q^{-1}\|\|P_k\|} V_k + 2\xi_{k+1}^T P_{k+1}^{-1} G_k \widetilde{\theta}_k \\ & + \xi_{k+1}^T P_{k+1}^{-1} \xi_{k+1}. \end{aligned} \tag{10.89}$$

By the elementary inequality $2|xy| \leq x^2 + y^2$ it follows that:

$$\begin{aligned} & 2|\xi_{k+1}^T P_{k+1}^{-1} G_k \widetilde{\theta}_k| \leq 2\|\xi_{k+1}^T P_{k+1}^{-\frac{1}{2}}\| \|P_{k+1}^{-\frac{1}{2}} G_k \widetilde{\theta}_k\| \\ \leq \; & 2\xi_{k+1}^T P_{k+1}^{-1} \xi_{k+1} (2 + \|Q^{-1}\| \|P_k\|) \\ & + \frac{\widetilde{\theta}_k^T G_k^T P_{k+1}^{-1} G_k \widetilde{\theta}_k}{2(2 + \|Q^{-1}\| \|P_k\|)}. \end{aligned} \tag{10.90}$$

Recall that by (10.83) and (10.88)

$$\widetilde{\theta}_k^T G_k^T P_{k+1}^{-1} G_k \widetilde{\theta}_k \leq V_k, \tag{10.91}$$

and (10.85) $P_{k+1} \geq RK_kK_k^T + Q$. Then by (10.86) it follows that

$$\begin{aligned}&\xi_{k+1}^T P_{k+1}^{-1}\xi_{k+1}\|P_{k+1}^{-\frac{1}{2}}(-K_kv_k + \Delta_{k+1}\|^2 \\ \leq\ & c(K_k^T P_{k+1}^{-1}K_k\|v_k\|^2 + \|\Delta_{k+1}\|^2) \leq c'Z_k^2 \end{aligned} \qquad (10.92)$$

where c and c' are constants.

Finally, substituting (10.90)-(10.92) into (10.89) yields

$$V_{k+1} \leq V_k - \frac{V_k}{2(2 + \|Q^{-1}\|\|P_k\|)} + c''\|P_k\|Z_k^2$$

where c'' is a constant. Hence (10.84) is true. Q.E.D.

Corollary 10.2. *Under the notations of Lemma 10.3 for any $q > 0$ there exist two constants $\mu \in (0,1)$ and $c \geq 1$ such that*

$$V_{k+1}^q \leq \left(1 - \frac{\mu}{1 + trP_k}\right) V_k^q + c(1 + trP_k)^{2q} Z_k^{2q}, \qquad (10.93)$$

where $Z_k = \|v_k\| + \|w_{k+1}\|$.

Proof. From Lemma 10.3 we know that

$$V_{k+1} \leq \left(1 - \frac{\mu_1}{1 + trP_k}\right) V_k + c_1(1 + trP_k)Z_k^2, \quad \forall k \geq 0 \qquad (10.94)$$

for some constants $\mu_1 \in (0, 1)$, $c_1 < \infty$.

We first consider the case where $q > 1$. Note that in this case x^q, $x > 0$, is a convex function. Hence it follows from (10.93) that

$$V_{k+1}^q \leq \left(1 - \frac{\mu_1}{1 + trP_k}\right) V_k^q + \frac{\mu_1}{1 + trP_k}\left[\frac{c_1(1 + trP_k)^2}{\mu_1} Z_k^2\right]^q,$$

which implies the desired result.

We now assume $q \in [0,1]$. In this case, the desired result can also easily be derived from (10.94) by applying the following elementary inequalities:

$$(x + y)^q \leq x^q + y^q, \qquad \forall x \geq 0,\ y \geq 0,\ q \in [0,1]$$

and

$$(1 - x)^q \leq 1 - qx, \qquad 0 < x < 1,\ q \in [0,1].$$

Q.E.D.

Lemma 10.4 *Let $\{P_k\}$ be defined by (10.67) with $\{\varphi_k\}$ satisfying the CR condition (10.46). Then for any $\mu \in (0,1)$ there exist constants $N > 0$, $\lambda \in (0,1)$ such that*

$$E\prod_{k=m}^{n}\left(1-\frac{1}{1+trP_k}\right) \le N\lambda^{n-m+1}, \quad \forall n \ge m \ge 0. \tag{10.95}$$

Proof. Let us denote

$$x_m = \frac{1}{\mu}(h+T_m)$$

where T_m is defined by (10.70). It follows from Lemma 10.2 that

$$x_{m+1} \le (1-a_{m+1})x_m + \frac{h+d}{\mu}.$$

Hence by noting (10.77)-(10.79) we see that Theorem 10.2 is applicable. So there are $N_0 > 0$ and $\lambda_0 \in (0,1)$ such that

$$E\prod_{k=m}^{n}\left(1-\frac{1}{x_{k+1}}\right) \le N_0\lambda_0^{n-m+1}, \quad \forall n \ge m \ge 0. \tag{10.96}$$

Clearly, for the final result (10.95) we need only to consider the case of $n-m > h$. Let i and j be two integers such that

$$ih \le n \le (i+1)h, \quad (j-1)h < m \le jh.$$

It then follows that

$$\begin{aligned}
E\prod_{k=m}^{n}\left(1-\frac{\mu}{1+trP_k}\right) &\le E\prod_{k=jh}^{ih}\left(1-\frac{\mu}{1+trP_k}\right)\\
&\le E\prod_{t=j}^{i}\left(1-\frac{\mu}{1+tr(P_{th})}\right)\\
&\le E\prod_{t=j}^{i}\left(1-\frac{1}{x_t}\right) \le N_0(\lambda_0)^{i-j+1}\\
&= N_0\left[(\lambda_0)^{1/h}\right]^{h(i-j)+h} \le N_0\left[(\lambda_0)^{1/h}\right]^{n-h-m-h+h}\\
&= \left[N_0(\lambda_0)^{-1-(1/h)}\right]\left[(\lambda_0)^{1/h}\right]^{n-m+1}.
\end{aligned}$$

Q.E.D.

Lemma 10.5 *Let $C_{nk} \in [0, 1]$, $n \geq k \geq 0$, be a double-indexed stochastic sequence satisfying*

$$E\,C_{nk} \leq N\lambda^{n-k}, \qquad \forall n \geq k \geq 0$$

for some constants $N > 0$, $\lambda \in (0, 1)$. If $\{x_k\}$ is a nonnegative stochastic sequence and

$$\sigma \stackrel{\Delta}{=} \sup_{k \geq 0} Ex_k \log^\beta(e + x_k) < \infty$$

for some $\beta > 1$, then

$$\sum_{k=0}^{n} EC_{nk}x_k \leq c\left(\sigma \log(e + \sigma^{-1})\right), \quad n \geq 0,$$

where c is a constant depending on β, N and λ.

Proof. Let us take a constant $d > 1$ such that $d\lambda < 1$. Then we have

$$\begin{aligned} EC_{nk}x_k &= EC_{nk}x_k I(x_k \leq \sigma d^{n-k}) + EC_{nk}x_k I(x_k > \sigma d^{n-k}) \\ &\leq \sigma N(d\lambda)^{n-k} + E\frac{x_k \log^\beta(e + x_k)}{\log^\beta(e + \sigma d^{n-k})} \\ &\leq \sigma N(d\lambda)^{n-k} + \frac{\sigma}{\log^\beta(e + \sigma d^{n-k})}. \end{aligned} \tag{10.97}$$

If $\sigma \geq 1$, then it is easy to see that

$$\sum_{k=0}^{n} \frac{1}{\log^\beta(e + \sigma d^{n-k})} \leq c', \quad \forall n \geq 0, \tag{10.98}$$

where c' is a constant independent of σ.

If $\sigma < 1$, then it is easy to see that $k_0 \leq (\log d)^{-1} \log(e + \sigma^{-1}) + 1$, where k_0 is the smallest integer such that $\sigma d^{k_0} \geq 1$. Thus we have

$$\begin{aligned} &\sum_{k=0}^{n} \frac{1}{\log^\beta(e + \sigma d^{n-k})} \\ = &\sum_{k=0}^{n-k_0+1} \frac{1}{\log^\beta(e + \sigma d^{n-k})} + \sum_{k=n-k_0}^{n} \frac{1}{\log^\beta(e + \sigma d^{n-k})} \\ \leq &\; c''(1 + \log(e + \sigma^{-1})), \end{aligned} \tag{10.99}$$

where c'' is some constant independent of σ.

Combining (10.97)-(10.99), we finally get

$$\begin{aligned} \sum_{k=0}^{n} EC_{nk}x_k &\leq \frac{N}{1 - d\lambda}\sigma + \sigma(c' + c'') + c''\sigma \log(e + \sigma^{-1}) \\ &\leq c\sigma \log(e + \sigma^{-1}) \end{aligned}$$

for some constant c. Q.E.D.

Lemma 10.6 *For any nonnegative numbers x, y, ε, α and σ_p the following inequalities hold:*

$$\begin{aligned} x^\alpha y &\le \sigma_p \exp(\varepsilon x) \\ &\quad + c_1 y \left[\log^\alpha(e+\sigma_p^{-1}) + \log^\alpha(e+y)\right], \qquad (10.100) \\ &\le \sigma_p \exp(\varepsilon x) \\ &\quad + 2c_1 \left[\log(e+\sigma_p^{-1})\right]^\alpha y \log^\alpha(e+y), \qquad (10.101) \end{aligned}$$

where c_1 can be taken as

$$c_1 = \left(\frac{1+\alpha}{\varepsilon}\right)^\alpha 4^{1+\alpha} \log^\alpha \left[e + \left(\frac{1+\alpha}{\varepsilon}\right)^\alpha\right].$$

Proof. Taking $\varphi(t) = \exp(t^{1/\alpha}) - 1$, and $\psi(t) = \log^\alpha(t+1)$ in the following Young's inequality (see, e.g. [Mi], [HLP]):

$$xy \le \int_0^x \varphi(t)dt + \int_0^y \psi(t)dt, \quad x \ge 0,\, y \ge 0, \quad \varphi(\psi(t)) = t.$$

we get

$$xy \le x\exp(x^{1/\alpha}) + y\log^\alpha(1+y).$$

Noting that

$$x \le \exp(\alpha x^{1/\alpha}), \quad \forall x \ge 0,\ \alpha > 0,$$

we have

$$xy \le \exp\left((\alpha+1)x^{1/\alpha}\right) + y\log^\alpha(1+y).$$

Replacing x and y in this inequality by $\left(\frac{\varepsilon x}{\alpha+1}\right)^\alpha$ and $\left(\frac{\alpha+1}{\varepsilon}\right)^\alpha \frac{y}{\sigma_p}$ respectively leads to

$$\begin{aligned} x^\alpha y &= \sigma_p \left\{ \left(\frac{\varepsilon x}{\alpha+1}\right)^\alpha \left(\frac{\alpha+1}{\varepsilon}\right)^\alpha \frac{y}{\sigma_p} \right\} \\ &\le \sigma_p \exp(\varepsilon x) + \left(\frac{\alpha+1}{\varepsilon}\right)^\alpha y \log^\alpha \left(e + \left(\frac{\alpha+1}{\varepsilon}\right)^\alpha \frac{y}{\sigma_p}\right) \\ &\le \sigma_p \exp(\varepsilon x) + \left(\frac{\alpha+1}{\varepsilon}\right)^\alpha y \\ &\quad \log^\alpha \left[\left(e + \left(\frac{\alpha+1}{\varepsilon}\right)^\alpha\right)\left(e + \frac{y}{\sigma_p}\right)\right] \\ &\le \sigma_p \exp(\varepsilon x) + \left(\frac{\alpha+1}{\varepsilon}\right)^\alpha y 2^\alpha \\ &\quad \left[\log^\alpha \left(e + \left(\frac{\alpha+1}{\varepsilon}\right)^\alpha\right) + \log^\alpha \left(e + \frac{y}{\sigma_p}\right)\right] \end{aligned}$$

$$\leq \sigma_p \exp(\varepsilon x) + \left(\frac{\alpha+1}{\varepsilon}\right)^{\alpha} 2^{\alpha+1} y$$

$$\log^{\alpha}\left(e + \left(\frac{\alpha+1}{\varepsilon}\right)^{\alpha}\right) \log^{\alpha}\left(e + \frac{y}{\sigma_p}\right)$$

$$\leq \sigma_p \exp(\varepsilon x) + \left(\frac{\alpha+1}{\varepsilon}\right)^{\alpha} 4^{\alpha+1} y \log^{\alpha}\left(e + \left(\frac{\alpha+1}{\varepsilon}\right)^{\alpha}\right)$$

$$\left[\log^{\alpha}\left(e + \sigma_p^{-1}\right) + \log^{\alpha}(e+y)\right].$$

Q.E.D.

We are now in a position to prove the following main results of this section.

Theorem 10.5 (Gu2) , *[ZGC]. Consider the time-varying model (10.1) and (10.2). Suppose that $\{v_k, \Delta_k\}$ is a stochastic sequence and satisfies for some $p > 0$ and $\beta > 1$,*

$$\sigma_p \triangleq \sup_{k\geq 0} E\left\{Z_k^p[\log(e+Z_k)]^{\beta+3p/2}\right\} < \infty, \tag{10.102}$$

and

$$E\left\{\|\tilde{\theta}_0\|^p[\log(e+\|\tilde{\theta}_0\|)]^{p/2}\right\} < \infty, \tag{10.103}$$

where $Z_k = \|v_k\| + \|\Delta_{k+1}\|$, $\tilde{\theta}_0 = \theta_0 - \hat{\theta}_0$, and v_k, Δ_k, θ_0 and $\hat{\theta}_0$ are respectively given by (10.1), (10.2) and (10.66). Then under the CR Condition (10.46), the estimation error $\{\theta_k - \hat{\theta}_k,\ k \geq 0\}$ generated by (10.66) and (10.67) is L_p-stable and

$$\limsup_{k\to\infty} E\|\theta_k - \hat{\theta}_k\|^p \leq A\left[\sigma_p \log^{1+3p/2}(e+\sigma_p^{-1})\right] \tag{10.104}$$

where A is a constant depending on h, a, M, M_0 and δ only.

Moreover, if $v_k \equiv 0$ and $\Delta_k \equiv 0$ (i.e., $\theta_k \equiv \theta_0$), then

$$E\|\theta_k - \hat{\theta}_k\|^p \longrightarrow 0 \quad as\ k \to \infty \tag{10.105}$$

and

$$E\|\theta_k - \hat{\theta}_k\|^q \longrightarrow 0 \quad exponentially\ fast \tag{10.106}$$

for any $q \in (0, p)$.

Remark 10.1. If in Theorem 10.5, $\{\varphi_k\}$ and $\{v_k, \Delta_k\}$ are assumed to be mutually independent, then for L_p-stability of $\{\hat{\theta}_k - \theta_k\}$, Condition (10.102) can be replaced by a weaker one:

$$\sup_{k\geq 0} EZ_k^p < \infty, \tag{10.107}$$

which is a natural condition for the desired L_p-stability. What condition (10.102) means is that if the independency between $\{\varphi_k\}$ and $\{v_k, \Delta_k\}$ is removed, then the L_p-stability of $\{\hat{\theta}_k - \theta_k\}$ is still preserved provided that the moment condition (10.107) is slightly strengthened.

Next, we present a result on time average of the estimation error $\{\hat{\theta}_k - \theta_k\}$.

Theorem 10.6 *Consider the time-varying model (10.1) and (10.2). Suppose that $\{v_k, \Delta_k\}$ is a stochastic sequence and for some $p > 0$,*

$$\varepsilon_p \triangleq \limsup_{k\to\infty} \frac{1}{k}\sum_{i=0}^{k-1}\{\|v_i\|^p + \|\Delta_{i+1}\|^p\} < \infty \ a.s. \tag{10.108}$$

Then under the CR condition (10.46), $\{\hat{\theta}_k - \theta_k, \quad k \ge 0\}$ is L_q-stable in the time average sense for any $q \in (0, p)$, and

$$\limsup_{k\to\infty} \frac{1}{k}\sum_{i=0}^{k}\|\hat{\theta}_i - \theta_i\|^q \le B(\varepsilon_p)^{q/p}, \tag{10.109}$$

where B is a constant depending on q, h, a, M, M_0 and δ, but independent of sample path.

Furthermore, if $v_k \equiv 0$, and $\theta_k \equiv \theta_0$, then

$$\hat{\theta}_k \to \theta_0 \quad a.s. \ \ exponentially\ fast. \tag{10.110}$$

We first prove Theorem 10.5 and then Theorem 10.6.

Proof of Theorem 10.5.
Set

$$f(x) = x\log^{p/2}(e + x), \quad p > 0, \quad x \ge 0. \tag{10.111}$$

Then from (10.100) (with $\alpha = \frac{p}{2}$) and the definition of V_k in (10.83) it follows that for any $p > 0$

$$\begin{aligned} \|\tilde{\theta}_n\|^p &\le \|P_n\|^{p/2}V_n^{p/2} \\ &\le \sigma_p \exp(\varepsilon\|P_n\|) \\ &\quad + c'\left[\log^{p/2}(e + \sigma_p^{-1})\, V_n^{p/2} + f(V_n^{p/2})\right] \end{aligned} \tag{10.112}$$

where and hereafter c' denotes a constant, which may be different from place to place, and where σ_p is defined by (10.102).

The first term on the right-hand side of (10.112) is easy to deal with, since by Theorem 10.4

$$\sup_n E\exp(\varepsilon\|P_n\|) < \infty, \qquad \forall\varepsilon \in (0, \varepsilon^*). \tag{10.113}$$

We now proceed to estimate $EV_n^{p/2}$ and $Ef(V_n^{p/2})$. By Corollary 10.2 we know that

$$V_{k+1}^{p/2} \le \left(1 - \frac{\mu}{1+trP_k}\right) V_k^{p/2} + c(1+trP_k)^p Z_k^p. \tag{10.114}$$

Note that $f(x)$ defined by (10.111) is convex. It follows from (10.114) that

$$f\left(V_{k+1}^{p/2}\right) \le \left(1 - \frac{\mu}{1+trP_k}\right) f\left(V_k^{p/2}\right) + \xi_k, \tag{10.115}$$

where

$$\xi_k = \frac{\mu}{1+trP_k} f\left(\frac{c(1+trP_k)^{p+1}}{\mu} Z_k^p\right).$$

It is easy to see that

$$\begin{aligned}
\xi_k &= c(1+trP_k)^p Z_k^p \log^{p/2}\left(e + \frac{c(1+trP_k)^{p+1}}{\mu} Z_k^p\right) \\
&\le c(1+trP_k)^p Z_k^p \log^{p/2}\left[(e+Z_k^p)\left(e + \frac{c(1+trP_k)^{p+1}}{\mu}\right)\right] \\
&\le 2^{p/2} c(1+trP_k)^p \\
&\quad \left\{f(Z_k^p) + Z_k^p \log^{p/2}\left(e + \frac{c(1+trP_k)^{p+1}}{\mu}\right)\right\} \\
&\le c'\left\{(1+trP_k)^p f(Z_k^p) + (1+trP_k)^{3p/2} Z_k^p\right\}.
\end{aligned} \tag{10.116}$$

By (10.101) we see that

$$\begin{aligned}
& (1+trP_k)^p f(Z_k^p) \\
\le\ & \sigma_p\, exp\{\varepsilon(1+trP_k)\} \\
& + c'\left[\log(e+\sigma_p^{-1})\right]^p f(Z_k^p) \log^p(e + f(Z_k^p))
\end{aligned}$$

and

$$\begin{aligned}
& (1+trP_k)^{3p/2} Z_k^p \\
\le\ & \sigma_p\, exp\{\varepsilon(1+trP_k)\} \\
& + c'\left[\log(e+\sigma_p^{-1})\right]^{3p/2} Z_k^p \log^{3p/2}(e + Z_k^p).
\end{aligned}$$

From this and (10.116) we arrive at

$$\begin{aligned}
\xi_k \le\ & c'\sigma_p\, exp\{\varepsilon(1+trP_k)\} \\
& + c'\left[\log(e+\sigma_p^{-1})\right]^{3p/2} Z_k^p \log^{3p/2}(e + Z_k^p).
\end{aligned} \tag{10.117}$$

Now, let us define $\Phi(n,k)$ as follows:

$$\Phi(n+1,k)=\left(1-\frac{\mu}{1+trP_n}\right)\Phi(n,k), \quad \Phi(k,k)=I. \tag{10.118}$$

Then by Lemma 10.4 we know that $\{\Phi(n+1,k)\}$ satisfies the same conditions as those for C_{nk} in Lemma 10.5.

By (10.117) and (10.118) it follows from (10.115) that for $\varepsilon \in (0, \varepsilon^*/r)$

$$\begin{aligned} f\left(V_{n+1}^{p/2}\right) &\le \Phi(n+1,0)f\left(V_0^{p/2}\right) \\ &+c'\sigma_p\sum_{k=0}^{n+1}\Phi(n+1,k)exp\{\varepsilon(1+trP_k)\} \\ &+c'\log^{3p/2}(e+\sigma_p^{-1})\sum_{k=0}^{n+1}\Phi(n+1,k)Z_k^p\log^{3p/2}(e+Z_k^p). \end{aligned} \tag{10.119}$$

By Lemma 10.4 it is seen that

$$E\Phi(n+1,k)\le N\lambda^{n-k}, \quad \forall n\ge k\ge 0 \tag{10.120}$$

for some constants $N>0$, $\lambda\in(0,1)$. Hence we have

$$E[\lambda_1^n\Phi(n+1,0)]\le N(\lambda\lambda_1)^n, \quad \forall\lambda_1\in\left(1,\frac{1}{\lambda}\right)$$

and

$$E\left(\sum_{n=0}^{\infty}\lambda_1^n\Phi(n+1,0)\right)=\sum_{n=0}^{\infty}E(\lambda_1^n\Phi(n+1,0))<\infty,$$

which implies

$$\lambda_1^n\Phi(n+1,0)\to 0 \quad a.s. \tag{10.121}$$

Hence the expectation of the first term on the right-hand side of (10.119) converges to zero by condition (10.103) and the dominated convergence theorem, while the other two terms can be estimated by applying Lemma 10.5. Therefore, we have

$$\limsup_{n\to\infty} Ef\left(V_{n+1}^{p/2}\right)\le c'\sigma_p\log^{3p/2+1}(e+\sigma_p^{-1}). \tag{10.122}$$

Next, we estimate $EV_n^{p/2}$. By (10.101) it is seen that

$$\begin{aligned}(1+trP_k)^pZ_k^p \;&\le\; \sigma_p\,exp\{\varepsilon(1+trP_k)\} \\ &+c'\left[\log(e+\sigma_p^{-1})\right]^p Z_k^p\log^p(e+Z_k^p).\end{aligned}$$

Hence, in exactly the same way as the proof of (10.122), from (10.114) we conclude that

$$\limsup_{n\to\infty} EV_{n+1}^{p/2} \le c'\sigma_p \log^{p+1}(e+\sigma_p^{-1}). \tag{10.123}$$

Finally, substituting (10.122) and (10.123) into (10.112) and taking account of (10.113) we obtain

$$\limsup_{n\to\infty} E\|\tilde{\theta}_n\|^p \le c'\sigma_p \log^{3p/2+1}(e+\sigma_p^{-1}).$$

This proves the first assertion (10.104).

We now prove (10.105) and (10.106). In this case $Z_k \equiv 0$, so by (10.114) and (10.115)

$$V_{n+1}^{p/2} \le \Phi(n+1,0)V_0^{p/2}, \tag{10.124}$$

$$f\left(V_{n+1}^{p/2}\right) \le \Phi(n+1,0) f\left(V_0^{p/2}\right). \tag{10.125}$$

Note that (10.112) is true for any $\sigma_p > 0$ (not necessarily the one defined by (10.102)). Consequently, by (10.124)-(10.125) and the dominated convergence theorem, we see from (10.112) that

$$\limsup_{n\to\infty} E\|\tilde{\theta}_n\|^p \le c'\sigma_p, \quad \sigma_p > 0.$$

Hence (10.105) follows by letting $\sigma_p \to 0$.

For the proof of (10.106), we take $s > 1$, $t > 1$ so that $\frac{1}{s}+\frac{1}{t}+\frac{q}{p}=1$. Then by the Hölder inequality we have

$$\begin{aligned} E\|\tilde{\theta}_n\|^q &\le E\|P_n\|^{q/2}V_n^{q/2} \\ &\le \|P_0^{-1}\|^{q/2}E\|P_n\|^{q/2}\Phi(n,0)\|\tilde{\theta}_0\|^q \\ &\le \|P_0^{-1}\|^{q/2}\left[E\|P_n\|^{qs/2}\right]^{1/s}\left[E\Phi(n,0)\right]^{1/t}\left[E\|\tilde{\theta}_0\|^p\right]^{q/p}. \end{aligned}$$

By virtue of Corollary 10.1, (10.120) and (10.103), $E\|\tilde{\theta}_n\|^q$ tends to zero exponentially fast. Q.E.D.

Proof of Theorem 10.6.

Let us take $s \in (q,p)$. By Corollary 10.1 and the Hölder inequality we have ($\frac{q}{s}+\frac{1}{t}=1$),

$$\limsup_{n\to\infty}\frac{1}{n}\sum_{k=0}^{n}\|\tilde{\theta}_k\|^q \le \limsup_{n\to\infty}\frac{1}{n}\sum_{k=0}^{n}\|P_k\|^{q/2}V_k^{q/2}$$

$$\le \limsup_{n\to\infty}\frac{1}{n}\sum_{k=0}^{n}\left(\frac{V_k^{s/2}}{1+trP_k}\right)^{q/s} \times (1+trP_k)^{(\frac{1}{2}+\frac{1}{s})q}$$

$$\le c'\left\{\limsup_{n\to\infty}\frac{1}{n}\sum_{k=0}^{n}\frac{V_k^{s/2}}{1+trP_k}\right\}^{q/s} \tag{10.126}$$

where c' is a constant.

Summing up both sides of (10.93) leads to

$$\sum_{k=0}^{n} \frac{V_k^{s/2}}{1+trP_k} \leq \mu^{-1} V_0^{s/2} + \mu^{-1} c \sum_{k=0}^{n} (1+trP_k)^s Z_k^s.$$

Then applying the Hölder inequality ($\frac{1}{u} + \frac{s}{p} = 1$) we see that

$$\limsup_{n\to\infty} \frac{1}{n} \sum_{k=0}^{n} \frac{V_k^{s/2}}{1+trP_k} \leq c' \left\{ \limsup_{n\to\infty} \frac{1}{n} \sum_{k=0}^{n} Z_k^p \right\}^{s/p}.$$

Hence from this, (10.126) and condition (10.108) it follows that

$$\limsup_{n\to\infty} \frac{1}{n} \sum_{k=0}^{n} \|\tilde{\theta}_k\|^q \leq c'(\varepsilon_p)^{q/p} \quad a.s.$$

This proves (10.109).

To prove (10.110), we note that by (10.76) of Theorem 10.4

$$\limsup_{n\to\infty} \frac{1}{n} exp\{\varepsilon \|P_n\|\} < \infty \quad a.s.$$

and hence

$$\limsup_{n\to\infty} \frac{\|P_n\|}{\log n} < \infty \quad a.s.$$

From this and (10.121) we see that

$$\|\tilde{\theta}_n\|^2 \leq V_n \|P_n\| \leq \Phi(n,0) V_0 \|P_n\|$$

which tends to zero exponentially fast. Q.E.D.

10.4 Analysis of LMS-Like Algorithms

A basic linear adaptive filtering algorithm used in adaptive signal processing is the following least mean-squares (LMS) algorithm [W]:

$$\hat{\theta}_{k+1} = \hat{\theta}_k + \mu_k \varphi_k (y_k - \varphi_k^\tau \hat{\theta}_k), \quad k \geq 0, \tag{10.127}$$

where $\mu_k \in (0,1)$ is a positive constant, often called the step-size, y_k and φ_k are respectively the noisy output and the measured signal. The LMS is so named because the increment of the algorithm (10.127) is opposite to the (stochastic) gradient of the mean square error

$$e_k = E|y_k - \varphi_k^\tau \theta|^2,$$

and (10.127) is a type of steepest descent algorithm that aims at recursively minimizing e_k.

The LMS algorithm is useful in many applications, for example, equalization and adaptive noise cancelling, and has naturally drawn much attention from researchers interested in its theoretical properties. As far as the tracking aspect is concerned, there is a vast literature on the tracking error analysis, for example, [WS], [Bi], [Be], [BR], [BA], [EM], [M], [So2] and [Ku], among others. Most of the works require some sort of stationarity and independence of the observations $\{y_k, \varphi_k\}$. Here such a restriction will not be made and only the upper bound of the tracking error is considered. The condition imposed on $\{\varphi_k\}$ is similar to the CR condition introduced in Section 10.2.

CR Condition for LMS

For the LMS algorithm with $\mu_n \in \mathcal{F}_n$ the regressor $\{\varphi_n, \mathcal{F}_n\}$ is said to satisfy CR condition if

$$\mu_m\|\varphi_m\|^2 \leq 1, \quad E\left\{\sum_{k=m+1}^{m+h} \mu_k\varphi_k\varphi_k^\tau|\mathcal{F}_m\right\} \geq \frac{1}{\alpha_m}I \quad a.s., \quad \forall m \geq 0, \tag{10.128}$$

where h is a positive integer and $\{\alpha_m, \mathcal{F}_m\}$ is a nonnegative sequence satisfying $\alpha_m \geq 1$, and

$$\alpha_{m+1} \leq a\alpha_m + \eta_{m+1}, \quad \forall m \geq 0, \quad E\alpha_0^{1+\delta} < \infty \tag{10.129}$$

where $a \in (0, 1)$ is a constant and $\{\eta_m, \mathcal{F}_m\}$ is a nonnegative sequence such that

$$\sup_{m\geq 0} E\left[\eta_{m+1}^{1+\delta}|\mathcal{F}_m\right] \leq M \quad a.s. \tag{10.130}$$

with $\delta > 0$ and $M < \infty$ being constants.

Clearly, if we take the step size μ_k as $\mu_k = \dfrac{1}{1+\|\varphi_k\|^2}$, then (10.128) coincides with the CR condition (10.46) introduced in Section 10.2.

We now present the main results of this section.

Theorem 10.7 *Consider the time-varying model (10.1) and (10.2). Suppose that Condition (10.128) holds and that for some $\alpha > 0$,*

$$\sigma_p \stackrel{\Delta}{=} \sup_{n\geq 0} E(|v_n|^\alpha + \|\Delta_n\|^\alpha) < \infty, \quad E\|\widetilde{\theta}_0\|^\alpha < \infty. \tag{10.131}$$

Then the tracking error $\widetilde{\theta}_k = \theta_k - \widehat{\theta}_k$ produced by the LMS algorithm (10.127) has the following property

$$\limsup_{n\to\infty} E\|\widetilde{\theta}_n\|^\beta \leq c(\sigma_\alpha)^{\frac{\beta}{\alpha}}, \quad \forall \beta \in (0, \alpha) \tag{10.132}$$

where c is a positive constant.

Moreover, if $v_n \equiv 0$ and $\Delta_n \equiv 0$, then

$$E\|\widetilde{\theta}_{n+1}\|^{\beta} \xrightarrow[n\to\infty]{} 0 \text{ exponentially fast, } \forall \beta \in (0,\alpha) \tag{10.133}$$

and

$$\widetilde{\theta}_{n+1} \xrightarrow[n\to\infty]{} 0 \quad \text{exponentially fast.} \tag{10.134}$$

Similarly, for the sample path average of the tracking error, we have the following result.

Theorem 10.8 *Consider the time-varying model (10.1) and (10.2). Suppose that Condition (10.128) holds and that for some $\alpha > 0$,*

$$E\|\widetilde{\theta}_0\|^{\alpha} < \infty, \quad \varepsilon_\alpha \triangleq \limsup_{n\to\infty} \frac{1}{n}\sum_{i=0}^{n}((|v_i|^{\alpha} + \|\Delta_i\|^{\alpha}) < \infty \quad a.s. \tag{10.135}$$

Then the tracking error $\widetilde{\theta}_k = \theta_k - \widehat{\theta}_k$ produced by the LMS algorithm (10.127) satisfies:

$$\limsup_{n\to\infty} \frac{1}{n}\sum_{i=0}^{n} \|\widetilde{\theta}_i\|^{\beta} \leq A(\varepsilon_\alpha)^{\frac{\beta}{\alpha}} \quad a.s. \tag{10.136}$$

for any $\beta \in (0,\alpha)$, provided that $\delta > \dfrac{2\beta}{\alpha-\beta}$ where δ is the constant appearing in (10.130), and A is a constant.

We now proceed to prove these theorems.

Recursively define

$$\Phi(n+1,m) = (I - \mu_n\varphi_n\varphi_n^{\tau})\Phi(n,m), \quad \Phi(m,m) = I, \quad \forall \; n \geq m \geq 0. \tag{10.137}$$

From (10.1), (10.2), (10.127) and (10.137) it is easy to see that

$$\begin{aligned}\widetilde{\theta}_{n+1} &= (I - \mu_n\varphi_n\varphi_n^{\tau})\widetilde{\theta}_n + \Delta_{n+1} - \mu_n\varphi_n v_n & (10.138)\\ &= \Phi(n+1,0)\widetilde{\theta}_0 + \sum_{i=0}^{n}\Phi(n+1,i+1)\xi_{i+1} & (10.139)\end{aligned}$$

where

$$\widetilde{\theta}_n \triangleq \theta_n - \widehat{\theta}_n \quad \text{and} \quad \xi_{n+1} \triangleq \Delta_{n+1} - \mu_n\varphi_n v_n. \tag{10.140}$$

From (10.139) we see that in order to prove the boundedness of $\widetilde{\theta}_{n+1}$ it is necessary to consider properties of the transition matrix $\Phi(n,m)$ first. This is done in the following two lemmas.

Lemma 10.7 *Under Condition (10.128) the following inequality holds*

$$\rho_k \leq 1 - \frac{1}{(1+h)^2\alpha_k}, \quad \forall k \geq 0, \tag{10.141}$$

where

$$\rho_k = \lambda_{max}\left(E[\Phi^\tau(k+h+1,\, k+1)\Phi(k+h+1,\, k+1)|\mathcal{F}_k]\right), \tag{10.142}$$

where $\lambda_{max}(X)$ denotes the maximum eigenvalue of a matrix X.

Proof. Let z_{k-1} be the unit eigenvector corresponding to the largest eigenvalue ρ_{k-1} of the matrix

$$E(\Phi^\tau(k+h,\, k)\Phi(k+h,\, k)|\mathcal{F}_{k-1}). \tag{10.143}$$

For any $j \geq k$ define z_j recursively by

$$z_j = (I - \mu_j\varphi_j\varphi_j^\tau)z_{j-1}. \tag{10.144}$$

It follows from (10.144) and the definition (10.137) that

$$z_{k+h-1} = \Phi(k+h,\, k)z_{k-1},$$

and hence

$$\begin{aligned} & E(\|z_{k+h-1}\|^2|\mathcal{F}_{k-1}) \\ = \ & z_{k-1}^\tau E(\Phi^\tau(k+h,\, k)\Phi(k+h,\, k)|\mathcal{F}_{k-1})z_{k-1} \\ = \ & \rho_{k-1}\|z_{k-1}\|^2 = \rho_{k-1}, \end{aligned}$$

i.e.

$$\rho_{k-1} = E(\|z_{k+h-1}\|^2|\mathcal{F}_{k-1}). \tag{10.145}$$

We now find connections between ρ_{k-1} and α_{k-1}.

Noticing (10.144), we have

$$z_j = z_{k-1} - \sum_{i=k}^{j}\mu_i\varphi_i\varphi_i^\tau z_{i-1}, \quad \forall j \in [k,\, k+h-1],$$

which leads to

$$\begin{aligned} E(\|z_{j-1} - z_{k-1}\|^2|\mathcal{F}_{k-1}) &= E\left[\left\|\sum_{i=k}^{j-1}\mu_i\varphi_i\varphi_i^\tau z_{i-1}\right\|^2 |\mathcal{F}_{k-1}\right] \\ \leq \ & E\left[\left(\sum_{i=k}^{j-1}\mu_i\|\varphi_i^\tau z_{i-1}\|^2\right)\left(\sum_{i=k}^{j-1}\mu_i\|\varphi_i\|^2\right)|\mathcal{F}_{k-1}\right] \\ \leq \ & hE\left(\sum_{i=k}^{j-1}\mu_i\|\varphi_i^\tau z_{i-1}\|^2|\mathcal{F}_{k-1}\right), \quad \forall j \in [k,\, k+h]. \end{aligned} \tag{10.146}$$

From (10.128) and the Minkowski inequality we conclude that

$$\begin{aligned}
\alpha_{k-1}^{-\frac{1}{2}} &\le \left(z_{k-1}^{\tau} E\left[\sum_{j=k}^{k+h-1} \mu_j \varphi_j \varphi_j^{\tau} | \mathcal{F}_{k-1} \right] z_{k-1} \right)^{\frac{1}{2}} \\
&= \left(E\left[\sum_{j=k}^{k+h-1} \mu_j |\varphi_j^{\tau} z_{k-1}|^2 | \mathcal{F}_{k-1} \right] \right)^{\frac{1}{2}} \\
&\le \left(E\left[\sum_{j=k}^{k+h-1} \mu_j |\varphi_j^{\tau} z_{j-1}|^2 | \mathcal{F}_{k-1} \right] \right)^{\frac{1}{2}} \\
&\quad + \left(E\left[\sum_{j=k}^{k+h-1} \|z_{j-1} - z_{k-1}\|^2 | \mathcal{F}_{k-1} \right] \right)^{\frac{1}{2}}.
\end{aligned}$$

From this and (10.146) it follows that

$$\alpha_{k-1}^{-\frac{1}{2}} \le (1+h) \left(E\left[\sum_{j=k}^{k+h-1} \mu_j |\varphi_j^{\tau} z_{j-1}|^2 | \mathcal{F}_{k-1} \right] \right)^{\frac{1}{2}}$$

i.e.

$$E\left[\sum_{j=k}^{k+h-1} \mu_j |\varphi_j^{\tau} z_{j-1}|^2 | \mathcal{F}_{k-1} \right] \ge \frac{1}{(1+h)^2 \alpha_{k-1}}. \tag{10.147}$$

Obviously, to complete the proof we should find the relationship between ρ_{k-1} and

$$E\left[\sum_{j=k}^{k+h-1} \mu_j |\varphi_j^{\tau} z_{j-1}|^2 | \mathcal{F}_{k-1} \right].$$

From (10.144) it is easy to see that

$$z_j^{\tau} z_j \le z_{j-1}^{\tau} z_{j-1} - \mu_j |\varphi_j^{\tau} z_{j-1}|^2,$$

which implies that

$$\begin{aligned}
\|z_{k+h-1}\|^2 &\le \|z_{k-1}\|^2 - \sum_{j=k}^{k+h-1} \mu_j |\varphi_j^{\tau} z_{j-1}|^2 \\
&= 1 - \sum_{j=k}^{k+h-1} \mu_j |\varphi_j^{\tau} z_{j-1}|^2.
\end{aligned} \tag{10.148}$$

Taking conditional expectations with respect to $\mathcal{F}_{k-1}$ for both sides of (10.148) and noticing (10.145) we have

$$\rho_{k-1} = E(\|z_{k+h-1}\|^2 | \mathcal{F}_{k-1}) \leq 1 - E\left[\sum_{j=k}^{k+h-1} \mu_j |\varphi_j^\tau z_{j-1}|^2 | \mathcal{F}_{k-1}\right],$$

which combining with (10.147) gives the desired result (10.141). Q.E.D.

Lemma 10.8 *Under Condition (10.128), there are constants c and $\gamma \in (0,1)$ such that*

$$E\|\Phi(n,m)\|^2 \leq c\gamma^{n-m}, \quad \forall n \geq m \geq 0, \tag{10.149}$$

where $\Phi(n,m)$ is defined by (10.137).

Proof. Let

$$k_0 = \min\{k: \ m \leq kh \leq n\}, \quad k_1 = \max\{k: \ m \leq kh \leq n\}. \tag{10.150}$$

It is clear that, if one of k_0 and k_1 exists, then the other one also exists and $k_1 \geq k_0 \geq 0$.

Noticing the inequalities

$$E\|\Phi(n,\ m)\|^2 \leq E\|\Phi(k_1 h,\ k_0 h)\|^2 \quad \text{and} \quad \gamma^{(k_1-k_0+2)h} \leq \gamma^{n-m} \tag{10.151}$$

which holds because $(k_1+1)h > n$ and $(k_0-1)h < m$, we find that for (10.149) it suffices to show that there exist constants $C \in (0,\infty)$ and $\gamma \in (0,1)$ such that

$$E\|\Phi(k_1 h,\ k_0 h)\|^2 \leq C\gamma^{(k_1-k_0+2)h}, \quad \forall k_1 \geq k_0. \tag{10.152}$$

We now proceed to prove (10.152).

Set

$$\rho_{k_1,k_0} = \lambda_{max}(E[\Phi^\tau(k_1 h, k_0 h)\Phi(k_1 h, k_0 h)]) \tag{10.153}$$

and let x_{k_1,k_0} be the unit eigenvector of the matrix $E[\Phi^\tau(k_1 h, k_0 h)\Phi(k_1 h, k_0 h)]$ corresponding to its largest eigenvalue ρ_{k_1,k_0}. Denote

$$z_k = \Phi(kh,(k-1)h)z_{k-1}, \quad \forall k \in [k_0+1, k_1], \quad z_{k_0} = x_{k_1,k_0}. \tag{10.154}$$

From (10.153)-(10.154) and Lemma 10.7 it follows that

$$z_k \in \mathcal{F}_{kh-1}, \quad \rho_{k_1,k_0} = E\|z_{k_1}\|^2, \tag{10.155}$$

$$E(\|z_k\|^2 | \mathcal{F}_{(k-1)h-1}) \leq \left(1 - \frac{1}{(1+h)^2 \alpha_{(k-1)h-1}}\right)\|z_{k-1}\|^2 \tag{10.156}$$

and

$$\|z_k\|^2 \le \|\Phi(kh,(k-1)h)\|^2\|z_{k-1}\|^2 \le \|z_{k-1}\|^2, \quad \forall k \in [k_0+1, k_1]. \tag{10.157}$$

Notice that from (10.129) we get

$$(1+h)^2\alpha_{(k+1)h-1} \le a^h(1+h)^2\alpha_{kh-1} + (1+h)^2\sum_{i=0}^{h-1} a^i \eta_{(k+1)h-i-1}.$$

Then applying Lemma 10.1 to the sequence $\{(1+h)^2\alpha_{kh-1}\}$ and noting $\alpha_{mh-1} \in \mathcal{F}_{mh-1} \triangleq \mathcal{Y}_m$ we know that there exist a constant $b \in (0,1)$ and two nonnegative adapted processes $\{y_k, \mathcal{Y}_k\}$ and $\{\delta_k, \mathcal{Y}_k\}$ depending only on $\{\eta_k\}$ such that

$$y_k \ge (1+h)^2\alpha_{kh-1}, \tag{10.158}$$

$$y_{k+1} = by_k + \delta_{k+1}, \quad \forall k \ge 0, \quad Ey_0^{1+\delta} < \infty \tag{10.159}$$

and

$$\sup_{k\ge 0} E\left[\delta_{k+1}^{1+\delta}|\mathcal{Y}_k\right] \le \overline{M}, \quad \sup_{k\ge 0} E[\delta_{k+1}|\mathcal{Y}_k] \le b, \tag{10.160}$$

where $\overline{M} < \infty$ is a constant.

We are now in a position to complete the proof of (10.152).

From (10.155)-(10.160) it is easy to see that

$$\begin{aligned}
\rho_{k_1,k_0} &= E\|z_{k_1}\|^2 \le Ey_{k_1}\|z_{k_1}\|^2 = E(by_{k_1-1} + \delta_{k_1})\|z_{k_1}\|^2 \\
&= bE[y_{k_1-1}E(\|z_{k_1}\|^2|\mathcal{F}_{(k_1-1)h-1})] + E\delta_{k_1}\|z_{k_1}\|^2 \\
&\le bE\left[y_{k_1-1}\left(1 - \frac{1}{y_{k_1-1}}\right)\|z_{k_1-1}\|^2\right] + E\delta_{k_1}\|z_{k_1-1}\|^2 \\
&= bE[y_{k_1-1}\|z_{k_1-1}\|^2] - bE\|z_{k_1-1}\|^2 \\
&\quad + E[\|z_{k_1-1}\|^2 E(\delta_{k_1}|\mathcal{F}_{(k_1-1)h-1})] \\
&= bE[y_{k_1-1}\|z_{k_1-1}\|^2] \le \dots \le b^{k_1-k_0}Ey_{k_0} \\
&= \left(Ey_0 + \frac{b}{1-b}\right)b^{k_1-k_0},
\end{aligned}$$

which implies (10.152), and hence (10.149) holds. Q.E.D.

Proof of Theorem 10.7.

We first show (10.133) and (10.134).

If $\frac{\alpha\beta}{\alpha-\beta} \geq 2$, then from $\|\Phi(n+1,i)\| \leq 1$ and (10.149) it follows that

$$E\|\Phi(n+1,i)\|^{\frac{\alpha\beta}{\alpha-\beta}} \leq E\|\Phi(n+1,i)\|^2 \leq c\gamma^{n+1-i}, \quad \forall n \geq i \geq 0. \tag{10.161}$$

If $\frac{\alpha\beta}{\alpha-\beta} < 2$, then by the Hölder inequality and (10.149), we have, for any $n \geq i \geq 0$,

$$\begin{aligned} E\|\Phi(n+1,i)\|^{\frac{\alpha\beta}{\alpha-\beta}} &\leq \left(E\|\Phi(n+1,i)\|^2\right)^{\frac{\alpha\beta}{2(\alpha-\beta)}} \\ &\leq \left(c\gamma^{n+1-i}\right)^{\frac{\alpha\beta}{2(\alpha-\beta)}}. \end{aligned} \tag{10.162}$$

Hence for any $\beta \in (0,\alpha)$ from (10.161)-(10.162) there always exist constants $c_1 \in (0,\infty)$ and $\gamma_1 \in (0,1)$ such that

$$E\|\Phi(n+1,i)\|^{\frac{\alpha\beta}{\alpha-\beta}} \leq c_1\gamma_1^{n+1-i}, \quad n \geq i \geq 0. \tag{10.163}$$

In the case where $v_n \equiv 0$ and $\Delta_n \equiv 0$, it follows from (10.139) that

$$\widetilde{\theta}_{n+1} = \Phi(n+1,0)\widetilde{\theta}_0. \tag{10.164}$$

Applying the Hölder inequality to (10.164) ($p = \frac{\alpha}{\beta}$, $q = \frac{\alpha}{\alpha-\beta}$) from (10.163) we see that (10.133) is true, i.e. there are constants $\xi \in (0,1)$ and $c_1 > 0$ such that

$$E\|\widetilde{\theta}_n\|^\beta \xi^{-n} < c_1 \quad \forall n \geq 0.$$

By the Borel-Cantelli lemma it is easy to see that

$$\lim_{n\to\infty} \|\widetilde{\theta}_n\|^\beta \xi^{-n/2} = 0 \quad a.s.$$

which proves (10.134).

It remains to prove (10.132).

From the definition (10.140) of ξ_n it is easy to see that there exists a constant N such that

$$\sup_{k\geq 0} E\|\xi_k\|^\alpha \leq N\sigma_\alpha. \tag{10.165}$$

For a given $\beta \in (0,\alpha)$ if $\beta \in (0,1]$, from (10.139) and the following elementary inequality

$$(x+y)^\beta \leq x^\beta + y^\beta, \quad \forall x \geq 0,\ y \geq 0,$$

it follows that

$$\begin{aligned} E\|\widetilde{\theta}_{n+1}\|^\beta &\leq E\|\Phi(n+1,0)\|^\beta \|\widetilde{\theta}_0\|^\beta \\ &\quad + \sum_{i=0}^{n} E\|\Phi(n+1,i+1)\|^\beta \|\xi_{i+1}\|^\beta, \end{aligned}$$

which by the Hölder inequality yields

$$E\|\widetilde{\theta}_{n+1}\|^\beta \le \left(E\|\Phi(n+1,0)\|^{\frac{\alpha\beta}{\alpha-\beta}}\right)^{\frac{\alpha-\beta}{\alpha}} \left(E\|\widetilde{\theta}_0\|^\alpha\right)^{\frac{\beta}{\alpha}} + \sum_{i=0}^{n} \left(E\|\Phi(n+1,i+1)\|^{\frac{\alpha\beta}{\alpha-\beta}}\right)^{\frac{\alpha-\beta}{\alpha}} \left(E\|\xi_{i+1}\|^\alpha\right)^{\frac{\beta}{\alpha}}. \quad (10.166)$$

If $\beta > 1$, from (10.139) by the Minkowski inequality and Hölder inequality we have

$$\begin{aligned} &\left(E\|\widetilde{\theta}_{n+1}\|^\beta\right)^{\frac{1}{\beta}} \\ \le\ & \left(E\|\Phi(n+1,0)\|^\beta\|\widetilde{\theta}_0\|^\beta\right)^{\frac{1}{\beta}} + \sum_{i=0}^{n}\left(E\|\Phi(n+1,i+1)\|^\beta\|\xi_{i+1}\|^\beta\right)^{\frac{1}{\beta}} \\ \le\ & \left\{\left(E\|\Phi(n+1,0)\|^{\frac{\alpha\beta}{\alpha-\beta}}\right)^{\frac{\alpha-\beta}{\alpha}}\left(E\|\widetilde{\theta}_0\|^\alpha\right)^{\frac{\beta}{\alpha}}\right\}^{\frac{1}{\beta}} \\ & + \sum_{i=0}^{n}\left\{\left(E\|\Phi(n+1,i+1)\|^{\frac{\alpha\beta}{\alpha-\beta}}\right)^{\frac{\alpha-\beta}{\alpha}}\left(E\|\xi_{i+1}\|^\alpha\right)^{\frac{\beta}{\alpha}}\right\}^{\frac{1}{\beta}} \\ =\ & \left(E\|\Phi(n+1,0)\|^{\frac{\alpha\beta}{\alpha-\beta}}\right)^{\frac{\alpha-\beta}{\alpha\beta}}\left(E\|\widetilde{\theta}_0\|^\alpha\right)^{\frac{1}{\alpha}} \\ & + \sum_{i=0}^{n}\left(E\|\Phi(n+1,i+1)\|^{\frac{\alpha\beta}{\alpha-\beta}}\right)^{\frac{\alpha-\beta}{\alpha\beta}}\left(E\|\xi_{i+1}\|^\alpha\right)^{\frac{1}{\alpha}}. \quad (10.167) \end{aligned}$$

By (10.163) from (10.166) and (10.167) the desired result (10.132) follows immediately. Q.E.D.

Proof of Theorem 10.8.
For any fixed $s = 0, 1, \ldots, h$ set

$$x_k(s) = \|\widetilde{\theta}_{k(h+1)+s}\|, \quad \forall k \ge 0. \quad (10.168)$$

Noticing from (10.138) and (10.140)

$$\widetilde{\theta}_{(k+1)(h+1)+s} = \Phi((k+1)(h+1)+s,\ k(h+1)+s)\widetilde{\theta}_{k(h+1)+s} + \sum_{i=k(h+1)+s}^{(k+1)(h+1)+s-1} \Phi((k+1)(h+1)+s,\ i+1)\xi_{i+1},$$

then by the boundedness of $\mu_n\|\varphi_n\|^2$ we have

$$\begin{aligned} x_{k+1}(s) \ \le\ & \|\Phi((k+1)(h+1)+s,\ k(h+1)+s)\widetilde{\theta}_{k(h+1)+s}\| \\ & + \bar{\xi}_{k+1}(s), \end{aligned} \quad (10.169)$$

where

$$\bar{\xi}_{k+1}(s) = \sum_{i=k(h+1)+s}^{(k+1)(h+1)+s-1} (|v_i| + \|w_{i+1}\|). \tag{10.170}$$

Let

$$a_{k+1}(s) = \\ = \begin{cases} 1 - \dfrac{\|\Phi((k+1)(h+1)+s,\ k(h+1)+s)\tilde{\theta}_{k(h+1)+s}\|}{\|\tilde{\theta}_{k(h+1)+s}\|}, \\ \quad \text{if } \ \|\tilde{\theta}_{k(h+1)+s}\| > 0 \\ 1, \quad \text{if } \ \|\tilde{\theta}_{k(h+1)+s}\| = 0 \end{cases} \tag{10.171}$$

It is clear that

$$a_k(s) \geq 0, \quad a_k(s) \in \mathcal{F}_{k(h+1)+s}, \quad a_k(s) \in [0,1] \tag{10.172}$$

and

$$x_{k+1}(s) \leq (1 - a_{k+1}(s))x_k(s) + \bar{\xi}_{k+1}(s). \tag{10.173}$$

We now show that

$$E[a_{k+1}(s)|\mathcal{F}_{k(h+1)+s}] \geq \frac{1}{2(1+h)^2\alpha_{k(h+1)+s}} \quad a.s. \tag{10.174}$$

Set $A_k(s) = \{\omega : \|\tilde{\theta}_{k(h+1)+s}\| = 0\}$. Obviously, we have $A_k(s) \in \mathcal{F}_{k(h+1)+s}$ and

$$\begin{aligned} & I_{A_k(s)} E[a_{k+1}(s)|\mathcal{F}_{k(h+1)+s}] \\ = \ & E[a_{k+1}(s) I_{A_k(s)} |\mathcal{F}_{k(h+1)+s}] = I_{A_k(s)}, \end{aligned} \tag{10.175}$$

which means that (10.174) is true on $A_k(s)$ since $\alpha_k \geq 1$, $\forall k \geq 0$.

From the Schwarz inequality and Lemma 10.7 we see that

$$\begin{aligned} & E[\|\Phi((k+1)(h+1)+s, \\ & k(h+1)+s)\tilde{\theta}_{k(h+1)+s}\||\mathcal{F}_{k(h+1)+s}] \\ \leq \ & \{\|\Phi((k+1)(h+1)+s, \\ & k(h+1)+s)\tilde{\theta}_{k(h+1)+s}\|^2|\mathcal{F}_{k(h+1)+s}\}^{\frac{1}{2}} \\ \leq \ & \left(1 - \frac{1}{(1+h)^2\alpha_{k(h+1)+s}}\right)^{\frac{1}{2}} \|\tilde{\theta}_{k(h+1)+s}\| \\ \leq \ & \left(1 - \frac{1}{2(1+h)^2\alpha_{k(h+1)+s}}\right) \|\tilde{\theta}_{k(h+1)+s}\|. \end{aligned}$$

Hence by (10.171) we derive

$$
\begin{aligned}
& I_{A_k^c(s)} E[a_{k+1}(s)|\mathcal{F}_{k(h+1)+s}] = E[I_{A_k^c(s)} a_{k+1}(s)|\mathcal{F}_{k(h+1)+s}] \\
\geq\ & I_{A_k^c(s)} \left(1 - \left(1 - \frac{1}{2(1+h)^2 \alpha_{k(h+1)+s}}\right)\right) \\
=\ & I_{A_k^c(s)} \frac{1}{2(1+h)^2 \alpha_{k(h+1)+s}},
\end{aligned}
$$

which together with (10.175) implies (10.174).

Finally, applying Theorem 10.3 to (10.173) we obtain

$$
\limsup_{n\to\infty} \frac{1}{n} \sum_{i=0}^{n} x_i^{\beta}(s) \leq A_s(\varepsilon_\alpha)^{\frac{\beta}{\alpha}}, \quad s = 0,\ 1,\ \ldots,\ h,
$$

where $A_s < \infty$, $s = 0, 1, \ldots, h$, are constants. This verifies the desired result (10.136). Q.E.D.

CHAPTER 11

Adaptive Control of Time-Varying Stochastic Systems

In many practical applications, the system structure is often unknown or just partially known; not only may it suffer from random disturbances but also its parameters may vary with time. The drifting or jumping phenomena of parameters in a system is one of the motivations for adaptive control since classical design methods may be difficult to use for such systems.

Over the past two decades, the area of adaptive control can roughly be divided into two directions: deterministic and stochastic. In the deterministic setup of adaptive control the system under study is normally assumed to be free of noise, or at most to have a uniformly bounded disturbance. When the adaptive controller is designed based on deterministic methods, optimality of performance cannot be guaranteed even for time-invariant plants with a uniformly bounded white noise. This is due to the fact that the algorithms used have the so-called *short memory property*, i.e., the adaptation gain is not vanishing. Nevertheless, algorithms of this kind have the merit that they may stabilize a time-varying system, as has been shown in, e.g., [TI], [MG].

In stochastic adaptive control the noise is an essential feature of the system, and it is not necessarily bounded; a standard example is the Gaussian white noise. In this case, especially for the constant parameter case, it is of interest not only to guarantee stability of the closed-loop system, but also to reject the noise optimally, or at least close to optimally. This is possible because the algorithms commonly used have the so-called *long memory property*, i.e., the adaptation gain tends to zero. This guarantees that no large deviations of the estimates can occur, at least for the constant parameter case. Indeed, it has been shown that in the constant parameter case the parameter estimates in a closed-loop adaptive system can be either nearly consistent [BKW] or strongly consistent [CG4]. However, it is this long memory property that prevents the adaptive law from being effective for general time-varying systems. Indeed, it has been found that with long memory algorithms it is difficult to deal with time-variations which are more general than those treated in [CCa1], [CG8]. For this reason it is believed that short memory algorithms may be more effective than the long memory ones in controlling time-varying systems.

However, the precise stability analysis for short-memory-algorithm-based stochastic adaptive controllers appears just recently. The stability of a first order adaptive control system is analyzed in [MC], where the noise is assumed to be Gaussian with known covariance. Convergence of control performance is established by using the Markov chain ergodic theory. The Gaussian assumptions are later removed in [GM] where the continuity of the performance with respect to the noise distribution is also addressed. By using a projected gradient algorithm, the stability of adaptive control for a general class of linear stochastic systems with deterministic or random time-varying coefficients is established in [Gu3]. The Markov chain ergodic theory can also be applied to the convergence study of control performance [MGu].

This chapter is based on [Gu3].

11.1 Preliminary Results

For derivation of the main results we need some inequalities and stability results for stochastic sequences, which we present in this section.

Lemma 11.1 *(Bellman-Gronwall). Let $\{x_k\}$, $\{f_k\}$ and $\{h_k\}$ be three non-negative sequences, and*

$$x_k \le f_k + \sum_{i=0}^{k-1} h_i x_i, \quad k \ge 0.$$

Then

$$x_k \le f_k + \sum_{i=0}^{k-1} \prod_{j=i}^{k-1} (1 + h_j) f_i, \quad k \ge 0.$$

Proof. Set

$$Y_k = \sum_{i=0}^{k} h_i x_i, \quad Y_{-1} = 0.$$

Then by the assumption we have

$$\begin{aligned} Y_k &= h_k x_k + \sum_{i=0}^{k-1} h_i x_i \le h_k \left(f_k + \sum_{i=0}^{k-1} h_i x_i \right) + \sum_{i=0}^{k-1} h_i x_i \\ &= (1+h_k)Y_{k-1} + h_k f_k. \end{aligned}$$

Hence it follows that

$$Y_k = \sum_{i=0}^{k} \prod_{j=i+1}^{k} (1+h_j) h_i f_i \le \sum_{i=0}^{k} \prod_{j=i}^{k} (1+h_j) f_i.$$

This in conjunction with the assumption yields the desired result.

Q.E.D.

Lemma 11.2 *Let $\{x_n, \mathcal{F}_n\}$ be an adapted sequence such that for some integer $r \ge 0$ and some $\alpha > 1$,*

$$\sup_n E\{|x_{n+1}|^\alpha | \mathcal{F}_{n-r}\} < \infty \quad a.s. \tag{11.1}$$

Then

$$\limsup_{n\to\infty} \frac{1}{N} \sum_{n=0}^{N} |x_n| < \infty \quad a.s.$$

Proof. We first note that for any fixed k, $0 \le k \le r$, the sequence

$$M_n = |x_{k+n(r+1)}| - E\left\{ |x_{k+n(r+1)}| \middle| \mathcal{F}_{k+(n-1)(r+1)} \right\}$$

is a martingale difference sequence with respect to $\{\mathcal{F}_{k+n(r+1)}\}$. Hence by (11.1) and Corollary 2.6, we know that

$$\frac{1}{N} \sum_{n=0}^{N} M_n \to 0 \quad a.s. \quad \text{as } N \to \infty.$$

Consequently, by (11.1) again we have

$$\limsup_{n\to\infty} \frac{1}{N} \sum_{n=0}^{N} |x_{k+n(r+1)}| < \infty \quad a.s. \quad \forall k \in [0, r].$$

Finally, the desired result follows by observing

$$\frac{1}{N} \sum_{n=0}^{N} |x_n| \le \frac{1}{N} \sum_{k=0}^{r} \sum_{n=0}^{[(N+1)/(r+1)]} |x_{k+n(r+1)}|,$$

where $[(N+1)/(r+1)]$ is the integer part of $(N+1)/(r+1)$. Q.E.D.

Lemma 11.3 *Let $\{x_n, \mathcal{F}_n\}$, $\{f_n, \mathcal{F}_n\}$ and $\{g_n, \mathcal{F}_n\}$ be three nonnegative sequences satisfying*

$$x_{n+1} \leq f_{n+1}x_n + g_{n+1}, \quad n \geq 0. \tag{11.2}$$

Assume that for some constants $\varepsilon_\alpha < 1$, $\alpha > 1$ and $c < \infty$,

$$\sup_n E\{(f_{n+1})^\alpha | \mathcal{F}_n\} \leq \varepsilon_\alpha \quad a.s., \quad \sup_n E\{(g_{n+1})^\alpha | \mathcal{F}_n\} \leq C. \tag{11.3}$$

Then

$$\sum_{n=0}^{N} x_n = O(N) \quad a.s. \quad as \quad N \to \infty. \tag{11.4}$$

Proof. Applying the Minkowski inequality to (11.2) and noting (11.3), we see that

$$\begin{aligned}
&\{E(x_{n+1})^\alpha\}^{1/\alpha} \\
\leq\ & \{E(f_{n+1}x_n)^\alpha\}^{1/\alpha} + \{E(g_{n+1})^\alpha\}^{1/\alpha} \\
=\ & \{E[E[(f_{n+1})^\alpha|\mathcal{F}_n](x_n)^\alpha]\}^{1/\alpha} + \{E(g_{n+1})^\alpha\}^{1/\alpha} \\
\leq\ & (\varepsilon_\alpha)^{1/\alpha}\{(x_n)^\alpha\}^{1/\alpha} + \sup_n \{E(g_{n+1})^\alpha\}^{1/\alpha}.
\end{aligned}$$

From this and the fact that $(\varepsilon_\alpha)^{1/\alpha} < 1$, it is easy to conclude that

$$\sup_n E(x_n)^\alpha < \infty. \tag{11.5}$$

Let us denote $M_n = x_n - E[x_n|\mathcal{F}_{n-1}]$; then by (11.5) and the martingale stability results (Corollary 2.6) it is evident that

$$\sum_{n=0}^{N} M_n = o(N) \quad a.s. \quad \text{as} \quad N \to \infty.$$

Thus by (11.3) and the recursion (11.2) we have (where ε_1 is defined as $(\varepsilon_\alpha)^{1/\alpha}$)

$$\begin{aligned}
\sum_{n=0}^{N} x_{n+1} &= \sum_{n=0}^{N} E[x_{n+1}|\mathcal{F}_n] + \sum_{n=0}^{N} M_{n+1} \\
&\leq \sum_{n=0}^{N} E[f_{n+1}|\mathcal{F}_n]x_n + \sum_{n=0}^{N} E[g_{n+1}|\mathcal{F}_n] + o(N) \\
&\leq \varepsilon_1 \sum_{n=0}^{N} x_n + O(N) \leq \varepsilon_1 \sum_{n=0}^{N} x_{n+1} + \varepsilon_1 x_0 + O(N).
\end{aligned}$$

Consequently the assertion (11.5) holds since $\varepsilon_1 < 1$. Q.E.D.

Lemma 11.4 *Assume that $\{f_n\}$ is a sequence of nonnegative random variables defined by*

$$f_n = \sum_{i=0}^{n-1} \lambda^{n-i} \prod_{j=i}^{n-1} x_j, \quad f_0 = 0,$$

where $\lambda \in (0,1)$ and $\{x_k, \mathcal{F}_k\}$ is a nonnegative adapted sequence satisfying $x_k \geq 1$ and

$$\left\{E[(x_{k+1})^{\alpha(r+1)}]|\mathcal{F}_{k-r}\right\}^{1/[\alpha(r+1)]} \leq C \quad a.s. \quad \lambda C < 1 \tag{11.6}$$

for some integer $r \geq 0$ and some constants $C > 0$ and $\alpha \geq 1$. Then

$$\sup_n \{E[f_n]^\alpha\}^{1/\alpha} \leq \lambda C^{r+2}(1-\lambda C)^{-1}. \tag{11.7}$$

Moreover, if in (11.6) $\alpha > 1$, then as $N \to \infty$

$$\sum_{n=0}^{N} f_n = O(N) \quad a.s. \tag{11.8}$$

Proof. By the Hölder inequality, we have

$$E\left\{\prod_{j=i}^{n-1} x_j\right\}^\alpha \leq E\left\{\prod_{j=i}^{i+r} \prod_{k=0}^{[(n-i)/(r+1)]} x^\alpha_{j+k(r+1)}\right\}$$

$$\leq \prod_{j=i}^{i+r}\left\{E \prod_{k=0}^{[(n-i)/(r+1)]} x^{\alpha(r+1)}_{j+k(r+1)}\right\}^{1/(r+1)}, \tag{11.9}$$

where $[(n-i)/(r+1)]$ is the integer part of $(n-i)/(r+1)$.

Note that for each i and j,

$$\begin{aligned}
& E \prod_{k=0}^{[(n-i)/(r+1)]} [x_{j+k(r+1)}]^{\alpha(r+1)} \\
= \ & E \prod_{k=0}^{[(n-i)/(r+1)]-1} x^{\alpha(r+1)}_{j+k(r+1)} \\
& \cdot E\{[x_{j+[(n-i)/(r+1)](r+1)}]^{\alpha(r+1)}|\mathcal{F}_{j+\{[(n-i)/(r+1)]-1\}(r+1)}\} \\
\leq \ & C^{\alpha(r+1)} E \prod_{k=0}^{(n-i)/(r+1)]-1} x^{\alpha(r+1)}_{j+k(r+1)} \leq \cdots \\
\leq \ & C^{\alpha(r+1)\{[(n-i)/(r+1)]+1\}} \leq C^{\alpha(n-i+r+1)}.
\end{aligned}$$

Substituting this into (11.9) we see that

$$E\left\{\prod_{j=i}^{n-1} x_j\right\}^{\alpha} \leq C^{\alpha(n-i+r+1)}.$$

Consequently by the definition of f_n and the Minkowski inequality

$$\begin{aligned}\{E[f_n]^{\alpha}\}^{1/\alpha} &\leq \sum_{i=0}^{n-1} \lambda^{n-i}\left\{E\prod_{j=i}^{n-1}(x_j)^{\alpha}\right\}^{1/\alpha} \\ &\leq C^{(r+1)}\sum_{i=0}^{n-1}(\lambda C)^{n-i} \leq \lambda C^{r+2}(1-\lambda c)^{-1}.\end{aligned}$$

We now prove (11.8). By the Hölder inequality

$$\begin{aligned}\sum_{n=0}^{N} f_n &= \sum_{n=0}^{N}\sum_{i=0}^{n-1}\lambda^{n-i}\prod_{j=i}^{n-1} x_j \\ &= \sum_{n=0}^{N}\sum_{i=0}^{n-1}\lambda^{n-i}\prod_{j=0}^{r}\prod_{k=0}^{[(n-i)/(r+1)]}[x_{i+j+k(r+1)}] \\ &\leq \prod_{j=0}^{r}\left\{\sum_{n=0}^{N}\sum_{i=0}^{n-1}\lambda^{n-i}\prod_{k=0}^{[(n-i)/(r+1)]}[x_{i+j+k(r+1)}]^{r+1}\right\}^{1/(r+1)}.\end{aligned} \tag{11.10}$$

Note that for each j

$$\begin{aligned}&\sum_{i=0}^{n-1}\lambda^{n-i}\prod_{k=0}^{[(n-i)/(r+1)]}[x_{i+j+k(r+1)}]^{r+1} \\ &\leq \sum_{s=0}^{r}\sum_{i=0}^{[n/(r+1)]}\lambda^{n-s-i(r+1)}\prod_{k=0}^{[(n-s)/(r+1)]-i}[x_{s+j+(i+k)(r+1)}]^{r+1},\end{aligned} \tag{11.11}$$

where by definition

$$\prod_{k=0}^{-1} = 1.$$

Let us denote

$$g_n = \sum_{i=0}^{[n/(r+1)]}\lambda^{n-s-i(r+1)}\prod_{k=0}^{[(n-s)/(r+1)]-i}[x_{s+j+(i+k)(r+1)}]^{r+1}, \quad 0 \leq s,\ j \leq r. \tag{11.12}$$

Similar to the proof of Lemma 11.2, we consider the following subsequence of $\{g_n\}$ for any fixed $t \in [0, r]$, $0 \le s,\ j \le r$:

$$\begin{aligned} g_{t+n(r+1)} &= \sum_{i=0}^{[t/(r+1)]+n} \lambda^{t-s+(n-i)(r+1)} \prod_{k=0}^{[(t-s)/(r+1)]-i+n} [x_{s+j+(i+k)(r+1)}]^{r+1} \\ &\le \sum_{i=0}^{n} \lambda^{t-s+(n-i)(r+1)} \prod_{k=0}^{n-i} [x_{s+j+(i+k)(r+1)}]^{r+1} \\ &\triangleq M_n. \end{aligned} \tag{11.13}$$

It is obvious that $\{M_n, \mathcal{G}_n\}$ is an adapted sequence, where $\mathcal{G}_n = \mathcal{F}_{s+j+n(r+1)}$. Note also that

$$M_n = [\lambda x_{s+j+n(r+1)}]^{r+1} M_{n-1} + \lambda^{t-s} [x_{s+j+n(r+1)}]^{r+1}$$

and that by the assumption (11.6)

$$\sup_n E\{[\lambda x_{s+j+n(r+1)}]^{(r+1)\alpha} | \mathcal{G}_{n-1}\} \le (\lambda C)^{\alpha(r+1)} < 1 \quad a.s.$$

Hence applying Lemma 11.3 we have

$$\begin{aligned} & \sum_{n=0}^{N} M_n = O(N) \quad a.s. \\ \Longrightarrow\ & \sum_{n=0}^{N} g_n = O(N) \quad a.s. \quad \text{(since in (11.13) } t \in [0, r] \text{ is arbitrary)} \\ \Longrightarrow\ & \sum_{n=0}^{N} \sum_{i=0}^{n-1} \lambda^{n-i} \prod_{k=0}^{[(n-i)/(r+1)]} [x_{i+j+k(r+1)}]^{r+1} = O(N) \quad a.s. \text{ (by (11.11))} \\ \Longrightarrow\ & \sum_{n=0}^{N} f_n = O(N) \quad a.s. \quad \text{(by (11.10))}. \end{aligned}$$

This completes the proof. Q.E.D.

We remark that f_k defined in Lemma 11.4 may be regarded as the solution of the following linear equation:

$$f_k = (\lambda x_{k-1}) f_{k-1} + 1, \quad k \ge 0, \quad f_0 = 0.$$

Lemma 11.5 *Let w and $\mathcal{F}$ be a random variable and a σ-algebra, respectively. If*

$$E\{\exp(w^2)|\mathcal{F}\} \le \exp(\delta) \quad a.s. \quad \textit{for some } \delta > 0,$$

then for any real number $a > 0$

$$E\{\exp(a|w|)|\mathcal{F}\} \le \exp\left\{a\delta^{1/2} + \left[\frac{1}{2} + 4a^2(1 + \exp(8a^2))\right]\delta\right\} \quad a.s.$$

We remark that the key point in the upper bound above is the dependence on δ. If $\delta < 1$, then the result stated above implies that

$$E\{\exp(a|w|)|\mathcal{F}\} \leq \exp\{f(a)\delta^{1/2}\},$$

where $f(\cdot)$ is a function defined by

$$f(x) = \frac{1}{2} + x + 4x^2[1 + \exp(8x^2)], \quad \forall x. \tag{11.14}$$

Proof. We first note that by the Jensen's inequality

$$E\{\exp(w^2)|\mathcal{F}\} \geq \exp\{E[w^2|\mathcal{F}]\} \quad a.s.$$

So it follows from the assumption that

$$E[w^2|\mathcal{F}] \leq \delta \quad a.s.$$

Next, we will use the following fact that can be proven in exactly the same way as that for Lemmas 4.1.1 of [St] (p.226): If $0 \leq Y \leq 1$ a.s., then

$$E\{\exp(Y)|\mathcal{F}\} \leq \exp\{E[Y|\mathcal{F}] + E[Y^2|\mathcal{F}]\}.$$

Applying this we have

$$E\{\exp[4a|w|I(4a|w| \leq 1)|\mathcal{F}]\} \leq \exp\{4a\delta^{1/2} + (4a)^2\delta\}.$$

Hence, by this inequality, the Schwarz inequality and the Markov inequality, we have (where $E^{\mathcal{F}}(\cdot)$ denotes $E(\cdot|\mathcal{F})$, for simplicity)

$$\begin{aligned}
& E^{\mathcal{F}}\exp(a|w|) = E^{\mathcal{F}}\exp\{a|w|[I(|w| \geq 2a) + I(|w| < 2a)]\} \\
\leq \; & E^{\mathcal{F}}\exp\left(\frac{w^2}{2}\right)\exp\{a|w|I(|w| < 2a)\} \\
\leq \; & \exp\left(\frac{\delta}{2}\right)\left\{E^{\mathcal{F}}\exp\left\{2a|w|\left[I\left(|w| \leq \frac{1}{4a}\right) + I\left(\frac{1}{4a} < |w| < 2a\right)\right]\right\}\right\}^{1/2} \\
\leq \; & \exp\left(\frac{\delta}{2}\right)\{E^{\mathcal{F}}\exp[4a|w|I(4a|w| \leq 1)] \\
& \cdot E^{\mathcal{F}}\exp\left[4a|w|I\left(\frac{1}{4a} < |w| < 2a\right)\right]\right\}^{1/2} \\
\leq \; & \exp\left(\frac{\delta}{2}\right)\{\exp[4a\delta^{1/2} + (4a)^2\delta]\}^{1/4} \\
& \cdot\left\{E^{\mathcal{F}}\exp\left[4a|w|I\left(\frac{1}{4a} < |w| < 2a\right)\right]\right\}^{1/4} \\
\leq \; & \exp\left[\frac{\delta}{2} + a\delta^{1/2} + 4a^2\delta\right]\left\{E^{\mathcal{F}}I\left(|w| \leq \frac{1}{4a}\right)\right.
\end{aligned}$$

$$+ \cdot E^{\mathcal{F}} \exp[4a|w|I(|w| < 2a)]I\left(|w| > \frac{1}{4a}\right)\Big\}^{1/4}$$

$$\leq \exp\left[a\delta^{1/2} + \left(\frac{1}{2} + 4a^2\right)\delta\right]\left\{1 + \exp(8a^2)P\left(|w| > \frac{1}{4a}|\mathcal{F}\right)\right\}^{1/4}$$

$$\leq \exp\left[a\delta^{1/2} + \left(\frac{1}{2} + 4a^2\right)\delta\right]\{1 + (4a^2)\delta\exp(8a^2)\}^{1/4}$$

$$\leq \exp\left\{a\delta^{1/2} + \left[\frac{1}{2} + 4a^2 + 4a^2\exp(8a^2)\right]\delta\right\}.$$

This completes the proof. Q.E.D.

11.2 Systems with Random Parameters

Let us consider the following linear time-varying stochastic model:

$$y_{k+1} = a_1(k)y_k + \dots + a_p(k)y_{k-p+1} + u_k + v_{k+1}, \quad y_k = u_k = v_k = 0, \quad \forall k < 0, \tag{11.15}$$

where y_k, u_k and v_k are the scalar output, input and random noise respectively, and $a_i(k)$, $1 \leq i \leq p$, are the unknown random time-varying parameters.

We now introduce the assumptions on the random noise sequence $\{v_k\}$.

Noise assumption. $\{v_k, \mathcal{F}_k\}$ is an adapted sequence where $\{\mathcal{F}_k\}$ is a nondecreasing family of σ-algebras, and for some integer $r \geq 0$ and deterministic positive constants ε and M_v:

$$E\{\exp[\varepsilon|v_{k+1}|^2]|\mathcal{F}_{k-r}\} \leq \exp\{M_v\} \quad a.s. \quad \forall k \geq 0. \tag{11.16}$$

Obviously, any sequence $\{v_k\}$ uniformly bounded in sample path satisfies this assumption. We note also that if $\{v_k\}$ is an r-dependent sequence (i.e. for any k, $\{v_i, i \leq k\}$ and $\{v_{i+r}, i > k\}$ are independent), then the assumption (11.16) reduces to

$$E\{\exp[\varepsilon|v_{k+1}|^2]\} \leq \exp\{M_v\} \quad a.s. \quad \forall k \geq 0. \tag{11.17}$$

Let us now give an example where the noise sequence $\{v_k\}$ is unbounded a.s.

Example 11.1. Let $\{v_k\}$ be the following time-varying moving average process:

$$v_k = e_k + c_1(k)e_{k-1} + \dots + c_r(k)e_{k-r}, \quad k \geq 0 \tag{11.18}$$

with deterministic coefficients $\{c_i(k)\}$ satisfying

$$\sum_{i=0}^{r} |c_i(k)|^2 \leq c < \infty \quad \forall k \geq 0, \quad (c_0(k) = 1) \tag{11.19}$$

and with $\{e_k\}$ being a Gaussian white noise sequence with variance $\sigma^2 > 0$. Then

$$\limsup_{k\to\infty} \frac{|v_k|}{(2\log k)^{1/2}} \geq \sigma \quad a.s. \tag{11.20}$$

and (11.16) holds for any

$$\varepsilon < \frac{1}{2c\sigma^2}, \quad M_v \geq \frac{\varepsilon c\sigma^2(r+1)}{1-2\varepsilon c\sigma^2}. \tag{11.21}$$

Proof. Property (11.20) follows from the conditional Borel-Cantelli lemma and the Gaussian assumption; details of the proof are omitted (see also [CT, p.64] for a related result). Here we need only to prove that (11.16) is true for any constants ε and M_v satisfying (11.21).

Apparently, $\{v_k\}$ is an r-dependent sequence, so we need only to verify (11.17). By elementary calculations, it is easy to verify that

$$\begin{aligned} E\exp\{\varepsilon|v_k|^2\} &\leq \{E\exp[\varepsilon c(e_1)^2]\}^{r+1} \\ &\leq \exp\left\{\frac{\varepsilon c\sigma^2}{1-2\varepsilon c\sigma^2}(r+1)\right\}. \end{aligned} \tag{11.22}$$

Hence by (11.21) and (11.22), we see that (11.17) is true.

We remark that in this example the constants ε and M_v depend only on the *upper bounds* of σ, c, and r.

For (11.15) we need the following assumption on the random parameter θ_k:

$$\theta_k = [a_1(k)\ \dots\ a_p(k)]^\tau. \tag{11.23}$$

Parameter assumption (random case). $\{\theta_k, \mathcal{F}_k\}$ defined in (11.23) is an adapted sequence satisfying

$$E\{\exp[M\|\theta_{k+1}\|^2]|\mathcal{F}_{k-m}\} \leq \exp\{M_\theta\} \quad a.s. \quad \forall k \geq 0, \tag{11.24}$$

$$E\{\exp[M\|\Delta_{k+1}\|^2]|\mathcal{F}_{k-m}\} \leq \exp\{\delta_\theta\} \quad a.s. \quad \forall k \geq 0, \tag{11.25}$$

where Δ_{k+1} is the parameter variation process:

$$\Delta_{k+1} = \theta_{k+1} - \theta_k, \quad k \geq 0, \tag{11.26}$$

and where $m \geq 0$ is an integer, M, M_θ and $\delta_\theta < 1$ are positive constants.

We now discuss this condition. Condition (11.24) means that the random process $\{\theta_k\}$ is bounded in an average sense and not necessarily bounded in sample path. In the main theorems to follow, we actually need the constant M to be suitably large and δ_0 is suitably small (see Remark 11.1). This means that the parameters are slowly varying in an average sense, and again, the variation is not necessarily small in the sample path. In particular, these conditions do not rule out occasional but possibly large jumps of the parameter process. Let us give a concrete example.

Example 11.2. Let the unknown parameter θ_k be a constant vector plus a p-dimensional moving average process:

$$\theta_k = \theta + \varepsilon_k + D_1\varepsilon_{k-1} + \ \dots\ + D_{m-1}\varepsilon_{k-m+1}, \quad \forall k \geq 0, \tag{11.27}$$

where D_i, $1 \leq i \leq m-1$, are deterministic matrices, and $\{\varepsilon_k\}$ is a Gaussian white noise sequence with covariance matrix $(\sigma_\varepsilon)^2 I$. Then for any $\sigma_\varepsilon > 0$,

$$\limsup_{k\to\infty} \|\theta_k\| = \infty \quad a.s., \quad \limsup_{k\to\infty} \|\theta_k - \theta_{k-1}\| = \infty \quad a.s. \tag{11.28}$$

Furthermore, the parameter assumption holds for all small σ_ε.

Proof. We need only to verify (11.24) and (11.25). Note that both the process $\{\theta_k\}$ and its variation process

$$\Delta_k = \varepsilon_k + (D_1 - 1)\varepsilon_{k-1} + \ \dots\ + (D_{m-1} - D_{m-2})\varepsilon_{k-m+1}, \tag{11.29}$$

are m-dependent sequences, so it suffices to verify (11.24) and (11.25) with conditional expectation replaced by expectation.

Similar to the proof of Example 11.1, we have for any constant $M > 0$,

$$E\{\exp[M\|\theta_k\|^2]\} \leq \exp\left\{2M\left[\|\theta\|^2 + \frac{pmd_0(\sigma_\varepsilon)^2}{1 - 4Md_0(\sigma_\varepsilon)^2}\right]\right\},$$

$$E\{\exp[M\|D_k\|^2]\} \leq \exp\left\{\frac{p(m+1)Md_1(\sigma_\varepsilon)^2}{1 - 2Md_1(\sigma_\varepsilon)^2}\right\},$$

where

$$d_0 = \sum_{i=0}^{m-1} \|D_i\|^2, \quad d_1 = 1 + \sum_{i=1}^{m} \|D_i - D_{i-1}\|^2, \quad (D_0 = I,\ D_m = 0).$$

Hence (11.24) and (11.25) hold. Q.E.D.

We now describe the estimation algorithm. Let $L > 0$ and $d > 0$ be two constants (which will be specified later). We define D as the following bounded domain:

$$D = \{x = (x_1, \ \dots\ x_p) \in R^p : |x_i| \leq L,\ 1 \leq i \leq p\} \tag{11.30}$$

and $\pi_D\{x\}$ as the nearest point in D from x (under the Euclidean norm).

The estimate for the unknown process $\{\theta_k\}$ is generated by the following projected version of the gradient algorithm:

$$\widehat{\theta}_{k+1} = \pi_D\left\{\widehat{\theta}_k + \frac{\varphi_k}{d + \|\varphi_k\|^2}(y_{k+1} - u_k - \varphi_k^\tau \widehat{\theta}_k)\right\} \tag{11.31}$$

with arbitrary initial condition $\theta_0 \in D$, where φ_k is defined by

$$\varphi_k = [y_k \ \dots \ y_{k-p+1}]^T, \quad k \geq 0. \tag{11.32}$$

We note that due to the special form of the domain D, the calculation of the projection in (11.31) is a very simple task.

Let $\{y_k^*\}$ be a bounded deterministic reference signal. The certainty equivalent adaptive (tracking) control law is

$$u_k = -\varphi_k^T \widehat{\theta}_k + y_{k+1}^*, \quad k \geq 0. \tag{11.33}$$

Stability of such a control is established in the following theorem.

Theorem 11.1 (Gu3) . *For the random parameter model (11.15), if the Noise assumption (11.16) and the Parameter assumptions (11.24) and (11.25) hold for suitably large M and small δ_θ, and if in the estimation algorithm (11.30)-(11.32) L and d are taken appropriately large, then under the adaptive control law (11.33) the closed-loop system is stable in the sense that*

$$\limsup_{n\to\infty} E\{|y_n|^\beta + |u_n|^\beta\} < \infty, \tag{11.34a}$$

$$\limsup_{N\to\infty} \frac{1}{N} \sum_{n=0}^{N} \{|y_n|^2 + |u_n|^2\} < \infty \ a.s., \tag{11.34b}$$

where $\beta > 2$ is a constant depending on M, δ_0, L, and d.

Remark 11.1. One may ask how large (small) the constant M (δ_θ) should be. Later, we shall prove that Theorem 11.1 is true when M and δ_0 satisfy the following inequality:

$$M \geq 3(m+1)2^7 \beta p^{3/2} \lambda^{-p}$$

and

$$\delta_0 < \min\left\{1, \left[\frac{\beta(m+1)}{2}(\log\lambda^{-1})\right]^2 \left[\frac{f(2^7(m+1)\beta p^{3/2} L_0 \lambda^{-p} M^{-1/2})}{2} + \frac{2^5 \beta p \lambda^{-p}(m+1)}{M}\right]^{-2}\right\} \tag{11.35}$$

for some $\lambda \in (0,1)$ and $\beta > 2$, where the function $f(\cdot)$ is defined by (11.14), and L_0 denotes

$$L_0 = \left\{\frac{4M_0}{M} + \lambda^p[96(m+1)\beta p]^{-1} \cdot \left|\log\left[\frac{(m+1)\beta}{2}\log(\lambda^{-1})\right]\right|\right\}^{1/2}. \tag{11.36}$$

Moreover, in the practical implementation of the algorithm, it is desirable to know the values of L and d. It will also be shown in the proof of Theorem 11.1 that one way to choose L and d is

$$L = L_0,$$

$$d > 16p(\varepsilon\lambda^p)^{-1}\max\{8M_v(\log\lambda^{-1})^{-1},\ 4\beta(r+1)\}. \tag{11.37}$$

The proof of the Theorem 11.1 is divided into several lemmas.

Lemma 11.6 *For system (11.15) and algorithm (11.30)-(11.32), the following inequality holds for any $k \geq 0$,*

$$\begin{aligned}\alpha_k \;\leq\; & 2(\|\widetilde{\theta}_k\|^2 - \|\widetilde{\theta}_{k+1}\|^2) + (8/d)|v_{k+1}|^2 \\ & +4\left\{2p^{1/2}L + \|\Delta_{k+1}\|\right\}\|\Delta_{k+1}\| \\ & +12\left\{p^{1/2}L + \|\theta_k\|\right\}\|\theta_k\| I(\theta_k \notin D),\end{aligned}$$

where $I(A)$ is the indicator function of a set A, Δ_{k+1} is defined by (11.26), and

$$\alpha_k = \frac{\|\varphi_k^\tau \widetilde{\theta}_k\|^2}{d + \|\varphi_k\|^2}, \quad \widetilde{\theta}_k = \theta_k - \widehat{\theta}_k. \tag{11.38}$$

Proof. Let us denote $\overline{\theta}_k = \theta_k I(\theta_k \in D)$. We have

$$\|\widehat{\theta}_{k+1} - \overline{\theta}_k\|^2 \leq 4pL^2$$

and

$$\begin{aligned}\|\theta_{k+1} - \overline{\theta}_k\| &\leq \|\theta_{k+1} - \theta_k\| + \|\theta_{k+1} I(\theta_k \in D)\| \\ \leq\; & \|\Delta_{k+1}\| + \|\theta_k I(\theta_k \in D)\|.\end{aligned}$$

So we have

$$\begin{aligned}\|\widehat{\theta}_{k+1} - \theta_{k+1}\|^2 \;=\; & \|\widehat{\theta}_{k+1} - \overline{\theta}_k + \overline{\theta}_k - \theta_{k+1}\|^2 \\ \leq\; & \|\widehat{\theta}_{k+1} - \overline{\theta}_k\|^2 + 4p^{1/2}L\{\|\Delta_{k+1}\| + \|\theta_k I(\theta_k \in D)\|\} \\ & +2\|\Delta_{k+1}\|^2 + 2\|\theta_k I(\theta_k \notin D)\|^2 \\ \leq\; & \|\widehat{\theta}_{k+1} - \overline{\theta}_k\|^2 + 2\{2p^{1/2}L + \|\Delta_{k+1}\|\}\|\Delta_{k+1}\| \\ & +2\{2p^{1/2}L + \|\theta_k\|\}\|\theta_k\| I(\theta_k \notin D).\end{aligned} \tag{11.39}$$

But by (11.3) and the properties of the projection we know that

$$\|\overline{\theta}_k - \widehat{\theta}_{k+1}\|^2 \;\leq\; \left\|\overline{\theta}_k - \widehat{\theta}_k - \frac{\varphi_k}{d + \|\varphi_k\|^2}[\varphi_k^\tau(\theta_k - \widehat{\theta}_k) + v_{k+1}]\right\|^2$$

$$
\begin{aligned}
&= \left\| \left(I - \frac{\varphi_k \varphi_k^T}{d + \|\varphi_k\|^2} \right) \widetilde{\theta}_k - \left\{ \theta_k I(\theta_k \notin D) - \frac{\varphi_k v_{k+1}}{d + \|\varphi_k\|^2} \right\} \right\|^2 \\
&\le \|\widetilde{\theta}_k\|^2 - \frac{\|\varphi_k^T \widetilde{\theta}_k\|^2}{d + \|\varphi_k\|^2} + 2\|\theta_k\|^2 I(\theta_k \notin D) + \frac{2}{d}\|v_{k+1}\|^2 \\
&\quad + 2\|\widetilde{\theta}_k\| \|\theta_k\| I(\theta_k \notin D) + 2\frac{\|\varphi_k^T \widetilde{\theta}_k\| |v_{k+1}|}{d + \|\varphi_k\|^2}.
\end{aligned}
$$

Applying the following elementary inequality

$$
2xy \le \frac{1}{2}x^2 + 2y^2 \quad \forall x,\ y
$$

with

$$
x = \frac{\|\varphi_k^T \widetilde{\theta}_k\|}{(d + \|\varphi_k\|^2)^{1/2}}, \quad y = \frac{|v_{k+1}|}{(d + \|\varphi_k\|^2)^{1/2}}
$$

to the last term, we then see that

$$
\begin{aligned}
\|\overline{\theta}_k - \widehat{\theta}_{k+1}\|^2 &\le \|\widetilde{\theta}_k\|^2 - \frac{1}{2}\frac{\|\varphi_k^T \widetilde{\theta}_k\|^2}{d + \|\varphi_k\|^2} + \frac{4}{d}\|v_{k+1}\|^2 \\
&\quad + 2\{(p^{1/2}L + \|\theta_k\|)\|\theta_k\| + \|\theta_k\|^2\} I(\theta_k \notin D) \\
&\le \|\widetilde{\theta}_k\|^2 - \frac{1}{2}\frac{\|\varphi_k^T \widetilde{\theta}_k\|^2}{d + \|\varphi_k\|^2} + \frac{4}{d}\|v_{k+1}\|^2 \\
&\quad + 2\{p^{1/2}L + 2\|\theta_k\|\}\|\theta_k\| I(\theta_k \notin D).
\end{aligned}
$$

Substituting this in (11.39) we have

$$
\begin{aligned}
\|\widetilde{\theta}_{k+1}\|^2 &\le \|\widetilde{\theta}_k\|^2 - \frac{1}{2}\frac{\|\varphi_k^T \widetilde{\theta}_k\|^2}{d + \|\varphi_k\|^2} + \frac{4}{d}\|v_{k+1}\|^2 \\
&\quad + 2\{2p^{1/2}L + \|\Delta_{k+1}\|\}\|\Delta_{k+1}\| \\
&\quad + 6\{p^{1/2}L + \|\theta_k\|\}\|\theta_k\| I(\theta_k \notin D),
\end{aligned}
$$

which is tantamount to the desired result. Q.E.D.

Lemma 11.7 *Let the closed-loop system be expressed by*

$$
y_{k+1} = \varphi_k^T \widetilde{\theta}_k + \overline{v}_{k+1}, \quad \overline{v}_{k+1} = v_{k+1} + y_{k+1}^*.
$$

Assume that there are constants $\lambda \in (0,1)$, $K_1 > 0$, $K_2 \ge 0$ and $K_3 \ge 0$ such that

$$
\sum_{i=0}^{n} \lambda^{n-i} \|\varphi_i\|^2 \le \sum_{i=0}^{n} \lambda^{n-i} \{K_1 (y_i)^2 + K_2 (\overline{v}_i)^2\} + K_3, \quad n \ge 0. \quad (11.40)
$$

Then for any $\beta \geq 2$

$$\{E\|\varphi_n\|^\beta\}^{1/\beta} \leq K_0 \left\{ 1 + (1-\lambda)^{-1/2} \left\{ E\left[\sum_{i=0}^{n} \lambda^{n-i} \prod_{j=i}^{n-1} (1 + 2K_1\lambda^{-1}\alpha_j)^2 \right]^{\beta/2} \right\}^{1/\beta} \right\}^{1/2},$$

$$\frac{1}{N}\sum_{n=0}^{N} \|\varphi_n\|^2 \leq O\left(\left\{ \frac{1}{N}\sum_{n=0}^{N}\sum_{i=0}^{n} \lambda^{n-i} \prod_{j=i}^{n-1}(1+2K_1\lambda^{-1}\alpha_j)^2 \right\}^{1/2} \right) + O(1), \tag{11.41}$$

where α_j *is defined in (11.38) and*

$$\begin{aligned} K_0 &= (1-\lambda)^{-1/2}\{[2K_1+K_2][\sigma_{2\beta}(\overline{v})]^2 + 2dK_1[2pL^2 \\ &\quad +2(\sigma_{2\beta}(\theta))^2]\}^{1/2} + \sqrt{K_3} \\ \sigma_{2\beta}(\overline{v}) &= \sup_k \{E|\overline{v}_k|^{2\beta}\}^{1/(2\beta)}, \quad \sigma_{2\beta}(\theta) = \sup_k\{E|\theta_k|^{2\beta}\}^{1/(2\beta)}. \end{aligned}$$

Proof. By the assumption it follows that

$$\begin{aligned} \|\varphi_n\|^2 &\leq \sum_{i=0}^{n} \lambda^{n-i}\|\varphi_i\|^2 \\ &\leq \sum_{i=0}^{n} \lambda^{n-i}\{K_1[2\|\varphi_{i-1}^\tau \widetilde{\theta}_{i-1}\|^2 + 2(\overline{v}_i)^2] + K_2(\overline{v}_i)^2\} + K_3 \\ &= 2K_1\sum_{i=0}^{n-1}\lambda^{n-i-1}\alpha_i(\|\varphi_i\|^2+d) + (2K_1+K_2)\sum_{i=0}^{n}\lambda^{n-i}(\overline{v}_i)^2 + K_3 \\ &\leq 2K_1\sum_{i=0}^{n-1}\lambda^{n-i-1}\alpha_i\|\varphi_i\|^2 + (2K_1+K_2)\sum_{i=0}^{n}\lambda^{n-i}(\overline{v}_i)^2 \\ &\quad +2dK_1\sum_{i=0}^{n-1}\lambda^{n-i-1}\|\widetilde{\theta}_i\|^2 + K_3. \end{aligned}$$

So by Lemma 11.1 with $x_i = \lambda^{-i}\|\varphi_i\|^2$, it is seen that

$$\|\varphi_n\|^2 \leq \xi_n + \sum_{i=0}^{n-1}\lambda^{n-i}\left[\prod_{j=i}^{n-1}(1+2K_1\lambda^{-1}\alpha_j)\right]\xi_i \tag{11.42}$$

where

$$\xi_i = (2K_1+K_2)\sum_{k=0}^{i}\lambda^{i-k}(\overline{v}_k)^2 + 2dK_1\sum_{k=0}^{i-1}\lambda^{i-k-1}\|\widetilde{\theta}_k\|^2 + K_3.$$

Applying the Minkowski inequality and the Schwarz inequality to (11.42), we get

$$
\begin{aligned}
&\{E\|\varphi_n\|^\beta\}^{2/\beta} \\
\le\ & \{E|\xi_n|^{\beta/2}\}^{2/\beta} + \left\{ E\left[\sum_{i=0}^{n-1} \lambda^{n-i} \prod_{j=i}^{n-1} (1+2K_1\lambda^{-1}\alpha_j)\xi_i\right]^{\beta/2}\right\}^{2/\beta} \\
\le\ & \{E|\xi_n|^{\beta/2}\}^{2/\beta} + \left\{ E\left[\sum_{i=0}^{n-1} \lambda^{n-i} \prod_{j=i}^{n-1} (1+2K_1\lambda^{-1}\alpha_j)^2\right]^{\beta/4}\right. \\
& \left.\left[\cdot\sum_{i=0}^{n-1} \lambda^{n-i}\|\xi_i\|^2\right]^{\beta/4}\right\}^{2/\beta} \\
\le\ & \{E|\xi_n|^{\beta/2}\}^{2/\beta} + \left\{ E\left[\sum_{i=0}^{n-1} \lambda^{n-i} \prod_{j=i}^{n-1} (1+2K_1\lambda^{-1}\alpha_j)^2\right]^{\beta/2}\right. \\
& \left.\cdot E\left[\sum_{i=0}^{n-1} \lambda^{n-i}\|\xi_i\|^2\right]^{\beta/2}\right\}^{1/\beta} \\
\le\ & \{E|\xi_n|^{\beta/2}\}^{2/\beta} + \left\{ E\left[\sum_{i=0}^{n-1} \lambda^{n-i} \prod_{j=i}^{n-1} (1+2K_1\lambda^{-1}\alpha_j)^2\right]^{\beta/2}\right\}^{1/\beta} \\
& \left\{\sum_{i=0}^{n-1} \lambda^{n-i} E[\|\xi_i\|^\beta]^{2/\beta}\right\}^{1/2} \\
\le\ & \left\{1+(1-\lambda)^{-1/2}\left\{ E\left[\sum_{i=0}^{n-1} \lambda^{n-i} \prod_{j=i}^{n-1} (1+2K_1\lambda^{-1}\alpha_j)^2\right]^{\beta/2}\right\}^{1/\beta}\right\} \\
& \cdot \sup_{0\le i\le n}\{E\|\xi_i\|^\beta\}^{1/\beta}.
\end{aligned}
$$

Again by the Minkowski inequality,

$$
\begin{aligned}
\{E\|\xi_i\|^\beta\}^{1/\beta} \le\ & (2K_1+K_2)\sum_{k=0}^{i} \lambda^{i-k}\{E\|\overline{v}_k\|^{2\beta}\}^{1/\beta} \\
& + 2dK_1 \sum_{k=0}^{i-1} \lambda^{i-k-1}\{E\|\widetilde{\theta}_k\|^{2\beta}\}^{1/\beta} + K_3
\end{aligned}
$$

$$\leq (1-\lambda)^{-1}\{[2K_1+K_2][\sigma_{2\beta}(\overline{v})]^2 \\ +2dK_1[2pL^2+2(\sigma_{2\beta}(\theta))^2]\}+K_3. \qquad (11.43)$$

Hence the first assertion of the lemma is true, while the second assertion can easily be proved by following the similar argument and by using (11.42), Lemma 11.2 and the Schwarz inequality. Q.E.D.

Lemma 11.8 *For system (11.15) if the adaptive control (11.33) is applied, then φ_n defined by (11.32) has the following properties:*

$$\{E\|\varphi_n\|^\beta\}^{1/\beta} \leq K_0\left\{1+(1-\lambda)^{-1/2}\prod_{k=1}^{4}\{E[I_k(n)]^{\beta/2}\}^{1/(4\beta)}\right\}^{1/2} \qquad (11.44)$$

$$\frac{1}{N}\sum_{n=0}^{N}\|\varphi_n\|^2 \leq O\left(\prod_{k=1}^{4}\left\{\frac{1}{N}\sum_{n=0}^{N}I_k(n)\right\}^{1/8}\right)+O(1) \quad a.s. \qquad (11.45)$$

where $I_k(n)$, $k=1, \ldots, 4$, are defined as

$$I_1(n) = \sum_{i=0}^{n-1}\lambda^{n-i}\exp\{8\beta_1\|\widetilde{\theta}_i\|^2\}, \quad \beta_1 = 4K_1\lambda^{-1}, \qquad (11.46)$$

$$I_2(n) = \sum_{i=0}^{n-1}\lambda^{n-i}\prod_{j=i}^{n-1}\exp\left\{\frac{32\beta_1}{d}\|v_{j+1}\|^2\right\} \qquad (11.47)$$

$$I_3(n) = \sum_{i=0}^{n-1}\lambda^{n-i}\prod_{j=i}^{n-1}\exp\left\{16\beta_1(2p^{1/2}L+\|\Delta_{j+1}\|)\|\Delta_{j+1}\|\right\}, \qquad (11.48)$$

$$I_4(n) = \sum_{i=0}^{n-1}\lambda^{n-i}\prod_{j=i}^{n-1}\exp\left\{48\beta_1(p^{1/2}L+\|\theta_j\|)\|\theta_j\|I(\theta_j\notin D)\right\}. \qquad (11.49)$$

Proof. To apply Lemma 11.7 we need to verify (11.40).

By the definition (11.32) for φ_n, it is easy to see that for any $\lambda\in(0,1)$

$$\sum_{i=0}^{n}\lambda^{n-i}\|\varphi_i\|^2 \leq p\lambda^{-(p-1)}\sum_{i=0}^{n}\lambda^{n-i}y_i^2.$$

Hence (11.40) holds with $K_1 = p\lambda^{-(p-1)}$, $K_2=0$, and $K_3=0$.

Note that the closed-loop system of Theorem 11.1 is

$$y_{k+1} = \varphi_k^\tau\theta_k+\overline{v}_{k+1}, \quad \overline{v}_{k+1} = v_{k+1}+y^*_{k+1}. \qquad (11.50)$$

By Lemma 11.6 and the inequality $\log(1+x) \le x$, for all $x \ge 0$, we have

$$\prod_{j=i}^{n-1}(1+2K_1\lambda^{-1}\alpha_j)^2 = \exp\left\{\sum_{j=i}^{n-1} 2\log(1+2K_1\lambda^{-1}\alpha_j)\right\}$$

$$\le \exp\left\{\beta_1 \sum_{j=i}^{n-1}\alpha_j\right\}, \quad (\beta_1 = 4K_1\lambda^{-1}),$$

$$\le \exp\{2\beta_1\|\widetilde{\theta}_i\|^2\}$$

$$\exp\left\{\beta_1\sum_{j=i}^{n-1}\left[\frac{8}{d}\|v_{j+1}\|^2 + 4(2p^{1/2}L + \|\Delta_{j+1}\|)\|\Delta_{j+1}\|\right.\right.$$

$$\left.\left. +12(p^{1/2}L + \|\theta_j\|)\|\theta_j\|I(\theta \notin D)\right]\right\}$$

$$\le \exp\{2\beta_1\|\widetilde{\theta}_i\|^2\}\prod_{j=i}^{n-1}\exp\left\{\frac{8\beta_1}{d}\|v_{j+1}\|^2\right\}$$

$$\cdot\prod_{j=i}^{n-1}\exp\left\{4\beta_1(2p^{1/2}L + \|\Delta_{j+1}\|)\|\Delta_{j+1}\|\right\}$$

$$\cdot\prod_{j=i}^{n-1}\exp\left\{12\beta_1(p^{1/2}L + \|\theta_j\|)\|\theta_j\|I(\theta \notin D)\right\}. \qquad (11.51)$$

Consequently, by the Hölder inequality and (11.46)-(11.49)

$$E\left[\sum_{i=0}^{n-1}\lambda^{n-i}\prod_{j=i}^{n-1}(1+2K_1\lambda^{-1}\alpha_j)^2\right]^{\beta/2}$$

$$\le E\left\{\prod_{k=1}^{4} I_k(n)\right\}^{\beta/8} \le \left\{\prod_{k=1}^{4} E[I_k(n)]^{\beta/2}\right\}^{1/4}.$$

Substituting this into Lemma 11.7, we see that (11.44) is true, while (11.45) can be proved in a similar way by using (11.51) and the Hölder inequality. The details will not be repeated. Q.E.D.

We now proceed to analyze the quantities $I_k(n)$, $k = 1, \ldots, 4$, appearing in (11.46)-(11.49) by using Lemma 11.4. For this we need the following lemma.

Lemma 11.9 *Under conditions of Theorem 11.1, the following inequalities hold (where $\beta_1 = 4p\lambda^{-p}$):*

(*i*) $$\sup_{(j,\omega)} E\{\exp[(16\beta\beta_1(r+1)/d)\|v_{j+1}\|^2]|\mathcal{F}_{j-r}\} < \lambda^{-\beta(r+1)/2},$$

$$\tag{11.52}$$

$$(ii) \quad \sup_{(j,\omega)} E\{\exp\{8\beta\beta_1(m+1)(2p^{1/2}L + \|\Delta_{j+1}\|)\|\Delta_{j+1}\|\}|\mathcal{F}_{j-m}\} < \lambda^{-\beta(m+1)/2}, \tag{11.53}$$

$$(iii) \quad \sup_{(j,\omega)} E\{\exp[24\beta\beta_1(m+1)(p^{1/2}L + \|\theta_{j+1}\|) \|\theta_{j+1}\| I(\theta_{j+1} \notin D)]|\mathcal{F}_{j-m}\} < \lambda^{-\beta(m+1)/2}, \tag{11.54}$$

where j takes nonnegative integer values and ω is the sampling point.

Proof. We need only to prove the lemma for M, δ_θ, L and d satisfying conditions of Remark 11.1.

(i) By the Noise assumption (11.16), the choice of d in (11.37) and the Hölder inequality, we have (note that $\beta_1 = 4p\lambda^{-p}$)

$$\begin{aligned} & E\{\exp[(16\beta\beta_1(r+1)/d)\|v_{j+1}\|^2]|\mathcal{F}_{j-r}\} \\ \leq \; & \exp\{2^6 p\lambda^{-p}\beta(r+1)M_v/(\varepsilon_0 d)\} < \lambda^{-\beta(r+1)/2}. \end{aligned}$$

(ii) By Lemma 11.5 and the Parameter assumption (11.25)

$$\begin{aligned} & E\{\exp\{2^5\beta\beta_1(m+1)p^{1/2}L\|\Delta_{j+1}\|\}|\mathcal{F}_{j-m}\} \\ \leq \; & E\{\exp\{[2^7(m+1)\beta p^{3/2}\lambda^{-p}LM^{-1/2}][M^{1/2}\|\Delta_{k+1}\|]\}|\mathcal{F}_{k-m}\} \\ \leq \; & \exp\{f(2^7(m+1)\beta p^{3/2}\lambda^{-p}LM^{-1/2})\delta_\theta^{1/2}\}, \end{aligned} \tag{11.55}$$

where the function $f(\cdot)$ is defined by (11.14).

Again, by the Parameter assumption (11.25) and the Hölder inequality,

$$\begin{aligned} & E\{\exp\{2^4\beta\beta_1(m+1)\|\Delta_{j+1}\|^2\}|\mathcal{F}_{j-m}\} \\ \leq \; & \exp\{2^6\beta p\lambda^{-p}(m+1)\delta_\theta/M\}. \end{aligned}$$

Combining this with (11.55) we have via the Schwarz inequality

$$\begin{aligned} & E\{\exp\{8\beta\beta_1(m+1)(2p^{1/2}L + \|\Delta_{j+1}\|)\|\Delta_{j+1}\|\}|\mathcal{F}_{j-m}\} \\ \leq \; & \exp\{[f(2^7(m+1)\beta p^{3/2}L\lambda^{-p}M^{-1/2})/2 \\ & +2^5\beta p\lambda^{-p}(m+1)/M]\delta_\theta^{1/2}\} \\ < \; & \lambda^{-\beta(m+1)/2}, \end{aligned}$$

where the last inequality is derived from (11.35).

(iii) We now proceed to prove (11.54). Let us denote $b = 192\beta(m+1)p^{3/2}\lambda^{-p}$. Then by the Parameter assumption (11.24) and the Markov inequality,

$$E\{\exp[48\beta\beta_1(m+1)p^{1/2}L\|\theta_{j+1}\|I(\theta_{j+1} \notin D)]|\mathcal{F}_{j-m}\}$$

$$
\begin{aligned}
&= E\{\exp[bL\|\theta_{j+1}\|I(\theta_{j+1} \notin D)]|\mathcal{F}_{j-m}\} \\
&= E\{I(\theta_{j+1} \in D)|\mathcal{F}_{j-m}\} + E\{\exp[bL\|\theta_{j+1}\|I(\theta_{j+1} \notin D)]|\mathcal{F}_{j-m}\} \\
&\leq 1 + \{E[\exp[2bL\|\theta_{j+1}\|)|\mathcal{F}_{j-m}]\}^{1/2}\{P(\|\theta_{j+1}\| > L|\mathcal{F}_{j-m})\}^{1/2} \\
&= 1 + \{E[\exp(2bL\|\theta_{j+1}\|)|\mathcal{F}_{j-m}]\}^{1/2} \\
&\quad \{P[\exp(2bL\|\theta_{j+1}\|) > \exp(2bL^2)|\mathcal{F}_{j-m}]\}^{1/2} \\
&\leq 1 + E[\exp(2bL\|\theta_{j+1}\|)|\mathcal{F}_{j-m}]/\exp(bL^2) \\
&\leq 1 + \exp(-bL^2/2)E[\exp(2b\|\theta_{j+1}\|^2)|\mathcal{F}_{j-m}] \\
&\leq 1 + \exp(-bL^2/2)\exp(2bM_\theta/M) \\
&\leq \exp\{\exp[192(m+1)\beta p^{3/2}\lambda^{-p}(2M_\theta/M - L^2/2)]\} \\
&\leq \exp\{\exp[192(m+1)\beta p\lambda^{-p}(2M_\theta/M - L^2/2)]\}, \qquad (11.56)
\end{aligned}
$$

where for the last inequality we have used the fact that $2M_\theta/M - L^2/2 \leq 0$, which is seen from (11.36) and the choice $L = L_0$.

Similarly, we have ($c = 192(m+1)\beta p\lambda^{-p}$),

$$
\begin{aligned}
&E\{\exp[48\beta\beta_1(m+1)\|\theta_{j+1}\|^2 I(\theta_{j+1} \notin D)]|\mathcal{F}_{j-m}\} \\
&= E\{\exp[c\|\theta_{j+1}\|^2 I(\theta_{j+1} \notin D)]|\mathcal{F}_{j-m}\} \\
&\leq 1 + \{E\{\exp[2c\|\theta_{j+1}\|^2]|\mathcal{F}_{j-m}]\}^{1/2}\{P(\|\theta_{j+1}\|^2 > L^2|\mathcal{F}_{j-m})\}^{1/2} \\
&\leq 1 + E\{\exp[2c\|\theta_{j+1}\|^2]|\mathcal{F}_{j-m}\}/\exp\{cL^2\} \\
&\leq 1 + \exp(2cM_\theta/M - cL^2) \\
&\leq \exp\{\exp[192(m+1)\beta p\lambda^{-p}(2M_\theta/M - L^2)]\}. \qquad (11.57)
\end{aligned}
$$

Combining (11.56) and (11.57), we obtain via the Schwarz inequality

$$
\begin{aligned}
&E\{\exp[24(m+1)\beta\beta_1(p^{1/2}L + \|\theta_{j+1}\|) \\
&\quad \|\theta_{j+1}\|I(\theta_{j+1} \notin D)]|\mathcal{F}_{j-m}\} \\
&\leq \exp\{\exp[192(m+1)\beta p\lambda^{-p}(2M_\theta/M - L^2/2)]\} \\
&< \lambda^{-\beta(m+1)/2},
\end{aligned}
$$

where the last inequality is obtained from (11.36). This completes the proof. Q.E.D.

Proof of Theorem 11.1. By Lemma 11.4 (with $\alpha = \beta/2 > 1$) and Lemma 11.9 we know that the quantities $I_K(n)$, $k = 2, \ldots, 4$, defined in Lemma 11.8 satisfy

$$
\sup_n E[I_k(n)]^{\beta/2} < \infty \quad \text{and} \quad \sum_{n=0}^{N} I_k(n) + O(N) \qquad (11.58)
$$

where $k = 2, 3, 4$, while for $I_1(n)$ we note that

$$
\exp\{8\beta_1\|\tilde{\theta}_i\|^2\} \leq \exp\{16\beta_1 pL^2\}\exp\{16\beta_1\|\theta_i\|^2\}.
$$

Then by the Parameter assumption (11.24) and Lemma 11.2, it is easy to see that (11.58) is also true for $k = 1$. Hence by Lemma 11.8 we get

$$\sup_n E\|\varphi_n\|^\beta < \infty \quad \text{and} \quad \sum_{n=0}^{N} \|\varphi_n\|^2 = O(N) \quad \text{as } N \to \infty.$$

Combining this with (11.33) we immediately conclude that Theorem 11.1 holds.
Q.E.D.

11.3 Systems with Deterministic Parameters

In this section we consider the following linear time-varying stochastic model:

$$\begin{aligned} & y_{k+1} = a_1(k)y_k + \ldots + a_s(k)y_{k-s+1} \\ & +b_1(k)u_k + \ldots + b_t(k)u_{k-t+1} + v_{k+1}, \quad k \geq 0, \\ & y_k = u_k = v_k = 0, \quad \forall k < 0, \end{aligned} \tag{11.59}$$

where $a_i(k)$, $b_j(k)$, $1 \leq i \leq s$, $1 \leq j \leq t$, are the unknown deterministic time-varying parameters.

If we introduce

$$\varphi_k = [y_k \ldots y_{k-s+1}, u_k \ldots u_{k-t+1}]^\tau, \quad \theta_k = [a_1(k) \ldots a_s(k), b_1(k) \ldots b_t(k)]^\tau, \tag{11.60}$$

then (11.59) can be rewritten as

$$y_{k+1} = \varphi_k^\tau \theta_k + v_{k+1}, \quad k \geq 0. \tag{11.61}$$

The assumption on the noise sequence $\{v_k\}$ is the same as that in the last section (see (11.16)), while here the parameter assumptions are different from there.

Parameter assumption (deterministic case). (i) There is a positive constant $b_1 > 0$ such that

$$b_1(k) \geq b_1, \quad \forall k \geq 0, \tag{11.62}$$

and the model (11.59) is uniformly stably invertible in the sense that there are two constants $A > 0$, $\rho \in (0,1)$ such that

$$|u_k|^2 \leq A \sum_{i=0}^{k+1} \rho^{k+1-i} \{|y_i|^2 + |v_i|^2\} \quad \forall k. \tag{11.63}$$

(ii) The parameter is slowly varying in the sense that

$$\|\theta_k\| \leq M_1, \quad \|\theta_{k+1} - \theta_k\| \leq \delta_1, \quad \forall k \geq 0, \tag{11.64}$$

where

$$M_1 < \infty,$$

$$\delta_1 < \min\{1,\ \log(\lambda^{-1})[24K_1\lambda^{-1}(4(s+t)^{1/2}M_1+1)]^{-1}\}, \tag{11.65}$$

$$K_1 = [(t+s)M_1^2/b_1^2+1]\{s\lambda^{-s+1}+2(t-1)\lambda^{-t+1}A\lambda^2/(\lambda-\rho)\}$$

and $\lambda \in (\rho, 1)$ is some constant.

We remark that since $\{\theta_k\}$ is bounded, the assumption (11.63) is implied by the uniformly asymptotic stability of the following time-varying polynomial:

$$B_k(z) = b_1(k) + b_2(k)z + \dots + b_t(k)z^{t-1}, \tag{11.66}$$

which in the constant parameter case is the standard minimum phase condition.

Let us introduce the following bounded domain:

$$\begin{aligned} D \;=\; & \{x = (x_1, \dots, x_{s+t}) \in R^{s+t} : \\ & |x_i| \le L,\ 1 \le i \le s+t,\ x_{s+1} \ge b_1\}. \end{aligned} \tag{11.67}$$

The estimation is carried out by the following projected gradient algorithm:

$$\hat{\theta}_{k+1} = \pi_D\left\{\hat{\theta}_k + \frac{\varphi_k}{d+\|\varphi_k\|^2}(y_{k+1} - \varphi_k^T\hat{\theta}_k)\right\}, \tag{11.68}$$

where the initial condition $\hat{\theta}_0 \in D$, and φ_k is defined as in (11.60).

Let $\{y_k^*\}$ be a given bounded deterministic reference signal (without loss of generality assume $\|y_k^*\| \le M_1$), and let the adaptive control u_k at any time k be solved from the following equation:

$$\varphi_k^T\hat{\theta}_k = y_{k+1}^*. \tag{11.69}$$

Similar to Theorem 11.1, we have the following result [Gu3].

Theorem 11.2 *For the deterministic parameter model (11.59) suppose that the Noise assumption (11.16) and the Parameter assumptions (11.62)-(11.65) hold, and that in the estimation algorithm (11.67) and (11.68) L is taken as M_1 appearing in (11.64) and*

$$d > 36K_1(\lambda\varepsilon)^{-1}\max\{\beta(r+1),\ 2M_v[\log(\lambda^{-1})]^{-1}\} \tag{11.70}$$

for some $\beta > 2$. Then under the adaptive control (11.69) the closed-loop system is stable in the sense that

$$\limsup_{n\to\infty} E\{|y_n|^\beta + |u_n|^\beta\} < \infty, \tag{11.71a}$$

$$\limsup_{N\to\infty} \frac{1}{N}\sum_{n=0}^{N}\{|y_n|^2 + |u_n|^2\} < \infty \text{ a.s.} \tag{11.71b}$$

We separate the proof into several lemmas. Similar to Lemma 11.6, we have the following result.

Lemma 11.10 *For system (11.61) and algorithm (11.68) if* $L \geq \|\theta_k\|$ *and* $\theta_k \in D\ \forall k$, *then*

$$\begin{aligned}\alpha_k &\leq 2(\|\widetilde{\theta}_k\|^2 - \|\widetilde{\theta}_{k+1}\|^2) + (6/d)|v_{k+1}|^2 \\ &\quad +2\|\Delta_{k+1}\|[4(s+t)^{1/2}L + \|\Delta_{k+1}\|].\end{aligned}$$

where $\Delta_{k+1} = \theta_{k+1} - \theta_k$, $\alpha_k = \|\varphi_k^T\widetilde{\theta}_k\|^2/(d + \|\varphi_k\|^2)$ *and* $\widetilde{\theta}_k = \theta_k - \widehat{\theta}_k$.

Proof. The proof is similar to that of Lemma 11.6. By $\theta_k \in D$ and by the properties of projection it is known that

$$\begin{aligned}&\|\theta_k - \widehat{\theta}_{k+1}\|^2 \\ \leq &\left\|\theta_k - \widehat{\theta}_k - \frac{\varphi_k}{d + \|\varphi_k\|^2}[\varphi_k^T(\theta_k - \widehat{\theta}_k) + v_{k+1}]\right\|^2 \\ \leq &\left\|\left(I - \frac{\varphi_k\varphi_k^T}{d + \|\varphi_k\|^2}\right)\widetilde{\theta}_k - \frac{\varphi_k v_{k+1}}{d + \|\varphi_k\|^2}\right\|^2 \\ \leq &\|\widetilde{\theta}_k\|^2 - \frac{\|\varphi_k^T\widetilde{\theta}_k\|^2}{d + \|\varphi_k\|^2} + \frac{\|v_{k+1}\|^2}{d} + 2\frac{\|\varphi_k^T\widetilde{\theta}_k\||v_{k+1}|}{d + \|\varphi_k\|^2} \\ \leq &\|\widetilde{\theta}_k\|^2 - \frac{\|\varphi_k^T\widetilde{\theta}_k\|^2}{d + \|\varphi_k\|^2} + \frac{\|v_{k+1}\|^2}{d} + \frac{1}{2}\frac{\|\varphi_k^T\widetilde{\theta}_k\|^2}{d + \|\varphi_k\|^2} \\ &+2\frac{|v_{k+1}|^2}{d + \|\varphi_k\|^2} \\ \leq &\|\widetilde{\theta}_k\|^2 - \frac{1}{2}\frac{\|\varphi_k^T\widetilde{\theta}_k\|^2}{d + \|\varphi_k\|^2} + 3\frac{\|v_{k+1}\|^2}{d}.\end{aligned}$$

By this and $\|\theta_{k+1} - \widehat{\theta}_{k+1}\| \leq L + (s+t)^{1/2}L$ we have

$$\begin{aligned}&\|\widetilde{\theta}_{k+1}\|^2 = \|\theta_{k+1} - \widehat{\theta}_{k+1}\|^2 \\ = &\|\theta_k - \widehat{\theta}_{k+1} - \Delta_{k+1}\|^2 \\ \leq &\|\theta_k - \widehat{\theta}_{k+1}\|^2 + 2\|\Delta_{k+1}\| \cdot \|\theta_k - \widehat{\theta}_{k+1}\| + \|\Delta_{k+1}\|^2 \\ \leq &\|\theta_k - \widehat{\theta}_{k+1}\|^2 + 2\|\Delta_{k+1}\|(L + (s+t)^{1/2}L) + \|\Delta_{k+1}\|^2 \\ \leq &\|\widetilde{\theta}_k\|^2 - \frac{1}{2}\alpha_k + \frac{3}{d}|v_{k+1}|^2 + 4\|\Delta_{k+1}\|(s+t)^{1/2}L + \|\Delta_{k+1}\|^2.\end{aligned}$$

Hence the desired result is true. Q.E.D.

Lemma 11.11 *Under conditions of Theorem 11.2, the property (11.40) holds with*

$$K_1 = s\lambda^{-s+1}\left[1 + \frac{(s+t)M_1^2}{b_1^2}\right] + K_2,$$

$$K_2 = 2(t-1)\lambda^{-t+1}\left[1+\left(\frac{M_1}{b_1}\right)^2(t+s)\right]\frac{A\lambda^2}{\lambda-\rho},$$

$$K_3 = \frac{s+t}{1-\lambda}\left(\frac{M_1}{b_1}\right)^2 + 2M_1K_2.$$

Furthermore

$$\{E\|\varphi_n\|^\beta\}^{1/\beta} \leq O\left(\left\{1+(1-\lambda)^{-1/2}\prod_{k=1}^{3}\{E[J_k(n)]^{\beta/2}\}^{1/(3\beta)}\right\}\right) \quad (11.72)$$

$$\frac{1}{N}\sum_{n=0}^{N}\|\varphi_n\|^2 \leq O\left(\prod_{k=1}^{3}\left\{\frac{1}{N}\sum_{n=0}^{N}J_k(n)\right\}^{1/6}\right)+O(1) \quad (11.73)$$

where $J_k(n)$, $k = 1, 2, 3$, *are defined as*

$$J_1(n) = \sum_{i=0}^{n-1}\lambda^{n-i}\exp\{6\delta_1\beta_1(n-i)[4(s+t)^{1/2}L+\delta_1]\},$$

$$\beta_1 = 4K_1\lambda^{-1}, \quad (11.74)$$

$$J_2(n) = \sum_{i=0}^{n-1}\lambda^{n-i}\exp\{6\beta_1\|\widetilde{\theta}_i\|^2\}, \quad (11.75)$$

$$J_3(n) = \sum_{i=0}^{n-1}\lambda^{n-i}\prod_{j=i}^{n-1}\exp\left\{\frac{18\beta_1}{d}\|v_{j+1}\|^2\right\}. \quad (11.76)$$

Proof. Let us write $\widehat{a}_i(k)$, $\widehat{b}_i(k)$ as the estimates for $a_i(k)$, $b_i(k)$ given by $\widehat{\theta}_k$. Then from (11.69) it is clear that

$$u_k = -\frac{1}{\widehat{b}_1(k)}\{\widehat{a}_1(k)y_k + \ldots + \widehat{a}_s(k)y_{k-s+1} + \widehat{b}_2(k)u_{k-1} + \ldots + \widehat{b}_t(k)u_{k-t+1} - y^*_{k+1}\}.$$

Since $\widehat{\theta}_k \in D$ and $L = M_1$, it follows from the Schwarz inequality that

$$|u_k|^2 \leq (s+t)\left(\frac{M_1}{b_1}\right)^2\left\{\sum_{j=0}^{s-1}|y_{k-j}|^2 + \sum_{j=1}^{t-1}|u_{k-j}|^2 + 1\right\}.$$

Therefore, by the definition of φ_i in (11.60) we have

$$\sum_{i=0}^{n}\lambda^{n-i}\|\varphi_i\|^2 = \sum_{i=0}^{n}\lambda^{n-i}\sum_{j=0}^{s-1}|y_{i-j}|^2 + \sum_{i=0}^{n}\lambda^{n-i}\sum_{j=1}^{t-1}|u_{i-j}|^2 + \sum_{i=0}^{n}\lambda^{n-i}|u_i|^2$$

$$\leq \left[1+(s+t)\frac{M_1^2}{b_1^2}\right]\sum_{i=0}^{n}\lambda^{n-i}\sum_{j=0}^{s-1}|y_{i-j}|^2 + \left[1+(s+t)\frac{M_1^2}{b_1^2}\right]\sum_{i=0}^{n}\lambda^{n-i}\sum_{j=1}^{t-1}|u_{i-j}|^2 + \frac{s+t}{1-\lambda}\frac{M_1^2}{b_1^2}$$

$$\leq s\lambda^{-s+1}\left[1+\frac{(s+t)M_1^2}{b_1^2}\right]\sum_{i=0}^{n}\lambda^{n-i}|y_i|^2 + (t-1)\lambda^{-t+1}\left[1+\left(\frac{M_1}{b_1}\right)^2(s+t)\right]\sum_{i=0}^{n-1}\lambda^{n-i}|u_i|^2 + \frac{s+t}{1-\lambda}\left(\frac{M_1}{b_1}\right)^2. \tag{11.77}$$

Note that by the assumption (11.63) we find

$$\sum_{i=0}^{n-1}\lambda^{n-i}|u_i|^2 \leq A\sum_{i=0}^{n-1}\lambda^{n-i}\sum_{j=0}^{i+1}\rho^{i+1-j}\{|y_j|^2+|v_j|^2\}$$

$$\leq A\sum_{j=0}^{n}\sum_{i=j-1}^{n-1}\lambda^{n-i}\rho^{i+1-j}\{|y_j|^2+|v_j|^2\}$$

$$= A\lambda\sum_{j=0}^{n}\sum_{i=j-1}^{n-1}\left(\frac{\rho}{\lambda}\right)^{i+1-j}\{|y_j|^2+|v_j|^2\}$$

$$= \frac{A\lambda^2}{\lambda-\rho}\sum_{j=0}^{n}\lambda^{n-j}\{|y_j|^2+|v_j|^2\}$$

$$\leq \frac{A\lambda^2}{\lambda-\rho}\sum_{j=0}^{n}\lambda^{n-j}\{|y_j|^2+2|\overline{v}_j|^2\} + \frac{2A\lambda^2M_1}{\lambda-\rho}.$$

Substituting this into (11.77) we see that (11.40) holds with K_1, K_2 and K_3 defined in the lemma. Hence Lemma 11.7 is applicable.

Now, by Lemma 11.10 ($\beta_1 = 4K_1\lambda^{-1}$) it is seen that

$$\prod_{j=i}^{n-1}(1+2K_1\lambda^{-1}\alpha_j)^2$$

$$\begin{aligned}&\le \exp\{2\delta_1\beta_1(n-i)[4(s+t)^{1/2}L+\delta_1]\}\\&\cdot\exp\{2\beta_1\|\widetilde{\theta}_i\|^2\}\prod_{j=i}^{n-1}\exp\left\{\frac{6\beta_1}{d}\|v_{j+1}\|^2\right\}.\end{aligned}$$

Hence the result of the lemma follows by applying Lemma 11.7 and the Hölder inequality. Q.E.D.

Proof of Theorem 11.2

For the desired stability by Lemma 11.11 we need only to consider the boundedness of $J_1(n)$ and $J_3(n)$. $J_1(n)$ can easily be dealt with by the assumption (11.65). So we need only to consider $J_3(n)$. For the boundedness of

$$E[J_3(n)]^{\beta/2} \quad \text{and} \quad \frac{1}{N}\sum_{n=0}^{N} J_3(n)$$

we may apply Lemma 11.4. For this we need only to verify that ($\alpha = \beta/2$)

$$\left\{E\left[\left(\exp\left\{\frac{18\beta_1}{d}\|v_{j+1}\|^2\right\}\right)^{\alpha(r+1)}|\mathcal{F}_{j-r}\right]\right\}^{1/\alpha(r+1)} < \lambda^{-1}. \tag{11.78}$$

By (11.70), the Noise assumption (11.16) and the Hölder inequality we derive

$$\begin{aligned}&\left\{E\left[\exp\left(18\alpha(r+1)\beta_1/d|v_{j+1}|^2\right)|\mathcal{F}_{j-r}\right]\right\}^{1/\alpha(r+1)}\\&\le \exp\left\{\frac{18M_v\beta_1}{d\varepsilon}\right\} = \exp\{72M_vK_1(\lambda\varepsilon)^{-1}/d\} < \lambda^{-1}.\end{aligned}$$

Hence the proof is complete. Q.E.D.

A Remark on Robustness

Let us assume that in addition to the random noise $\{v_k\}$ there is unmodeled dynamics $\{\eta_k\}$ existing in the system (11.61):

$$y_{k+1} = \varphi_k^T\theta_k + v_{k+1} + \eta_k.$$

We assume that the unmodeled dynamics $\{\eta_k\}$ depends on the previous input-output data, and has the following time-varying upper bound [IT], [CG8]:

$$|\eta_k| \le \varepsilon^* m_k, \quad m_k = \gamma m_{k-1} + \|\varphi_k\|, \quad m_0 > 0, \quad k \ge 0,$$

where $\varepsilon^* > 0$, $\gamma \in (0,1)$.

Similar to the normalization idea used in [IT], we replace the quantity $d + \|\varphi_k\|^2$ in (11.68) by $d + (m_k)^2$, and consider the following algorithm:

$$\widehat{\theta}_{k+1} = \pi_D \left\{ \widehat{\theta}_k + \frac{\varphi_k}{d + (m_k)^2} (y_{k+1} - \varphi_k^\tau \widehat{\theta}_k) \right\}.$$

Then stability of the closed-loop system under the certainty equivalent adaptive control (11.69) can also be established, provided that ε^* is appropriately small. The proof is essentially the same as that for Theorem 11.2.

CHAPTER 12

Continuous-Time Stochastic Systems

12.1 The Model

We start with the following standard linear state space model:

$$dx_t = Ax_t dt + Bu_t dt + Ddw_t \tag{12.1}$$

$$dy_t = Cx_t dt + dw_t \tag{12.2}$$

where y_t and u_t are the scalar output and input, respectively, x_t is the p-dimensional state vector and w_t is a standard Wiener process (Brownian motion). A, B, C, D and F are real matrices of compatible dimensions.

Without loss of generality, we assume that (A, B, C) has the following observable canonical form:

$$A = \begin{bmatrix} -a_1 & 1 & & 0 \\ \vdots & 0 & \ddots & \\ \vdots & \vdots & \ddots & 1 \\ -a_p & 0 & \cdots & 0 \end{bmatrix}, \quad B = \begin{bmatrix} b_1 \\ \vdots \\ \vdots \\ b_p \end{bmatrix}$$

$$C = [1\ 0 \ldots 0], \quad D \triangleq [d_1 \ldots d_p]^\tau .$$

Thus, by setting $x_t = [x_t^{(1)} \ldots x_t^{(p)}]^\tau$ we see from (12.1) and (12.2) that

$$dx_t^{(1)} = -a_1 x_t^{(1)} dt + x_t^{(2)} dt + b_1 u_t dt + d_1 dw_t \tag{12.3}$$

$$dx_t^{(2)} = -a_2 x_t^{(1)} dt + x_t^{(3)} dt + b_2 u_t dt + d_2 dw_t \tag{12.4}$$

$$\ldots\ldots$$

$$dx_t^{(p-1)} = -a_{p-1} x_t^{(1)} dt + x_t^{(p)} dt + b_{p-1} u_t dt + d_{p-1} dw_t \tag{12.5}$$

$$dx_t^{(p)} = -a_p x_t^{(1)}dt + b_p u_t dt + d_p dw_t \tag{12.6}$$
$$dy_t = x_t^{(1)}dt + dw_t. \tag{12.7}$$

Let us further assume that $y_0 = 0$ and $x_0 = 0$. Denote the integral operator S as $Sf_t = \int_0^t f_\lambda d\lambda$ for integrable function f_λ. Then we have from (12.6)

$$x_t^{(p)} = -a_p S x_t^{(1)} + b_p S u_t + d_p w_t.$$

Substituting this into (12.5) and applying the operator S, we get

$$\begin{aligned} x_t^{(p-1)} &= -a_{p-1}Sx_t^{(1)} - a_p S^2 x_t^{(1)} + b_p S^2 u_t + d_p S w_t \\ &\quad + b_{p-1} S u_t + d_{p-1} w_t \\ &= (-a_{p-1}S - a_p S^2)x_t^{(1)} + (b_{p-1}S + b_p S^2)u_t + (d_{p-1} + d_p S)w_t. \end{aligned}$$

Continuing this procedure, we finally get from (12.3)

$$\begin{aligned} x_t^{(1)} &= (-a_1 S - \ \dots \ - a_p S^p)x_t^{(1)} \\ &\quad +(b_1 S + \ \dots \ + b_p S^p)u_t \\ &\quad +(d_1 + d_2 S + \ \dots \ + d_p S^{p-1})w_t. \end{aligned}$$

From this and (12.7) it follows that

$$\begin{aligned} &(1 + a_1 S + \ \dots \ + a_p S^p)y_t \\ &= (b_1 S + \ \dots \ + b_p S^p) S u_t \\ &\quad + [1 + (a_1 + d_1)S + \ \dots \ + (a_p + d_p)S^p]\, w_t \end{aligned}$$

or

$$A(S)y_t = B(S)Su_t + C(S)w_t, \tag{12.8}$$

where $A(S)$, $B(S)$ and $C(S)$ are polynomials defined in an obvious way.

This model is studied in [Che1]. Since (12.8) has the same form as the ARMAX model investigated in preceding chapters, it may be referred to as the continuous-time ARMAX model.

As is noted in [Mo4], it is usually preferable to introduce a prefilter $D^{-1}(S)$ which is exponentially stable, giving rise to prefiltered variables $(D^{-1}(S)y_t,\ D^{-1}(S)u_t,\ D^{-1}(S)w_t)$. Thus, if we multiply $D^{-1}(S)$ on both sides of (12.8) and make the following change of variables

$$(D^{-1}(S)y_t, D^{-1}(S)u_t, D^{-1}(S)w_t) \to (y_t, u_t, v_t)$$

then (12.8) becomes

$$A(S)y_t = SB(S)u_t + C(S)v_t, \quad v_t = D^{-1}(S)w_t.$$

Of course, the observed variables (y_t, u_t) in this model are different from those in (12.8).

In this chapter, we consider a class of models that is more general than (12.8), dealing with systems which contain both modeled and unmodeled dynamics.

To be precise, let $\{\mathcal{F}_t\}$ be a family of nondecreasing σ-algebras defined on a probability space $(\Omega, \mathcal{F}, P)$, and let the system be described by the following stochastic integral equation:

$$\begin{aligned}
&[I + \mu_1 S H_1(S)]A(S)y_t \\
= \; & [I + \mu_2 H_2(S)]SB(S)u_t + [I + \mu_3 S H_3(S)]C(S)v_t \\
& + \mu_4 S\xi_t(y, u), \quad t \geq 0, \quad y_0 = 0, \quad u_0 = 0, \quad \xi_0 = 0, \qquad (12.9)
\end{aligned}$$

where S denotes the integral operator (e.g., $Sy_t = \int_0^t y_z dz$), and y_t and u_t adapted to $\{\mathcal{F}_t\}$ are m-dimensional output and l-dimensional input, respectively. The quantities μ_i, $i = 1, \ldots, 4$, are small constants, $H_i(S)$, $i = 1$, 2, 3, are unmodeled matrix transfer functions, and $\xi_t(y, u)$, dependent on the previous observation $\{y_s,\, u_s,\, 0 \leq s \leq t\}$, is an unknown nonanticipative measurable process characterizing the unmodeled dynamics. Finally, v_t is the system noise generated via a known filter $D^{-1}(S)$ from a standard Wiener process $\{w_t, \mathcal{F}_t\}$:

$$D(S)v_t = w_t, \quad t \geq 0. \qquad (12.10)$$

Assume that $A(S)$, $B(S)$, and $C(S)$ are matrix polynomials in S, with unknown coefficients but known upper bounds for the true orders:

$$\begin{aligned}
A(S) &= I + A_1 S + \ldots + A_p S^p, \quad p \geq 0, && (12.11)\\
B(S) &= B_1 + B_2 S + \ldots + B_q S^{q-1}, \quad q \geq 1, && (12.12)\\
C(S) &= I + C_1 S + \ldots + C_r S^r, \quad r \geq 0 && (12.13)\\
D(S) &= I + D_1 S + \ldots + D_r S^r. && (12.14)
\end{aligned}$$

The polynomials $A(S)$, $B(S)$ and $C(S)$ characterize the modeled part which corresponds to (12.8). The model (12.9) may be rewritten as

$$\begin{aligned}
A(S)y_t &= SB(S)u_t + C(S)v_t + \eta_t, && (12.15)\\
\eta_t &= \mu_4 S\xi_t(y, u) - \mu_1 S H_1(S)A(S)y_t \\
&\quad + \mu_2 S H_2(S)B(S)u_t + \mu_3 S H_3(S)C(S)v_t. && (12.16)
\end{aligned}$$

The quantity η_t is called the unmodeled dynamics. Of course, it is assumed that the output $\{y_t\}$ can be uniquely determined by the process $\{u_t, w_t\}$ via (12.9) or (12.15).

12.2 Parameter Estimation

Denote the collection of unknown matrix coefficients of $A(S)$, $B(S)$ and $C(S)$ by θ:

$$\theta^{\tau} = [-A_1 \ldots -A_p \quad B_1 \ldots B_q \quad C_1 \ldots C_r]. \tag{12.17}$$

In the sequel, θ is estimated by the continuous-time least squares algorithm

$$\begin{aligned} d\theta_t &= P_t\varphi_t D(S)(dy_t^{\tau} - \varphi_t^{\tau}\theta_t dt), \quad \theta_0 = 0, && (12.18)\\ dP_t &= -P_t\varphi_t\varphi_t^{\tau}P_t dt, \quad P_0 = aI \quad (a = \dim \text{ of } \varphi_t), && (12.19)\\ \varphi_t &= [y_t^{\tau}\ Sy_t^{\tau} \ldots S^{p-1}y_t^{\tau} \quad u_t^{\tau} \ldots S^{q-1}u_t^{\tau} \quad \widehat{v}_t^{\tau} \ldots S^{r-1}\widehat{v}_t^{\tau}]^{\tau} && (12.20)\\ \widehat{v}_t &= y_t - S\theta_t^{\tau}\varphi_t. && (12.21)\end{aligned}$$

Obviously, if $r = 0$, then (12.18) and (12.19) can be expressed as

$$\begin{aligned} \theta_t &= P_t\int_0^t \varphi_s dy_s^{\tau} + P_t(P_0)^{-1}\theta_0, && (12.22)\\ P_t &= \left(\int_0^t \varphi_s\varphi_s^{\tau}ds + a^{-1}I\right)^{-1}, && (12.23)\end{aligned}$$

and the right-hand side of (12.22) is completely determined by the observations $\{u_s, y_s, s \le t\}$.

In the general $r > 0$ case, however, the regressor φ_t depends on $\{\theta_s, s \le t\}$. Then (12.18)-(12.21) constitute a system of nonlinear stochastic differential equations for θ_t.

The existence of the global solution is not obvious (see [LSh, Chap. 4], for example). Henceforth, we assume that the stochastic differential equation (12.18)-(12.21) has a unique strong solution $\{\theta_t, t \ge 0\}$.

Set

$$\varphi_t^0 = [y_t^{\tau}\ Sy_t^{\tau} \ldots S^{p-1}y_t^{\tau} \quad u_t^{\tau}\ Su_t^{\tau} \ldots S^{q-1}u_t^{\tau} \quad v_t^{\tau} \ldots S^{r-1}v_t^{\tau}]^{\tau} \tag{12.24}$$

$$\begin{aligned} \widetilde{\varphi}_t &= [0 \ldots 0 \quad 0 \ldots 0 \quad \widetilde{v}_t^{\tau} \ldots S^{r-1}\widetilde{v}_t^{\tau}]^{\tau}, \quad \widetilde{v}_t = v_t - \widehat{v}_t && (12.25)\\ Y_t &= [y_t^{\tau} \ldots S^{p-1}y_t^{\tau}]^{\tau} \quad U_t = [u_t^{\tau} \ldots S^{q-1}u_t^{\tau}]^{\tau} && (12.26)\\ V_t &= [v_t^{\tau} \ldots S^{r-1}v_t^{\tau}]^{\tau} \quad \widehat{V}_t = [\widehat{v}_t^{\tau} \ldots S^{r-1}\widehat{v}_t^{\tau}]^{\tau}, \quad \widetilde{V}_t = V_t - \widehat{V}_t. && (12.27)\end{aligned}$$

Then it follows that

$$\varphi_t = [Y_t^{\tau}\ U_t^{\tau}\ \widehat{V}_t^{\tau}]^{\tau}, \quad \varphi_t^0 = [Y_t^{\tau}\ U_t^{\tau}\ V_t^{\tau}]^{\tau}, \quad \widetilde{\varphi}_t = [0\ 0\ \widetilde{V}_t^{\tau}]^{\tau}. \tag{12.28}$$

Furthermore, we set

$$F_d = \begin{bmatrix} -D_1 & \dots & \dots & -D_r \\ I & 0 & \dots & 0 \\ 0 & \dots & \dots & \dots \\ \dots & \dots & \dots & \dots \\ 0 & \dots & I & 0 \end{bmatrix}, \quad F_c = \begin{bmatrix} -C_1 & \dots & \dots & -C_r \\ I & 0 & \dots & 0 \\ 0 & \dots & \dots & \dots \\ \dots & \dots & \dots & \dots \\ 0 & \dots & I & 0 \end{bmatrix}. \tag{12.29}$$

By use of these types of matrices it is easy to represent an input-output equation in the state space form. For example, from (12.10) and (12.27) we may write V_t as

$$dV_t = F_d V_t dt + [I\ 0\ \dots\ 0]^\tau dw_t. \tag{12.30}$$

In what follows similar representations will be used without explanation. We need the following assumptions.

Assumption 1. There is a real number $\varepsilon \geq 0$ such that

$$\int_0^t \|\dot{\eta}_t\| ds \leq \varepsilon r_t, \quad t \geq 0, \tag{12.31}$$

where

$$\begin{aligned} \dot{\eta}_t &= \mu_4 \xi_t(y,u) - \mu_1 H_1(S)A(S)y_t \\ &\quad + \mu_2 H_2(S)B(S)u_t + \mu_3 H_3(S)C(S)v_t \end{aligned} \tag{12.32}$$

and

$$r_t = e + \int_0^t \|\varphi_s\|^2 ds. \tag{12.33}$$

We remark that in the ideal case where there is no unmodeled dynamics (i.e. $\eta_t \equiv 0$), then Assumption 1 is automatically satisfied. More discussions on this Assumption can be found in [CG10].

Assumption 2. $D(S)$ is stable and the transfer matrix $D(S)C^{-1}(S) - I/2$ is SPR.

Theorem 12.1 (CG10) . *For the model (12.15) and (12.16) let Assumptions 1 and 2 hold. Then the estimation error $\widetilde{V}_t$ defined by (12.27) satisfies as $t \to \infty$*

$$\int_0^t \|\widetilde{V}_\lambda\|^2 d\lambda = O(\log r_t) + O(\varepsilon r_t). \tag{12.34}$$

Furthermore, if the system is persistently excited in the sense that

$$K \triangleq \limsup_{t\to\infty} r_t/\lambda_{\min}(t) < \infty,$$

then

$$\limsup_{t\to\infty} \|\theta_t - \theta\| \leq \alpha K\varepsilon \quad a.s. \tag{12.35}$$

where $\alpha \in (0,\infty)$ is a constant, ε and r_t are defined in Assumption 1, and $\lambda_{\min}(t)$ denotes the minimum eigenvalue of P_t^{-1}.

We remark that similar results hold also for the discrete-time case, see [CG8]. If there is no unmodeled dynamics, that is, $\eta_t \equiv 0$ in (12.15), then similar results to those in Section 4.2 are also obtainable.

Theorem 12.2 (CM) . *Consider the model (12.15) with $\eta_t \equiv 0$ and the algorithm (12.18). If Assumption 2 holds, then (i) the noise estimation error $\widetilde{V}_t$ defined by (12.27) as $t \to \infty$ satisfies*

$$\int_0^t \|\widetilde{V}_\lambda\|^2 d\lambda = O(\log r_t) \quad a.s.; \tag{12.36}$$

(ii) the parameter estimation error $\widetilde{\theta}_t = \theta_t - \theta$ satisfies

$$\|\widetilde{\theta}_t\|^2 = O\left(\frac{\log r_t}{\lambda_{\min}(t)}\right) \quad a.s. \tag{12.37}$$

Before proving the theorems, we first present several lemmas.

Lemma 12.1 *Let $E(S)$ and $F(S)$ be matrix polynomials in the integral operator S, such that the transfer matrix $F(S)E^{-1}(S)$ is stable and proper. Then*

$$\int_0^t \|F(S)E^{-1}(S)x_z\|^2 dz \leq c \int_0^t \|x_z\|^2 dz$$

for any square integrable function $\{x_t\}$, where c is a constant depending on $E(S)$ and $F(S)$ only.

Proof. Let us write

$$E(S) = I + E_1 S + \ldots + E_d S^d, \quad F(S) = I + F_1 S + \ldots + F_d S^d$$

and set

$$z_t = E^{-1}(S)x_t, \quad Z_t = [z_t^\tau \; \ldots \; S^{d-1}z_t^\tau]^\tau.$$

Similar to (12.30) we have

$$Z_t = F_e S Z_t + [x_t^\tau \; 0 \ldots 0]^\tau \tag{12.38}$$

with

$$F_e = \begin{bmatrix} -E_1 & \ldots & \ldots & -E_d \\ I & 0 & \ldots & 0 \\ 0 & \ldots & \ldots & \ldots \\ \ldots & \ldots & \ldots & \ldots \\ 0 & \ldots & I & 0 \end{bmatrix}.$$

The linear differential equation (12.38) has the solution

$$Z_t = F_e \int_0^t \exp\{F_e(t-s)\}[x_s^\tau\ 0...0]^\tau ds + [x_t^\tau\ 0...0]^\tau.$$

Since F_e is stable, there are constants $c_1 \geq 1$ and $\rho > 0$ such that

$$\|\exp\{F_e t\}\| \leq c_1 e^{-\rho t} \quad \forall t \geq 0$$

where and hereafter c_i, $i = 1, 2, ...,$ denote constants.

It then follows that

$$\begin{aligned}
&\int_0^t \|Z_z\|^2 dz \\
\leq\ & 2\|F_e\|^2 \int_0^t \left\{ \left\| \int_0^z \exp\{F_e(z-s)\}[x_s^\tau\ 0...0]^\tau ds \right\|^2 + \|x_z\|^2 \right\} dz \\
\leq\ & 2(c_1)^2 \|F_e\|^2 \int_0^t \left\{ \int_0^z \exp[-\rho(z-s)] ds \right. \\
& \left. \int_0^z \exp[-\rho(z-s)] \|x_s\|^2 ds + \|x_z\|^2 \right\} dz \\
\leq\ & 2(c_1)^2 \rho^{-1} \|F_e\|^2 \int_0^t \left\{ \int_0^z \exp[-\rho(z-s)] \|x_s\|^2 ds + \|x_z\|^2 \right\} dz \\
\leq\ & 2(c_1)^2 \rho^{-1} \|F_e\|^2 \left\{ \int_0^t \|x_z\|^2 dz + \rho^{-1} \int_0^t \|x_s\|^2 ds \right\} \\
\leq\ & 2(c_1)^2 \rho^{-1} \|F_e\|^2 (1+\rho^{-1}) \int_0^t \|x_s\|^2 ds. \qquad (12.39)
\end{aligned}$$

Furthermore, by (12.38) it follows that

$$SZ_t = (F_e)^{-1} \left\{ Z_t - \begin{bmatrix} x_t \\ 0 \\ \vdots \\ 0 \end{bmatrix} \right\}$$

and

$$\int_0^t \|SZ_z\|^2 dz \leq c_2 \int_0^t \|x_z\|^2 dz \qquad (12.40)$$

by (12.39).

Finally, the lemma follows from (12.39) and (12.40):

$$\int_0^t \|F(S)E^{-1}(S)x_s\|^2 ds = \int_0^t \|z_s + F_1 S z_s + ... + F_d S^d z_s\|^2 ds$$

$$= \int_0^t \|[I\ 0 \dots 0]z_s + [F_1 \dots F_d]SZ_s\|^2 ds \le c \int_0^t \|x_s\|^2 ds.$$

Q.E.D.

Lemma 12.2 *Let $\{\varphi_t, \mathcal{F}_t\}$ be a measurable process of dimension a (not necessarily defined by (12.19)), and P_t be defined by (12.19) or (12.23). Then*

$$\int_0^t \varphi_s^T P_s \varphi_s ds = \log \det(P_t)^{-1} - a \log a^{-1}. \tag{12.41}$$

Proof. Let the elements of P_t^{-1} be $x_{ij}(t)$, and the algebraic complement corresponding to $x_{ij}(t)$ be $A_{ij}(t)$. Then from the identity

$$\det P_t^{-1} = \sum_j x_{ij}(t) A_{ij}(t), \quad \forall i$$

we know that

$$\frac{d}{dt}[\det P_t^{-1}] = \sum_{i,j} \frac{\partial(\det P_t^{-1})}{\partial x_{ij}} \cdot \frac{dx_{ij}(t)}{dt}$$

$$= \sum_{i,j} A_{ij}(t) \frac{dx_{ij}(t)}{dt} = tr\left\{(\mathrm{Adj} P_t^{-1})\left(\frac{dP_t^{-1}}{dt}\right)\right\}$$

$$= tr\left\{[\det P_t^{-1}]P_t\left(\frac{dP_t^{-1}}{dt}\right)\right\} = [\det P_t^{-1}] tr\left\{P_t \frac{dP_t^{-1}}{dt}\right\}.$$

Hence, it follows that

$$\int_0^t \varphi_s^T P_s \varphi_s ds = \int_0^t tr[P_s \varphi_s \varphi_s^T] ds = \int_0^t tr\left[P_s \frac{dP_s^{-1}}{ds}\right] ds$$

$$= \int_0^t \frac{d(\det P_s^{-1})}{\det P_s^{-1}} = \log \det P_t^{-1} - a \log a^{-1}.$$

Q.E.D.

Lemma 12.3 *Let $\{x_t, \mathcal{F}_t\}$ be an adapted process such that*

$$\int_0^t |x_s|^2 ds < \infty \quad a.s. \quad \forall t. \tag{12.42}$$

If $\{w_t, \mathcal{F}_t\}$ is a Wiener process, then as $t \to \infty$

$$\int_0^t x_s dw_s = O\left(\sqrt{a(t) \log\log[a(t) + e]}\right) \quad a.s. \tag{12.43}$$

where $\{a(t), \mathcal{F}_t\}$ is an adapted process defined by

$$a(t) = \int_0^t x_s^2 ds. \tag{12.44}$$

Proof. By noting that $a(t)$ is the intrinsic time of the stochastic integral $\int_0^t x_s dw_s$, we know from Theorem 1.6 in [Go, p.56] that there is a probability space $(\widetilde{\Omega}, \widetilde{\mathcal{F}}, \widetilde{P})$ and a Brownian motion $(\widetilde{w}_t, \widetilde{\mathcal{F}}_t)$ on it such that

$$\int_0^t x_s dw_s = \widetilde{w}_{a(t)}. \tag{12.45}$$

Now, on the set $\{\lim_{t\to\infty} a(t) < \infty\}$ it is easy to see from (12.45) that (12.43) holds. So, we need only to consider the set $\{\lim_{t\to\infty} a(t) = \infty\}$.

By the law of the iterated logarithm for Brownian motion (see Theorem 1.14), we know that as $t \to \infty$,

$$|\widetilde{w}_t| = O(\sqrt{t \log\log t}) \quad \text{a.s.}$$

Combining this and (12.45) we conclude that (12.43) holds almost surely on $\{\lim_{t\to\infty} a(t) = \infty\}$. This completes the proof. Q.E.D.

Lemma 12.4 *If F_d defined by (12.29) is stable, then*

$$\frac{1}{t}\int_0^t V_s V_s^\tau ds \to R \quad a.s. \quad as \quad t \to \infty \tag{12.46}$$

where V_t is defined in (12.27) and

$$R \triangleq \int_0^\infty \exp\{F_d\lambda\} \begin{bmatrix} I & 0 \\ 0 & 0 \end{bmatrix} \exp\{F_d^\tau \lambda\} d\lambda.$$

Proof. Since F_d is stable, there exists a positive definite matrix $P > 0$ such that

$$PF_d + F_d^\tau P = -I.$$

By this and the Ito formula (Theorem 1.15) we see from (12.30) that

$$\begin{aligned} d[V_t^\tau P V_t] &= V_t^\tau (PF_d + F_d^\tau P) V_t dt \\ &\quad + tr\left\{ \begin{bmatrix} I \\ 0 \\ \vdots \\ 0 \end{bmatrix} [I\ 0 \ldots 0] P dt + 2V_t^\tau P \begin{bmatrix} I \\ 0 \\ \vdots \\ 0 \end{bmatrix} dw_t \right\} \\ &= -\|V_t\|^2 dt + tr \begin{bmatrix} I & 0 \\ 0 & 0 \end{bmatrix} P dt + 2V_t^\tau P \begin{bmatrix} I \\ 0 \\ \vdots \\ 0 \end{bmatrix} dw_t. \end{aligned}$$

So it follows by applying Lemma 12.3

$$\begin{aligned}
&V_t^T P V_t + \int_0^t \|V_s\|^2 ds \\
&= tr\begin{bmatrix} I & 0 \\ 0 & 0 \end{bmatrix} Pt + O\left(\left\{\int_0^t \|V_s\|^2 ds\right\}^{1/2+\eta}\right), \quad \eta > 0.
\end{aligned} \tag{12.47}$$

Consequently, we conclude that

$$\int_0^t \|V_s\|^2 ds = O(t) \quad \text{a.s.} \tag{12.48}$$

Again, by the Ito's formula we get

$$\begin{aligned}
V_t V_t^T &= \left(\int_0^t V_s V_s^T ds\right) F_d^T + F_d\left(\int_0^t V_s V_s^T ds\right) + \begin{bmatrix} I & 0 \\ 0 & 0 \end{bmatrix} t \\
&\quad + \int_0^t \begin{bmatrix} I \\ 0 \\ \vdots \\ 0 \end{bmatrix} dw_s V_s^T + \int_0^t V_s dw_s^T [I \; 0 \ldots 0]
\end{aligned}$$

and hence

$$\begin{aligned}
&\int_0^t V_s V_s^T ds \\
&= \int_0^t \exp[F_d(t-z)] \int_0^z \left\{ \begin{bmatrix} I \\ 0 \\ \vdots \\ 0 \end{bmatrix} dw_s V_s^T + V_s dw_s^T [I \; 0 \ldots 0] \right\} \\
&\qquad \exp[F_d^T(t-z)] dz \\
&\quad + \int_0^t \exp[F_d(t-z)] \begin{bmatrix} I & 0 \\ 0 & 0 \end{bmatrix} z \exp[F_d^T(t-z)] dz.
\end{aligned} \tag{12.49}$$

We now consider the first term on the right-hand side of (12.49). By Lemma 12.3, (12.48) and the stability of F_d, it is easy to see that there is a constant $\rho > 0$ such that

$$\left\| \int_0^t \exp[F_d(t-z)] \int_0^z \left\{ \begin{bmatrix} I \\ 0 \\ \vdots \\ 0 \end{bmatrix} dw_s V_s^T + V_s dw_s^T [I \; 0 \ldots 0] \right\} \exp[F_d^T(t-z)] dz \right\|$$

$$= O\left(\int_0^t \exp[-2\rho(t-z)]\left\{\int_0^z \|V_s\|^2 ds\right\}^{1/2+\eta} dz\right)$$

$$= O\left(\int_0^t \exp[-2\rho(t-z)]z^{1/2+\eta}dz\right) = O(t^{1/2+\eta}) \quad \forall\eta > 0.$$

Hence the lemma follows immediately from this and (12.49).

Lemma 12.5 *Under the conditions of Theorem 12.1 there is a constant $k_0 > 0$ such that*

$$tr\widetilde{\theta}_t^\tau P_t^{-1}\widetilde{\theta}_t \le O(1) + O\left(\left\{\int_0^t \|g_s\|^2 ds\right\}^{1/2+\eta}\right) + O(\log r_t)$$
$$+2\left\{-\left(k_0 - \frac{c}{2}\right)\int_0^t \|g_s\|^2 ds + \frac{1}{2c}\int_0^t \|D(S)C^{-1}(S)\dot{\eta}_s\|^2 ds\right\},$$
$$\forall\eta > 0, \quad c > 0,$$

where $\widetilde{\theta}_t = \theta - \theta_t$, $g_t = \widetilde{\theta}_t^\tau \varphi_t$.

Proof. By (12.15) and (12.24) it is easy to see that

$$dy_t = \theta^\tau \varphi_t^0 dt + dv_t + \dot{\eta}_t dt = \theta^\tau \widetilde{\varphi}_t dt + \theta^\tau \varphi_t dt + dv_t + \dot{\eta}_t dt$$

and hence

$$\begin{aligned}\theta^\tau \widetilde{\varphi}_t dt &= dy_t - \theta_t^\tau \varphi_t dt + (\theta_t - \theta)^\tau \varphi_t dt - dv_t - \dot{\eta}_t dt\\ &= d\widehat{v}_t - \widetilde{\theta}_t^\tau \varphi_t dt - dv_t - \dot{\eta}_t dt\\ &= -d\widetilde{v}_t - \widetilde{\theta}_t^\tau \varphi_t dt - \dot{\eta}_t dt \qquad (12.50)\end{aligned}$$

or

$$C(S)\left(\frac{d\widetilde{v}_t}{dt}\right) = -g_t - \dot{\eta}_t \quad \text{or} \quad \left(\frac{d\widetilde{v}_t}{dt}\right) = -C^{-1}(S)(g_t + \dot{\eta}_t). \qquad (12.51)$$

Let us now set

$$f_t = \left\{\frac{[C(S) - D(S)]}{S}\right\}\widetilde{v}_t + \frac{g_t}{2}; \qquad (12.52)$$

it then follows that

$$f_t = \left[D(S)C^{-1}(S) - \frac{I}{2}\right]g_t + \left\{\frac{[D(S)C^{-1}(S) - I]}{S}\right\}\eta_t. \qquad (12.53)$$

From this and Assumption 2 there are constants $k_0 > 0$, $k_1 > 0$ such that

$$\int_0^t g_s^\tau\{f_s + [I - D(S)C^{-1}(S)]\dot{\eta}_s - k_0 g_s\}ds + k_1 > 0. \qquad (12.54)$$

From (12.18), (12.51) and (12.52) it follows that

$$\begin{aligned} d\widetilde{\theta}_t &= -P_t\varphi_t D(S)[dy_t^\tau - \varphi_t^\tau \theta_t dt] = -P_t\varphi_t D(S)[dv_t - d\widetilde{v}_t]^\tau \\ &= -P_t\varphi_t[dw_t^\tau - d\widetilde{v}_t - D_1\widetilde{v}_t dt - \ \dots \ - D_r S^{r-1}\widetilde{v}_t dt]^\tau \\ &= -P_t\varphi_t\left[g_t dt + \dot{\eta}_t dt + \frac{C(S)-D(S)}{S}\widetilde{v}_t dt + dw_t\right]^\tau \\ &= -P_t\varphi_t\left(f_t dt + \frac{1}{2}g_t dt + \dot{\eta}_t dt + dw_t\right)^\tau . \end{aligned} \tag{12.55}$$

Applying Ito's formula (Theorem 1.15) we obtain

$$\begin{aligned} d[tr\widetilde{\theta}_t^\tau P_t^{-1}\widetilde{\theta}_t] &= -2g_t^\tau[f_t dt + \dot{\eta}_t dt + dw_t] + \varphi_t^\tau P_t\varphi_t dt \\ &= -2g_t^\tau\{f_t + [I - D(S)C^{-1}(S)]\dot{\eta}_t - k_0 g_t\}dt \\ &\quad +2g_t^\tau[I - D(S)C^{-1}(S)]\dot{\eta}_t dt - 2k_0\|g_t\|^2 dt \\ &\quad -2g_t^\tau\dot{\eta}_t dt - 2g_t^\tau dw_t + \varphi_t^\tau P_t\varphi_t dt; \end{aligned}$$

then by (12.54)

$$\begin{aligned} 0 &\le tr\widetilde{\theta}_t^\tau P_t^{-1}\widetilde{\theta}_t \\ &\le tr\widetilde{\theta}_0^\tau P_0^{-1}\widetilde{\theta}_0 + \int_0^t \varphi_s^\tau P_s^{-1}\varphi_s ds + 2k_1 \\ &\quad +2\left\{-k_0\int_0^t \|g_s\|^2 ds\right. \\ &\quad \left. -\int_0^t g_s^\tau D(S)C^{-1}(S)\dot{\eta}_s ds - \int_0^t g_s^\tau dw_s\right\}. \end{aligned} \tag{12.56}$$

Noticing the following simple facts:

$$\begin{aligned} &2\int_0^t g_s^\tau D(S)C^{-1}(S)\dot{\eta}_s ds \\ \le\ & c\int_0^t \|g_s\|^2 ds + \frac{1}{c}\int_0^t \|D(S)C^{-1}(S)\dot{\eta}_s\|^2 ds, \quad \forall c > 0 \end{aligned}$$

and

$$\log\det P_t^{-1} = O(\log r_t),$$

and applying Lemmas 12.2 and 12.3, we can easily conclude the lemma from (12.56). Q.E.D.

Proof of Theorems 12.1 and 12.2.

Since $D(S)C^{-1}(S)$ is strictly positive real, $C(S)$ is necessarily stable. Then by Lemma 12.1 and Assumption 1, it follows that

$$\int_0^t \|D(S)C^{-1}(S)\dot{\eta}_s\|^2 ds \le \varepsilon c_0 r_t, \quad \text{for some } c_0 > 0.$$

Taking $c < 2k_0$ in Lemma 12.5 we see that

$$0 \le tr\widetilde{\theta}_t^T P_t^{-1}\widetilde{\theta}_t \;\le\; O(1) - \left(k_0 - \frac{c}{2}\right)\int_0^t \|g_s\|^2 ds$$
$$+O(\varepsilon r_t) + O(\log r_t). \tag{12.57}$$

We now show that

$$r_t \xrightarrow[t\to\infty]{} \infty \quad \text{a.s.} \tag{12.58}$$

By (12.57),

$$\int_0^t \|g_s\|^2 ds = O(\varepsilon r_t) + O(\log r_t). \tag{12.59}$$

From (12.51) it follows that

$$\widetilde{V}_t = -\int_0^t \exp\{F_c(t-s)\}[g_s + \dot{\eta}_s]ds.$$

Then by (12.59) and Assumption 1, we have for some $\rho > 0$ and $c_1 > 0$

$$\int_0^t \|\widetilde{V}_z\|^2 dz \le \int_0^t \left\| \int_0^z \exp\{F_c(z-s)\}[g_s + \dot{\eta}_s]ds \right\|^2 dz$$
$$\le \; (c_1)^2 \int_0^t \left\{ \int_0^z \exp[-\rho(z-s)][\|g_s\| + \|\dot{\eta}_s\|]ds \right\}^2 dz$$
$$\le \; 2(c_1)^2 \int_0^t \int_0^z \exp[-\rho(z-s)]ds$$
$$\int_0^z \exp[-\rho(z-s)][\|g_s\|^2 + \|\dot{\eta}_s\|^2]dsdz$$
$$\le \; 2\rho^{-1}(c_1)^2 \int_0^t \int_0^z \exp[-\rho(z-s)]dz[\|g_s\|^2 + \|\dot{\eta}_s\|^2]ds$$
$$\le \; 2\rho^{-2}(c_1)^2 \int_0^t [\|g_s\|^2 + \|\dot{\eta}_s\|^2]ds$$
$$\le \; 2\rho^{-2}(c_1)^2\{O(\log r_t) + O(\varepsilon r_t) + \varepsilon r_t\}$$
$$= O(\log r_t) + O(\varepsilon r_t). \tag{12.60}$$

Assume the converse were true, i.e., r_t was bounded in t; then from (12.60) it would follow that $\int_0^t \|\widetilde{V}_z\|^2 dz$ would be bounded. But by (12.28) and (12.33) it is clear that

$$r_t \ge \int_0^t \|\widehat{V}_z\|^2 dz = \int_0^t \|V_z\|^2 dz + \int_0^t \|\widetilde{V}_z\|^2 dz - 2\int_0^t V_z^T \widetilde{V}_z dz.$$

From this and the boundedness of r_t and $\int_0^t \|\widetilde{V}_z\|^2 dz$, it follows that $\int_0^t \|V_z\|^2 dz$ is bounded. This contradicts Lemma 12.4. Hence $r_t \to \infty$ a.s., and (12.58) holds.

By (12.57) it follows that for large t

$$tr\widetilde{\theta}_t^T \widetilde{\theta}_t \le \frac{tr(\widetilde{\theta}_t^T P_t^{-1} \widetilde{\theta}_t)}{\lambda_{\min}(t)}$$

$$\le \frac{1}{\lambda_{\min}(t)} \Big\{ O(1) - \Big(k_0 - \frac{c}{2}\Big) \int_0^t \|g_s\|^2 ds$$

$$+ O(\varepsilon r_t) + O(\log r_t) \Big\}. \tag{12.61}$$

Finally, Theorem 12.1 follows from (12.60) and (12.61) by taking $c < 2k_0$ and noting (12.58); while Theorem 12.2 follows from (12.60) and (12.61) by taking $c < 2k_0$ and $\varepsilon = 0$. Q.E.D.

12.3 Adaptive Control

Let $\{u_t^*\}$ be a bounded deterministic and differentiable reference signal with $u_0^* = 0$. Our objective here is to design the adaptive control u_t, so that the output $\{y_t\}$ tracks the output of the following reference model:

$$E(S) y_t^* = u_t^*$$

where $E(S) = I + E_1 S + \ldots + E_p S^p$ is a stable matrix polynomial in S.

Similar to (12.26), we set

$$Y_t^* = [y_t^* \ \ldots S^{p-1} y_t^*]^T.$$

By a representation similar to (12.30), it is easy to see that $\{Y_t^*\}$ is a bounded sequence.

From now on we assume that the upper bound for the order of the polynomial $A(S)$ is equal to that of $C(S)$, i.e., $p = r$.

Similar to the discrete-time case, we need the following standard minimum phase condition.

Assumption 3. $B(S)$ is stable.

Let us define the adaptive control u_t via the following equation:

$$\theta_t^T \varphi_t = \frac{dy_t^*}{dt}. \tag{12.62}$$

This together with (12.15), (12.18) and (12.19) form a system of nonlinear stochastic differential equations, for which the existence and uniqueness of the strong solution are assumed.

Theorem 12.3 (CG10) . *Consider the system (12.9)-(12.14) with $p = r$, and the estimation algorithm (12.18)-(12.21). If Assumptions 1-3 hold, and the control law is defined from (12.62), then there exists $\varepsilon_1 > 0$ such that whenever ε in (12.31) lies in the internal $[0, \varepsilon_1)$, the following properties hold:*

$$\limsup_{T\to\infty} \frac{1}{T} \int_0^T (\|Y_t\|^2 + \|U_t\|^2) dt < \infty \quad a.s. \tag{12.63}$$

and

$$\limsup_{T\to\infty} \frac{1}{T} \int_0^T (\|Y_t - Y_t^*\|^2 dt = trR + \delta \quad a.s. \tag{12.64}$$

where $|\delta| = O(\varepsilon^{1/2})$, and R is defined in Lemma 12.4.

Proof. From (12.21) and (12.62) it is easy to see that

$$y_t - y_t^* = \widehat{v}_t, \quad Y_t^* - Y_t = \widetilde{V}_t - V_t. \tag{12.65}$$

Note that

$$r_T = \int_0^T (\|Y_t\|^2 + \|U_t\|^2 + \|\widehat{V}_t\|^2) dt + e. \tag{12.66}$$

We have by (12.46), (12.60) and (12.65) that

$$\int_0^T \|Y_t\|^2 dt = O(T) + \varepsilon c_3 r_T + O(\log r_T), \quad c_3 > 0. \tag{12.67}$$

From this, (12.15) and the stability of $B(S)$ it follows that

$$\int_0^T \|U_t\|^2 dt = O(T) + \varepsilon c_4 r_T + O(\log r_T), \quad c_4 > 0. \tag{12.68}$$

Noting also that by (12.60) and Lemma 12.4, we have

$$\begin{aligned}\int_0^T \|\widehat{V}_t\|^2 dt &\leq O(T) + 2\int_0^T \|\widetilde{V}_t\|^2 dt \\ &\leq O(T) + \varepsilon c_5 r_T + O(\log r_T).\end{aligned}$$

Hence, combining (12.66)-(12.68) leads to

$$r_t \leq O(t) + \varepsilon c_6 r_t + O(\log r_t), \quad c_6 > 0,$$

which yields

$$\limsup_{t\to\infty} \frac{r_t}{t} < \infty \quad \text{for any } \varepsilon \in [0, \varepsilon_1)$$

with $\varepsilon_1 = 1/c_6$. Thus (12.63) is true.

We now proceed to prove (12.64). From (12.60) we have for any $\varepsilon \in [0, \varepsilon_1)$

$$\frac{1}{T}\int_0^T \|\widetilde{V}_t\|^2 dt = O\left(\frac{\log T}{T}\right) + O(\varepsilon); \tag{12.69}$$

then by (12.65)

$$\begin{aligned}
&\frac{1}{T}\int_0^T (Y_t - Y_t^*)(Y_t - Y_t^*)^\tau dt \\
= &\frac{1}{T}\int_0^T (V_t - \widetilde{V}_t)(V_t - \widetilde{V}_t)^\tau dt \\
= &\frac{1}{T}\int_0^T V_t V_t^\tau dt + \frac{1}{T}\int_0^T \widetilde{V}_t \widetilde{V}_t^\tau dt - \frac{1}{T}\int_0^T (V_t \widetilde{V}_t^\tau + \widetilde{V}_t V_t^\tau) dt \\
= &R + \left[\frac{1}{T}\int_0^T V_t V_t^\tau dt - R\right] + O\left(\frac{\log T}{T}\right) + O(\varepsilon) \\
&+ O\left(\left\{\frac{\log T}{T} + \varepsilon\right\}^{1/2}\right).
\end{aligned} \tag{12.70}$$

Hence (12.64) is also true. Q.E.D.

Remark. If the initial value of the reference signal is not zero, i.e., $u_0^* \neq 0$, then we may replace (12.62) by

$$\theta_t^\tau \varphi_t = \frac{dz_t^*}{dt},$$

where $z_t^* = E^{-1}(S)\{u_t^* - \exp(-t^2)u_0^*\}$. In this case, Theorem 12.3 is true for $\{z_t^*\}$, which approximates $\{y_t^*\}$ exponentially.

We now consider the case where $\eta_t \equiv 0$ in (12.15). We shall get the convergence rate for the tracking errors. It is worth noting that the completely similar results for the discrete-time case are not available yet.

Theorem 12.4 *Consider the system described by (12.15) with $\eta_t = 0$ and $p = r$, and estimation algorithm (12.18). If Assumptions 2 and 3 are satisfied, and if the adaptive control is defined from (12.62), then*

$$\|R_T - R\|^2 = O\left(\frac{\log T}{T}\right) \quad a.s. \text{ as } T \to \infty, \tag{12.71}$$

where R is given in Lemma 12.4 and

$$R_T = \frac{1}{T}\int_0^T (Y_t - Y_t^*)(Y_t - Y_t^*)^\tau dt.$$

Proof. We first consider the convergence rate of $\frac{1}{T}\int_0^T V_t V_t^\tau dt$. From (12.47) it is clear that

$$\limsup_{T\to\infty} \frac{1}{T}\int_0^T \|V_s\|^2 ds \le 2tr \begin{bmatrix} I & 0 \\ 0 & 0 \end{bmatrix} P.$$

Then Lemma 12.3 implies

$$\limsup_{T\to\infty} \frac{1}{(T\log\log T)^{1/2}} \left\| \int_0^T V_t dw_t^\tau \right\| < \infty \quad \text{a.s.,}$$

and hence

$$\left\| \int_0^t \exp[F_d(t-z)] \int_0^z \left\{ \begin{bmatrix} I \\ 0 \\ \vdots \\ 0 \end{bmatrix} dw_s V_s^\tau + V_s dw_s^\tau [I\ 0 \ldots 0] \right\} \exp[F_d^\tau (t-z)] dz \right\|$$
$$= O(\{t\log\log t\}^{1/2}) \quad \text{a.s.}$$

Consequently, it follows from (12.49) that

$$\left\| \frac{1}{T}\int_0^T V_s V_s^\tau ds - \int_0^\infty \exp\{F_d s\} \begin{bmatrix} I & 0 \\ 0 & 0 \end{bmatrix} \exp\{F_d^\tau s\} ds \right\|$$
$$= O\left(\left\{ \frac{\log\log T}{T} \right\}^{1/2} \right). \tag{12.72}$$

Setting $\varepsilon = 0$ in (12.69) and (12.70) and using (12.72), we see that

$$\begin{aligned} R_t &= \int_0^\infty \exp\{F_d s\} \begin{bmatrix} I & 0 \\ 0 & 0 \end{bmatrix} \exp\{F_d^\tau s\} ds \\ &\quad + O\left(\left\{ \frac{\log\log T}{T} \right\}^{1/2} \right) + O\left(\frac{\log T}{T} \right) + O\left(\left\{ \frac{\log T}{T} \right\}^{1/2} \right) \\ &= \int_0^\infty \exp\{F_d s\} \begin{bmatrix} I & 0 \\ 0 & 0 \end{bmatrix} \exp\{F_d^\tau s\} ds + O\left(\left\{ \frac{\log T}{T} \right\}^{1/2} \right), \end{aligned}$$

which verifies (12.71). The proof is complete. Q.E.D.

References

[A1] H. Akaike, Stochastic theory of minimal realization, *IEEE Trans. Autom. Control*, **19**(1974), 667-674.

[A2] H. Akaike, A new look at the statistical model identification, *IEEE Trans. Autom. Control*, **19**(1974), 716-723.

[An1] B.D.O. Anderson, A system criterion for positive real matrices, *SIAM J. Control and Optimization*, **5**(1967), 2, 171-182.

[AM1] B.D.O. Anderson and J.B. Moore, *Linear Optimal Control*, Prentice-Hall, Englewood Cliffs, NJ, 1971.

[AM2] B.D.O. Anderson and J.B. Moore, *Optimal Filtering*, Prentice-Hall, Englewood Cliffs, NJ, 1979.

[An2] T.W. Anderson, *The Statistical Analysis of Time Series*, Wiley, New York, 1971.

[As] K.J. Åström, *Introduction to Stochastic Control*, Academic Press, New York, 1970.

[AW] K.J. Åström and B. Wittenmark, On self-tuning regulators, *Automatica*, **9**(1973), 195-199.

[BKW] A. Becker, P.R. Kumar and C.Z. Wei, Adaptive control with the stochastic approximation algorithm: geometry and convergence, *IEEE Trans. Autom. Control*, **30**(1985), 4, 330-338.

[BR] A. Benveniste and G. Ruget, A measure of the tracking capacity of recursive stochastic algorithms with constant gains, *IEEE Trans. Autom. Control*, **27**(1982), 639-649.

[Be] A. Benveniste, Design of adaptive algorithms for tracking of time-varying systems, *Adaptive Control and Signal Processing*, **1**(1987), 3-29.

[Bi] R. Bitmead, Convergence in distribution of LMS-type adaptive parameter estimates, *IEEE Trans. Autom. Control*, **28**(1983), 54-60.

[BA] R. Bitmead and B.D.O. Anderson, Performance of adaptive estimation algorithm in dependent random enviroment, *IEEE Trans. Autom. Control*, **25**(1980), 788-794.

[BJ] G.E.P. Box and G. Jenkins, *Time Series Analysis, Forecasting and Control*, Holden-Day, San-Francisco, 1970.

[Br] B.M. Brown, Martingale central limit theorems, *Ann. Math. Statist.*, **42**(1971), 59-66.

[Ca] P.E. Caines, *Linear Stochastic Systems*, Wiley, New York, 1988.

[CL] P.E. Caines and S. Lafortune, Adaptive control with recursive identification for stochastic linear systems, *IEEE Trans. Autom. Control*, **30**(1985), 185-189.

[Che1] H.F. Chen, Strong consistency and convergence rate of least squares identification, *Scientia Sinica (Series A)*, **25**(1982), 7, 771-784.

[Che2] H.F. Chen, Recursive system identification and adaptive control by use of the modified least squares algorithm, *SIAM J. Control and Optimization*, **22**(1984), 5, 758-776.

[Che3] H.F. Chen, *Recursive Estimation and Control for Stochastic Systems*, Wiley, New York, 1985.

[CCa1] H.F. Chen and P.E. Caines, On the adaptive control of a class systems with random parameters and disturbances, *Automatica*, **21**(1985), 6, 737-741.

[CCa2] H.F. Chen and P.E. Caines, Strong consistency of stochastic gradient algorithm of adaptive control, *IEEE Trans. Autom. Control*, **30**(1985), 2, 189-192.

[CG1] H.F. Chen and L. Guo, The limit of stochastic gradient algorithm for identifying systems excited not persistently, *Kexue Tongbao (Science Bulletin)*, (1986), 1, 6-9.

[CG2] H.F. Chen and L. Guo, Strong consistence of parameter estimates in optimal adaptive tracking systems, *Scientia Sinica (Series A)*, **29**(1986), 11, 1145-1156.

[CG3] H.F. Chen and L. Guo, Convergence rate of least squares identification and adaptive control for stochastic systems, *Int. J. Control*, **44**(1986), 5, 1459-1476.

[CG4] H.F. Chen and L. Guo, Asymptotically optimal adaptive control with consistent parameter estimates, *SIAM J. Control and Optimization*, **25**(1987), 3, 558-575.

[CG5] H.F. Chen and L. Guo, Optimal adaptive control and consistent parameter estimates for ARMAX model with quadratic cost, *SIAM J. Control and Optimization*, **25**(1987), 4, 845-867.

[CG6] H.F. Chen and L. Guo, Consistent estimation of the order of the stochastic control systems, *IEEE Trans. Autom. Control*, **32**(1987), 6, 531-535.

[CG7] H.F. Chen and L. Guo, Stochastic adaptive control for a general quadratic cost, *J. Sys. Sci. & Math. Scis.*, **7**(1987), 4, 287-302.

[CG8] H. F. Chen and L. Guo, A robust adaptive controller, *IEEE Trans. Autom. Control*, **33**(1988), 11, 1035-1043.

[CG9] H. F. Chen and L. Guo, The limit passage in the stochastic adaptive LQ control problem, *Control Theory and Applications*, **6**(1989), 1, 51-56.

[CG10] H. F. Chen and L. Guo, Continuous time stochastic adaptive tracking: robustness and asymptotic properties, *SIAM J. Control & Optimization*, **28**(1990), 3, 513-529.

[CG11] H. F. Chen and L. Guo, Adaptive control with recursive identification for stochastic linear systems, in *Control and Dynamic Systems* (ed. C.T. Leondes), **Vol.26**(2), 1987, Academic Press, INC.

[CKS] H. F. Chen, P.R. Kumar and J.H. Van Schuppen, On Kalman filter for conditionally Gaussian systems with random matrices, *Systems and Control Letters*, **13**(1989), 513-529.

[ChM] H.F. Chen and J.B. Moore, Convergence rate of continuous time ELS parameter estimation, *IEEE Trans. Autom. Control*, **32**(1987), 3, 267-269.

[CZ1] H. F. Chen and J.F. Zhang, Convergence rate in stochastic adaptive tracking, *Int. J. Control*, **49**(1989), 6, 1915-1935.

[CZ2] H. F. Chen and J.F. Zhang, Stochastic adaptive control for systems with noise being an ARMA process, *Syst. Sci. & Math. Scis.*, **2**(1989), 1, 40-53.

[CZ3] H. F. Chen and J.F. Zhang, Identification of linear systems without assuming stability and minimum-phase, *Science in China (Series A)*, **33**(1990), 6, 641-653.

[CZ4] H. F. Chen and J.F. Zhang, Identification and adaptive control for systems with unknown orders, time-delay and coefficients (Uncorrelated noise case), *IEEE Trans. Autom. Control*, **35**(1990), 8, 866-877.

[Cho] Y.S. Chow, Local convergence of martingales and the law of large numbers, *Ann. Math. Stat.*, **36**(1965), 552-558.

[CT] Y.S. Chow and H. Teicher, *Probability Theory: Independence, Interchangeability, Martingales*, Springer-Verlag, New York, 1978.

[Chu1] K.L. Chung, *Markov Chains with Stationary Transition Probabilities*, Springer-Verlag, New-York, 1967.

[Chu2] K.L. Chung, *A Course in Probability Theory*, Harcourt, Brace and World, New York, 1968.

[Cl] D.W. Clarke, Self-tuning control of non-minimum phase systems, *Automatica*, **20**(1984), 501-518.

[CMT] D.W. Clarke, C. Mohtadi and P.S. Tuffs, Generalized predictive control — Part I and Part II, *Automatica*, **23**(1987), 137-160.

[Da] M.H.A. Davis, *Linear Estimation and Stochastic Control*, Chapman and Hall, New York, 1977.

[Do] J. L. Doob, *Stochastic Processes*, Wiley, New York, 1953.

[Du1] J. Durbin, Efficient estimation of parameters in moving-average model, *Biometrika*, **46**(1959), 306-316.

[Du2] J. Durbin, The fitting of time series model, *Int. Statist. Rev.*, **28**(1961), 233-344.

[Dy] E. B. Dynkin, *Markov Processes*, Plenum, New York, 1963.

[EM] E. Eweda and O. Macchi, Tracking error bounds of adaptive nonstationary filtering, *Automatica*, **21**(1985), 3, 293-302.

[GL] S. Gannarsson and L. Ljung, Frequency domain tracking characteristics adaptive algorithms, *IEEE Trans. Acoustics, Speech, and Signal Processing*, **37**(1989), 1072-1089.

[GSk] I.I. Gihman and A.V. Skorohod, *The Theory of Stochastic Processes III*, Springer-Verlag, 1979.

[Gl] K. Glover, All optimal Hankel-norm approximations of linear multivariable systems and their L^{∞} error bounds, *Int. J. Control*, **39**(1984), 1115-1193.

[Go] G.L. Gong, *Introduction to Stochastic Differential Equations* (in Chinese), Beijing University Press, 1987.

[GRC1] G.C. Goodwin, P.J. Ramadge and P.E. Caines, Discrete time multivariable adaptive control, *IEEE Trans. Autom. Control*, **25**(1980), 3, 449-456.

[GRC2] G.C. Goodwin, P.J. Ramadge and P.E. Caines, Discrete time stochastic adaptive control, *SIAM J. Control and Optimization*, **19**(1981), 6, 829-853.

[Gu1] L. Guo, *Identification and Adaptive Control for Dynamic Systems*, Ph. D. Desertation, Institute of Systems Science, The Chinese Academy of Sciences, Beijing, 1986.

[Gu2] L. Guo, Estimating time-varying parameters by Kalman filter based algorithm: stability and convergence, *IEEE Trans. Autom. Control*, **35**(1990), 2, 141-147.

[Gu3] L. Guo, On adaptive stabilization of time-varying stochastic systems, *SIAM J. Control and Optimization*, **28**(1990), 6, 1432-1451.

[GC1] L. Guo and H.F. Chen, Convergence and optimality of self-tuning tracker, *Science in China (Series A)* (to appear).

[GC2] L. Guo and H.F. Chen, Åström-Wittenmark self-tuning regulator revisited and ELS-based tracker, *IEEE Trans. Autom. Control* (to appear).

[GC3] L. Guo and H.F. Chen, Convergence rate of ELS based adaptive trackers, *Syst. Sci. & Math. Scis.*, **1**(1988), 2, 131-138.

[GCZ] L. Guo, H.F. Chen and J.F. Zhang, Consistent order estimation for linear stochastic feedback control systems (CARMA model), *Automatica*, **25**(1989), 1, 147-151.

[GH] L. Guo and D. Huang, Least squares identification for ARMAX models without the positive real condition, *IEEE Trans. Autom. Control*, **34**(1989), 1094-1098.

[GHH] L. Guo, D. Huang and E.J. Hannan, On ARX(∞) approximation, *J. of Multivariate Analysis*, **32**(1990), 1, 17-47.

[GM] L. Guo and S.P. Meyn, Adaptive control for time-varying systems: A combination of martingale and Markov chain techniques, *Adaptive Control and Signal Processing*, **3**(1989), 1, 1-14.

[GMo] L. Guo and J.B. Moore, Stochastic system identification via adaptive spectral factorization, *Syst. Sci. & Math. Scis.* (to appear).

[GXM] L. Guo, L. Xia and J.B. Moore, Tracking randomly varying parameters analysis of a standard algorithm, *Proc. 27th IEEE Conf. on Decision and Control*, Austin, Taxas, (1988), 1514-1591; see also *Mathematics of Control, Signals and Systems* (1990).

[GS] G.C. Goodwin and K.S. Sin, *Adaptive Filtering, Prediction and Control*, Prentice-Hall, Englewood Cliffs, NS, 1984.

[HH] P. Hall and C.C. Heyde, *Martingale Limit Theory and Its Applications*, Academic Press, New York, 1980.

[H] E.J. Hannan, Rational transfer function approximation, *Statist. Sci.*, **2**(1987), 2, 135-161.

[HDe] E.J. Hannan and M. Deistler, *The Statistical Theory of Linear Systems*, Wiley, New York, 1988.

[HQ] E.J. Hannan and B.G. Quinn, The determination of the order of an autoregression, *J.R. Stat. Soc. Set-B*, **41**(1979), 190-195.

[HR] E.J. Hannan and J. Rissanen, Recursive estimation of ARMA order, *Biometrika*, **69**(1982), 81-94.

[Hu1] D.W. Huang, Convergence rate of sample autocorrelations and autocovariances for stationary time series, *Scientia Sinica (Series A)*, **31**(1988), 4, 406-424.

[Hu2] D.W. Huang, Recursive method for ARMA model estimation, *Acta Math. Appli. Sinica*, **11**(1988), No.3.

[Hu3] D.W. Huang, Estimating the orders and parameters for ARUMA model, *The Annals of Statistics* (to appear).

[HG] D.W. Huang and L. Guo, Estimation of nonstationary ARMAX models based on Hannan-Rissanen method, *The Annals of Statistics* (to appear).

[HLP] G.H. Hardy, J.E. Litllewood and G. Polya, *Inequalities*, Cambridge Univ. Press, Cambridge, 1934.

[HDa] E.M. Hermerly and M.H.A. Davis, Recursive order estimation of stochastic control systems, *Mathematical Systems Theory*, **22**(1989), 323-346.

[IT] P.A. Ioannou and K. Tsakalis, Robust discrete-time adaptive control, in *Adaptive and Learning Systems; Theory and Applications* (ed. K.S. Narendra), Plenum Press, New York, (1985), 73-85.

[IW] N. Ideda and S. Watanabe, *Stochastic Differential Equations and Diffusion Processes*, North-Holland Publishing Company, 1981.

[J] A.H. Jazwinski, *Stochastic Processes and Filtering Theory*, Academic Press, New York, 1970.

[Ki] T.Kailath, *Linear Systems*, Prentice-Hall, Englewood Cliffs, NJ, 1980.

[Ka1] R.E. Kalman, On the general theory of control systems, *Proceedings of the* 1st *IFAC Congress* (Moscow, 1960), Butterworth, London, **1**(1961), 481-492.

[Ka2] R.E. Kalman, A new approach to linear filtering and prediction problems, *ASME Trans. Part D*, **82**(1960), *J. Basic Eng.*, 35-45.

[KFA] R.E. Kalman, P.L. Falb and M.A. Arbib, *Topic in Mathematical System Theory*, McGraw-Hill, New York, 1969.

[KG] G. Kitagawa and W. Gersh, A smoothness priors time varying AR coefficient modelling of non-stationary covariance time series, *IEEE Trans. Autom. Control*, **30**(1985), 48-56.

[Ku1] P.R. Kumar, A survey of some results in stochastic adaptive control, *SIAM J. Control and Optimization*, **23**(1985), 3, 329-380.

[Ku2] P.R. Kumar, Convergence of adaptive control schemes using least squares parameter estimates, *IEEE Trans. Autom. Control*, **35**(1990), 4, 416-424.

[KP] P.R. Kumar and L. Praly, Self-tuning trackers, *SIAM J. Control and Optimization*, **25**(1987), 1053-1071.

[KV] P.R. Kumar and P.P. Varayia, *Stochastic Systems: Estimation, Identification, Adaptive Control*, Prentice-Hall, Englewood Cliffs, NJ, 1986.

[Ku] H.J. Kushner, *Approximation and Weak Convergence Methods for Random Processes*, MIT Press, Cambridge, Massachusetts, 1984.

[LW1] T.L. Lai and C.Z. Wei, Least squares estimates in stochastic regression models with application to identification and control of dynamic systems, *Ann. Stat.*, **10**(1983), 1, 154-166.

[LW2] T.L. Lai and C.Z. Wei, Extended least squares and their applications to adaptive control and prediction in linear systems, *IEEE Trans. Autom. Control*, **31**(1986), 898-906.

[LW3] T.L. Lai and C.Z. Wei, Asymptotically efficient self-tuning regulators, *SIAM J. Control and Optimization*, **25**(1987), 2, 466-481.

[LY] T.L. Lai and Z. Ying, Parallel recursive algorithms in asymptotically efficient adaptive control of linear stochastic systems, *Tech. Report* (1989), Dept. of Statistics, Stanford University.

[La] I.D. Landau, *Adaptive Control: The Model Reference Approach*, Marcel Dekker, New York, 1979.

[LM] G. Ledwich and J.B. Moore, Multivariable Self-Tuning Filters, Differential Games and Control Theory, II, *Lecture Notes on Pure and Applied Math.*, New York: Marcel-Dekker, (1977), 345-376.

[LSh] R.S. Liptser and A.N. Shiryayev, *Statistics of Random Processes*, Springer-Verlag, New York, 1977.

[Lj1] L. Ljung, Consistency of least squares identification method, *IEEE Trans. Autom. Control*, **21**(1976), 779-781.

[Lj2] L. Ljung, On positive real functions and the convergence of some recursive schemes, *IEEE Trans. Autom. Control*, **22**(1977), 539-551.

[LSo] L. Ljung and T. Söderström, *Theory and Practice of Recursive Identification*, MIT Press, Cambridge, MA, 1983.

[Lo] M. Loève, *Probability Theory*, Springer-Verlag, New York, 1977-1978.

[Ma] O. Macchi, Optimization of adaptive identification for time-varying filters, *IEEE Trans. Autom. Control*, **31**(1986), 283-287.

[MC1] S.P. Meyn and P.E. Caines, The zero divisor problem of multivariable stochastic adaptive control, *Systems and Control Letters*, **6**(1985), 4, 235-238.

[MC2] S.P. Meyn and P.E. Caines, A new approach to stochastic adaptive control, *IEEE Trans. Autom. Control*, **32**(1987),220-226.

[MGu] S.P. Meyn and L. Guo, Stability, convergence and performance of an adaptive control algorithm applied to a randomly varying systems, *IEEE Trans. Autom. Control* (to appear).

[MG] R.H. Middleton and G.C. Goodwin, Adaptive control of time-varying linear systems, *IEEE Trans. Autom. Control*, **33**(1988), 2, 150-155.

[MGHM] R.H. Middleton, G.C. Goodwin, D.J. Hill and D.Q. Mayne, Design issues in adaptive control, *IEEE Trans. Autom. Control*, **33**(1988), 50-58.

[Mi] D.S. Mitrinović, *Analytic Inequalities*, Springer-Verlag, 1970.

[Mo1] J.B. Moore, On strong consistency of least squares algorithm, *Automatica*, **14**(1978), 505-509.

[Mo2] J.B. Moore, Persistence of excitation in extended least squares, *IEEE Trans. Autom. Control*, **28**(1983), 60-68.

[Mo3] J.B. Moore, Side-stepping the positive real condition for stochastic adaptive schemes, *Ricerche di Automatica*, **8**(1982), 501-523.

[Mo4] J.B. Moore, Convergence of continuous time stochastic ELS parameter estimation, *Stochastic Processes and Their Applications*, **27**(1988), 195-215.

[N] J. Neveu, *Discrete Parameter Martingales*, North-Holland, Amsterdam, 1975.

[Ri1] J. Rissanen, Modeling by shortest data description, *Automatic*, **14**(1978), 467-471.

[Ri2] J. Rissanen, Stochastic completely and modeling, *The Annals of Statistics*, **14**(1986), 1080-1100.

[Ro] Y.A. Rozanov, *Stationary Random Processes*, A. Feinstein, translator, Holden-Day, Sanfrancisco, 1967.

[Sc] R. Scattlini, A multivariable self-tuning controller with integral action, *Automatica*, **22**(1986), 619-627.

[SK] U. Shaked and P.R. Kumar, Minimum variance control of discrete time multivariable ARMAX systems, *SIAM J. Control and Optimization*, **24**(1986), 396-411.

[Sh] A.N. Shiryayev, *Probability*, Springer-Verlag, New York, 1984.

[SG] K.S. Sin and G.C. Goodwin, Stochastic adaptive control using a modified least squares algorithm, *Automatica*, **18**(1982), 315-321.

[So1] V. Solo, The convergence of AML, *IEEE Trans. Autom. Control*, **24**(1979), 958-962.

[So2] V. Solo, The limiting behavior of LMS, *IEEE Trans. Acoustics, Speech, and Signal Processing*, **37**(1989), 1909-1922.

[S] G.W. Stewart, *Introduction to Matrix Computation*, Academic Press, 1973.

[St] W.F. Stout, *Almost Sure Convergence*, Academic Press, New York, 1974.

[Sw] G. Schwarz, Estimating the dimension of a model, *The Annals of Statistics*, **6**(1978), 416-464.

[TI] K. Tsakalis and P.A. Ioannou, Adaptive control of linear time-varying plants, in *Proc. IFAC Workshop on Adaptive Control and Signal Processing*, Lund, Sweden, 1986.

[W] B. Widrow, J.M. McColl, M.G. Larimore and C.R. Johnson, Jr., Stationary and non-stationary characteristics of the LMS adaptive filter, *Proc. IEE*, **64**(1976), 1151-1162.

[WS] B. Widrow and S. Stearns, *Adaptive Signal Processing*, Englewood Cliffs, NJ, Pretice-Hall, 1985.

[Wol] W.A. Wolovich, *Linear Multivariable Systems*, Springer-Verlag, New York, 1974.

[We] C.Z. Wei, Adaptive prediction by least squares predictors in stochastic regression models with application to time series, *The Annals of Statistics*, **15**(1987), 1667-1682.

[Won] M.W. Wonham, *Linear Multivariable Control — A Geometric Approach*, Springer-Verlag, New York, 1979.

[ZGC] J.F. Zhang, L. Guo and H.F. Chen, L_p-stability of estimation errors of Kalman filter for tracking time-varying parameters, *Int. J. Of Adaptive Control and Signal Processing*, (1991).

Index